ADVANCED PHOTONIC SCIENCES

Edited by **Mohamed Fadhali**

Advanced Photonic Sciences
Edited by Mohamed Fadhali

Published by InTech
Janeza Trdine 9, 51000 Rijeka, Croatia

Publishing Process Manager Vedran Greblo

Technical Editor Teodora Smiljanic

Cover Designer InTech Design Team

Additional hard copies can be obtained from orders@intechopen.com

Advanced Photonic Sciences, Edited by Mohamed Fadhali
p. cm.
ISBN 978-953-51-0153-6

For EU product safety concerns:
IN TECH d.o.o., Prolaz Marije Krucifikse Kozulić 3, 51000 Rijeka, Croatia,
info@intechopen.com or visit our website at intechopen.com.

Meet the editor

Dr Mohamed Fadhali, PhD (Physics, UTM, Malaysia), MSc (Physics, CUSAT, India), BSc (Honor-Applied Physics / Laser, UT, Iraq), is an Assistant Professor of Photonics and Laser Physics at the Department of Physics, Ibb University, Yemen. He has about 12 years of teaching and research experience in laser physics and optics. Dr. Fadhali is the author and co-author of about 22 peer-reviewed papers in international journals and more than 30 proceedings and conference presentations. He also has two years of experience of working as an Optical Communication Engineer in Public Telecommunication Corporation, Yemen. He is a member of SPIE and OSA. He is a reviewer in many international peer-reviewed journals. Dr Fadhali has adequate experience in laser weld machines and their peripherals, vacuum machines and pulsed laser deposition. He is very familiar with most of photonics/optoelectronics systems applied for optical communications and biophotonics. He has adequate experience in computer applications and programming as well as MATLAB coding & simulations for a variety of applications. Dr Fadhali was invited as guest speaker at the Advanced Photonic Science Institute (APSI), UTM, Malaysia, August 2008. He was later invited as a visiting lecturer at (APSI) (September 2008 – February 2008). He held many invited talks locally and internationally. Dr Fadhali was awarded the SPIE award for the best paper (ICPN2007, Pattaya, Thailand) and the best research paper award for professionals (CUTSE 2008, Malaysia).

Contents

Preface

The new emerging field of photonics has significantly attracted the interest of many societies, professionals and researchers around the world. The great importance of this field is due to its applicability and possible utilization in almost all scientific and industrial areas. One can say that photonics is currently at the forefront of science and technology, both on a basic and an advanced level.

The scientific revolution in fiber optic communication, nanotechnology, materials fabrication and optical diagnostic techniques made these technologies some of the leading photonic technologies of the era. These vast technological developments with wide promising applications made photonics the field of choice for further advancement of human life.

The comprehensive applicability of photonics technology appeares in the following areas:

- Information Technology and Telecommunications
- Optics for Military and Surveillance applications
- Optics in Manufacturing
- Manufacturing Optical Components and Systems
- Health Care and the Life Sciences
- Optical Sensing, Lighting, Imaging and Energy

The advanced research in photonics is still very extensive and hence lot of research problems related to integrated photonics and photonic material fabrications are still under investigation and may give amazing results in the future. The advanced research in photonics has no limitations, it may be linked to all fields of Science and Engineering. Therefore, every application or technique employing photons in some way can be considered a branch of photonics. Consequentially, photonics is usually defined as the technology of generating and harnessing light and other forms of radiant energy whose quantum unit is the photon.

This book contains 16 chapters presenting quite diverse research topics in photonics. An attempt has been made to provide researchers in this field with a comprehensive reference point for advanced research in various topics related to photonics, such as integrated photonics, with many remarkable works to pave the way for further innovations in the future.

This book is divided into three general sections: integrated photonics, photonic materials, and photonic applications. The section dealing with integrated photonics presents information on integrated photonics in silicon and germanium, experimental engineering of photonic quantum entanglement, photonic band gap and utilizing photonic quantum angular momentum for information processing. The photonic materials section concentrates on the analysis related to new lead fluoride nanophotonic materials, Fianite in photonics, hybrid polyfluorene for optoelectronic devices materials and plasmonics for green technologies. The last section contains information on lightwave analysis in some applications, high resolution photonic spectroscopy, photonic imaging, time resolved photonic devices, near field optical microscopy and LIDAR atmospheric sensing.

Dr. Mohamed Fadhali
Department of Physics, Faculty of Science Ibb University, Ibb
Yemen

Part 1

Integrated Photonics

1

Germanium-on-Silicon for Integrated Silicon Photonics

Xiaochen Sun
Massachusetts Institute of Technology
USA

1. Introduction

To meet the unprecedented demands for data transmission speed and bandwidth silicon integrated photonics that can generate, modulate, process and detect light signals is being developed. Integrated silicon photonics that can be built using existing CMOS fabrication facilities offers the tantalizing prospect of a scalable and cost-efficient solution to replace electrical interconnects. Silicon, together with commonly used dielectric materials in CMOS processes such as silicon dioxide, is a great material system for optical confinement and wave transmission in near infrared range. However, silicon is not a good choice for active photonic devices due to its transparency in such wavelength range. Germanium and GeSi alloy, the materials that have long been adopted to improve the performance of silicon transistors in many ways, have been showing their potential as the building blocks of such active integrated optical devices. This chapter discusses the research of using germanium and GeSi for silicon-integrated photodetection and light source in the contexts of material physics and growth, device design and fabrication.

This introduction section briefly introduces the background of integrated silicon photonics and some germanium properties which are important for photonic applications. Next section focuses on waveguide-integrated germanium photodetectors which can readily be integrated with silicon waveguide on mature silicon or silicon-on-insulator (SOI) platform. The physics and design considerations of these devices are presented with details. The fabrication processes of these devices are also discussed with some extent. Next section includes a newly developed field that using germanium for light sources in silicon photonics applications such as light-emitting-diodes (LEDs) and lasers. There has been a few breakthroughs in this topic including the author's work of epitaxial germanium LEDs and optically pumped lasers operating at room temperature. The physics of this unusual concept of using indirect band gap material for light emission is discussed in details and some important results are presented.

1.1 Integrated silicon photonics

Silicon integrated circuits (ICs) had been developed in an extraordinary pace for almost four decades before 2005. It is known as Moore's law that the number of transistors in an integrated circuit doubles roughly every eighteen months (Moore, 1965). The scalability is the main reason of the tremendous success of many silicon IC based technologies (Haensch et al., 2006), such as silicon complimentary metal oxide semiconductor (Si-CMOS) technology.

The scalability of Si-CMOS technology is not only about the shrinkage of the dimensions of the devices, but also a number of other factors for maintaining the power density while boosting the performance. However, many issues against further scaling have been found including applied voltage barrier (threshold voltage) and passive component heating due to sub-threshold leakage current. In 2005, "for the first time in thirty five years, the clock speed of the fastest commercial computer chips has not increased" (Muller, 2005) because of the above reasons.

In order to solve these problems in data transmission speed scaling and passive component heat dissipation, optical interconnects relying on silicon photonics are proposed as a promising solution. Silicon photonics offers a platform for the monolithic integration of optics and microelectronics on the same silicon chips (Lipson, 2004), aiming for many applications including the optical interconnects (Haurylau et al., 2006). At present, electrical circuit speed or propagation delay is limited by interconnect RC delay which increases with the scaling of device dimensions. Optical interconnects which use phonon as information carrier are not subject to the RC delay in the first place. The weak electrical interaction between photons and guiding media inherently minimize heat generation in propagation. The interaction among photons is also weak at lower optical power so that multiple transmission using different wavelengths can co-exist in the same propagation channel. This wavelength multiplexing (WDM) technique dramatically increase the aggregated bandwidth of the transmission system. To achieve such optical interconnects using compatible silicon processing techniques is the major task of monolithic integrated silicon photonics (Aicher & Haberger, 1999; OĆonnor et al., 2007; O'Connor et al., 2006). The compatibility with silicon enables a cost-effective, scalable and manufacturable solution for implementing silicon photonics. To address its importance and rapid growth, a roadmap of silicon photonics has been developed by academic and industrial professionals and released on a yearly basis (Kirchain & Kimerling, 2007).

1.2 Germanium epitaxy on silicon

The world's first transistor is made from Germanium by John Bardeen, Walter Brattain and William Shockley at Bell Labs in December, 1947. Beginning with this invention, a revolution of semiconductor electronics quickly started and have profoundly changed our life in many ways. But germanium, the first used semiconductor material, soon gave its crown to another semiconductor in the same elementary group - silicon - in this great revolution. Silicon has superior properties than germanium in many ways: chemical and mechanical stability, stable oxides and etc. Based on silicon, integrated circuit (IC) was realized and appeared in numerous electronic devices that we use and carry everyday.

Silicon was so successful that germanium was forgotten by most researchers for decades soon after the first silicon transistor was invented and commercialized from Texas Instrument in 1952. However, germanium has its own advantages, e.g. higher carrier mobility than silicon, thereby some researchers investigated ways to integrate germanium on silicon substrates. As semiconductor devices generally require single crystalline germanium thin films, the attempts of heterogeneous epitaxy of germanium on silicon emerged. People soon found a great difficulty of growing germanium thicker than its critical thickness because there is 4% mismatch between the lattice constants of germanium and silicon. Although the earliest work of the epitaxy of germanium-silicon (GeSi) alloy on silicon dates back to 1962 (Miller & Grieco, 1962), good quality pure germanium films or high germanium content GeSi films have been grown on silicon since late 1970s after the invention of molecular beam epitaxy (MBE) (Bean et al., 1984; Garozzo et al., 1982; Kasper & Herzog, 1977; Tsaur et al., 1981).

Despite the importance of the studies of germanium epitaxy by MBE, it is recognized by many researchers that MBE is not likely to be a good choice for massive device manufacturing. The complexity of MBE equipment and the low yield of single wafer MBE process eventually gave way to chemical vapor deposition (CVD) for growing germanium and GeSi films on silicon substrate. A variety of CVD growth techniques have been developed for germanium and GeSi epitaxy since 1980s and the crystalline quality is equal to, if not better than, what can be achieved in MBE process.

The development of germanium and GeSi CVD epitaxy followed the work of CVD silicon epitaxy demanded by advanced metal-oxide-semiconductor field-effect transistor (MOSFET) devices. Like silicon epitaxy, early GeSi epitaxy in CVD systems was performed at atmospheric pressure. This type of growth usually requires a hydrogen prebake with an option of hydrochloride (HCl) vapor etch at very high temperature (>1100°C) to volatilize or dissolve native SiO_2 or carbon on the silicon surface (Raider et al., 1975; Vossen et al., 1984). The growth temperature is carefully chosen (e.g. around 800oC) to balance between a reasonable germanium growth rate and the prevention of relaxation of metastable strained layer. A typical CVD reactor of such kind used by some researchers for GeSi growth (Kamins & Meyer, 1991) is made by ASM International. Atmospheric pressure CVD has insurmountable shortcomings which make it less popular in growing high quality silicon and germanium epitaxial films nowadays. The atmospheric pressure chamber condition can not avoid impurities contaminations from ambient and the need for very high temperature bake and growth causes autodoping issue, a phenomenon where doped regions existent in a substrate transfer substantial amounts of dopant into the epitaxial layer.

To overcome these shortcomings, ultra-high vacuum chemical vapor deposition (UHVCVD) was introduced by Meyerson and co-workers at IBM T. J. Watson Research Center in 1986 (Meyerson, 1986). In a UHVCVD system, the base pressure of the chamber is usually in the range of $10^{-8} \sim 10^{-9}$ torr when idling and 10^{-3} torr during growth. At this ultra high vacuum environment, the contamination can be well controlled at a very low level. For example, the system induced background partial pressure of contaminants such as water vapor, oxygen, and hydrocarbons is limited to values in the range of 10^{-11} torr (Meyerson, 1992). As a result of the superior condition of ambient and substrate surface, the germanium and GeSi epitaxial growth is performed at relatively low temperatures (400$\sim$700°C) without requiring extensive high temperature hydrogen pre-bake. UHVCVD technique is suitable for manufacturing owing to its high throughput multiple-wafer growth process. The early work of GeSi UHVCVD epitaxy focused on high silicon content GeSi alloy films driven by the demands for mobility enhancement by strained GeSi in CMOS transistors (Meyerson, 1990). There are numerous work on this type of GeSi epitaxy, however it is not our primary interest in the photonic applications described in this chapter. There was some UHVCVD germanium epitaxy reports in the early 1990s (Kobayashi et al., 1990), while the extensive development of pure germanium epitaxy by UHVCVD began in the early to late 1990s (Currie et al., 1998; Luan et al., 1999) as a result of the booming photonics research.

On the other hand, germanium can also be grown with very high carrier (mainly hydrogen gas) flow rate at relaxed chamber base pressure. Rapid growth rate and surface passivation as a result of high carrier flow rate reduces of the chance of impurities contamination. The base pressure during growth e.g. tens of Torrs, is much higher than that in the UHVCVD case while lower than atmospheric pressure thereby this approach is called reduced pressure chemical vapor deposition (RPCVD). Epitaxial germanium film with comparable threading dislocation density has been successfully grown on silicon using this approach (Hartmann et al., 2004). This approach quickly became popular as the maturely developed epitaxy reactors for (high silicon content) SiGe growth can be directly used with modified processes. It's also an approach compatible with manufacturing owing to its mature technique and rapid growth rate.

Besides the widely used UHVCVD and RPCVD approaches, there are a few other ways to grow germanium and high germanium content GeSi epitaxial films on silicon. Similarly to graded GeSi layers, GeSn can also be used as a buffer layer for germanium growth (Fang et al., 2007) though the research of the GeSn material system is still at its early stage. Epitaxial germanium has also been grown on silicon using a newly developed low energy plasma enhance chemical vapor deposition (LEPECVD) approach with low dislocation density (Osmond et al., 2009). The lack of germanium growth selectivity (between on silicon and on dielectrics) of this approach limits its use in many types of photonic devices. Besides direct epitaxy on silicon, an alternative way to form crystalline germanium on dielectrics through a so-called rapid melt growth (RMG) was introduced (Liu, Deal & Plummer, 2004). In this approach, amorphous germanium is deposited on top of dielectrics with a small window to expose crystalline silicon below the dielectrics. The germanium film is then capped with oxides and melt for a short time by rapid thermal annealing (RTA). Single crystalline germanium can be grown from the exposed silicon seed window through liquid-phase epitaxy (LPE) process. The RTA temperature profile is critical to the success of this approach though good quality of germanium film can be achieved.

As stated previously, the biggest challenge to grow high quality epitaxial germanium films with sufficient thickness is the lattice constant mismatch between germanium and silicon. To solve this problem a few approaches have been developed. A straightforward solution is to grow a graded GeSi layer or layers with the composition gradually changed from silicon to germanium. This approach was at first attempted in MBE growth (Fitzgerald et al., 1991) and was later introduced to UHVCVD growth (Samavedam & Fitzgerald, 1997) and RPCVD. It has been demonstrated that with a sufficiently thick graded layer high quality and low dislocated germanium films can be successfully grown on silicon (Currie et al., 1998).

A thick graded GeSi layer complicates epitaxial growth and device fabrication, so an alternative approach which uses a low temperature germanium layer (called buffer layer or seed layer) was developed (Colace et al., 1997). The low temperature buffer layer is used to kinetically prevent germanium from islanding and to plastically release lattice strain energy with misfit dislocations at Ge/Si interface when germanium thickness is beyond Stranski-Krastanov (S-K) critical thickness. In addition, another benefit of using low temperature is to make growing surface hydrogen act as surfactant to reduce island nucleation (Eaglesham et al., 1993). A thicker germanium film is then homo-epitaxially grown on the relaxed Ge buffer at higher temperatures for better growth rate. Therefore this approach is sometimes called two-step growth. The quality of germanium films by this type of growth is generally worse than the graded GeSi layer case, while the dislocations in the epitaxial film can be greatly minimized by a proper post annealing process (Luan et al., 1999).

Graded GeSi layer and low temperature Ge layer can both be used as buffer layer in germanium epitaxy and the required thickness of graded GeSi layer can be greatly reduced. Overgrowing germanium on a patterned silicon/SiO_2 surface is another way to reduce dislocations resulted from Ge/Si lattice mismatch (Langdo et al., 2000). The "epitaxial necking" process as a result of selective germanium growth on silicon and SiO_2 terminates dislocations on SiO_2 sidewalls and leave the overgrown germanium above Si/SiO_2 layer with reduced dislocation densities. The disadvantage of this approach is the difficulty of surface morphology control and the inconvenience of using the fine patterned Si/SiO_2 surface in many cases.

1.3 Germanium band structure

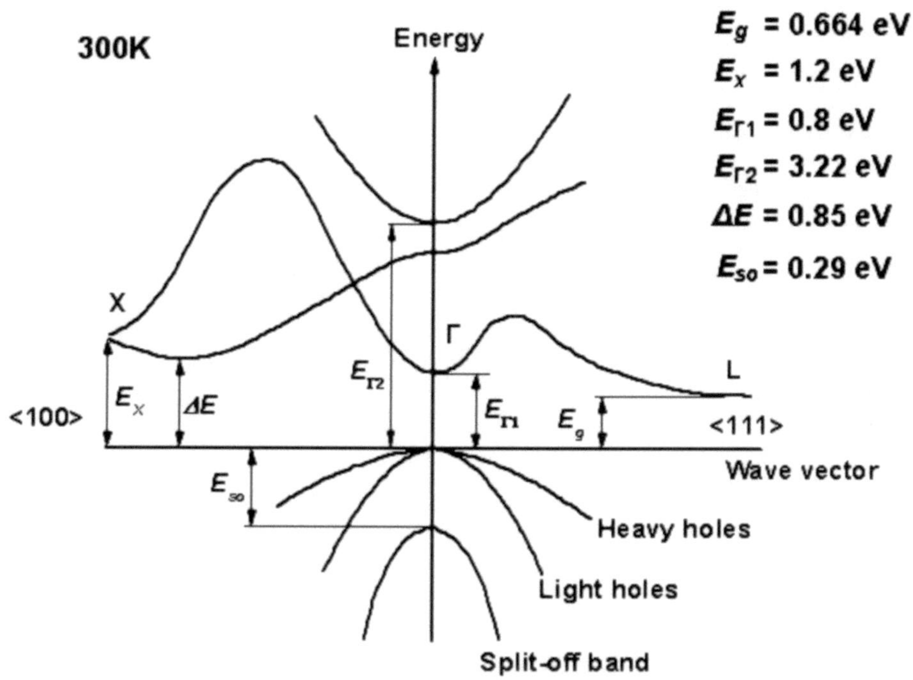

Fig. 1. Ge band structure at 300K.(M.Levinstein et al., 1996)

The electronic band structure of bulk germanium at room temperature is shown in Fig. 1. The valence band is composed of a light-hole band, a heavy-hole band, and a split-off band from spin-orbit interaction. The light-hole band and the heavy-hole band are degenerate at wave vector $\mathbf{k} = 0$ or Γ point which is at the maximal energy of the valence band. The minimal energy of the conduction band is located at $\mathbf{k} =< 111 >$ or L point. The energy difference between the conduction band at L point and the valence band at Γ point determines the narrowest band gap in germanium: $E_g = 0.664$ eV. This type of band gap is called an indirect band gap since the referred energies do not occur at the same $\mathbf{k}$. On the other hand, the two energy gaps between the two local minima of the two conduction bands and the maximum of the valence band at the same Γ point, i.e. $E_{\Gamma1}$ and $E_{\Gamma2}$ in Fig. 1, are called direct band gaps. Because the energy gap $E_{\Gamma2}$ is much larger than $E_{\Gamma1}$ and E_g, there is barely any electrons at such high energy levels so that it has negligible effect on the light-matter interaction in most cases. Therefore, people usually refer the direct band gap to $E_{\Gamma1}$ thus the author denotes $E_{g\Gamma} \equiv E_{\Gamma1}$ and $E_{gL} \equiv E_g$ in the chapter. The part of the conduction band near Γ point is usually called direct valley or Γ valley while the part near L point is called indirect valley or L valley. In germanium crystal, the energy is 4-fold degenerate with regard to the changes of the secondary total angular-momentum quantum number at L point, four L valleys are considered in the following calculations. [1]

1.4 Germanium optical absorption

In the study of germanium in optoelectronic and photonic applications, one of the most important properties is the optical absorption. The band-to-band optical absorption is a process that transferring the energy of an incoming photon to an electron in valence band and make the electron jump to conduction band and leave a hole in valence band. When this process occurs in a crystalline material, both energy and momentum are conserved for the system. Therefore, the rate of such event is much lower in an indirect band-to-band transition compared to a direct band-to-band transition due to the need of one or more phonons to conserve momentum. The band structure described earlier indicates that the optical absorption of bulk germanium is substantial at any wavelengths less than 1.55 μm (corresponding to 0.80 eV direct band gap energy) and the direct optical absorption is what matters in most of the applications studied here.

The direct gap absorption of a semiconductor can theoretically modeled by solving the electron-photon scattering in crystalline potential with Fermi's golden rule. The detailed mathematical derivation is skipped the only the result is shown here:

$$\alpha(h\nu) = \frac{e^2 h c \mu_0}{2{m_e}^2} \frac{|p_{cv}|}{n} \frac{1}{h\nu} \rho_r(h\nu - E_g), \tag{1}$$

where $|p_{cv}|$ is related to an element of optical transition operator matrix and n is index of refraction, both of which are material properties and may be considered as constants within a small range of wavelengths. $h\nu$ and E_g are the energy of photon and band gap respectively. ρ_r is the joint density of states of the conduction band (Γvalley) and the valence band. A quadratic approximation is usually adapted to describe the density of states near an extremum of any energy band in semiconductor and ρ_r is subsequently calculated

$$\rho_r(h\nu - E_g) = 2\pi \left(\frac{2m_r}{h^2}\right)^{3/2} \sqrt{h\nu - E_g}, \tag{2}$$

where $m_r = m_c m_v/(m_c + m_v)$ is the reduced effective mass as an expression of the effective masses of conduction band (m_c) and the valence band (m_v).

In a concise form, the direct gap absorption can be approximately written as

$$\alpha(h\nu) = A \frac{\sqrt{h\nu - E_g}}{h\nu}, \tag{3}$$

where A is a constant usually determined from experiments.

Some experimental optical absorption data of intrinsic and unstrained germanium at photon energies around its direct band gap (Braunstein et al., 1958; Dash & Newman, 1955; Frova & Handler, 1965; Hobden, 1962) are shown in Fig. 2. The sharp absorption drop at a photon energy of 0.8 eV from all the data sources indicates the energy of the direct gap. The data from (Dash & Newman, 1955) and (Hobden, 1962) show good agreement at photon energies more

[1] Degeneracy with regard to the changes of electron spin quantum number is not explicitly accounted here though it is considered in density of states calculations.

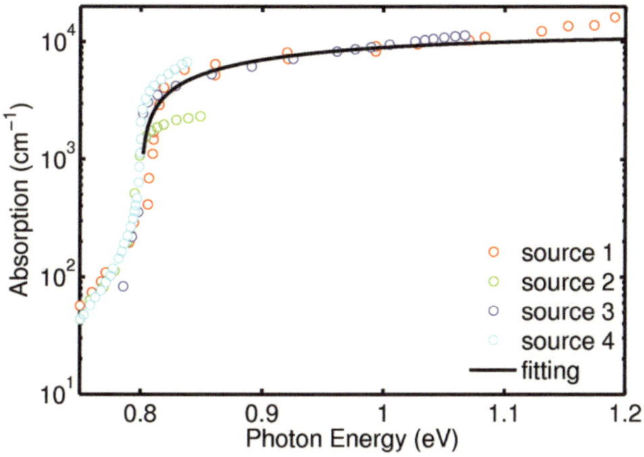

Fig. 2. Comparison of germanium absorption from different published sources. The experimental data (Source 1 to 4 in the figure) are obtained from (Braunstein et al., 1958; Dash & Newman, 1955; Frova & Handler, 1965; Hobden, 1962), respectively. The fitting based on the theoretical model is drawn with a black solid line.

than the direct gap. A calculation based on the above theoretical model (Equation 3) with $A = 2.0 \times 10^4$ eV$^{1/2}$/cm is shown with black solid line and fits well with the experimental data.

2. Waveguide integrated germanium photodetector

2.1 Tensile strain engineering of germanium

For an integrated optical interconnect and transmission system on silicon platform, any wavelength above silicon absorption edge (about 1.1 μm) can be adopted for photon carriers as silicon is the guiding material. Shorter wavelengths can also be used if other compatible materials (e.g. silicon nitride) are used for wave guiding however silicon is a preferred material for the miniaturization reason owing to the high refractive index contrast between silicon and cladding materials (e.g. silicon oxide). But wavelengths near 1.55 μm (fiber optics C-band) are commonly used because the communication design tool boxes at this wavelength band are mature in fiber optics technology. But germanium has weak absorption at 1.55 μm which is about its direct band gap energy as described in the band diagram earlier. Germanium strain engineering has been adopted to address this issue.

Semiconductor band structure is associated with the crystal structure which can be altered by the existence of strain. This effect can be calculated using a strain-modified $k \cdot p$ method Chuang (1995). Pikus-Bir Hamiltonian and Luttinger-Kohn's model are used in the method to describe the degenerate bands in germanium. This calculation shows that strain changes the energy levels of the direct Γ valley, the indirect L valleys, the light-hole band, and the heavy-hole band relative to vacuum level.[2] As a result, the direct band gap and indirect band gap are changed and the light-hole and the heavy-hole bands become non-degenerate with separation at Γ point. A band structure comparison of unstrained germanium and 0.2% tensile-strained germanium is shown in Fig. 3.

---

[2] The energy levels of other bands such as spin-orbit split-off band are also changed.

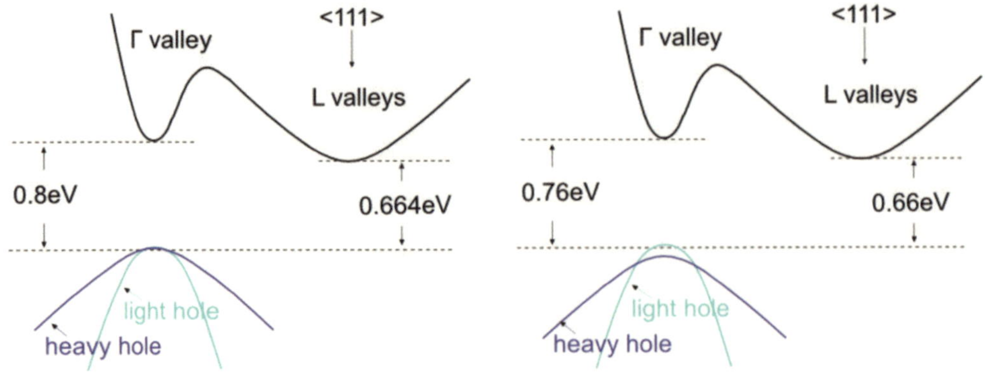

(a) (b)

Fig. 3. Comparison of the band structures of (a) unstrained germanium and (b) strained germanium.

The direct band gap and indirect band gap under strain can be calculated using this method[3]

$$E_{g\Gamma hh} = a_{c\Gamma}(\epsilon_{xx} + \epsilon_{yy} + \epsilon_{zz}) + P + Q \tag{4}$$

$$E_{g\Gamma lh} = a_{c\Gamma}(\epsilon_{xx} + \epsilon_{yy} + \epsilon_{zz}) + P - Q/2 + E_{so}/2 - \sqrt{E_{so}^2 + 2E_{so}Q + 9Q^2} \tag{5}$$

$$E_{gLhh} = a_{cL}(\epsilon_{xx} + \epsilon_{yy} + \epsilon_{zz}) + P + Q \tag{6}$$

$$E_{gLlh} = a_{cL}(\epsilon_{xx} + \epsilon_{yy} + \epsilon_{zz}) + P - Q/2 + E_{so}/2 - \sqrt{E_{so}^2 + 2E_{so}Q + 9Q^2} \tag{7}$$

where

$$P = -a_v(\epsilon_{xx} + \epsilon_{yy} + \epsilon_{zz}) \tag{8}$$

$$Q = -b(\epsilon_{xx}/2 + \epsilon_{yy}/2 - \epsilon_{zz}) \tag{9}$$

. $E_{g\Gamma hh}$ and $E_{g\Gamma lh}$ are energy gaps between the Γ valley and the heavy-hole band and the light-hole band respectively. E_{gLhh} and E_{gLlh} are energy gaps between the L valley and the heavy-hole band and the light-hole band respectively. ϵ_{xx}, ϵ_{yy}, and ϵ_{zz} are strain components. E_{so} is the energy difference between valence bands and spin-orbit split-off band at Γ point. $a_{c\Gamma}$, a_{cL}, a_v and b are deformation potentials for Γ valley, L valleys, the average of three valence bands (light-hole, heavy-hole and spin-orbit split-off) and a strain of tetragonal symmetry. They are material properties which can be either calculated from first-principle calculation or determined by experiments.

For epitaxial germanium films, the strain is usually induced by in-plane stress from adjacent layers. If a biaxial stress is applied, i.e.

$$\epsilon_{xx} = \epsilon_{yy}, \tag{10}$$

the relation of the stress tensors for the isotropic material can be determined from Hooke's law in tensor form:

$$\epsilon_{zz} = -2C_{12}/C_{11}\epsilon_{xx}, \tag{11}$$

[3] Shear strain $\epsilon_{xy(x\neq y)}$, which is negligible in the thin film material, is not considered here.

where C_{11} and C_{12} are the elements of the elastic stiffness tensor.

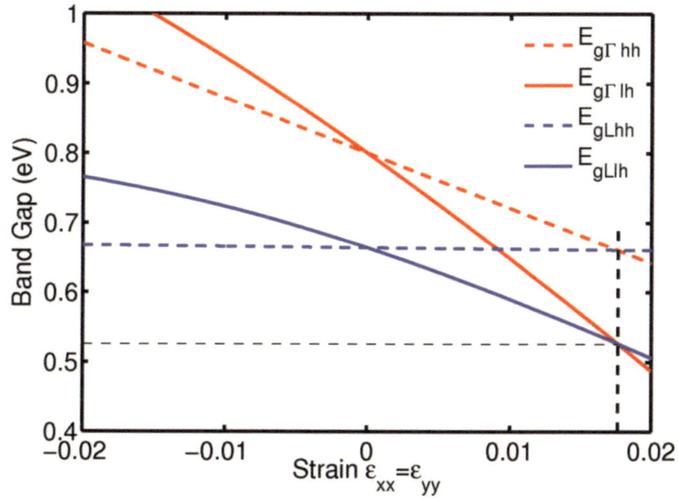

Fig. 4. The direct and the indirect band gaps of germanium with in-plane strain. $E_{g\Gamma hh}$ and $E_{g\Gamma lh}$ are energy gaps between the Γ valley and the heavy-hole band and the light-hole band respectively. E_{gLhh} and E_{gLlh} are energy gaps between the L valley and the heavy-hole band and the light-hole band respectively.

The direct and the indirect band gaps of strained germanium can be obtained using Eq. 4−7 with the experimental results of $a_{c\Gamma} = -8.97$ eV and $b = -1.88$ eV from reference (Liu, Cannon, Ishikawa, Wada, Danielson, Jongthammanurak, Michel & Kimerling, 2004) $C_{11} = 128.53$ GPa and $C_{12} = 48.26$ GPa from reference (Madelung & et al, 1982), and the calculated results of a_{cL}=-2.78 eV and a_v=1.24 eV from reference (de Walle, 1989). These energy gaps are calculated and shown in Fig. 4. We can see both the direct band gap and the indirect band gap shrink with tensile strain (positive ϵ_{xx}) and the direct band gap decreases faster than the indirect band gap does due to $|a_{c\Gamma}| > |a_{cL}|$. The direct band gap becomes equal to the indirect band gap at $\epsilon \approx 1.8\%$ where Ge becomes a direct band gap material. The energy gaps from the conduction band (equal for both direct Γ valley and indirect L valley) to the light-hole and the heavy-hole band are 0.53 eV and 0.66 eV respectively.It is noted that there are two kinds of energy gaps at both Γ point and L point due to the separation of the light-hole band and the heavy-hole band. The optical band gap is determined by the smaller gap with respect to the light-hole band. We refer the band gap to this energy gap in the following discussion unless explicitly stated otherwise.

The absorption edge of tensile strain germanium moves towards lower energy because of the shrinkage of the direct band gap. Since the light-hole band and the heavy-hole band separate under strain, two optical transitions corresponding to the two energy gaps contribute the overall optical absorption. Therefore the absorption spectrum of tensile-strain germanium can be expressed as

$$\alpha(h\nu) = A\left(k_1\frac{\sqrt{h\nu - E_{g\Gamma lh}}}{h\nu} + k_2\frac{\sqrt{h\nu - E_{g\Gamma hh}}}{h\nu}\right), \qquad (12)$$

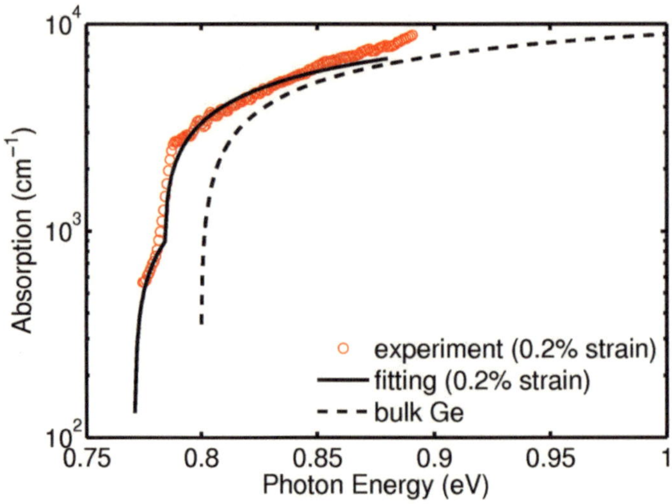

Fig. 5. The experimental result and the theoretical fitting of the optical absorption of 0.2% tensile-strained germanium. The optical absorption of unstrained germanium is shown with a black dash line for comparison.

where $k_1 = m_{rlh}{}^{3/2}/(m_{rlh}{}^{3/2} + m_{rhh}{}^{3/2})$ and $k_2 = m_{rhh}{}^{3/2}(m_{rlh}{}^{3/2} + m_{rhh}{}^{3/2})$ are coefficients attributing to the difference between the two transitions due to the different reduced effective masses. k_1 and k_2 are normalized so that $k_1 + k_2 = 1$. $k_1 = 0.68168$ and $k_2 = 0.31832$ are calculated for germanium. The measured absorption spectrum of 0.2% tensile-strained germanium is shown in Fig. 5 and $A = 1.9 \times 10^4$ $\mathrm{eV}^{1/2}/\mathrm{cm}$ is calculated by fitting the experimental result using Eq. 12. The A of strained germanium is approximately the same as that of unstrained germanium which underlies that applied strain is too small to affect the optical transition matrix of Fermi's golden rule.

Introduction of tensile strain in germanium is not intuitive in germanium-on-silicon epitaxy. It is usually understood that strain is induced from the lattice mismatch between epitaxial and substrate materials. In such way, epitaxial germanium is compressively strained as the lattice constant of germanium (5.658 Å) is greater than that of silicon (5.431 Å). However, the above analysis is only applied to the case that an epitaxial layer is thinner than its critical thickness on a certain substrate. Beyond this thickness, the elastic strain of the epitaxial film releases plastically by introducing misfit dislocations near material interface region. Another possible result beyond critical thickness is that three dimensional growth (islanding) occurs to balance elastic strain energy with surface tension.

In the two-step growth described earlier, the buffer layer is grown at low temperature to prevent such islanding process by reducing the mobility of adsorbed atoms on growing surface. Thus the following germanium layer grown at higher temperatures layer is on a strain-relaxed surface of the buffer layer, the thickness which is more than critical thickness. The epitaxial germanium and the silicon substrate shrink as the substrate temperature is cooled down to room temperature. Germanium shrinks more due to its larger coefficient of thermal expansion than that of silicon therefore in-plain tensile strain is accumulated in the germanium film as illustrated in Fig. 6.

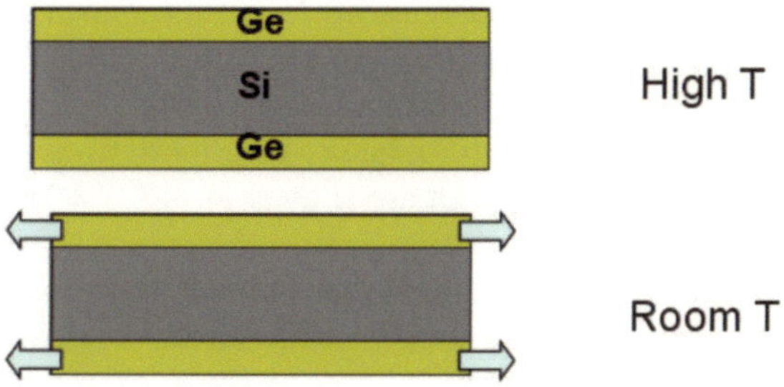

Fig. 6. Illustration of tensile strain formation in germanium on silicon epitaxy.

As silicon substrate is always much thicker than germanium epitaxial layer, the tensile strain $\epsilon_{//}$ in germanium caused by in-plain stress can be obtained using

$$\epsilon_{//} = \int_{T_0}^{T_1} (\alpha_{\mathrm{Ge}} - \alpha_{\mathrm{Si}}) dT, \tag{13}$$

where, T_0 and T_1 are room temperature and growth temperature respectively, and $\alpha_{\mathrm{Ge}} = \alpha_{\mathrm{Ge}}(T)$ and $\alpha_{\mathrm{Si}} = \alpha_{\mathrm{Si}}(T)$ are thermal expansion coefficients of germanium and silicon respectively. In general, these coefficients are a function of temperature and the relations are experimentally determined (Singh, 1968) and (Okada & Tokumaru, 1984):

$$\alpha_{\mathrm{Ge}}(T) = 6.050 \times 10^{6} + 3.60 \times 10^{9} T - 0.35 \times 10^{-12} T^2 (^oC^{-1}), \tag{14}$$

$$\alpha_{\mathrm{Si}}(T) = 3.725 \times 10^{-6} \times [1 - \exp(-5.88 \times 10^{-3} \times (T + 149.15))] + 5.548 \times 10^{-10} T (^oC^{-1}). \tag{15}$$

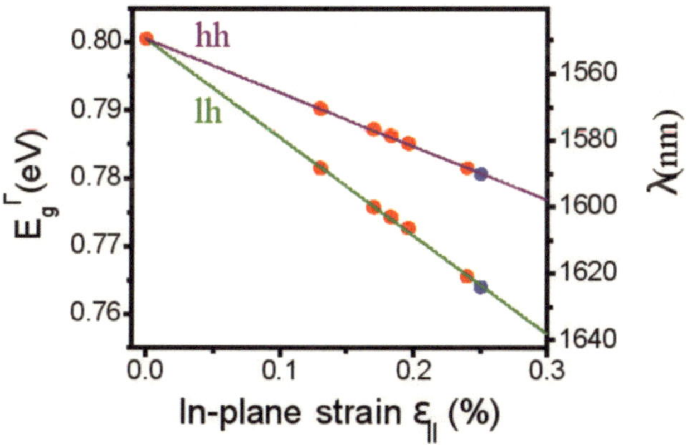

Fig. 7. Ge direct energy gaps versus tensile strain measured by X-ray diffraction (XRD) and photoreflectance (PR) experiments. (Liu, Cannon, Ishikawa, Wada, Danielson, Jongthammanurak, Michel & Kimerling, 2004)

From the above formula, different tensile strain can be achieved with different growth temperature and the experimental results of germanium grown at various temperatures are also plotted in Fig. 7. The energy gaps are measured by photoreflectance (PR) experiment and the tensile strain is obtained from crystal lattice constant measurement by X-ray diffraction

(XRD) (Liu, Cannon, Ishikawa, Wada, Danielson, Jongthammanurak, Michel & Kimerling, 2004). The direct-gap deformation potentials of germanium can then be calculated from fitting the experimental data. The calculated two direct energy gaps, i.e. from the direct Γ valley to the light-hole band (indicated by lh) and to the heavy-hole band (indicated by hh) respectively, of germanium as a function of tensile strain are shown in Fig. 7 with the fitted deformation potentials.

If post-growth thermal annealing is performed, germanium epitaxial layer is further relaxed at the annealing temperature therefore more tensile-strain is possibly obtained at room temperature. In practice, it is found that the tensile strain in germanium cease to increase at temperatures beyond about 750 °C and the measured strain is usually less than the theoretical values calculated from above formula (Cannon et al., 2004). It is believed that the existence of residual compressive strain retrains the full relaxation of germanium.

2.2 Germanium photodetector

Germanium has been investigated as an efficient photodetector on silicon platform since the germanium MBE was realized. Although the early work of germanium on silicon photodetector was in 1980s (Luryi et al., 1984), it only became extensively researched since late 1990s aligned with the boom of CVD epitaxy of germanium on silicon (Colace et al., 2000; 1998; Dehlinger et al., 2004; Hartmann et al., 2004; Liu, Michel, Giziewicz, Pan, Wada, Cannon, Jongthammanurak, Danielson, Kimerling, Chen, Ilday, Kartner & Yasaitis, 2005; Luan et al., 2001).

Early epitaxial germanium photodetectors are mostly vertical P-N or P-I-N junction devices designed for surface optical incidence for the ease of fabrication. There is a trade-off between optical absorption and carrier transit time for this type of photodetector and many of devices are designed for a 3dB bandwidth less or around 10GHz for 10 Gb/s applications (Colace et al., n.d.; Liu, Michel, Giziewicz, Pan, Wada, Cannon, Jongthammanurak, Danielson, Kimerling, Chen, Ilday, Kartner & Yasaitis, 2005; Morse et al., 2006). A lateral P-I-N junction design can be adopted to reduce carrier transit time for very high speed applications though the photo responsivity is usually compromised (Dehlinger et al., 2004). Therefore, waveguide coupled germanium photodetectors are favored in high speed applications especially for devices working at 1.55 μm in which longer germanium absorption length is required due to weaker absorption.

The speed of photodetectors is characterized by its 3dB bandwidth which is usually limited by a combination of Resistive-capacitive (RC) delay and carrier transit time. The RC delay limited 3dB bandwidth is

$$f_{RC} = \frac{1}{2\pi RC}. \tag{16}$$

It be can proved that the carrier transit time limited 3dB bandwidth is

$$f_{tr} = 0.44 \times \frac{v_d}{d}, \tag{17}$$

where v_d is the average drift velocity of carriers (electrons or holes) and d is the junction depletion width in germanium. v_d is generally a function of electric field (and temperature) but it approaches the saturation velocity v_{sat} when applied electric field is sufficiently large. The experimentally measured electron drift velocity in germanium as a function of electric field (Jacoboni et al., 1981) is shown in Fig. 8. This relation can be described by an empirical formula:

$$v_d = \frac{1}{(1/(v_{sat})^n + 1/(\mu E)^n)^{1/n}}, \tag{18}$$

where $\mu = 3.7 \times 10^3 cm^2/V \cdot s$ is the electron mobility and $n = 2.5$ is a coefficient determined empirically. The saturation velocity of electrons in germanium is about 6×10^6 cm/s while the value is a little less for holes.

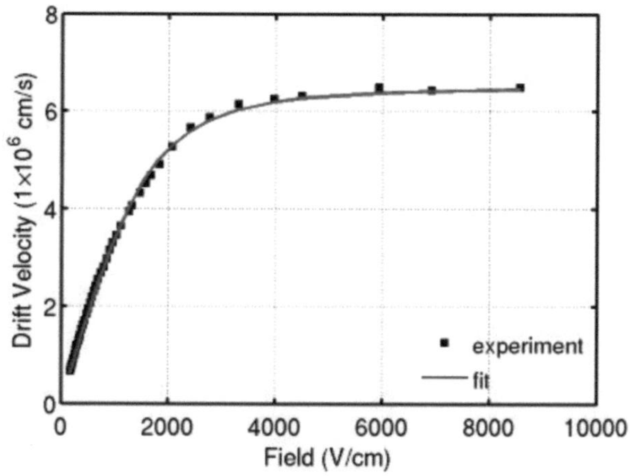

Fig. 8. Experimental result and theoretical fitting of the electron drift velocity in germanium versus electric field.

When both RC delay and transit time are considered the overall 3dB bandwidth is

$$f_{3dB} = \frac{1}{\sqrt{1/f_{RC}{}^2 + 1/f_{tr}{}^2}}. \tag{19}$$

Based on this formula, the calculated 3dB bandwidth contour of a germanium diode with respect to its depletion width and device area is shown in Fig. 9. In general, the transit time limits the maximal thickness of depletion width which is close to the germanium thickness for a fully depleted vertical P-I-N junction diode. The RC delay, on the other hand, limits the maximal device area. In the calculation, a termination resistance of 50 Ω is used for the compatibility of common impedance and no parasitics is considered. When the series resistance and parasitic capacitance (e.g. contact pad capacitance) are taken into account, the 3dB bandwidth is reduced.

Dark current is another important parameter of photodetectors. Germanium photodetector usually suffers from a larger dark current as a result of its narrow indirect band gap and poor surface passivation capability. A narrow band gap leads to higher density of intrinsic carriers and higher trap assisted generation rate through Shockley-Read-Hall (SRH) process. The surface passivation of germanium, i.e. termination of the dangling bonds or surface states, is generally more different than silicon as germanium oxide is much less robust than silicon dioxide. Therefore people use other materials such as hydrogenate amorphous silicon (aSi:H), hydrogen-rich silicon nitride (Si_3N_4), germanium oxynitride ($Ge(ON)_x$) and etc. to passivate germanium surface. The dark current associated with material quality (e.g. defects)

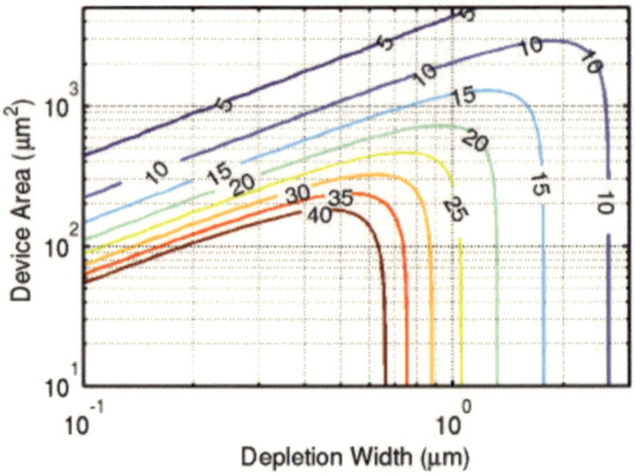

Fig. 9. Calculated 3dB bandwidth contour of a germanium diode with respect to its depletion width and device area in ideal conditions.

is sometimes called bulk leakage while that with surface condition is called surface leakage. The former one is scaled with the area of devices while the latter one is scaled with the perimeter.

As surface leakage highly depends on surface passivation processes, bulk leakage is a good measure of material quality. In many epitaxial germanium photodetectors with relatively larger size, bulk leakage dominates overall dark current. As we mentioned earlier in this chapter, misfit dislocations exist at germanium and silicon interface as a result of lattice mismatch and stress relaxation. These misfit dislocations along with other crystal imperfections are important sources of threading dislocations in germanium. These threading dislocations introduce deep level traps (SRH recombination-generation centers) which result in the majority of bulk leakage current in many cases. The effect of threading dislocation on leakage current can be obtained by calculating generation current induced by deep level traps:

$$J_{gen}/d = \frac{en_i}{\tau}, \tag{20}$$

where J_{gen}/d is the generation current per unit length, e is elementary charge, n_i is intrinsic carrier density and the minority lift time τ is equal to

$$\tau = \frac{1}{\sigma v_{th} N_D N_{TD}}, \tag{21}$$

where σ is trap capture cross-section, v_{th} is carrier thermal velocity, N_D is threading dislocation density and N_{TD} is the number of traps per unit length of dislocation. By substituting the expression of τ the current density per length

$$J_{gen}/d = en_i\sigma v_{th} N_D N_{TD}. \tag{22}$$

The above formula can be applied to either electrons or holes though one of them usually dominates the generation process. In germanium, $v_{th} = 2.3 \times 10^7 \text{cm/s}$ for electrons and $v_{th} = 1.7 \times 10^7 \text{cm/s}$ for holes. The trap capture cross section is different for different types of traps and it usually between 1×10^{-14}cm^{-2} to 1×10^{-15}cm^{-2}. The leakage current density per unit length as a function of threading dislocation density is calculated by using $\sigma = 1 \times 10^{-15}$cm^{-2} for $N_{TD} = 1 \times 10^{7}$cm^{-1} and $N_{TD} = 1 \times 10^{8}$cm^{-1}, respectively. The results are shown in Fig. 10. A few experimental results (Colace et al., n.d.; Fama et al., n.d.; Liu, Cannon, Wada, Ishikawa, Jongthammanurak, Danielson, Michel & Kimerling, 2005; Samavedam et al., 1998; Sutter et al., 1994) are also presented in the figure for comparison. For the experimental results, the dark current density is divided by the germanium thickness as an approximating for the depletion width. All the experimental results fall in the two calculation curves and show good agreement with the calculated relation.

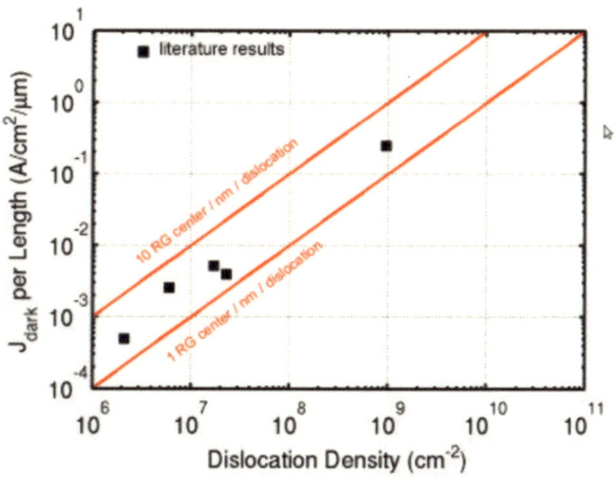

Fig. 10. Summary of dark current density (at -1V) bias versus measured threading dislocation density from various literature (Colace et al., n.d.; Fama et al., n.d.; Liu, Cannon, Wada, Ishikawa, Jongthammanurak, Danielson, Michel & Kimerling, 2005; Samavedam et al., 1998; Sutter et al., 1994). Theoretical calculation with two conditions of generation-recombination center per unit length are shown in two solid lines.

2.3 Integration of germanium photodetector and waveguide

Waveguide coupled photodetector is adopted to break the trade-off between optical absorption and device speed as stated earlier. It also allows the realization of planar integration of photodetectors with electronics on silicon substrate. Therefore many research work on silicon waveguide coupled germanium photodetectors emerged soon after the surge of surface incidence epitaxial germanium photodetector research. The research started with polycrystalline germanium on silicon waveguide (Colace et al., 2006) followed by numerous work of waveguide coupled epitaxial germanium photodetectors around the world. In most of these work, silicon-on-insulator is the choice for substrate to form silicon waveguide with optical confinement by buried oxide and surrounding low refractive index materials (Feng et al., 2009; 2010; Masini et al., 2007; Vivien et al., 2009; 2007; Wang, Loh, Chua, Zang, Xiong, Loh, Yu, Lee, Lo & Kwong, 2008; Wang, Loh, Chua, Zang, Xiong, Tan, Yu, Lee, Lo & Kwong, 2008; Yin et al., 2007) while silicon nitride can also be adopted as waveguide to couple light into germanium (Ahn et al., 2007). Most of these photodetectors are vertical or lateral P-I-N diodes while quantum well germanium photodetectors are also investigated (Fidaner et al., 2007). Among these devices, more than 30GHz 3dB bandwidth have been achieved at reasonable bias (usually -1V to -2V) with good responsivity (e.g. 0.9-1.0 A/W). These high speed photodetectors can be potential substitutes for commercially available photodetectors based on III-V materials for up to 25Gb/s digital transmission applications.

In waveguide coupled photodetectors, one of the critical design roles is to optimize waveguide to germanium optical coupling. The required length of a photodetector is determined by the coupling efficiency for an evanescent coupling. Therefore the device speed is also affected by coupling design as the capacitance of the device is scaled with its dimension. In general, the germanium is on top of an input silicon waveguide for evanescent coupling. This design is usually adopted for small silicon waveguides such as channel waveguides or small rib waveguides on SOI substrates with relatively thin device silicon layer. However, for a large rib waveguide the coupling efficiency is very sensitive to the thickness of germanium layer as shown later. Therefore it is sometimes necessary to make a mode converter to push guided light down to the slab part of the rib waveguide and couple the light from the slab to germanium photodetectors. The two different designs are schematically shown in Fig. 11.

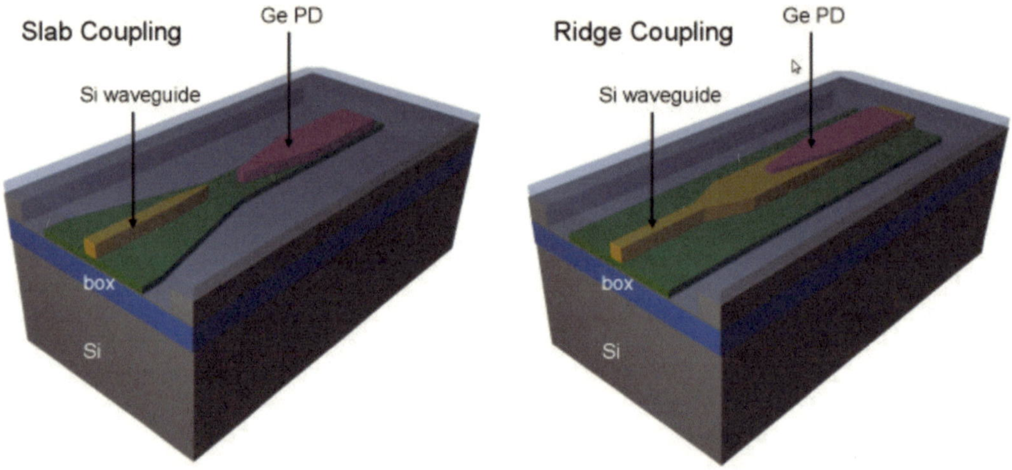

Fig. 11. Two designs of waveguide to germanium photodetector coupling on SOI substrate.

The coupling between waveguide and germanium can be simulated by numerical methods such as finite-difference time-domain (FDTD) method. A sample simulation results is shown in Fig. 12. In the figure, the propagated electric field shows the process of evanescent coupling. By calculating the absorption in germanium in this coupling process, the ideal responsivity can be obtained. Simulated ideal responsivity as a function of germanium thickness for waveguide-coupled photodetector with two silicon thicknesses (0.55μm and 0.8μm) are shown in Fig. 13. The responsivity is generally a periodic function of germanium thickness as a result of coupling resonance conditions. But the sensitivity of peak responsivity on germanium thickness is higher for thicker silicon thickness. It is the reason of reducing silicon thickness for large silicon rib waveguides to avoid the sensitivity of the performance of germanium photodetectors due to processing variations. The reduction of the coupling sensitivity on the polarization state of guided modes is another design consideration. The peak coupling conditions of a transverse-electric (TE) mode and a transverse-magnetic (TM) mode are not usually aligned. Optimizing TE and TM modes or creating multiple modes in the waveguides with some mode transformers can minimize the polarization sensitivity.

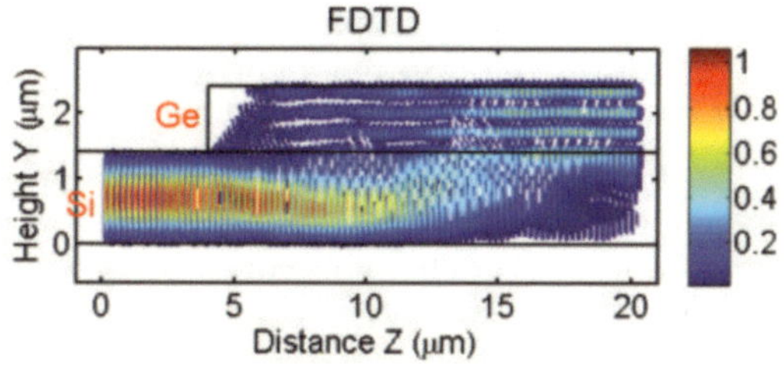

Fig. 12. FDTD simulation of light coupling from silicon waveguide to germanium photodetector.

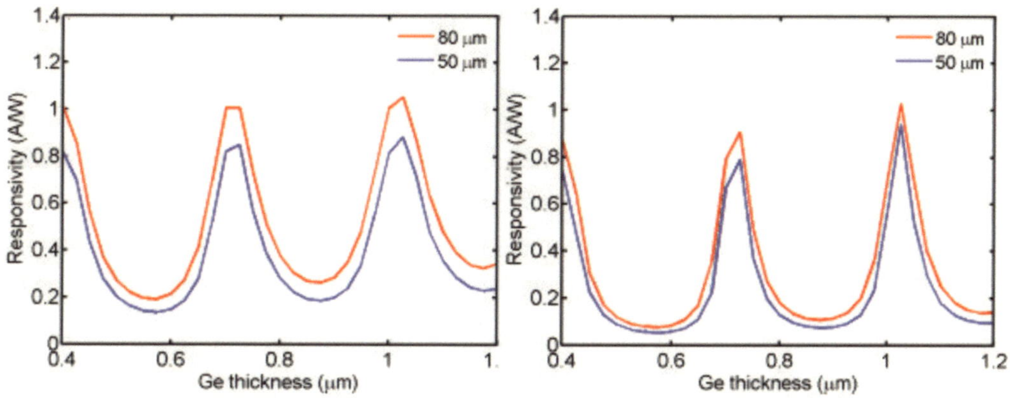

(a) (b)

Fig. 13. Simulated ideal responsivity versus germanium thickness for waveguide-coupled photodetector with silicon thickness of (a) 0.55μm and (b) 0.8μm, respectively.

3. Germanium light source

3.1 Silicon based light source

Photonic-electronic monolithic integration requires a silicon-based light source. A silicon-based laser is arguably the most challenging element in silicon photonics because both silicon and SiGe (including pure germanium) are inefficient light emitters due to their indirect band structure. Many approaches have been investigated to solve this challenge including the efforts on porous silicon (Gelloz & Koshida, 2000; Koshida & Koyama, 1992), silicon nanostructures (Irrera et al., 2003) and SiGe nanostructures (Peng et al., 1998), silicide (Leong et al., 1997), erbium doped silicon (Zhang et al., 1994). However, all above approaches are challenged by the lack of sufficient gain to surpass material loss for laser action.

Erbium-doped silicon dielectrics which benefit from less energy back transfer than erbium-doped silicon are potential gain media . Erbium atoms which are capable of light emission at 1.54 μm have been incorporated into silicon oxide (Adeola et al., 2006; Fujii et al., 1997; Kik et al., 2000) or silicon nitride (Makarova et al., 2008; Negro et al., 2008) matrix materials. Silicon nanocrystals are sometimes introduced in these materials as recombination sensitizers. The optical gain of such extrinsic light-emitting materials is generally very small (Han et al., 2001) due to limited erbium solubility and energy up-conversion. Therefore lasing can only occur in extremely low loss resonators such as toroidal structures (Polman et al., 2004). These dielectric materials also suffer from the difficulty of electrical injection. Appreciable concentration of carriers only presents under very high electric field via effects like tunneling process (Iacona et al., 2002; Nazarov et al., 2005).

Non-linear optical effects can also produce net gain in silicon. Based on stimulated Raman scattering (SRS) effect, optically pumped silicon waveguide lasers have been realized with pulse (Boyraz & Jalali, 2004; Rong et al., 2004) and continuous-wave operation (Rong et al., 2005). Similarly, the inevitable requirement of optical injection makes them not suitable for integrated photonics.

Following a hybrid approach, researchers successfully integrated III-V semiconductor lasers on silicon substrates. The integration can be accomplished by directly growing GaAs/InGaAs on silicon with graded Si_xGe_{1-x} buffer layers (Groenert et al., 2003) or bonding III-V materials on silicon surface for laser fabrication (Fang et al., 2006; Park et al., 2005).

Electronic-photonic monolithic integration requires a light source with the capability of electrical injection and room temperature operation based on silicon compatible materials and CMOS-compatible processes. Germanium is a very promising material for making such a light source if it can be engineered for more efficient direct gap light emission and net optical gain.

3.2 Germanium band structure engineering

The inefficiency of light emission from germanium comes from is its indirect band structure. To understand how we can engineer its band structure for more efficient light emission and even net optical gain we start with the carrier recombination analysis in germanium.

The band structure and the carrier distribution of germanium at equilibrium are drawn in the Fig. 14 (a). Most of the thermally activated electrons occupy the lowest energy states in the indirect L valleys governed by Fermi distribution:

$$f(E) = \frac{1}{1 + (E - E_f)/k_B T}, \tag{23}$$

where E_f is equilibrium Fermi level. For a direct gap material InGaAs shown in Fig. 14 (b), most of the electrons occupy the direct Γ valley.

The carrier occupancy states determine the light emitting properties of a material. The band-to-band optical transition requires excess free carriers which are injected electrically or optically. At steady state, the carriers obey quasi Fermi distribution with respect to carrier quasi Fermi level:

$$f_c(E) = \frac{1}{1 + (E - E_{fc})/k_B T}, \text{ and} \tag{24}$$

$$f_v(E) = \frac{1}{1 + (E - E_{fv})/k_B T}, \tag{25}$$

where E_{fc} and E_{fv} are the quasi Fermi levels of electrons and holes. A single quasi Fermi level exists for electrons or holes in steady state despite of the multi-valley band structure because of very fast inter-valley scattering process.

The carrier distribution under injection for germanium and InGaAs are shown in Fig. 15. Despite germanium is an indirect band gap material, there are still some electrons pumped to the Γ valley owing to the small energy difference (0.136 eV) between the direct band gap and the indirect band gap. The excess electrons in the Γ valley recombine radiatively with the holes in the valence band. This light emission process is as efficient as the direct

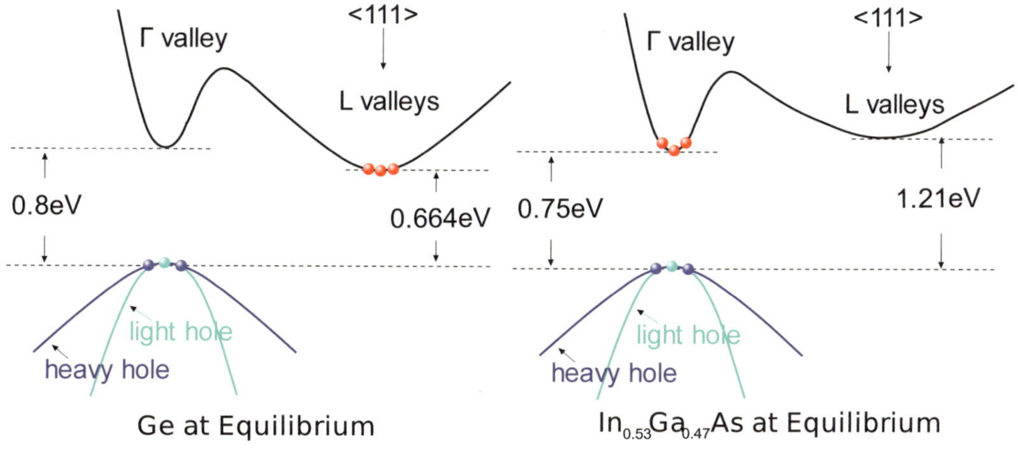

(a) (b)

Fig. 14. Comparison of the band structure and the carrier distribution between (a) Ge and (b) $In_{0.53}Ga_{0.47}As$ at equilibrium.

radiative recombination in direct band gap materials. But the overall light emission from germanium is weak because most of the injected electrons are in the *L* valleys. Most of these electrons recombine non-radiatively as indirect phonon-assisted radiative recombination is very slow. For III-V direct gap materials, injected electrons are in the direct valley so overall light emission is efficient.

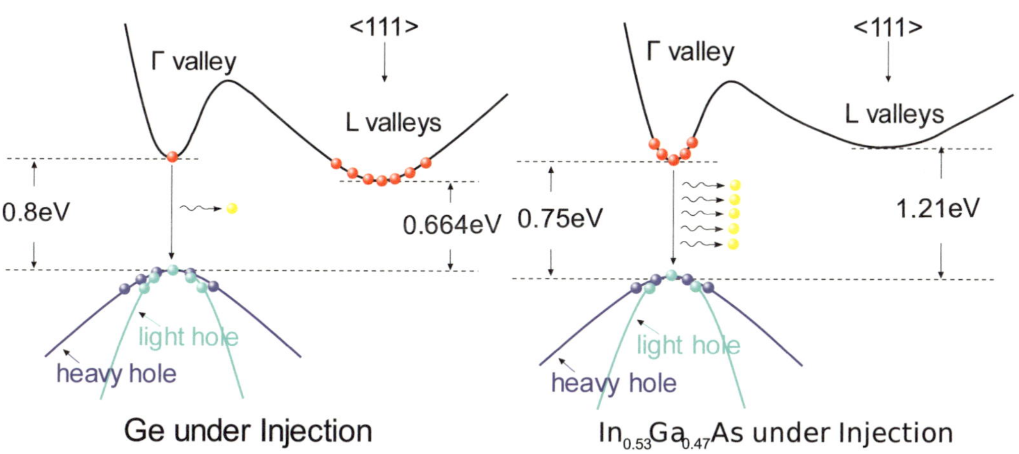

(a) (b)

Fig. 15. Comparison of the carrier distribution and the light emission process between (a) Ge and (b) $In_{0.53}Ga_{0.47}As$ under injection.

To improve the light emission efficiency in germanium, more injected electrons are required to be pumped into Γ valley. In other words, the objective is to make germanium a direct gap material. In the germanium photodetector section, it has shown that tensile strain can used to reduce direct band gap for more optical absorption. It should also be noted that the difference of direct and indirect band gaps are also reduced under tensile strain. Based on the calculation a tensile strain of about 1.8% is required to make germanium direct band gap. It is difficult to achieve such high strain in epitaxial germanium. By using a thermal mismatch approach to stress epitaxial germanium on silicon as explained earlier, the maximal tensile strain is below 0.20~0.25%. The energy difference between the direct and the indirect gaps at this condition is only reduced by 20 meV therefore the majority of the injected electrons still occupy the indirect L valleys.

N-type doping can be used to solve this problem. When the indirect L valleys are filled with the extrinsic electrons thermally activated from n-type donors, the Fermi level is raised to push more injected electrons into the direct Γ valley. An example of the Fermi level versus active n-type doping concentration for 0.25% tensile-strained germanium is shown in Fig. 16. The Fermi level becomes equal to the bottom of Γ valley at a doping concentration of 7×10^{19} cm^{-3}, where most states of the indirect L valleys are filled by electrons and Ge behaves as a direct band gap material.

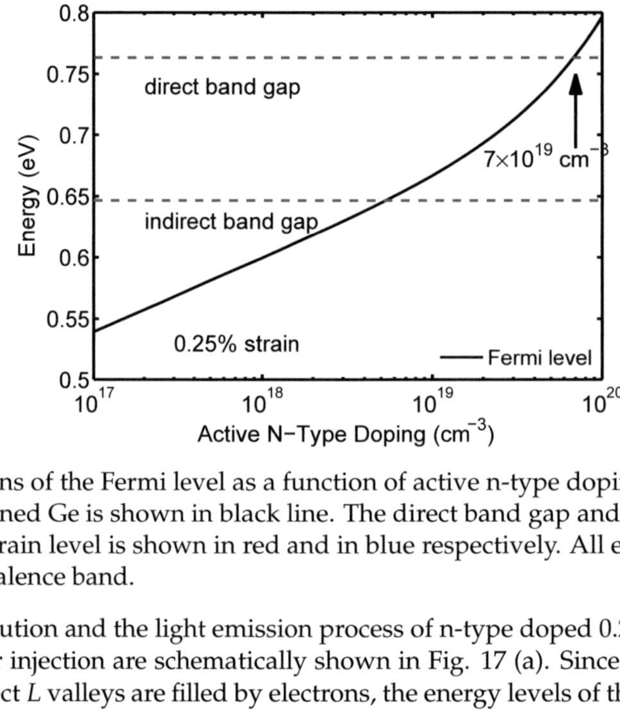

Fig. 16. Calculations of the Fermi level as a function of active n-type doping concentration in 0.25% tensile-strained Ge is shown in black line. The direct band gap and the indirect band gap at the same strain level is shown in red and in blue respectively. All energies are referred to the top of the valence band.

The carrier distribution and the light emission process of n-type doped 0.25% tensile-strained germanium under injection are schematically shown in Fig. 17 (a). Since the lower quantum states of the indirect L valleys are filled by electrons, the energy levels of the available states in both the direct Γ valley and the indirect L valleys are equal in energy. The excess electrons are injected into both valleys and the electrons in the direct Γ valley contribute to the direct gap light emission. The heavy n-type doping introduces free carrier absorption which negatively affects the occurrence of net gain, therefore the amount of doping should be carefully modeled and optimized.

3.3 Germanium gain modeling

Direct band-to-band optical transition includes three processes: optical absorption, stimulated emission, and spontaneous emission. The rate of these electron-photon scattering processes can be described by the product of scattering strength and carrier occupation probabilities. We

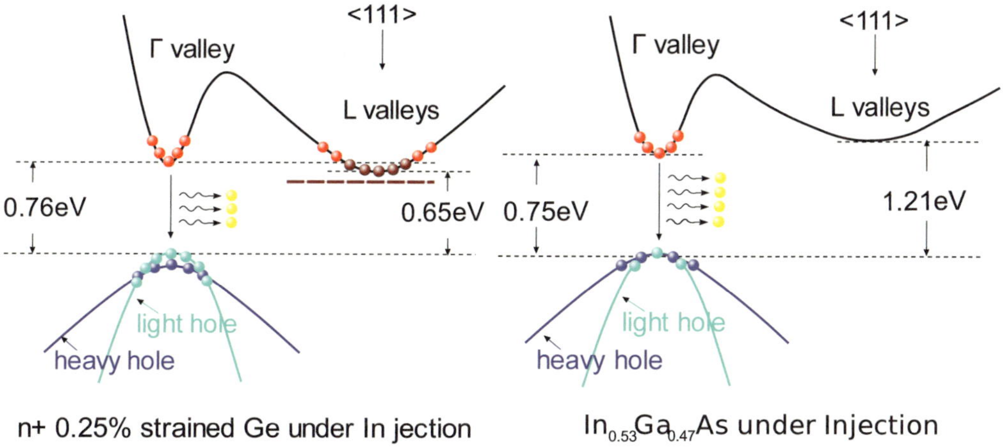

Fig. 17. Comparison of the carrier distribution and the light emission process between (a) n-type doped 0.25% tensile-strained Ge and (b) $In_{0.53}Ga_{0.47}As$ under injection.

denote the scattering strength of absorption, stimulated emission, and spontaneous emission at photon energy $h\nu$ are $\alpha(h\nu)$, $e_{st}(h\nu)$, and $e_{sp}(h\nu)$ respectively. $\alpha(h\nu)$ is the same absorption coefficient that we calculated earlier in the photodetector section. The rate equation of carriers N related to radiative recombinations is

$$\frac{dN}{dt} = \alpha f_v(1-f_c)N_{ph} - e_{st}f_c(1-f_v)N_{ph} - e_{sp}f_c(1-f_v) \tag{26}$$

where N_{ph} is the number of photons and f_c and f_v are the occupation probabilities of the electron with respect to the electron quasi Fermi level at E_1 and the hole with respect to the hole quasi Fermi level at E_2, respectively. And

$$E_1 = E_c + \frac{h\nu - E_g}{1 + m_c/m_v} \text{and} \tag{27}$$

$$E_2 = E_v - \frac{h\nu - E_g}{1 + m_v/m_c} \tag{28}$$

are the energy levels associated with the optical transition at photon energy $h\nu$ and are related by $E_1 - E_2 = h\nu$. Since quasi Fermi level depends on the injection level, f_c (f_v) is an implicit function of the density of electrons (holes) which include equilibrium electrons (holes) and injected electrons (holes).

The meaning of $f_c(1-f_v)$ is the joint probability of the existence of an electron at E_1 in the conduction band and the absence of a hole at E_2 in the valence band, which is the condition satisfied when an optical transition occurs at a given photon energy $h\nu$. It represents both stimulated emission and spontaneous emission. Optical absorption is the opposite process of the stimulated emission therefore $f_v(1-f_c)$ is used instead.

Optical gain $g(h\nu)$ is determined by the competition of the stimulated emission and the absorption:

$$g(h\nu) = e_{st}f_c(1-f_v) - \alpha f_v(1-f_c). \tag{29}$$

A detailed balance analysis proves that all the three scattering strength coefficients, α, e_{st}, and e_{sp}, are equal at any photon energy $h\nu$. Therefore optical gain can be rewritten as

$$g(h\nu) = \alpha(f_c - f_v). \tag{30}$$

$(f_c - f_v)$ is called population inversion factor. It is negative at equilibrium or low injection indicating a net optical loss (absorption). It becomes positive at high injection, when the population of the electrons inverts indicating a net optical gain. By using the absorption data presented in the photodetector section, optical gain can be calculated with this formula.

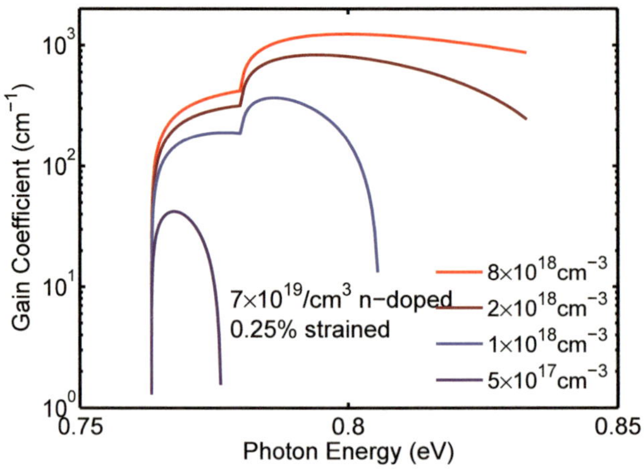

Fig. 18. Calculated optical gain spectrum of Ge with 0.25% tensile strain and $7 \times 10^{19} cm^{-3}$ n-type doping at various injection levels with respect to photon energies close to its direct band gap energy.

As demonstrated earlier, a combination of 0.25% tensile strain and 7×10^{19} cm^{-3} n-type doping results in an effective direct band gap germanium. The optical gain spectrum of such engineered germanium is shown in Fig. 18. The optical gain occurs at 0.76 eV which is the energy gap between the direct Γ valley and the light-hold band under 0.25% tensile strain. Since the effective masses are very light for these bands, population inversion occurs at low injection levels of $\sim 10^{17}$ cm^{-3}. As the injection level increases, the separation of the electron and the hole quasi Fermi levels becomes larger than the energy gap between the direct Γ valley and the heavy-hole band. Thus the optical gain contributed by electron heavy-hole recombination occurs, which can be seen from the fast raise of the optical gain at 0.78 eV at injection levels above 10^{18} cm^{-3}. A peak gain over 1000 cm^{-1} around 0.8 eV (1550 nm) is achieved at injection level of 8×10^{18} cm^{-3}.

The occurrence of optical gain in a material does not necessarily lead to lasing which requires that the optical gain overcomes optical losses from all sources. The material related optical loss is dominated by free carrier absorption. Free carrier absorption is a process that an electron or a hole absorbs the energy of a photon and moves to an empty higher energy state without inter-band recombination. Free carrier absorption increases with wavelength and becomes significant at high carrier densities. When a material is under carrier injection for population inversion, free carrier absorption caused by large amount of injected carriers competes against optical gain. Free carrier absorption is usually the major obstacle for lasing in a gain medium.

In n-typed doped germanium, additional free carrier absorption exists due to the existence of extrinsic electrons.
Free carrier absorption α_{fc} can be expressed in the following empirical formula:

$$\alpha_{\mathrm{fc}}(\lambda) = k_e n_c \lambda^{a_e} + k_h p_v \lambda^{a_h}, \tag{31}$$

where n_c and p_v are the densities of electrons and holes and k_e, k_h, a_e and a_h are the constants of a material. a_e and a_h are usually between 1.5 and 3.5. By fitting the free carrier absorption data in n^+Ge Spitzer et al. (1961) and p^+Ge Newman & Tyler (1957) in a carrier density range of $10^{19} - 10^{20}$ cm^{-3} at room temperature, we obtain

$$\alpha_{\mathrm{fc}}(\lambda) = -3.4 \times 10^{-25} n_c \lambda^{2.25} - 3.2 \times 10^{-25} p_v \lambda^{2.43}, \tag{32}$$

where α_{fc} is in unit of cm^{-1}, n_c and p_v in units of cm^{-3}, and λ in units of nm.

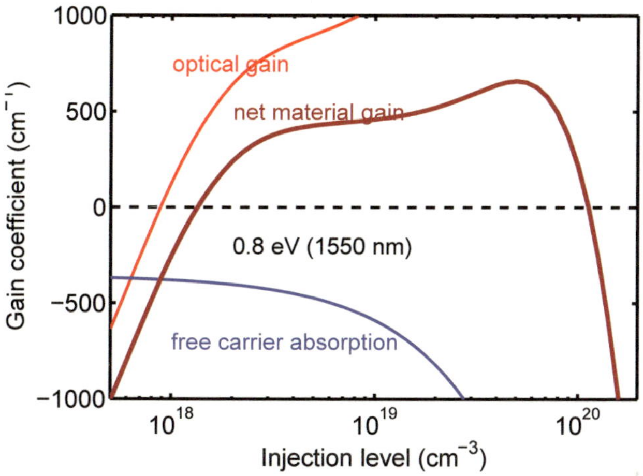

Fig. 19. The optical gain, the free carrier absorption, and the net material gain of Ge with 0.25% tensile strain and 7×10^{19} cm^{-1} n-type doping at various injection levels at the photon energy of 0.8 eV (1550 nm).

The optical gain, the free carrier absorption, and the net material gain which is the difference between first two are calculated with respect to injection level at the photon energy of 0.8 eV (1550 nm) for Ge with 0.25% tensile strain and 7×10^{19} cm^{-3} n-type doping. The results are shown in Fig. 19. The free carrier absorption is significant even at low injections since the material is heavily doped. The optical gain overcomes the free carrier loss above the injection level of 1.2×10^{18} cm^{-3} where germanium becomes a gain medium. At very high carrier injection levels ($> 10^{20}$ cm^{-3}), the free carrier absorption exceeds the optical gain leading to net loss. Between a large injection range of 10^{18} cm^{-3} and 10^{20} cm^{-3}, the tensile-strained n^+Ge is a gain medium. For photon energies other than 0.8 eV, the net gain range varies but the same characteristic holds.
The comparison of the net material gain versus injection level of germanium with both tensile strain and n-type doping, with either of the two, and with neither of two are shown in Fig. 20. Since the optical gain spectrum varies with strain, the net gain for each condition is calculated at the photon energy where maximal gain is achieved. Net gain can not be achieved for intrinsic Ge no matter with or without 0.25% tensile strain because the free carrier absorption

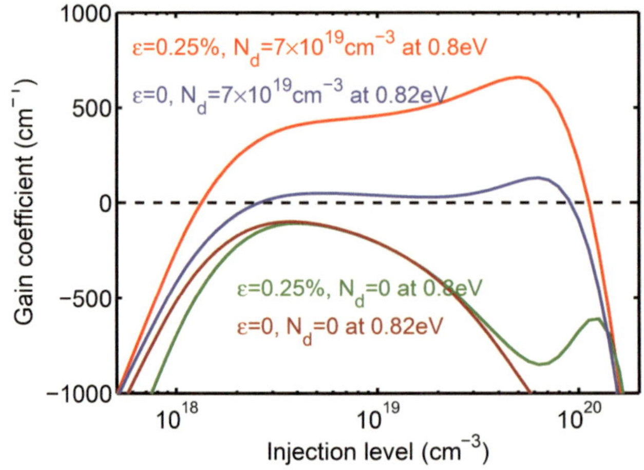

Fig. 20. Comparison of the calculated net material gain versus injection level of the Ge with both tensile strain and n-type doping, with either of two, and with neither of two at peak gain photon energy respectively.

always exceeds the optical gain. It is the reason that net gain has not been experimentally observed from germainium in the history. With n-type doping, the net gain can be achieved in both tensile-strained and unstrained cases. Tensile strain increases the population of the injected electrons in the direct Γ valley leading to a lower net gain threshold as well as higher net gain above threshold.

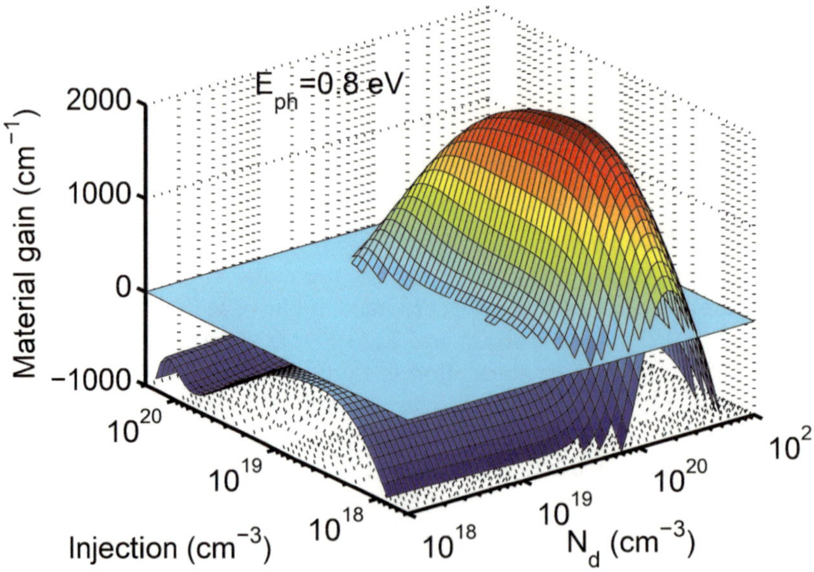

Fig. 21. Dependence of the calculated net material gain of Ge on n-type doping level (N_d) and carrier injection level at the photon energy of 0.8 eV.

The dependence of the net gain on n-type doping level and on injection level is calculated and shown in Fig. 21. There exists a range of n-type doping concentration and carrier injection level, in which net gain occurs. Doping concentration higher than a few 10^{19} cm^{-3} is required for net gain occurrence at photon energy of 0.8 eV. Net loss occurs at either very high doping concentrations or very high injection levels because the free carrier loss trumps the optical gain. The intercept of the 3-D plot at zero gain surface is a threshold boundary surrounding the net gain region. The threshold boundary varies at photon energies other than 0.8 eV.

3.4 Germanium photoluminescence and electroluminescence

Photoluminescence (PL) and electroluminescence (EL) are commonly used material characterization experiments. The intensity and spectral content of the luminescence is a direct measure of various important material properties. The carrier recombination characteristics of germanium can be studied with these experiments.
Germanium is a multi-valley indirect band gap material therefore both the direct and the indirect band-to-band radiative recombination exist. As discussed earlier indirect transition is an inefficient process as most injected electron-hole pairs recombine non-radiatively before the occurrence of radiative recombination. Therefore indirect PL can only be observed in ultra high quality germanium at cryogenic temperatures at which non-radiative recombinations are greatly suppressed. On the contrary, direct transition is a fast process with radiative recombination rate 4-5 orders of magnitude higher than that of the indirect transition(Haynes & Nilsson, 1964). However, the lack of sufficient injected electrons in the direct Γ valley results in weak overall light emission. The tensile strain and n-type doping techniques improve the electron concentration in the direct valley hence more efficient luminescence is expected.

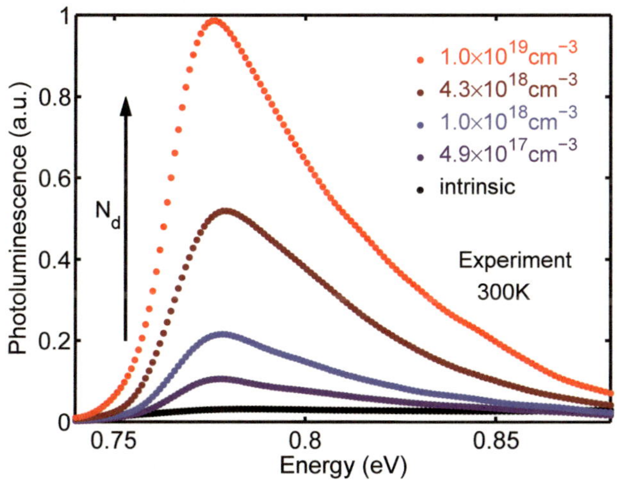

Fig. 22. PL spectra of tensile-strained Ge film with various n-type doping levels. The PL of a 1×10^{19} cm^{-3} doped Ge film is over 50 times brighter than that of an undoped one. (Sun et al., 2009a)

The effect of n-type doping and tensile-strain on PL is investigated using a series of epitaxial germanium samples with doping concentrations from less than $1 \times 10^{16} cm^{-3}$ (approximately undoped) to 2×10^{19} cm^{-3}. The doping is achieved by in situ phosphorous incorporation during germanium epitaxy. Some PL spectrum examples are shown in Fig. 22. All PL spectra are measured at room temperature. The doping dependence of PL intensities agrees with the analysis earlier that direct gap light emission is enhanced with n-type doping owing to the indirect valley states filling with extrinsic electrons. The PL of a 1×10^{19} cm^{-3} doped epitaxial germanium is over 50 times brighter than that of an undoped one. All spectra exhibit the same spectral characteristic. The same PL peak position implies that phosphorous doping has negligible effect on band structure or tensile strain in germanium.

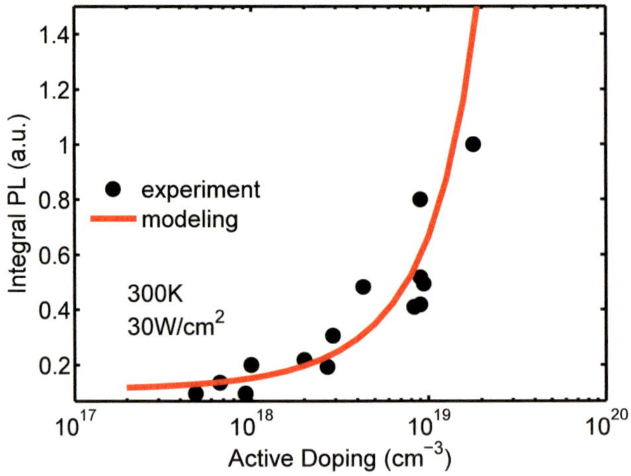

Fig. 23. A summary of integral PL intensity versus active doping concentration. The theoretical calculation is represented in red solid line and describes the trend of the experimental data.(Sun et al., 2009a)

A plot of the enhancement of integral PL intensity with active doping concentration is shown in Fig. 23. The integral PL-doping relation can be calculated by

$$I(N_d) \propto n_e^{\Gamma}(N_d)n_h \propto n_e^{\Gamma}(N_d) \tag{33}$$

It shows the direct gap integral PL intensity is proportional to the product of the electron concentration in the Γ valley and the hole concentration at quasi-equilibrium under excitation. The hole concentration remains the same for each sample at the same excitation level therefore the PL intensity is only determined by the electron concentration in the direct Γ valley. This concentration is a function of n-type doping concentration. Higher n-type doping results in more injected electrons in the direct Γ valley. A theoretical calculation based on this analysis is shown in Fig. 23 with red solid line, exhibiting good agreement with the experimental data. Ion-implantation is an effective way to achieve high doping concentration at the cost of crystalline damage. The lattice damages introduce defects acting as non-radiative recombination centers and reduce lifetime in materials. Fig. 24 shows the PL from epitaxial germanium implanted with phosphorus at three doses aiming for doping concentrations of 1×10^{18} cm^{-3}, 1×10^{19} cm^{-3}, and 1×10^{20} cm^{-3}, respectively. All samples were annealed at 800 ^{o}C after implantation for dopant activation and subsequent Hall effect measurement confirmed over 80% dopants were activated. PL increases with the doping level in implanted germanium as expected. But all the implanted samples show weaker PL than the in situ doped one confirming the negative effect of lattice damage on radiative recombination. In general, in situ doping is a superior approach to achieve n-type doping in epitaxial germanium for light emission applications.

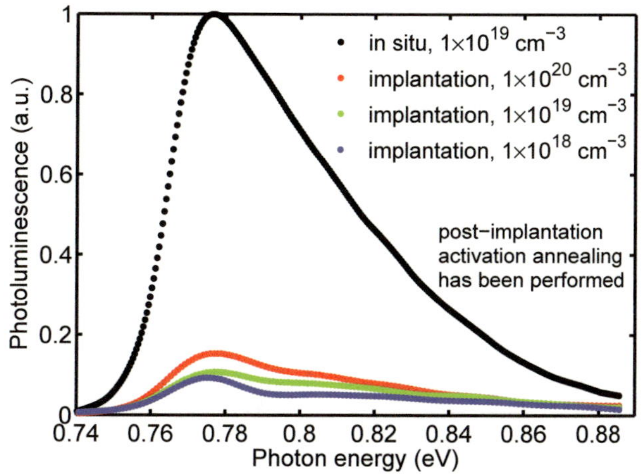

Fig. 24. PL Comparison of in situ doped Ge and implanted Ge. Post-implantation annealing has been performed for implanted Ge films for dopant activation.

Compared to PL, EL emission is excited electrically other than optically. EL is usually described as an electrical-optical phenomenon in which a material emits light in response to an electric current passed through it, or to a strong electric field. Due to different injection mechanisms, the characteristics observed in EL are not necessary the same as in PL.

Si/Ge/Si structure forms a natural hetetojunction for a light-emitting diode. In the EL experiment on such Si/Ge/Si diodes, the onset of EL at forward bias has been observed at room temperature. In author's own work, EL was observed at 0.5 V forward bias or 1.3 mA injection current for a 20 μm by 100 μm diode.

The EL spectrum measured at room temperature at 50 mA forward current is shown in Fig. 25 (a). The EL peak is located at 0.76 eV corresponding to the direct band-to-band optical transition in 0.2% tensile-strained Ge. The full width at half maximum (FWHM) of the peak is about 60 meV ($\sim$2 kT) consistent with the direct band-to-band transition model. The sharp peaks are from the Fabry-Perot resonance of the air gap between device surface and the flat facet of the fiber used for light collection. These resonances are reproducible in the experiments and can be filtered by fast Fourier transformation (FFT). After the FFT filtering, the smoothed curve represents the "real" EL characteristics which is shown with red solid line. The EL spectrum is consistent with the room temperature photoluminescence (PL) spectrum measured from a 0.2% tensile-strained epitaxial germanium sample shown in Fig. 25 (b). The small red shift in the EL peak position compared to PL peak position is a result of heating from current injection which slightly reduces the band gap.

The injection dependence of the direct gap EL emission was measured. The results show unique direct gap light emission characteristics of germanium which is different from that of a direct gap semiconductor.

A superlinear relation of the integral direct gap EL intensity with the injected electrical current is shown in Fig. 26. In a direct band gap semiconductor, the relation is expected to be linear at lower injection levels and to roll over at higher injecion levels due to increasingly significant non-radiative recombinations. This superlinear dependence on injection is a unique feature of direct gap light emission in germanium. As discussed earlier, direct light emission intensity is

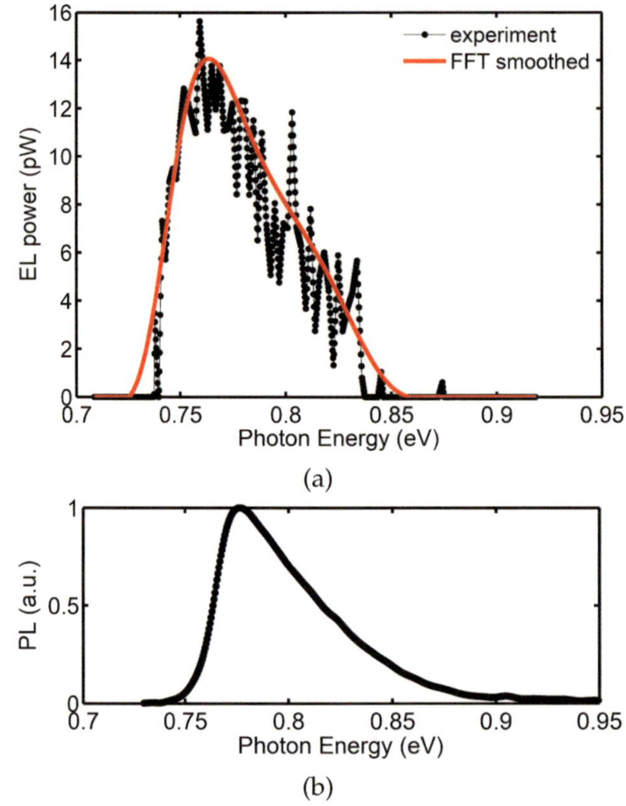

Fig. 25. (a) Direct gap EL spectrum of a 20 μm by 100μm 0.2% tensile-strained Si/Ge/Si light-emitting diode measured at room temperature. The periodic sharp peaks are due to Fabry-Perot resonances. (b) Room temperature direct gap photoluminescence of a 0.2% tensile-strained Ge film epitaxially grown on Si. (Sun et al., 2009b)

determined by the injected electrons in the direct Γ valley $n_{e\Gamma}$ and can be further expressed as

$$P_{\Gamma} \propto n_{e\Gamma} = n_{\text{tot}} \cdot f_{\Gamma}, \tag{34}$$

where n_{tot} is the total injected electron density and f_{Γ} is the fraction of the electrons injected into the direct Γ valley. The total injected electron scales linearly with the injected electrical current. The fraction term also increases with the injection level due to the increase of the electron quasi Fermi level leading to larger portion of electrons in the direct Γ valley. The multiplication of the two terms results in a superlinear behavior with injection current. A theoretical calculation based on this analysis is shown in solid line in Fig. 26. It agrees well with the experimental data. The small difference between the theoretical and the experimental result is due to the small deviation from the ideal square-root density of states near the band edge.

3.5 Germanium net gain and lasing

It has been theoretically shown that germanium can be engineered by tensile strain and n-type doping for better direct gap light emission at room temperature. The direct gap PL and EL

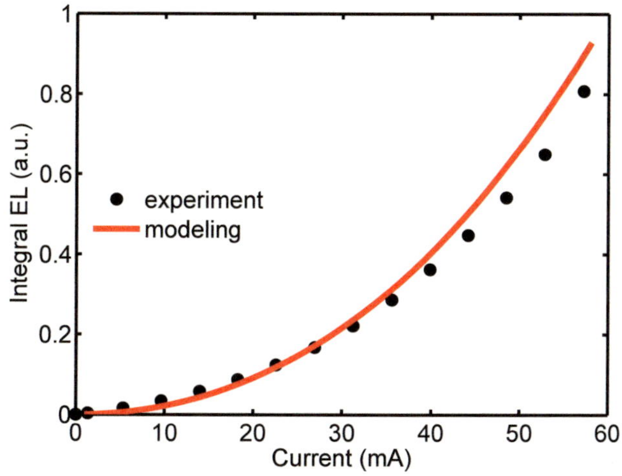

Fig. 26. Integral direct gap EL intensity of a 0.2% tensile-strained Si/Ge/Si light emitting diode. The EL intensity increases superlinearly with electrical current. The theoretical calculation (solid line) agrees well with the experimental result.

experiment greatly support this theory. Investigating the capability of net gain in germanium and to achieving lasing are next tasks.

A non-degenerate pump-probe experiment with a tunable laser as probe and a high power laser as pump is used for optical gain measurement. The fiber outputs of the pump laser and the probe laser are combined through a wavelength division multiplexing (WDM) coupler and illuminate to the top of n-type doped, tensile-strained epitaxial germanium. The transmitted optical power is measured by an integral optical sphere for transmissivity calculation with respect to incident light. The probe light is modulated at a few kHz and is filtered by a lock-in amplifier for precise transmission measurement.

The measured transmissivity is correlated with the optical gain or absorption under optical pumping:

$$g(h\nu) = \alpha(f_c - f_v). \tag{35}$$

The relation between transmissivity and optical gain/absorption can be modeled by transfer matrix method (TMM). The detailed calculated is omitted here.

The inversion factor $(f_c - f_v)$ ranges from -1 to 1. In the absence of pumping, $(f_c - f_v) = -1$, i.e. $g(h\nu) = -\alpha$ is pure absorption. As the material is increasingly pumped, $(f_c - f_v)$ increases from -1 towards 0. The optical absorption becomes less which is called optical bleaching effect. At sufficiently high pumping levels, the material becomes transparent and a gain medium for the inversion factor equals to and more than 0, respectively.

For heavily n-type doped germanium, the measured net gain/absorption includes both optical gain/absorption and free carrier absorption:

$$g_{\text{tot}}(h\nu) = -\alpha_{\text{tot}}(h\nu) = \alpha(f_c - f_v) - \alpha_{\text{fc}}, \tag{36}$$

where α_{fc} represents free carrier absorption. The existence of free carrier absorption increases the injection level required for transparency or net gain. The effective inversion factor can be expressed by

$$(f_c - f_v) - \frac{\alpha_{\text{fc}}}{\alpha} = -\frac{\alpha_{\text{tot}}}{\alpha}, \tag{37}$$

which can be obtained from the experimental results.
A tensile-strained n^+ doped germanium mesa was fabricated for pump-probe measurement. The germanium is 1.0×10^{19} cm^{-3} n-type doped and 0.2% tensile-strained. The transmissivity spectra were measured at 0 and 60 mW optical pumping powers, respectively. The absorption is then calculated by using transfer matrix method with the consideration of Kramer-Kronig relations.

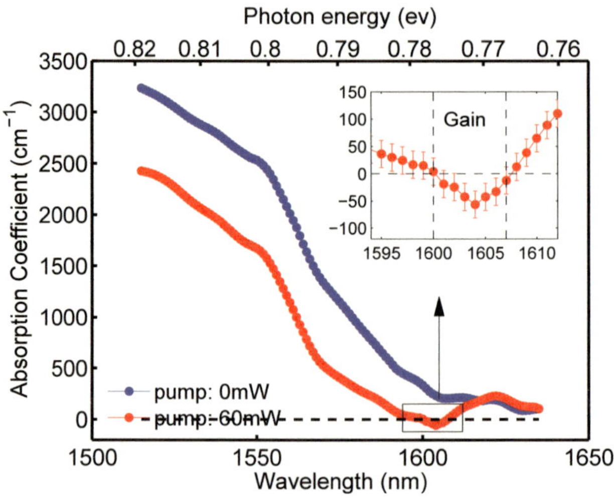

Fig. 27. The absorption spectra of the n+ Ge mesa under 0 and 60 mW optical pumping, respectively. Negative absorption corresponding to the onset of net gain was observed in wavelength between 1600 nm and 1608 nm. The error bars in the inset represent the transmissivity measurement errors. Liu et al. (2009).

Fig. 27 shows the absorption spectra of the n^+Ge mesa under pumping powers of 0 and 60W, respectively. The absorption at photon energies above 0.77 eV (at wavelengths below 1610 nm) decreases significantly upon optical pumping. Negative absorption corresponding to the onset of optical gain is observed in the wavelength range of 1600-1608 nm, as shown in the inset of Fig. 27. The maximum gain coefficient is $g_{tot} = -\alpha_{tot} = 50 \pm 25$ cm^{-1} at 1605 nm. The error bars represent the transmissivity measurement errors.
Like PL in doped germanium, the optical bleaching effect also increases with the n-type doping concentration. The effective inversion factor is calculated from the experimental results for germanium with different doping concentrations.
The comparison of the effective inversion factor spectra of tensile-strained Ge for both blanket film and mesa samples with various n-type doping concentrations is shown in Fig. 28. The optical bleaching effect (inversion factor more than -1) can be seen from all samples. The bleaching increases with n-type doping concentration confirming the theory of the effect of n-type doping on optical gain. The Ge mesa sample exhibits a positive inversion factor at the direct band edge (1600-1608 nm) underlying the occurrence of net gain. The effective inversion factors at longer wavelengths are less than -1 for all samples because free carrier absorption overcomes optical bleaching. The optical bleaching in the Ge mesa sample is more than any Ge film samples as a result of the lateral confinement of injected carriers by the mesa structure.
With the observation of net gain in tensile-strained n-doped germanium, a optically pumped Fabry-Perot germanium laser was realized (Liu et al., 2010). The waveguide laser was excited

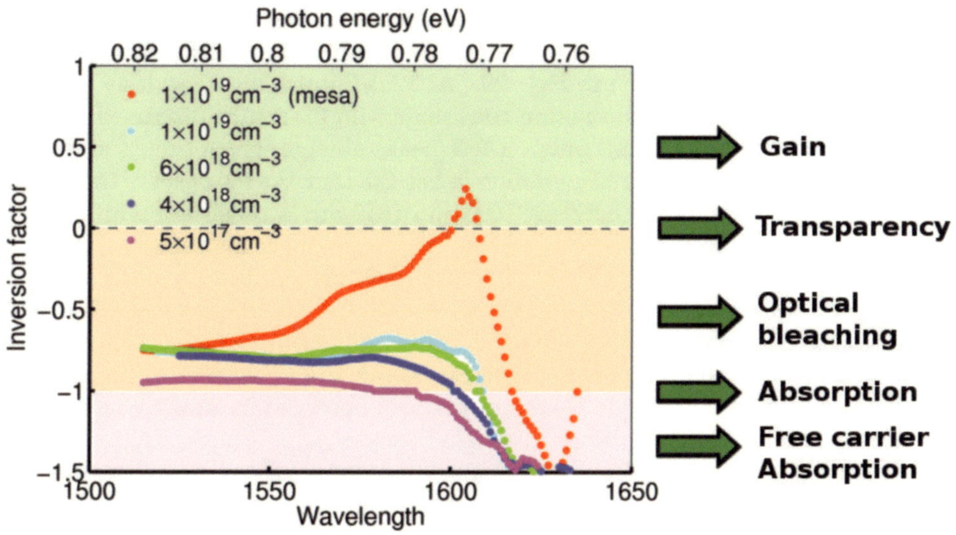

Fig. 28. Comparison of the effective inversion factors of tensile-strained Ge film and mesa with various n-type doping concentrations. The observed optical bleaching effect increases with n-type doping concentration.

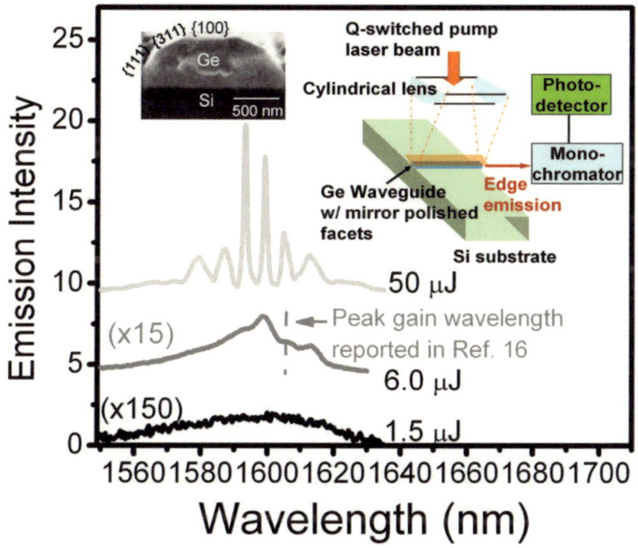

Fig. 29. Edge emission spectra of a Ge waveguide with mirror polished facets under excitation. The three spectra at 1.5, 6.0 and 50 μJ/pulse pumping power correspond to spontaneous emission, threshold for lasing, and laser emission, respectively. The inset shows a cross-sectional SEM picture of the Ge waveguide and a schematic drawing of the experimental setup for optical pumping.

by a 1064 nm Q-switched laser through a cylindrical focusing lens. The lasing measurement is schematically shown in the inset of Fig. 29. The light emission spectra of the laser under different injection levels are shown in Fig. 29. At 1.5 μJ/pulse of pump laser power, the spectrum is a typical spontaneous emission consistent with PL results discussed earlier. As the pump power increases to 6.0 μJ/pulse, a few peaks emerge which occurs at the pump power corresponding to the threshold condition in Fig. 30. It marks the onset of transparency. The occurrence of emission peaks between 1600 and 1610 nm is consistent with the optical gain spectrum peaked at 1605 nm shown earlier. As pump power increases to 50 μJ/pulse, the widths of the emission peaks at 1594, 1599 and 1605 nm significantly decrease and the polarization became to predominant TE other than a mixture of TE and TM at lower injections. These results represent a typical lasing behavior. The multiple emission peaks are most likely due to multiple guided modes in the germanium waveguide as a result of high refractive index contrast. A similar multimode behavior has been observed in an early work on III-V semiconductor lasers (Miller et al., 1977).
Fig. 30 shows the integral edge emission intensity as a function of pump power. An obvious threshold behavior is observed. The threshold pumping power is about 5 μJ/pulse. The absorbed pump power density at the threshold is about 30 kW/cm^2 by considering various optical losses of the incident pump light. The threshold is expected to further decrease with increased n-type doping concentration based on the calculation earlier. With lower injection threshold requirement, an electrically pumped germanium laser diode can be realized. As shown in the EL experiment discussion, Ge/Si/Ge heterojunction may be suitable diode structure for such lasers which will eventually complete the integrated silicon photonic circuits for the next generation of data communication and interconnects.

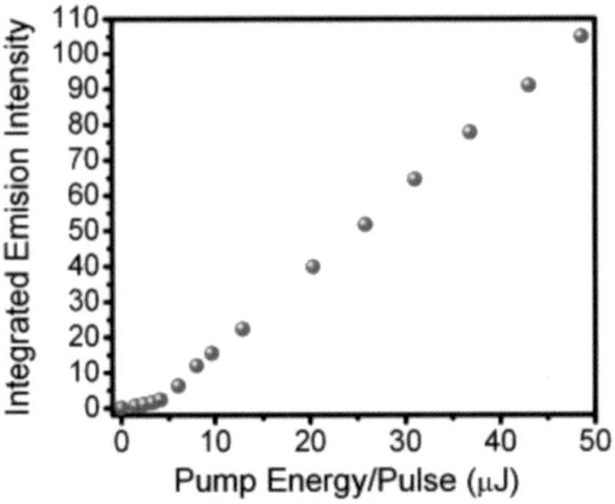

Fig. 30. Integral edge emission power versus optical pump power showing a lasing threshold behavior.

4. References

Adeola, G. W., Jambois, O., Miska, P., Rinnert, H. & Vergnat, M. (2006). Luminescence efficiency at 1.5 μm of er-doped thick sio layers and er-doped sio / sio2 multilayers, *Appl. Phys. Lett.* 89: 101920.

Ahn, D., yin Hong, C., Liu, J., Giziewicz, W., Beals, M., Kimerling, L. C. & Michel, J. (2007). High performance, waveguide integrated Ge photodetectors, *Opt. Express* 15(7): 3916–3921.

Aicher, W. & Haberger, K. (1999). Influence of optical interconnects at the chip and board levels, *Opt. Eng.* 38: 313–322.

Bean, J. C., Sheng, T. T., l. C. Feldman, Fiory, A. T. & Lynch, R. T. (1984). Pseudomorphic growth of gexsi1-x on silicon by molecular beam epitaxy, *Appl. Phys. Lett.* 44: 102.

Boyraz, O. & Jalali, B. (2004). Demonstration of a silicon Raman laser, *Opt. Express* 12(21): 5269–5273.

Braunstein, R., Moore, A. R. & Herman, F. (1958). Intrinsic optical absorption in germanium-silicon alloys, *Phy. Rev.* 109(3): 695–710.

Cannon, D. D., Liu, J., Ishikawa, Y., Wada, K., Danielson, D. T., Jongthammanurak, S., Michel, J. & Kimerling, L. C. (2004). Tensile strained epitaxial Ge films on Si(100) substrates with potential application in L-band telecommunications, *Appl. Phys. Lett* 84(6): 906–908.

Chuang, S. L. (1995). *Physics of Optoelectronic Devices*, John Wiley & Sons. Inc, chapter 4, p. 144.

Colace, L., Balbi, M., Masini, G., Assanto, G., Luan, H.-C. & Kimerling, L. C. (n.d.). Ge on si p-i-n photodiodes operating at 10 gbit/s, *Appl. Phys. Lett.* 88: 101111.

Colace, L., Masini, G., Altieri, A. & Assanto, G. (2006). Waveguide photodetectors for the near-infrared in polycrystalline germanium on silicon, *IEEE Photon. Technol. Lett.* 18: 1094.

Colace, L., Masini, G., Assanto, G., Luan, H.-C., Wada, K. & Kimerling, L. C. (2000). Efficient high-speed near-infrared ge photodetectors integrated on si substrates, *Appl. Phys. Lett.* 76(10): 1231.

Colace, L., Masini, G., Galluzi, F., Assanto, G., Capellini, G., Gaspare, L. D., & Evangelisti, F. (1997). Ge/si (001) photodetector for near infrared light, *Solid State Phenom.* 54: 55.

Colace, L., Masini, G., Galluzzi, F., Assanto, G., Capellini, G., Gaspare, L. D., Palange, E. & Evangelisti, F. (1998). Metal-semiconductor-metal near-infrared light detector based on eptixial ge/si, *Appl. Phys. Lett.* 72(24): 3175–3177.

Currie, M. T., Samavedam, S. B., Langdo, T. A., Leitz, C. W. & Fitzgerald, E. A. (1998). Controlling threading dislocation densities in ge on si using graded sige layers and chemical-mechanical polishing, *Appl. Phys. Lett.* 72: 1718–1720.

Dash, W. C. & Newman, R. (1955). Intrinsic optical absorption in single-crystal germanium and silicon at 77Âřk and 300Âřk, *Phy. Rev.* 99(4): 1151–1155.

de Walle, C. G. V. (1989). Band lineups and deformation potentials in the model-solid theory, *Phys. Rev. B* 39(3): 1871–1883.

Dehlinger, G., Koester, S. J., Schaub, J. D., Chu, J. O., Ouyang, Q. C. & Grill, A. (2004). High-Speed Germanium-on-SOI Lateral PIN Photodiodes, *IEEE Photon. Technol. Lett.* 16(11): 2547–2549.

Eaglesham, D. J., Unterwald, F. C. & Jacobson, D. C. (1993). Growth morphology and the equilibrium shape: the role of surfactants in ge/si island formation, *Phys. Rev. Lett.* 70(7): 966–969.

Fama, S., Colace, L., Masini, G., Assanto, G. & Luan, H.-C. (n.d.). High performance germanium-on-silicon detectors for optical communications, *Appl. Phys. Lett.* 81(4): 586.

Fang, A. W., Park, H., Cohen, O., Jones, R., Paniccia, M. J. & Bowers, J. E. (2006). Electrically pumped hybrid AlGaInAs-silicon evanescent laser, *Opt. Express* 14(20): 9203–9210.

Fang, Y.-Y., Tolle, J., Roucka, R., Chizmeshya, A. V. G., Kouvetakis, J., D'Costa, V. R. & Menendez, J. (2007). Perfectly tetragonal, tensile-strained ge on $ge_{1-y}sn_y$ buffered si(100), *Appl. Phys. Lett.* 90: 061915.

Feng, D., Liao, S., Dong, P., Feng, N.-N., Liang, H., Zheng, D., Kung, C.-C., Fong, J., Shafiiha, R., Cunningham, J., Krishnamoorthy, A. V., & Asghari, M. (2009). High-speed ge photodetector monolithically integrated with large cross-section silicon-on-insulator waveguide, *Appl. Phys. Lett.* 95: 261105.

Feng, N.-N., Dong, P., Zheng, D., Liao, S., Liang, H., Shafiiha, R., Feng, D., Li, G., Cunningham, J. E., V.Krishnamoorthy, A. & Asghari, M. (2010). Vertical p-i-n germanium photodetector with high external responsivity integrated with large core si waveguides, *Optics Express* 18(1): 96.

Fidaner, O., Okyay, A. K., Roth, J. E., Schaevitz, R. K., Kuo, Y.-H., Saraswat, K. C., James S. Harris, J. & Miller, D. A. B. (2007). GeâĂŞsige quantum-well waveguide photodetectors on silicon for the near-infrared, *IEEE Photon. Technol. Lett.* 19(20): 1631.

Fitzgerald, E. A., Xie, Y.-H., Green, M. L., Brasen, D., Kortan, A. R., Michel, J., Mii, Y.-J. & Weir, B. E. (1991). Totally relaxed $ge_x si_{1-x}$ layers with low threading dislocation densities grown on si substrates, *Appl. Phys. Lett.* 59(7): 811–813.

Frova, A. & Handler, P. (1965). Franz-keldysh effect in the space-charge region of a germanium p-n junction, *Phy. Rev.* 137(6A): A1857–A1861.

Fujii, M., Yoshida, M., Kanzawa, Y., Hayashi, S. & Yamamoto, K. (1997). 1.54mum photoluminescence of er^{3+} doped into sio_2 films containing si nanocrystals: Evidence for energy tranfer from si nanocrystals to er^{3+}, *Appl. Phys. Lett.* 71(9): 1198–1200.

Garozzo, M., Conte, G., Evangelisti, F. & Vitali, G. (1982). Heteroepitaxial growth of ge on <111> si by vacuum evaporation, *Appl. Phys. Lett.* 41: 1070.

Gelloz, B. & Koshida, N. (2000). Electroluminescence with high and stable quantum efficiency and low threshold voltage from anodically oxidized thin porous silicon diode, *J. Appl. Phys.* 88(7): 4319–4324.

Groenert, M. E., Leitz, C. W., Pitera, A. J., Yang, V., Lee, H., Ram, R. & Fitzerald, E. A. (2003). Monolithic integration of room-temperature cw GaAs/AlGaAs lasers on Si substrates via relaxed graded GeSi buffer layers, *J. Appl. Phys.* 93(1): 362–367.

Haensch, W., Nowak, E. J., Dennard, R. H., Solomon, P. M., Bryant, A., Dokumaci, O. H., Kumar, A., Wang, X., Johnson, J. B. & Fischetti, M. V. (2006). Silicon cmos devices beyond scaling, *IBM J. Res. & Dev.* 50: 339–361.

Han, H.-S., Seo, S.-Y. & Shin, J. H. (2001). Optical gain at 1.54 mum erbium-doped silicon nanocluster sensitized waveguide, *Appl. Phys. Lett.* 79(27): 4568–4570.

Hartmann, J. M., Abbadie, A., Papon, A. M., Holliger, P., Rolland, G., Billon, T., Fedeli, J. M., Rouviere, M., Vivien, L. & Laval, S. (2004). Reduced pressure-chemical vapor deposition of ge thick layers on si(001) for 1.3-1.55μm photodetection, *J. Appl. Phys.* 95: 5905.

Haurylau, M., Chen, G., Chen, H., Zhang, J., Nelson, N. A., Albonesi, D. H., Friedman, E. G. & Fauchet, P. M. (2006). On-chip optical interconnect roadmap: Challenges and critical directions, *IEEE J. Sel. Topic Quantum Electron.* 12(6): 1699–1705.

Haynes, J. R. & Nilsson, N. G. (1964). The direct radiative transitions in germanium and their use in the analysis of lifetime, *Proceedings of VIIth International Conference on Physics of Semiconductors*, Paris, p. 21.

Hobden, M. V. (1962). Direct optical transitions from the split-off valence band to the conduction band in germanium, *J. Phys. Chem. Solids* 23(6): 821–822.

Iacona, F., Pacifici, D., Irrera, A., Miritello, M., Franzo, G. & Priolo, F. (2002). Electroluminescence at 1.54 μm in er-doped si nanocluster-based devices, *Appl. Phys. Lett.* 81(17): 3242–3244.

Irrera, A., Pacifici, D., Miritello, M., Franzo, G., Priolo, F., Iacona, F., Sanfilippo, D., Stefano, G. D. & Fallica, P. G. (2003). Electroluminescence properties of light emitting devices based on silicon nanocrystals, *Physica E* 16(3-4): 395–399.

Jacoboni, C., Nava, F., Canali, C. & Ottaviani, G. (1981). Electron drift velocity and diffusivity in germanium, *Phys. Rev. B* 24(2): 1014–1026.

Kamins, T. I. & Meyer, D. J. (1991). *Appl. Phys. Lett.* 59: 178.

Kasper, E. & Herzog, H.-J. (1977). Elastic strain and misfit dislocation density in si0.92ge0.08 films on silicon substrates, *Thin Solid Films* 44: 357.

Kik, P. G., Brongersma, M. L. & Polman, A. (2000). Strong exciton-erbium coupling in si nanocrystal-doped sio_2, *Appl. Phys. Lett.* 76(17): 2325–2327.

Kirchain, R. & Kimerling, L. C. (2007). A roadmap for nanophotonics, *Nature Photonics* 1: 303–305.

Kobayashi, S., Cheng, M.-L., Kohlhase, A., Sato, T., Murota, J. & Mikoshiba, N. (1990). Selective germanium epitaxial growth on silicon using cvd technology with ultra-pure gases, *J. Crystal Growth* 99: 259.

Koshida, N. & Koyama, H. (1992). Visible electroluminescence from porous Si, *Appl. Phys. Lett.* 60(3): 347–349.

Langdo, T. A., Leitz, C. W., Currie, M. T., Fitzgerald, E. A., Lochtefeld, A. & Antoniadis, D. A. (2000). High quality ge on si by epitaxial necking, *Appl. Phys. Lett.* 75(25): 3700–3702.

Leong, D., Harry, M., Reeson, K. J. & Homewood, K. P. (1997). A silicon/iron-disilicide light-emitting diode operating at a wavelength of 1.5 μm, *Nature* 387(6634): 686–688.

Lipson, M. (2004). Overcoming the limitations of microelectronics using Si nanophotonics: solving the coupling, modulation and switching challenges, *Nanotechnology* 15(10): S622–S627.

Liu, J., Cannon, D. D., Wada, K., Ishikawa, Y., Jongthammanurak, S., Danielson, D. T., Michel, J. & Kimerling, L. C. (2005). Tensile strained ge p-i-n photodetectors on si platform for c and l band telecommunications, *Appl. Phys. Lett.* 87(1): 011110.

Liu, J., Cannon, D., Ishikawa, Y., Wada, K., Danielson, D. T., Jongthammanurak, S., Michel, J. & Kimerling, L. C. (2004). Deformation potential constants of biaxially tensile stressed ge epitaxial films on si(100), *Phy. Rev. B* 70(15): 155309.

Liu, J., Michel, J., Giziewicz, W., Pan, D., Wada, K., Cannon, D. D., Jongthammanurak, S., Danielson, D. T., Kimerling, L. C., Chen, J., Ilday, F. O., Kartner, F. X. & Yasaitis, J. (2005). High-performance, tensile-strained Ge p-i-n photodetectors on a Si platform, *Appl. Phys. Lett* 87(10): 103501.

Liu, J., Sun, X., Camacho-Aguilera, R., Kimerling, L. C. & Michel, J. (2010). Ge-on-si laser operating at room temperature, *Optics Lett.* 35(5): 679.

Liu, J., Sun, X., Kimerling, L. C. & Michel, J. (2009). Direct band gap photoluminescence and onset of optical gain of band-engineered ge-on-si at room temperature, *Optics Lett.* . accepted.

Liu, Y., Deal, M. D. & Plummer, J. D. (2004). High-quality single-crystal ge on insulator by liquid-phase epitaxy on si substrates, *Appl. Phys. Lett.* 84: 2563.

Luan, H.-C., Lim, D. R., Lee, K. K., Chen, K. M., Sandland, J. G., Wada, K. & Kimerling, L. C. (1999). High-quality Ge epilayers on Si with low threading-dislocation densities, *Appl. Phys. Lett.* 75(19): 2909–2911.

Luan, H.-C., Wada, K., Kimerling, L. C., Masini, G., Colace, L. & Assanto, G. (2001). High eÂĄciency photodetectors based on high quality epitaxial germanium grown on silicon substrates, *Optical Mater.* 17: 71.

Luryi, S., Kastalsky, A. & Bean, J. C. (1984). New infrared detector on a silicon chip, *IEEE Trans. Electron. Dev.* 31: 1135.

Madelung, O. & et al (eds) (1982). *Properties of Group IV Elements and III-V, II-VI, and I-VII Compounds*, Vol. 17a, Springer, Berlin.

Makarova, M., Sih, V., Warga, J., Li, R., Negro, L. D. & Vuckovic, J. (2008). Enhanced light emission in photonic crystal nanocavities with Erbium-doped silicon nanocrystals, *Appl. Phys. Lett.* 92(16): 161107.

Masini, G., Capellini, G., Witzens, J. & Gunn, C. (2007). A 1550nm, 10Gbps monolithic optical receiver in 130nm CMOS with integrated Ge waveguide photodetector, *4th IEEE International Conference on Group IV Photonics*, Tokyo, Japan, pp. 19–21.

Meyerson, B. S. (1986). Low-temperature silicon epitaxy by ultrahigh vacuum chemical vapor deposition, *Appl. Phys. Lett.* 48: 797–799.

Meyerson, B. S. (1990). Low-temperature si and sige epitaxy by ultrahigh-vacuum chemical deposition: process fundamentals, *IBM J. Res. Develop* 34(6): 806–815.

Meyerson, B. S. (1992). Uhvcvd growth of si and sige alloys: chmical, physics, and device applications, *Proc. IEEE* 80(10): 1592–1608.

Miller, K. J. & Grieco, M. J. (1962). Epitaxial silicon-germanium alloy films on silicon substrates, *J. Electrochem. Society* 109: 70.

Miller, R. C., Nordland, W. A., Logan, R. A. & Johnson, L. F. (1977). Optically pumped taper-coupled gaas-algaas laser with a second-order bragg reflector, *J. Appl. Phys* 49: 539.

M.Levinstein, Rumyantsev, S. & Shur, M. (eds) (1996). *Handbook Series on Semiconductor Parameters*, World Scientific.

Moore, G. (1965). Cramming more components onto integrated circuits, *Electronics* 38: 144âĂŞ–117.

Morse, M., Dosunmu, O., Sarid, G. & Chetrit, Y. (2006). Performance of ge-on-si p-i-n photodetectors for standard receiver modules, *IEEE Photon. Technol. Lett.* 18: 2442.

Muller, D. A. (2005). A sound barrier for silicon?, *Nature Mater.* 4: 645–647.

Nazarov, A., Sun, J. M., Skorupa, W., Yankov, R. A., Osiyuk, I. N., Tjagulskii, I. P., Lysenko, V. S. & Gebel, T. (2005). Light emission and charge trapping in er-doped silicon dioxide films containing silicon nanocrystals, *Appl. Phys. Lett.* 86: 151914.

Negro, L. D., Li, R., Warga, J. & Basu, S. N. (2008). Sensitized erbium emission from silicon-rich nitride/silicon superlattice structures, *Appl. Phys. Lett.* 92(18): 181105.

Newman, R. & Tyler, W. W. (1957). Effect of impurities on free-hole infrared absorption in p-type germanium, *Phys. Rev.* 105: 885–886.

OĆonnor, I., Tissafi-Drissi, F., Gaffiot, F., Dambre, J., Wilde, M. D., Campenhout, J. V., Thourhout, D. V., Campenhout, J. V. & Stroobandt, D. (2007). Systematic simulation-based predictive synthesis of integrated optical interconnect, *IEEE Trans. VLSI Sys.* 15: 927–940.

O'Connor, I., Tissafi-Drissi, F., Navarro, D., Mieyeville, F., Gaffiot, F., Dambre, J., Wilde, M. D., Stroobandt, D. & Briere, M. (2006). Integrated optical interconnect for on-chip data transport, *Circuits and Systems, 2006 IEEE North-East Workshop on*, Gatineau, Que., p. 209.

Okada, Y. & Tokumaru, Y. (1984). Precise determination of lattice parameter and thermal expansion coefficient of silicon between 300 and 1500 k, *J. Appl. Phys.* 56: 314.

Osmond, J., Isella, G., Chrastina, D., Kaufmann, R., Acciarri, M. & von Kanel, H. (2009). Ultralow dark current ge/si(100) photodiodes with low thermal budget, *Appl. Phys. Lett.* 94: 201106.

Park, H., Fang, A. W., Kodamaa, S. & Bowers, J. E. (2005). Hybrid silicon evanescent laser fabricated with a silicon waveguide and III-V offset quantum wells, *Opt. Express* 13(23): 9460–9464.

Peng, C. S., Huang, Q., Cheng, W. Q., Zhou, J. M., Zhang, Y. H., Sheng, T. T. & Tung, C. H. (1998). Optical properties of ge self-organized quantum dots in si, *Phys. Rev. B* 57: 8805–8808.

Polman, A., Min, B., Kalkman, J., Kippenberg, T. J. & Vahala, K. J. (2004). Ultralow-threshold erbium-implanted toroidal microlaser on silicon, *Appl. Phys. Lett.* 84(7): 1037–1039.

Raider, S. I., Flitsch, R. & Palmer, M. J. (1975). Oxide growth on etched silicon in air at room temperature, *J. Electrochem. Soc.* 12: 413.

Rong, H., Jones, R., Liu, A., Cohen, O., Hak, D., Fang, A. & Paniccia, M. (2005). A continuous-wave Raman silicon laser, *Nature* 433: 725–728.

Rong, H., Liu, A., Jones, R., Cohen, O., Hak, D., Nicolaescu, R., Fang, A. & Paniccia, M. (2004). An all-silicon Raman laser, *Nature* 433(1): 292–294.

Samavedam, S. B., Currie, M. T., Langdo, T. A. & Fitzgerald, E. A. (1998). High-quality germanium photodiodes integrated on silicon substrates using optimized relaxed graded buffers, *Appl. Phys. Lett.* 73(15): 2125.

Samavedam, S. B. & Fitzgerald, E. A. (1997). Novel dislocation structure and surface morphology effects in relaxed ge/si-ge(graded)/si structures, *J. Appl. Phys.* 81(7): 3108–3116.

Singh, H. P. (1968). Determination of thermal expansion of germanium, rhodium, and iridium by x-rays, *Acta Crystallogr., Sect. A: Cryst. Phys., Diffr., Theor. Gen. Crystallogr.* 24: 469–471.

Spitzer, W. G., Trumbore, F. A. & Logan, R. A. (1961). Properties of heavily doped n-type germanium, *J. Appl. Phys.* 32: 1822–1830.

Sun, X., Liu, J., Kimerling, L. C. & Michel, J. (2009a). Direct gap photoluminescence of n-type tensile-strained ge-on-si, *Appl. Phys. Lett.* . in review.

Sun, X., Liu, J., Kimerling, L. C. & Michel, J. (2009b). Room temperature direct band gap electroluminesence from ge-on-si light emitting diodes, *Optics Lett.* 34(8): 1198–1200.

Sutter, P., Kafader, U. & von Kanel, H. (1994). Thin film photodetectors grown epitaxially on silicon, *Solar Energy Mater. Solar Cell* 31: 541.

Tsaur, B.-Y., Geis, M. W., Fan, J. C. C. & Gale, R. P. (1981). Heteroepitaxy of vacuum-evaporated ge films on single-crystal si, *Appl. Phys. Lett.* 38: 779.

Vivien, L., Osmond, J., Fedeli, J.-M., Marris-Morini, D., cand Jean-Francois Damlencourt, P. C., Cassan, E., Y.Lecunff & Laval, S. (2009). 42 ghz p.i.n germanium photodetector integrated in a silicon-on-insulator waveguide, *Optics Express* 17(8): 6252.

Vivien, L., Rouviere, M., Fedeli, J.-M., Marris-Morini, D., Damlencourt, J.-F., Mangeney, J., Crozat, P., Melhaoui, L. E., Cassan, E., Roux, X. L., Pascall, D. & Lavall, S. (2007). High speed and high responsivity germanium photodetector integrated in a Silicon-On-Insulator microwaveguide, *Opt. Express* 15(15): 9843.

Vossen, J. L., Thomas, J. H., Maa, J. S. & ONeill, J. J. (1984). Preparation of surfaces for high quality interface formation, *J. Vac. Sci. Technol. A* 2: 212–215.

Wang, J., Loh, W. Y., Chua, K. T., Zang, H., Xiong, Y. Z., Loh, T. H., Yu, M. B., Lee, S. J., Lo, G.-Q. & Kwong, D.-L. (2008). Evanescent-coupled ge p-i-n photodetectors on si-waveguide with seg-ge and comparative study of lateral and vertical p-i-n configurations, *IEEE Electron Dev. Lett.* 29(5): 445.

Wang, J., Loh, W. Y., Chua, K. T., Zang, H., Xiong, Y. Z., Tan, S. M. F., Yu, M. B., Lee, S. J., Lo, G. Q. & Kwong, D.-L. (2008). Low-voltage high-speed (18 ghz/1 v) evanescent-coupled thin-film-ge lateral pin photodetectors integrated on si waveguide, *IEEE Photon. Technol. Lett.* 20(17): 1485.

Yin, T., Cohen, R., Morse, M. M., Sarid, G., Chetrit, Y., Rubin, D. & Paniccia, M. J. (2007). 31GHz Ge n-i-p waveguide photodetectors on Silicon-on-Insulator substrate, *Opt. Express* 15(21): 13965–13971.

Zhang, B., Michel, J., Ren, F. Y. G., Kimerling, L. C., Jacobson, D. C. & Poate, J. M. (1994). Room-temperature sharp line electroluminescence at λ=1.54 μm from an erbiqm-doped, silicon light-emitting diode, *Appl. Phys. Lett.* 64(21): 2842–2844.

2

Experimental Engineering of Photonic Quantum Entanglement

Stefanie Barz[1], Gunther Cronenberg[1,2] and Philip Walther[1]
[1]*Vienna Center for Quantum Science and Technology (VCQ), Faculty of Physics, University of Vienna, Vienna*
[2]*Atominstitut, Technische Universität Wien, Vienna*
Austria

1. Introduction

Entangled photons are a crucial resource for linear optical quantum communication and quantum computation. Besides the remarkable progress of photon state engineering using atomic memories (Kimble (2008); Yuan et al. (2008)) the majority of current experiments is based on the production of photon pairs in the process of spontaneous parametric down-conversion (SPDC), where the entangled photon pair is concluded from post-selection of randomly occurring coincidences. Here we present new insights into the heralded generation of photon states (Barz et al. (2010); Wagenknecht et al. (2010)) that are maximally entangled in polarization (Schrödinger (1935)) with linear optics and standard photon detection from SPDC (Kwiat et al. (1995)). We utilize the down-conversion state corresponding to the generation of three pairs of photons, where the coincident detection of four auxiliary photons unambiguously heralds the successful preparation of the entangled state (Śliwa & Banaszek (2003)). This controlled generation of entangled photon states is a significant step towards the applicability of a linear optics quantum network (Nielsen & Chuang (2000)), in particular for entanglement distribution (Bennett et al. (1996)), entanglement swapping (Kaltenbaek et al. (2009); Pan et al. (1998)), quantum teleportation (Bouwmeester et al. (1997)), quantum cryptography (Bennett & Brassard (1984); Ekert (1991); Jennewein et al. (2000)) and scalable approaches towards photonics-based quantum computing schemes (Browne & Rudolph (2004); Gottesman & Chuang (1999); Knill et al. (2001)).

2. Background

Photons are generally accepted as the best candidate for quantum communication due to their lack of decoherence and their possibility of photon broadcasting (Bouwmeester et al. (2000)). However, it has also been discovered that a scalable quantum computer can in principle be realized by using only single-photon sources, linear-optics elements and single-photon detectors (Knill et al. (2001)). Several proof-of-principle demonstrations for linear optical quantum computing have been given, including controlled-NOT gates (Gasparoni et al. (2004); O'Brien et al. (2003); Pittman et al. (2003; 2001); Sanaka et al. (2004)), Grover's search algorithm (Grover (1997); Kwiat et al. (2000); Prevedel et al. (2007)), Deutsch-Josza algorithm (Deutsch (1985); Tame et al. (2007)), Shor's factorization algorithm (Lanyon et al. (2007); Lu et al. (2007); Politi et al. (2009)) and the promising and new model of the one-way quantum computation (Chen et al. (2007); Kiesel et al. (2005); Prevedel et al. (2007); Vallone et al. (2008); Walther et al. (2005)). A main issue on the path of photonic quantum information processing is that the best current photon source, SPDC, is a process where the photons are created at random times (Zukowski et al. (1993)). All photons involved in a protocol need to be measured including a detection of the desired output state. This impedes the applicability of many of the beautiful proof-of-principle experiments, especially when multiple photon-pairs are involved (Bouwmeester et al. (2000)).

Other leading technologies in this effort are based on other physical systems including single trapped atoms and atomic ensembles (Kimble (2008); Yuan et al. (2008)), quantum dots (Michler et al. (2000)), or Nitrogen-Vacancy centers in diamond (Kurtsiefer et al. (2000)). Although these systems are very promising candidates, each of these quantum state emitters faces significant challenges for realizing heralded entangled states; typically due to low outcoupling efficiencies or the distinguishability in frequency.

Within linear optics several approaches exist to overcome the probabilistic nature originating from SPDC and to prepare two-photon entangled states conditioned on the detection of auxiliary photons (Eisenberg et al. (2004); Hnilo (2005); Śliwa & Banaszek (2003); Kok & Braunstein (2000b); Pittman et al. (2003); Walther et al. (2007)). It was shown that the production of one heralded polarization-entangled photon pair using only conventional down-conversion sources, linear-optics elements, and projective measurements is not possible with less than three initial pairs (Kok & Braunstein (2000a)). Here we describe an experimental realization for producing heralded two-photon entanglement along theses lines, suggested by Śliwa and Banaszek that relies on triple-pair emission from a single down-conversion source (Śliwa & Banaszek (2003)). This scheme shows significant advantages compared to other schemes where either several SPDC sources (Pittman et al. (2003)) or more ancilla photons (Walther et al. (2007)) are required.

Current down-conversion experiments allow for the simultaneous generation of up to six photons (Lu et al. (2009; 2007); Prevedel et al. (2009); Radmark et al. (2009); Wieczorek et al. (2009); Zhang et al. (2006)) with typical detection count rates, dependent on the experimental configuration, of about 10^{-3} to $10^{-1}\,s^{-1}$. In the demonstrated case the coincident detection of four photons is used to predict the presence of two polarization entangled photons in the output modes. The auxiliary photons thus herald the presence of a Bell state without performing a measurement on that state.

3. Theory and experimental design

Figure 1 gives a schematic diagram of the used setup to generate the heralded state

$$|\phi^+\rangle = \frac{1}{\sqrt{2}}\left(t_{1H}^\dagger t_{2H}^\dagger + t_{1V}^\dagger t_{2V}^\dagger\right)|vac\rangle\,, \tag{1}$$

where H and V denote horizontal and vertical polarization, respectively, whereas t_1 and t_2 correspond to the transmitted modes after the beam splitters. For generating the heralded state, $|\phi^+\rangle$, three photon pairs have to be emitted simultaneously into spatial modes a_1 and a_2, which can be expressed in terms of creation operators:

$$|\Psi_3\rangle = \frac{1}{12}\left(a_{1H}^\dagger a_{2V}^\dagger - a_{1V}^\dagger a_{2H}^\dagger\right)^3|vac\rangle\,. \tag{2}$$

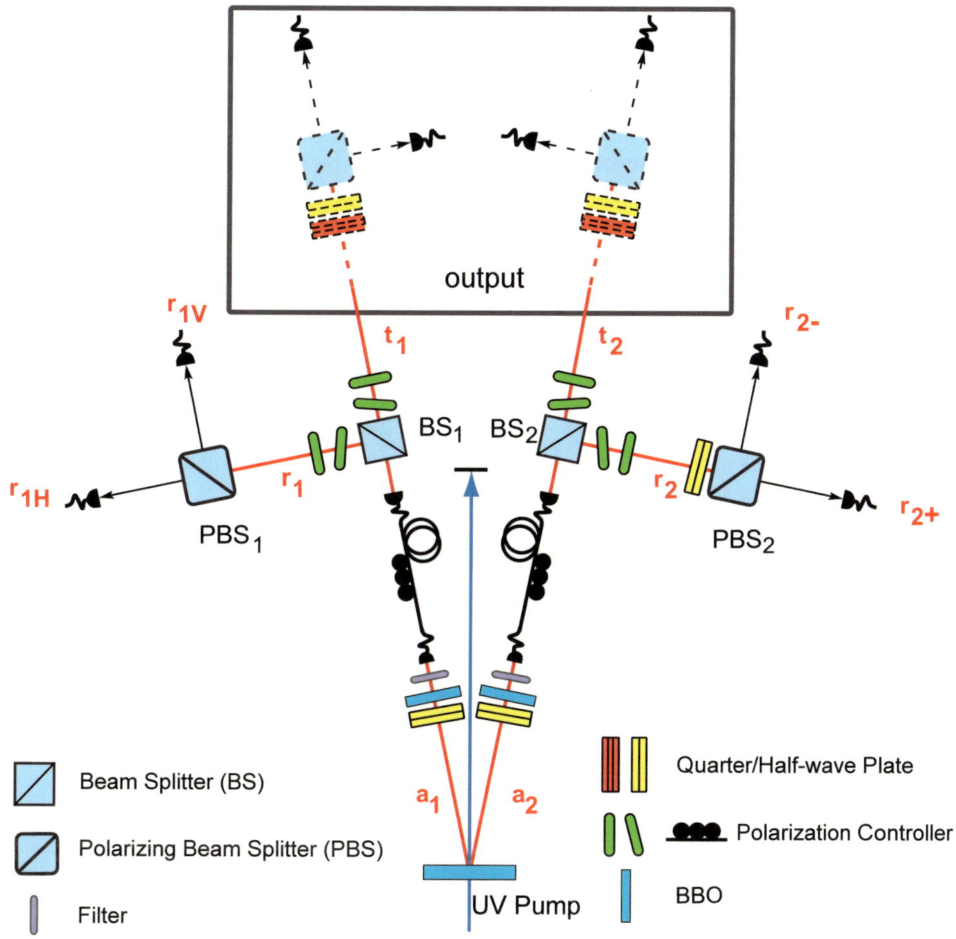

Fig. 1. Setup for the heralded generation of entangled photon pairs. Six photons are created simultaneously by exploiting higher-order emissions in a spontaneous parametric down-conversion process. The photons are brought to beam splitters and the reflected modes are analyzed in $|H/V\rangle$ basis and in $|\pm\rangle = \frac{1}{\sqrt{2}}(|H\rangle \pm |V\rangle)$ basis, respectively, using polarizing beam splitters and half-wave plates. State characterization of the heralded photon pair in the transmitted modes is performed via polarization analysis and the help of quarter-wave plates, half-wave plates and polarizing beam splitters.

These photons are guided to non-polarizing beam splitters (BS1 and BS2) with various splitting ratios. The scheme only succeeds when four photons, two photons at BS1 and BS2, respectively, are reflected, and detected in each of the output modes as four-fold coincidence. The two reflected photons of BS1 are projected onto the $|H/V\rangle$ basis for mode r_1, while the two reflected photons of BS2 are measured in the $|\pm\rangle = \frac{1}{\sqrt{2}}(|H\rangle \pm |V\rangle)$ basis for mode r_2. Only the case where one photon is present in each of the modes $r_{1H,1V}$ and $r_{2+,2-}$ is of interest for a successful heralding of the output state. Considering only these terms, the output state results in

$$|\Psi_3\rangle = C(\theta_1,\theta_2)\cdot\frac{1}{\sqrt{2}}\left(t^\dagger_{1H}t^\dagger_{2H}+t^\dagger_{1V}t^\dagger_{2V}\right)\cdot r^\dagger_{1H}r^\dagger_{1V}r^\dagger_{2+}r^\dagger_{2-}\,|vac\rangle \quad (3)$$

where $C(\theta_1,\theta_2)$ is a constant depending on transmission coefficients of beam splitters. The coincident detection of one and only one photon in the modes r_{1H}, r_{1V}, r_{2+} and r_{2-} heralds the presence of an entangled photon pair in $|\phi^+\rangle$ state in the output modes t_1, t_2.
In the present scheme such a case can only be achieved by three (or more)-pair emission from SPDC. The contribution from two-pair emission is suppressed by destructive quantum interference in the half-wave plate (HWP) rotation used for $r_{2+,2-}$. At this specific angle any possible four-photon state, emitted into the four modes, $r_{1H,1V}$ and $r_{2+,2-}$, will result only in a three-fold coincidence when projected on the $|H/V\rangle$ basis. Thus, these two photons will never contribute to a fourfold coincidence. This results from the quantum interference (see Figure 2) of

$$r^\dagger_{2+}r^\dagger_{2-} = r^\dagger_{2H}r^\dagger_{2H} - r^\dagger_{2V}r^\dagger_{2V}. \quad (4)$$

This quantum interference together with the use of number-resolving detectors ensures that the remaining two photons are found in the transmitted modes. If a high transmittance of the beam splitters is chosen, it still can be assumed that the two photons are transmitted even without the use of number-resolving detectors. The quantum interference can be seen when rotating the HWP in mode r_2. Figure 2 shows this dependency of the four-fold coincidences with a visibility of $(86.7 \pm 1.2)\%$

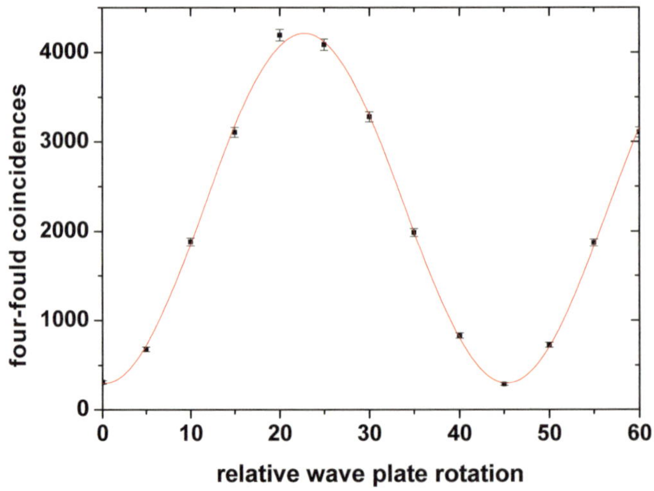

Fig. 2. Visibility of the four-fold coincidences at the reflected modes r_{1H}, r_{1V} and r_{2+}, r_{2-}. The quantum interference for the detected two-pair emission can be seen with respect to the half-wave plate (HWP) used for r_{2+}, r_{2-}. At the specific HWP rotation of 0° relative to the $|\pm\rangle$ basis, the curve shows a suppression of the corresponding four-photon detection. An angle of 22.5° relative to the $|\pm\rangle$ basis results in a measurement in the $|H/V\rangle$ and thus leads to a maximum of the four-fold rates. This quantum interference is a key feature of the experiment as it enables the triggering on the desired six-photon emission by measuring only four detection events.

4. Experimental setup

Six photons in the $|\Psi_3\rangle$-state are produced simultaneously by using higher-order emissions of a non-collinear type-II SPDC process. A mode-locked Mira HP Ti:Sa oscillator is pumped by a Coherent Inc. Verdi V-18 laser to reach output powers high enough to be able to exploit third-order SPDC emissions. The pulsed-laser output (τ = 200 fs, λ= 808 nm, 76 MHz) is frequency-doubled using a 2 mm thick Lithium triborate (LBO) crystal, resulting in UV pulses of 1.2 W cw-average. A stable source of UV-pulses is achieved by translating the LBO with a stepper-motor to avoid optical damage to the anti-reflection coating of the crystal (count-rate fluctuations less than 3 % over 24 h). Afterwards, dichroic mirrors are used to separate the up-converted light from the infrared laser light.

The UV beam is focused on a 2 mm-thick β-barium borate (BBO) crystal cut for non-collinear type-II parametric down-conversion. Half-wave plates and additional BBO-crystals compensate walk-off effects and allow the production of any Bell-state. Narrowband interference filters ($\Delta\lambda$ = 3 nm) are used to spatially and spectrally select the down-converted photons which are then coupled into single mode fibers that guide them to the analyzer setup. At this detection unit, the photon pairs are directed to non-polarizing beam splitters with different splitting ratios for different experiment runs. The reflected modes are analyzed in $|H/V\rangle$ basis and in $|\pm\rangle$ basis, respectively, as described above. This is implemented by using a half-wave plate (HWP) oriented at 45° followed by PBS2.

5. Methods

In the demonstrated case of using standard detectors (photo-avalanche diodes by PerkinElmer) the transmission of the non-polarizing beam splitters should ideally be as high as possible; i.e. that a measured four-photon coincidence corresponds to precisely four photons and thus heralds the desired state in the output modes t_1 and t_2. Obviously the trade-off for increasing this probability of heralding $|\phi^+\rangle$ - which in principle can be approximately unity - is a reduction in the four-fold coincidence rate for triggering this state. Therefore for demonstrating this dependency beam splitters with different transmission rates T, of 17 %, 50 % and 70 % are chosen. For each of this beam splitter ratios the density matrix ρ of the heralded entangled pair, triggered by the successful registration of a four-fold coincidence in modes $r_{1H}, r_{1V}, r_{2+}, r_{2-}$ is reconstructed. This density matrix is obtained by an over-complete tomographic set of measurements, where 36 combinations of the single photon projections $|H/V\rangle$, $|\pm\rangle$, and $|R/L\rangle = \frac{1}{\sqrt{2}}(|H\rangle \pm i\,|V\rangle)$, on each of the two photons in modes t_1 and t_2 were used. Successful projections are signaled by 6-photon coincidence measurements and the most likely physical density matrix for the 2-qubit output states is extracted using maximum-likelihood reconstruction (Banaszek et al. (1999); Hradil (1997); James et al. (2001)).

6. Results

The rate R of the four-fold and the six-fold coincidences are shown in Table 1. Figure 3 shows the probability of obtaining the heralded state, i.e. to find a photon pair in the output modes triggered by the four-fold coincidence, with respect to the beam splitter transmission.

The probabilities were $P_{17/83} = 2.5 \pm 0.2\,\%$, $P_{50/50} = 29.4 \pm 1.0\,\%$ and $P_{70/30} = 77.2 \pm 6.6\,\%$ for the different transmission rates (see Figure 3). Obviously, these obtained values for heralding

	four-fold rate /minute	six-fold rate /day
17/83	83	30
$30/70^{LP}$	22	5
50/50	14	60
70/30	0.4	5

Table 1. Overview over the experimental four-fold coincidence rates showing high- and low-power (LP) measurements.

$|\phi^+\rangle$ reflect the limitations of photon losses, mostly due to using standard detectors without the capability of resolving the photon number.

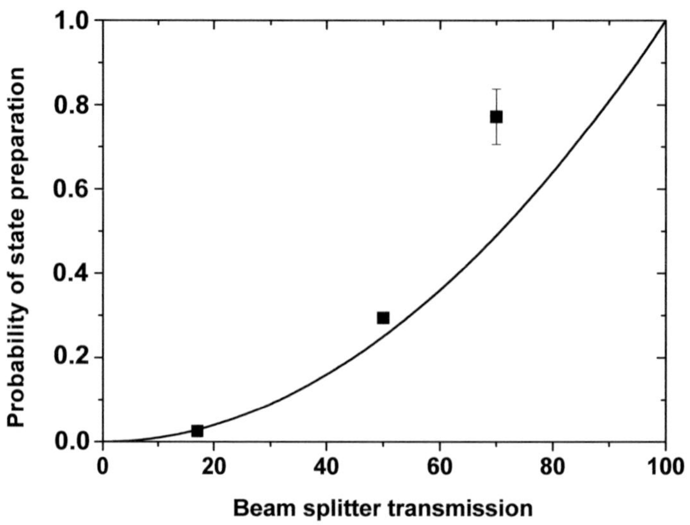

Fig. 3. Heralding efficiency. Probability of the heralded entangled photon-pair generation with respect to various beam splitter transmission rates in %. The deviation from the expected quadratic behavior (black line) originated from spurious high-order emissions, which increase the probability of preparing the entangled state $|\phi^+\rangle$ for higher beam splitter transmissions.

The corresponding fidelity, $F = \langle\phi^+|\rho|\phi^+\rangle$ of the heralded photon pair with the pure quantum state $|\phi^+\rangle$, was $F_{17/83} = 63.7 \pm 4.9\,\%$, $F_{50/50} = 57.5 \pm 3.4\,\%$, and $F_{70/30} = 61.9 \pm 7.7\,\%$ for the different beam splitters via local unitary transfomations (see Figure 4).

The corresponding density matrices are shown in Figure 5. Uncertainties in quantities extracted from these density matrices were calculated using a Monte Carlo routine and assumed Poissonian errors. As expected, these fidelities are basically independent of the beam splitter ratio. The small deviation can be explained by the typical variations of the quality for these custom-made beam splitters. The reduced fidelities to the real state $|\phi^+\rangle$, however, is a result of the eight-photon emission. At the given laser power the probability of obtaining a higher-order emission for a given six-fold coincidence is about 10 %.

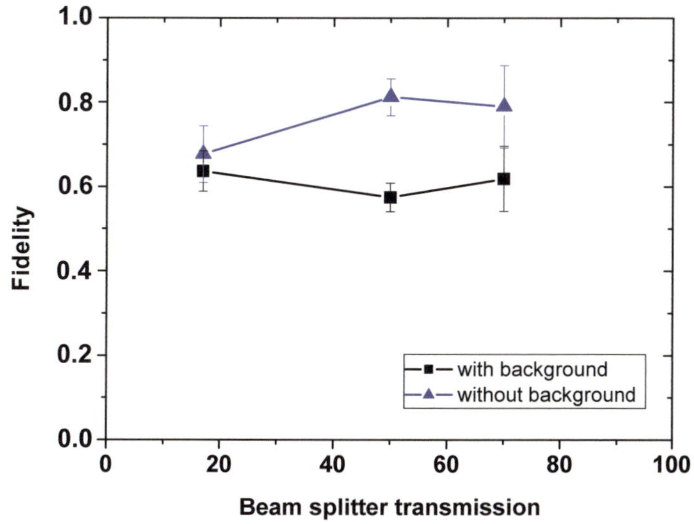

Fig. 4. Fidelities. The experimentally obtained quantum state fidelities with respect to the ideal state $|\phi^+\rangle$ is shown with (squares) and without (triangles) background for various beam splitter transmission rates in %. The background was assumed to be mostly state $|\psi^-\rangle$ originating from from eight-photon emissions.

This unwanted contribution adds a significant $|\psi^-\rangle$ component to the density matrices and thus reduces the overlap with the ideal state $|\phi^+\rangle$. A theoretical calculation based on the used experimental parameters allowed to estimate this unwanted background which is qualitatively different than the desired output state. When subtracting this $|\psi^-\rangle$ contribution from the measured density matrices the corrected fidelities became $F'_{17/83} = 67.7 \pm 6.7\,\%$, $F'_{50/50} = 81.2 \pm 4.4\,\%$, and $F'_{70/30} = 79.0 \pm 9.8\,\%$, which demonstrates the that laser systems with less peak power per pulse but much higher repetition rate could achieve such state fidelities. From this data, the tangle (Coffman et al. (2000)) is extracted as a measure of entanglement that ranges from 0 for separable states to 1 for maximally entangled states. The values are $t_{17/83} = 0.43 \pm 0.08$, $t_{50/50} = 0.37 \pm 0.03$ and $t_{70/30} = 0.45 \pm 0.11$ for the different beam splitter ratios. It is important to note that the noise is not intrinsic in the setup and is only due to practical drawbacks. For demonstrating that this limitation will be overcome in the near future, an additional experimental run with a reduced laser power of 620 mW and beam splitter transmissions of 30 % was performed. The post-selected density matrix of this state is shown in Figure 2b. The extracted state of $F_{30/70} = (84.2 \pm 8.5)\,\%$ and a tangle of $\tau_{30/70} = 0.55 \pm 0.19$ evidently demonstrate the generated entanglement for this measured state.

This state's density matrix as shown in Figure 6, if commonly used in the coincidence basis, would allow a violation of local realistic theories by almost 2 standard deviations as the it implies a maximum Clauser-Horne-Shimony-Holt (Clauser et al. (1969); Horodecki et al. (1995)) Bell parameter of $S = 2.36 \pm 0.22$.

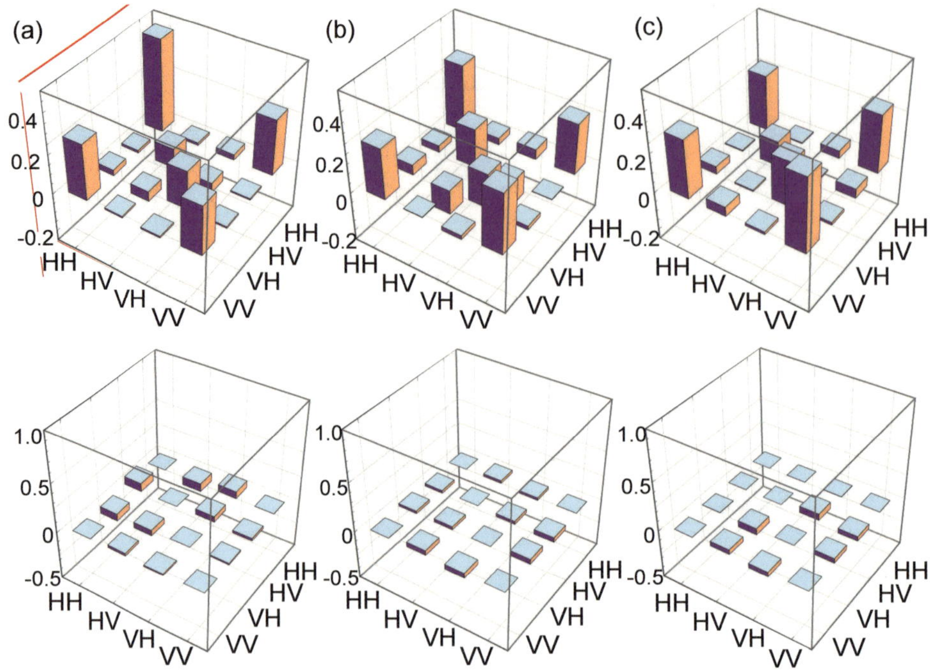

Fig. 5. The two-qubit density matrix for different beam splitter transmissions. Shown are the real (top) and imaginary (bottom) parts of the reconstructed density matrix for beam splitter transmissions of 17 % (a), 50 % (b), and 83 % (c). Large diagonal elements in the $|HH\rangle$ and $|VV\rangle$ positions along with large positive coherences indicate that this state has the qualities of the desired heralded entangled state $|\phi^+\rangle$. The real density matrix was reconstructed by way of a maximum likelihood method using six-photon coincidence rates obtained in 36 polarization projections. The experimentally measured density matrix has a fidelity of (63.7 ± 4.9)%(a), (57.5 ± 3.4)% (b), and (61.9 ± 7.7)% (c) with the ideal state $|\phi^+\rangle$ via local unitary transformations.

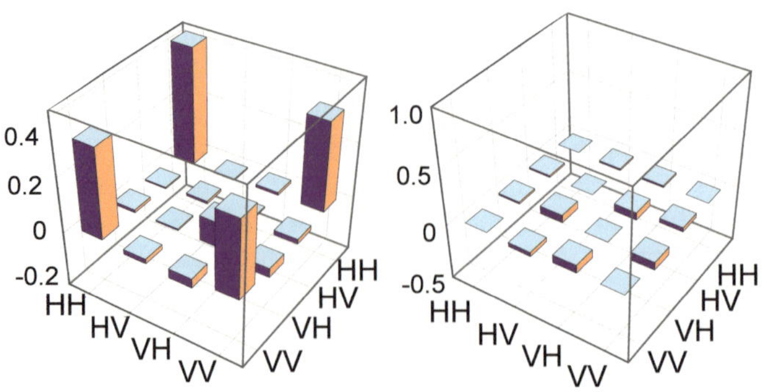

Fig. 6. Low-power density matrix. The reduction of the background is demonstrated when reducing the laser-power. The experimentally reconstructed real part (left) and imaginary part (right) of the two-qubit polarization density matrix is shown. The measurements were performed with a reduced laser-power of 0.62 Watt and a beam splitter transmission of 30%.

7. Conclusion

In conclusion, an efficient method for the generation of heralded polarization-entangled photon states, which are a crucial resource for photonic quantum computing, quantum communication and quantum metrology is demonstrated. This experiment uses currently available technologies - it relies only on linear optics, parametric down-conversion and standard photon detection - and is therefore of direct practical relevance. The performance of the photon-pair source was characterized by measuring the quantum state fidelity of the output states and by demonstrating the relation of the preparation efficiency with respect to the beam splitter transmission rate. A fidelity of better than 84% and a state preparation efficiency of 77% have been achieved. The feasibility of this experiment and the promising application for linear optics quantum information processing and quantum metrology makes it important and interesting for future quantum information experiments.

8. References

Banaszek, K., D'Ariano, G. M., Paris, M. G. A. & Sacchi, M. F. (1999). Maximum-likelihood estimation of the density matrix, *Phys. Rev. A* 61(1): 010304.

Barz, S., Cronenberg, G., Zeilinger, A. & Walther, P. (2010). Heralded generation of entangled photon pairs, *Nature Photonics* 4(8): 553–556.

Bennett, C. H. & Brassard, G. (1984). *Quantum Cryptography: Public Key Distribution and Coin Tossing*, Vol. 70, IEEE, Bangalore, India, p. 175.

Bennett, C. H., Brassard, G., Popescu, S., Schumacher, B., Smolin, J. A. & Wootters, W. K. (1996). Purification of noisy entanglement and faithful teleportation via noisy channels, *Phys. Rev. Lett.* 76: 722–725.

Bouwmeester, D., Ekert, A. & Zeilinger, A. (eds) (2000). *The Physics of Quantum Information*, Springer, Berlin.

Bouwmeester, D., Pan, J.-W., Mattle, K., Eibl, M., Weinfurter, H. & Zeilinger, A. (1997). Experimental quantum teleportation, *Nature* 390: 575–579.

Browne, D. E. & Rudolph, T. (2004). Efficient linear optical quantum computation, *quant-ph/0405157* .

Chen, K., Li, C.-M., Zhang, Q., Chen, Y.-A., Goebel, A., Chen, S., Mair, A. & Pan, J.-W. (2007). Experimental realization of one-way quantum computing with two-photon four-qubit cluster states, *Physical Review Letters* 99(12): 120503.

Clauser, J. F., Horne, M. A., Shimony, A. & Holt, R. A. (1969). Proposed experiment to test local hidden-variable theories, *Phys. Rev. Lett.* 23(15): 880–884.

Coffman, V., Kundu, J. & Wootters, W. K. (2000). Distributed entanglement, *Phys. Rev. A* 61(5): 052306.

Deutsch, D. (1985). Quantum theory, the church-turing principle and the universal quantum computer, *Proceedings of the Royal Society of London. A. Mathematical and Physical Sciences* 400(1818): 97–117.

Eisenberg, H. S., Khoury, G., Durkin, G. A., Simon, C. & Bouwmeester, D. (2004). Quantum entanglement of a large number of photons, *Phys. Rev. Lett.* 93(19): 193901.

Ekert, A. K. (1991). Quantum cryptography based on bell's theorem, *Phys. Rev. Lett.* 67: 661–663.

Gasparoni, S., Pan, J.-W., Walther, P., Rudolph, T. & Zeilinger, A. (2004). Realization of a photonic cnot gate sufficient for quantum computation, *Phys. Rev. Lett.* 92: 020504.

Gottesman, D. & Chuang, I. L. (1999). Demonstrating the viability of universal quantum computation using teleportation and single-qubit operations, *Nature* 402: 390–393.

Grover, L. K. (1997). Quantum mechanics helps in searching for a needle in a haystack, *Phys. Rev. Lett.* 79(2): 325–328.

Hnilo, A. A. (2005). Three-photon frequency down-conversion as an event-ready source of entangled states, *Phys. Rev. A* 71(3): 033820.

Horodecki, R., Horodecki, P. & Horodecki, M. (1995). Violating bell ineqaulity by mixed spin-$\frac{1}{2}$ states: necessary and sufficient condition, *Phys. Lett. A* 200: 340–344.

Hradil, Z. (1997). Quantum-state estimation, *Phys. Rev. A* 55(3): R1561–R1564.

Śliwa, C. & Banaszek, K. (2003). Conditional preparation of maximal polarization entanglement, *Phys. Rev. A* 67(3): 030101.

James, D., Kwiat, P., Munro, W. & White, A. (2001). Measurement of qubits, *Physical Review A* 64(5): 52312.

Jennewein, T., Simon, C., Weihs, G., Weinfurter, H. & Zeilinger, A. (2000). Quantum cryptography with entangled photons, *Phys. Rev. Lett.* 84: 4729–4732.

Kaltenbaek, R., Prevedel, R., Aspelmeyer, M. & Zeilinger, A. (2009). High-fidelity entanglement swapping with fully independent sources, *Physical Review A* 79(4): 40302.

Kiesel, N., Schmid, C., Weber, U., Tóth, G., Gühne, O., Ursin, R. & Weinfurter, H. (2005). Experimental analysis of a four-qubit photon cluster state, *Phys. Rev. Lett.* 95(21): 210502.

Kimble, H. (2008). The quantum internet, *Nature* 453: 1023–1030.

Knill, E., Laflamme, R. & Milburn, G. (2001). A scheme for efficient quantum computation with linear optics, *Nature* 409(6816): 46–52.

Kok, P. & Braunstein, S. (2000a). Limitations on the creation of maximal entanglement, *Physical Review A* 62(6): 64301.

Kok, P. & Braunstein, S. L. (2000b). Postselected versus nonpostselected quantum teleportation using parametric down-conversion, *Phys. Rev. A* 61(4): 042304.

Kurtsiefer, C., Mayer, S., Zarda, P. & Weinfurter, H. (2000). Stable solid-state source of single photons, *Phys. Rev. Lett.* 85(2): 290–293.

Kwiat, P. G., Mattle, K., Weinfurter, H., Zeilinger, A., Sergienko, A. V. & Shih, Y. (1995). New high-intensity source of polarization-entangled photon pairs, *Phys. Rev. Lett.* 75(24): 4337–4341.

Kwiat, P., Mitchell, J., Schwindt, P. & White, A. (2000). Grover's search algorithm: An optical approach, *J. Mod. Opt.* 47: 257–266.

Lanyon, B. P., Weinhold, T. J., Langford, N. K., Barbieri, M., James, D. F. V., Gilchrist, A. & White, A. G. (2007). Experimental demonstration of a compiled version of shor's algorithm with quantum entanglement, *Physical Review Letters* 99(25): 250505.

Lu, C.-Y., Browne, D. E., Yang, T. & Pan, J.-W. (2007). Demonstration of a compiled version of shor's quantum factoring algorithm using photonic qubits, *Physical Review Letters* 99(25): 250504.

Lu, C.-Y., Yang, T. & Pan, J.-W. (2009). Experimental multiparticle entanglement swapping for quantum networking, *Physical Review Letters* 103(2): 020501.

Lu, C., Zhou, X., Gühne, O., Gao, W., Zhang, J., Yuan, Z., Goebel, A., Yang, T. & Pan, J. (2007). Experimental entanglement of six photons in graph states, *Nature Physics* 3(2): 91–95.

Michler, P., Kiraz, A., Becher, C., Schoenfeld, W. V., Petroff, P. M., Zhang, L., Hu, E. & Imamoglu, A. (2000). A quantum dot single-photon turnstile device, *Science* 290(5500): 2282–2285.

Nielsen, M. A. & Chuang, I. L. (2000). *Quantum Computation and Quantum Information*, Cambridge University Press, Cambridge.

O'Brien, J. L., Pryde, G. J., White, A. G., Ralph, T. C. & Branning, D. (2003). Demonstration of an all-optical quantum controlled-not gate, *Nature* 426: 264–267.

Pan, J.-W., Bouwmeester, D., Weinfurter, H. & Zeilinger, A. (1998). Experimental entanglement swapping: Entangling photons that never interacted, *Phys. Rev. Lett.* 80: 3891–3894.

Pittman, T., Donegan, M., Fitch, M., Jacobs, B., Franson, J., Kok, P., Lee, H. & Dowling, J. (2003). Heralded two-photon entanglement from probabilistic quantum logic operations on multiple parametric down-conversion sources, *IEEE Journal of Selected Topics in Quantum Electronics* 9(6): 1478–1482.

Pittman, T., Fitch, M., Jacobs, B. & Franson, J. (2003). Experimental controlled-not logic gate for single photons, *Phys. Rev. A* 68: 032316.

Pittman, T., Jacobs, B. & Franson, J. (2001). Probabilistic quantum logic operations using polarizing beam splitters, *Phys. Rev. A* 64: 062311.

Politi, A., Matthews, J. C. F. & O'Brien, J. L. (2009). Shor's quantum factoring algorithm on a photonic chip, *Science* 325(5945): 1221–.

Prevedel, R., Cronenberg, G., Tame, M. S., Paternostro, M., Walther, P., Kim, M. S. & Zeilinger, A. (2009). Experimental realization of dicke states of up to six qubits for multiparty quantum networking, *Physical Review Letters* 103(2): 020503.

Prevedel, R., Walther, P., Tiefenbacher, F., Böhi, P., Kaltenbaek, R., Jennewein, T. & Zeilinger, A. (2007). High-speed linear optics quantum computing using active feed-forward, *Nature* 445(7123): 65–69.

Radmark, M., Zukowski, M. & Bourennane, M. (2009). Experimental high fidelity six-photon entangled state for telecloning protocol, *quant-ph/09061530* .

Sanaka, K., Jennewein, T., Pan, J.-W., Resch, K. & Zeilinger, A. (2004). Experimental nonlinear sign shift for linear optics quantum computation, *Phys. Rev. Lett.* 92: 017902.

Schrödinger, E. (1935). Die gegenwärtige Situation in der Quantenmechanik, *Naturwissenschaften* 23(49): 823–828.

Tame, M. S., Prevedel, R., Paternostro, M., Böhi, P., Kim, M. S. & Zeilinger, A. (2007). Experimental realization of deutsch's algorithm in a one-way quantum computer, *Physical Review Letters* 98(14): 140501.

Vallone, G., Pomarico, E., Martini, F. D. & Mataloni, P. (2008). One-way quantum computation with two-photon multiqubit cluster states, *Physical Review A (Atomic, Molecular, and Optical Physics)* 78(4): 042335.

Wagenknecht, C., Li, C., Reingruber, A., Bao, X., Goebel, A., Chen, Y., Zhang, Q., Chen, K. & Pan, J. (2010). Experimental demonstration of a heralded entanglement source, *Nature Photonics* 4(8): 549–552.

Walther, P., Aspelmeyer, M. & Zeilinger, A. (2007). Heralded generation of multiphoton entanglement, *Physical Review A (Atomic, Molecular, and Optical Physics)* 75(1): 012313.

Walther, P., Resch, K., Rudolph, T., Schenck, E., Weinfurter, H., Vedral, V., Aspelmeyer, M. & Zeilinger, A. (2005). Experimental one-way quantum computing, *Nature* 434(7030): 169–176.

Wieczorek, W., Krischek, R., Kiesel, N., Michelberger, P., Tóth, G. & Weinfurter, H. (2009). Experimental entanglement of a six-photon symmetric dicke state, *Physical Review Letters* 103(2): 020504.

Yuan, Z., Chen, Y., Zhao, B., Chen, S., Schmiedmayer, J. & Pan, J. (2008). Experimental demonstration of a bdcz quantum repeater node, *Nature* 454(7208): 1098–1101.

Zhang, Q., Goebel, A., Wagenknecht, C., Chen, Y., Zhao, B., Yang, T., Mair, A., Schmiedmayer, J. & Pan, J. (2006). Experimental quantum teleportation of a two-qubit composite system, *Nature Physics* 2(10): 678–682.

Zukowski, M., Zeilinger, A., Horne, M. A. & Ekert, A. K. (1993). "Event-ready-detectors" Bell experiment via entanglement swapping, *Phys. Rev. Lett.* 71(26): 4287–4290.

3

Photonic Band Gap Engineered Materials for Controlling the Group Velocity of Light

Andrea Blanco[1] and Joseba Zubía[2]
[1]*Tecnalia Research & Innovation*
[2]*University of the Basque Country*
Spain

1. Introduction

The power of creating engineered materials in which one can customize the propagation of light unleashes a huge range of fascinating applications. Commonly, these materials are based on the appearance of a forbidden frequency range, after which they are named photonic band gap materials or, more generally, photonic crystals. Photonic crystals are artificial, periodic arrangements of dielectric media with sub-wavelength periodicity lengths. As a consequence of the periodic dielectric function, in analogy with natural crystals where atoms or molecules are periodically spaced, a range of forbidden propagating energies may appear under the proper conditions, giving rise to a photonic band gap. Exploiting the effects associated to the photonic band gap, one can dramatically alter the flow of electro-magnetic waves (Joannopoulos et al., 2008).

One of the optical properties that can be radically modified in such materials is the group velocity of light pulses. Usually, the unmatched velocity of light (in vacuum c=299.792.458 m/s) is an advantage. Thanks to this fact, light confined within the core of an optical fiber could give a complete round to the equatorial circumference of Earth in 200 milliseconds (light in glass travels at c/1.5). In other cases, where processing the information carried by light pulses is needed, the huge speed of light entails a problem. In quantum optics, for instance, the quantum state of light can be stored for longer if light is slowed down. As another example, buffering and storing information on a photonic chip is crucial in all-optical communications routers and optical computing, the same as electronic RAMs are essential in conventional networks and computers. Furthermore, countless applications arise from the increase of energy density, and consequent stronger light-matter interaction, due to slower light propagation. Higher energy density implies enhanced nonlinearities, and therefore more efficient, compact, and low-power consuming nonlinear optical schemes, as Raman amplification or optical regenerators (Krauss, 2008). Stronger light-matter interaction is also beneficial for sensing applications, especially for biosening, and more efficient photovoltaic cells.

Unluckily, light is inherently difficult to control, not to mention how hard it is to store it. This is not only due to its speed, but also to the very nature of its basic unit, the photon, being massless and having no electric charge. A number of ingenious and diverse schemes to diminish the speed of light in a controllable fashion has been proposed, giving rise to the intriguing field of slow light. All slow light schemes are based in the same basic principle, in spite of their interdisciplinary nature: the existence of a single sharp resonance or multiple resonances (Khurgin & Tucker, 2009). As it will be justified, photonic crystal devices are, to our understanding, the most promising approach for the development of practical photonic devices based on slow light.

This chapter begins by reviewing the motivation and principles of slow light, followed by a description and assessment of the different techniques to diminish the speed of light and the applications of this phenomenon. In a subsequent section the particularly promising approach of slow-light photonic crystal arrangements will be addressed. In that section, together with the fundamentals of photonic band gap materials, their simulation methods and fabrication techniques, simulations results and analysis on our designs of slow-light photonic crystal arrangements will be presented. The chapter will be concluded with an assessment of photonic band gap engineered materials as a technology for developing slow-light based devices for telecommunications and microwave photonics applications.

2. Slow light: From theory to practice

Being able of governing something as essentially free as light is captivating in itself. Along this section, the physical principles behind the drastic reduction of group velocity and the techniques to achieve it will be presented. However, what is more appealing about slow light is its significant number of applications. In practice, unfortunately, slow light entails severe difficulties and a noteworthy research effort is being done by many groups to cover the gap between theory and practical applications.

2.1 Principles of slow light

The term slow lights refers to the phenomenon of light propagation in media with reduced group velocity. It is thence convenient to throw some light on the concepts of group velocity, phase velocity, and front velocity. Fig. 1 serves as a reference for those concepts.

The group velocity is the speed at which the wave envelope, the overall shape of an amplitude varying wave, propagates. This is frequently considered to be the velocity of the information transported by light signals, which is correct for most cases. Nevertheless, as it will be discussed bellow, there are some exceptions to this statement. The group velocity of a wave in a certain medium corresponds to the velocity of light in vacuum, c, divided by the group index of the medium, n_g, and it is given by the equation

$$v_g = \frac{c}{n_g} = \frac{d\omega}{dk} \tag{1}$$

where ω is the angular frequency and k is the angular wave number. The variation ω with k is called the dispersion relation, and thence, the group velocity is given by the slope of the dispersion relation.

The phase velocity is the speed at which the phase of every particular frequency component of the wave travels and it is given by

$$v_p = \frac{\omega}{k} \tag{2}$$

It can be seen that for media in which ω varies linearly with k the phase and the group velocity coincide. However, in general, for dispersive media, phase and group velocities differ and both vary with frequency.

There is one last concept that it is worth to mention, the front velocity. The front velocity is the speed of the pulse leading edge, i.e. the velocity of the earliest part of a pulse. This is truly the velocity of the signal, i.e. of the information carried by the light wave. According to the principle of Einstein causality, information can never travel faster than light in vacuum and therefore the signal velocity it is always slower than c. This concept has led to discrepancies among the scientific community in the past, when some groups claimed measured superluminal *signal* velocity (Heitmann & Nimtz, 1994; Steinberg et al., 1993), menacing the validity of the special theory of relativity. In those experiments, the reported measured velocity corresponded to that of the peak of the photon's wave and that of the envelope of the signal, but not to the signal velocity itself. This is explained by the reshaping of the pulse along its propagation, so that its peak moves towards its leading edge. The velocity of the peak may be, in this case, faster than the velocity of light in vacuum, but the signal (front) velocity will remain bellow c. In other words, even if the group velocity appears to be higher than c, causality still holds.

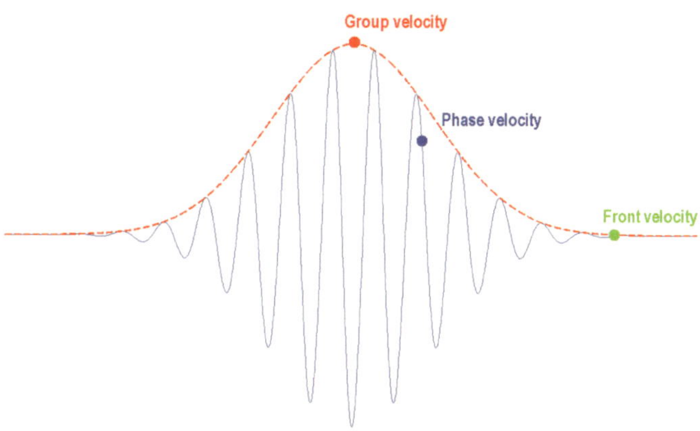

Fig. 1. Representation of the points of measurement for the group, phase and front velocities

Coming back to the point of slow light, it has been defined as the phenomenon of light propagation at reduced group velocity. Bearing in mind Eq. 1, it is straightforward to see that slow light relies on increasing the group index of the medium, which is given by

$$n_g = n + \omega \frac{dn}{d\omega} \tag{3}$$

The refractive index of the material n varies with frequency and can be slightly altered by making use of a plurality of phenomena, such as electro-optic or thermo-optic effects. Nonetheless, when aiming at a significant change in the group index, it is the second term in Eq. 3, $dn/d\omega$, the one that dominates, i.e. the dispersive behaviour of the media. This is the reason why sharp spectral resonances are always behind slow light techniques.

By making use of different approaches, that will be briefly reviewed in next section, extremely low group velocities have been achieved, as low as 17 m s^{-1} (Hau et al., 1999).

Nonetheless, from the point of view of practical applications, a more useful figure of merit (FOM) than the absolute value of the group velocity itself is the number of pulse widths that can be delayed by the system. This FOM is given by the delay-bandwidth product, being the bandwidth inversely proportional to the duration of the pulse, and it gives an idea of the information storage capacity of the system (Boyd & Narum, 2007). The delay-bandwidth product is limited by two major impediments: the pulse distortion, mainly caused by group velocity dispersion, and the propagation loss. These two factors present different origin and significance for each slow-light approach but it is a common feature to all of them that higher delays come at the expense of bandwidth reductions and, usually, of higher loss.

2.2 Techniques to achieve slow light

In Fig. 2 we present a classification of the different approaches to slow light.

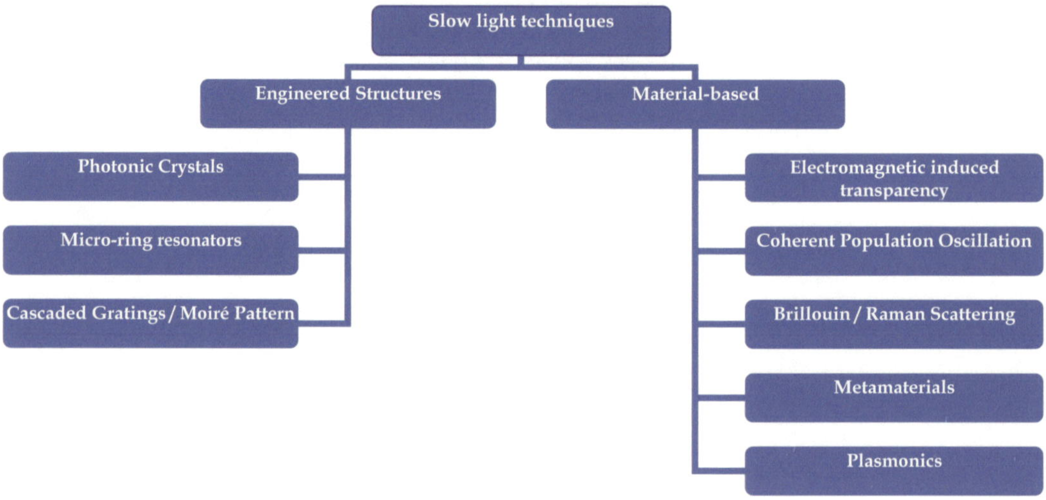

Fig. 2. Classification of the proposed approaches to slow light

How to control something that has no mass and no electric charge? These fundamental properties make the task of stopping light very challenging. However, far from being intimidated, science community has dedicated a great number of research efforts to 'trap' light. More specifically, these efforts try, not to store the photons, but the information that they carry.

In Fig. 2 slow light techniques have been separated into two subclasses: those based in modifying the waveguide dispersion by using engineered structures, such as gratings, photonic crystals or micro-ring resonators; and those based upon the change of material dispersion to large and positive values involving the use and modification of properties inherent to the material, e.g. the transmission spectrum or the scattering effects.

2.2.1 Material-based slow light generation

Within material-based methods, the exploitation of atomic resonances was the first to be discovered. Dramatic modifications of the velocity of light are often based on this principle. Highly dispersive media present extreme values of the group velocity and thus they are well suited for realizing this kind of systems, as in electromagnetically induced transparency (EIT), and coherent population oscillations (CPO). EIT and CPO have been used to create narrow transparency windows in absorbing materials. These techniques, not only allow making controllable the degree of slowing (or speeding), but they also alleviate other important issues related to strong atomic resonances: excessive absorption and dispersion.

The electromagnetically induced transparency is a quantum effect that permits the propagation of light through a medium otherwise opaque. In (Hau et al., 1999), the extremely low speed of light of 17 $m \cdot s^{-1}$ was achieved by exploiting this technique. The biggest disadvantages of EIT arise because the method has a highly limited operation bandwidth owing to its narrow transparency window and to higher-order dispersion effects. Moreover EIT relies on delicate interference between two quantum amplitudes and thus the presence of collisions or any other dephasing effect can destroy the interference. On the face of it, EIT requires cryogenic temperatures and atomic media, preventing its practical use.

The quantum coherence technique of coherent population oscillations is studied as an alternative to EIT, due to its larger bandwidth and because it is highly insensitive to dephasing effects. The CPO method relies on creating a spectral hole due to population oscillations. Slow light propagation with a group velocity as low as 57.5m/s was observed employing CPO at room temperature in a ruby crystal (Bigelow et al., 2003). However, significant delays can only be achieved for signals of a few kb/s.

More recently, stimulated Brillouin and Raman scattering have also been proposed as material-based methods for slowing light. Brillouin scattering arises from the interaction of light with propagating density waves or acoustic phonons. Raman scattering arises from the interaction of light with the vibrational or rotational modes of the molecules in the scattering medium. The main advantage of these techniques is that they make use of optical fibers, which is an unmatched medium in terms of low attenuation levels. They are very adequate to provide practical moderate delays as in (Gonzalez-Herraez et al., 2005). Unluckily, Brillouin presents a limited bandwidth due to its narrow gain linewidth while Raman presents reduced slow-light efficiency.

The innovative realm of metamaterials has also been explored for developing photonic buffers. Metamaterials are artificial composites with dramatically different electromagnetic properties. The most noticeable approach to storing light in metamaterials is presented in (Tsakmakidis & Hess, 2007). So far, this theoretical approach is unfeasible due to the material losses, and, difficulty of fabrication in a light wavelength scale.

Recently, novel approaches to slow light based on plasmonics have been proposed, as in (Søndergaard & Bozhevolnyi, 2007). The main argument for using plasmonics is that it overcomes the diffraction limit affecting photonics, i.e. surface plasmon polaritons (SPPs) enable focusing light in nanoscale while photonics is size-limited to a wavelength scale. The main limitation to plasmonics today is that plasmons tend to dissipate after only a few millimetres. For sending data longer distances, the technology would need a great breakthrough. SPPs are also very sensitive to surface roughness and they are difficult to excite efficiently, although some "tricks" (e.g. Kretschmann geometry) have been already proposed for their excitation.

2.2.2 Engineered structures for slow-light generation

Differing from the previously explained material-based methods, the following slow-light mechanisms use strong spatial resonances between electromagnetic waves travelling along special structures. The structure-based slow-light approaches can be materialized in various types of arrangements, such as Fabry-Perot resonators, cascaded fiber Bragg gratings (FBGs), photonic crystals defects, or ring resonators. In general, these approaches outperform material-based ones for high-bandwidth signals, due to the fact that material resonances have generally a narrow linewidth and are more limited by dispersion effects (Melloni et al. 2010).

The election of a slow-light technique strongly depends on the targeted application. Photonic crystals constitute, to our understanding, the most promising approach in order to build devices for applications as the ones outlined bellow: small operating power (dozens of µW per cell), compact footprint (cell size around 10µm²), and fast access to information (tens to hundreds of picoseconds). Contributing to its practicality are also the facts that they can be operated at room temperature offering wide bandwidth. Finally, their fabrication processes can be made compatible with CMOS technology, enabling the use of silicon industry mass manufacturing facilities, and thus reducing mass fabrication costs and simplifying its integration with electronic circuitry.

2.3 Practical applications of slow light

The very first consequence of slow light is light pulse delaying. This fact unleashes in itself a series of valuable applications as optical buffers for optical packet switching (OPS) (Blanco et al. 2009a; Tucker, 2009), tuneable delay-lines for optical synchronization and correlation (Willner et al., 2009), or photonic true-time delay beamforming for phased array antennas, radio-over-fiber and analog-to-digital conversion (Capmany & Novak, 2007).

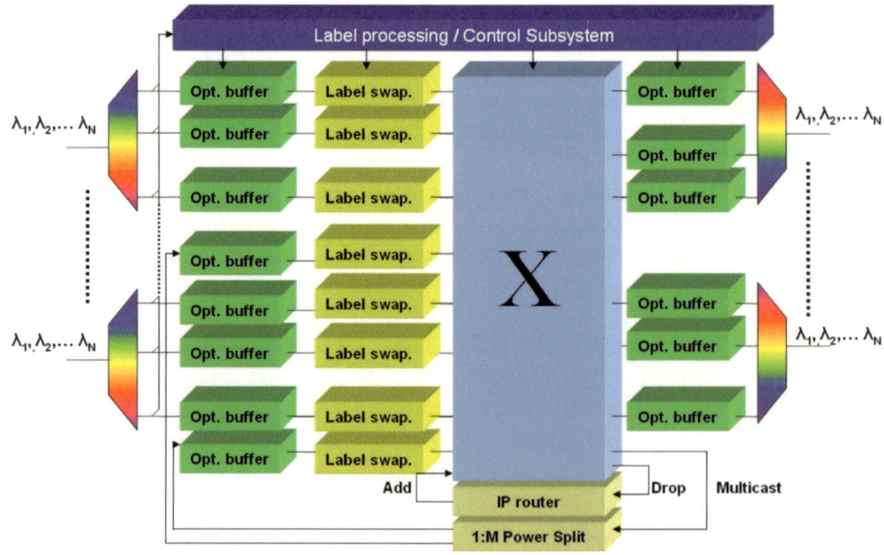

Fig. 3. Schematics of a proposed optical packet switching router design, including optical buffers and optical labels; a photonic switch fabric is represented by X (Blanco et al., 2009b)

To illustrate the use of optical buffers, Fig.3 shows the schematics of a proposed design for an OPS core router. Optical packets are wavelength-demultiplexed and immediately tapped: a copy of the packet is passed to the control subsystem while the other copy must remain "stored" in the optical domain. This latter copy must be released as soon as control decisions are made to ensure efficiency. Once the packets have been optically switched to the appropriate output, optical buffers are needed to resolve possibly arising collisions.

Not so obvious applications arise from the slow-light-based enhancement of light-matter interaction. Two facts explain this effect: on the one hand, slowly travelling photons are more likely to interact with the surrounding matter simply because it takes longer for them to go through it; on the other hand, at the very moment a light pulse enters a slow light device, its leading edge starts propagating at extraordinarily low speeds while the rest of the pulse is still propagating at normal velocities. This fact generates an accordion effect that implies spatial pulse compression and consequently an increase of local energy density and nonlinear effects. In Fig. 4 one of our simulations serves to illustrate the pulse compression and energy density increase suffered by a gaussian pulse entering in a photonic crystal slow light region from a conventional ridge waveguide.

Optical sensing, especially biosensing, is one of the application fields most benefitted from the strengthening of light-matter interaction, since its principle relies on identifying substance changes on the basis of light-matter relations. Higher standards of sensitivity and resolution are enabled by the use of slow light in sensing devices (Biallo et al., 2007; Pedersen et al., 2008). Fig. 5 illustrates the concept of optical biosensing using a slow-light photonic crystal waveguide to improve sensitivity to the biomarkers bounded to its surface.

Distributed Raman amplification is one of the nonlinear systems benefiting from slow light, with an expected efficiency improvement by a factor of 66.000 (McMillan et al., 2006). In quantum optics, as another example, slowing light down provides a mean to achieve sufficiently long storage time of quantum states to enable quantum operations (Dutton et al. 2004). Finally, it is worth to mention the role that slow light may play at improving the conversion efficiency of next generation solar cells. It is well known that recent thin film and organic photovoltaic (PV) technologies lack of a sufficient solar light-to-electrical energy conversion efficiency. By making photons travel slower within the active zone of the PV cell, up to a 50% increase in photon absorption (El Daif et al., 2010).

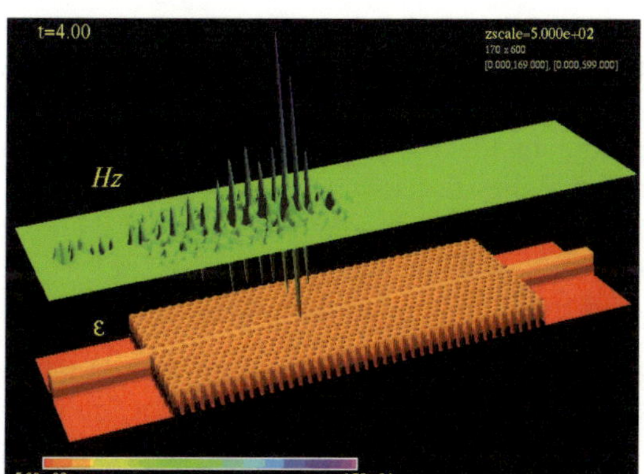

Fig. 4. Slow light photonic crystal waveguide (bottom inset), and transverse magnetic field component of the optical wave (top inset) to illustrate pulse compression

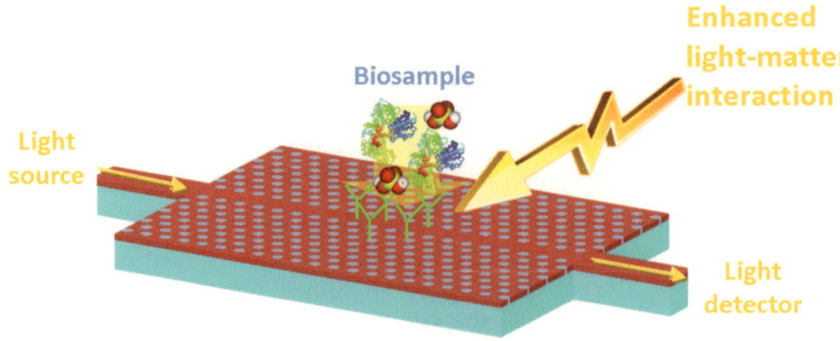

Fig. 5. Illustration of an optical biosensor based on a slow light photonic crystal waveguide.

3. Photonic band gap engineered materials for slow light applications

Dielectric materials having periodic dielectric constant present singular properties for electromagnetic waves propagation. The most remarkable of these properties is the appearance of the so called photonic band gap, a range of frequencies unable to propagate through the material. Many profitable effects stem from the existence of photonic band gaps: from the possibility of tight light confinement to the opportunity of achieving extremely low group velocities.

Photonic crystals are periodic, artificial, dielectric materials, which under certain conditions, present a photonic band gap. Biological photonic crystals are found in nature: in butterflies' wings, in peacocks' feathers, in comb-jellyfishes... The characteristics of photonics crystals confer these species their peculiar and outstanding colouration. Engineered or artificial photonic crystals can be designed with tailored electromagnetic and propagation properties, giving rise to a huge range of possibilities and practical applications. The next points give a flavour on the principles of design of photonic crystal devices with tailored group velocity.

3.1 Principles of photonic crystals

A photonic band gap engineered material, or a photonic crystal possessing a band gap, can be built from the periodic arrangement of different dielectric media in one, two or three dimensions. It has to be noticed that, even under the appropriate conditions of index contrast, lattice period and geometry, the photonic band gap will only appear in the plane of periodicity. Therefore, only a three-dimensional photonic crystal will be able to localize light in three dimensions by means of the photonic band gap, while light modes in one- or two-dimensional arrangements will only be localized in one or two dimensions respectively. Examples of one-, two-, and three-dimensional photonic crystals are shown in Fig. 6.

To understand how light propagates within a photonic crystal one has to resort to the macroscopic Maxwell equations and specialise to the particular case of mixed dielectric media in the absence of free currents or charges and in which the structure does not vary with time. This development has already been done in several excellent books and we refer the reader to (Joannopoulus et al., 2008) or (Sakoda, 2005), just to cite two of them, for a comprehensive understanding. In the following, the main results of this development,

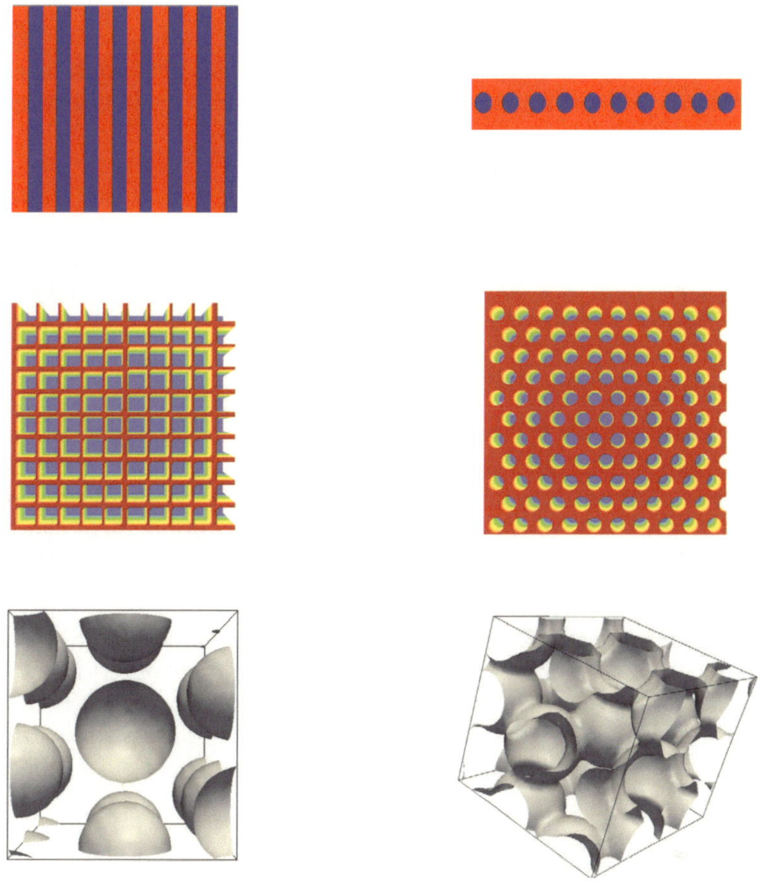

Fig. 6. Examples of one-, two- and three-dimensional photonic crystals.

indispensable to understand the basic principles of photonic crystals, are summarized. The resultant master equation is the following:

$$\nabla \times \left(\frac{1}{\varepsilon(r)} \nabla \times H(r) \right) = \left(\frac{\omega}{c} \right)^2 H(r) \tag{4}$$

Given a known spatial dielectric constant arrangement, *ε(r)*, one can find the magnetic field spatial profile of the modes allowed by the structure, *H(r)*, and their corresponding frequencies, ω, by solving Eq. 4 subject to the following transversality requirements:

$$\nabla \cdot H(r) = 0 \tag{5}$$

$$\nabla \cdot (\varepsilon(r) E(r)) = 0 \tag{6}$$

Subsequently one can obtain the electric field by using the following expression:

$$E(r) = \frac{i}{\omega \varepsilon_0 \varepsilon(r)} \nabla \times H(r) \tag{7}$$

The solutions to the master equation *H(r)* can be written as Bloch states, i.e. as a product of a plane wave and a periodic function, with a period equal to the lattice period of the photonic crystal:

$$H_k(r) = e^{ikr} u_k(r) \tag{8}$$

All information about the spatial profile is given by the wave vector, *k,* and the periodic function, $u_k(r)$. For a given k, an infinite family of modes with discretely spaced frequencies $\omega_n(k)$ satisfy the master equation in Eq. 4 in particular these functions $\omega_n(k)$, represented in the *band structure* diagram, provide us with most of the valuable information about the optical properties of a given photonic crystal.

From this moment on, we will restrict our attention to two-dimensional structures, mainly because the state of the art fabrication techniques do not allow for reliable and cost-affordable realization of three-dimensional photonic crystals. Additionally, 2D structures are suitable for photonics on-chip integration. 3D light confinement can be obtained in 2D photonic crystals by using the photonic band gap effect to confine light laterally and total internal reflection for vertical confinement.

An ideal 2D photonic crystal is homogeneous in *z* (real 2D photonic crystal devices will be patterned in slabs or membranes to ensure vertical confinement). Modes will oscillate in that direction with its wave vector k_z unrestricted. We can particularize Eq. 8 for this case, expressing the Bloch modes of a two-dimensional photonic crystal structure as a function of the mode index, *n*, the in-plane and off-plane wave vectors, $k_{||}$ and k_z, and the projection of *r* in the *xy* plane, *ρ*:

$$H_{(n,k_z,k_{||})}(r) = e^{ik_{||}\rho} e^{ik_z r} u_{(n,k_z,k_{||})}(\rho) \tag{9}$$

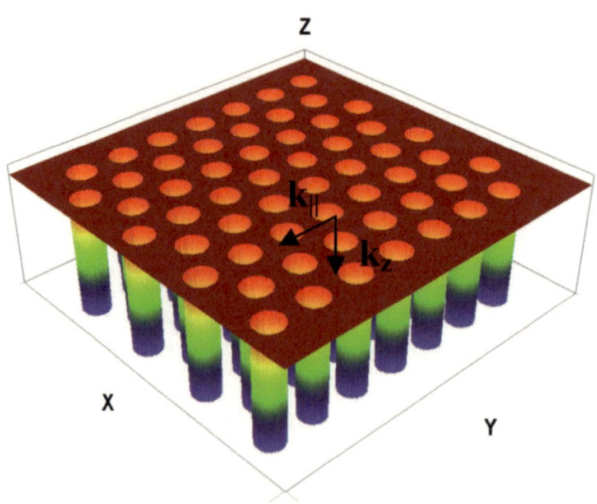

Fig. 7. Modes propagating parallel to the plane of periodicity xy (k_z=0), can be classified in transverse electric (TE) and transverse magnetic (TM) in two dimensional photonic crystals

Modes propagating parallel to the periodicity plane have kz=0 and posess mirror symmetry through that plane. This allows us to classify modes in transverse electric (TE) and transverse magnetic (TM), having the electric field (E) in plane and the magnetic field (H) polarized in z and vice versa.

In Fig.8 the computed TE band structure for a two-dimensional photonic crystal made of air holes on silicon substrate is represented, accompanied by its transmission diagram. It is noticeable that transmission is zero at the photonic band gap frequencies. This type of arrangement, under certain conditions on index contrast and holes radii, presents a photonic band gap for TE and TM polarizations. Nonetheless, typically only TE polarized light is used since TE band gap is larger and therefore confinement will be stronger.

Several concepts must be clarified to be able to interpret this diagram correctly. First, notice that the wave vector in the plane of periodicity, $k_{||}$, is indexed in four points (Γ, M, K, Γ), representing the limits of the irreducible Brillouin zone for this particular structure. A complete description of Brillouin zone is given in the Appendix B of (Joannopoulus et al., 2008). Here we will just mention that the Brillouin zone is a primitive cell in the reciprocal lattice, the Fourier transform of the spatial function of the original lattice. The irreducible Brillouin zone is the first Brillouin zone reduced by all symmetries in the point group of the lattice, i.e. all periodicity- and symmetry-imposed redundancies in *k* are avoided. Secondly, the frequency is given in dimensionless units ωa/2πc, or equivalently a/λ. By normalizing all magnitudes, the frequency and the holes radii in this case, by the lattice period *a* it is possible to explode the inherent scalability of photonic crystals. If the given structure had a=1μm, the middle of the band gap ω_m=0.3 would be at a wavelength λ=1/0.3= 3.33 μm. If of one wanted to set the middle of the band gap at the third window of communications, λ=1550nm, it would suffice with setting a=0,3·1550=465nm.

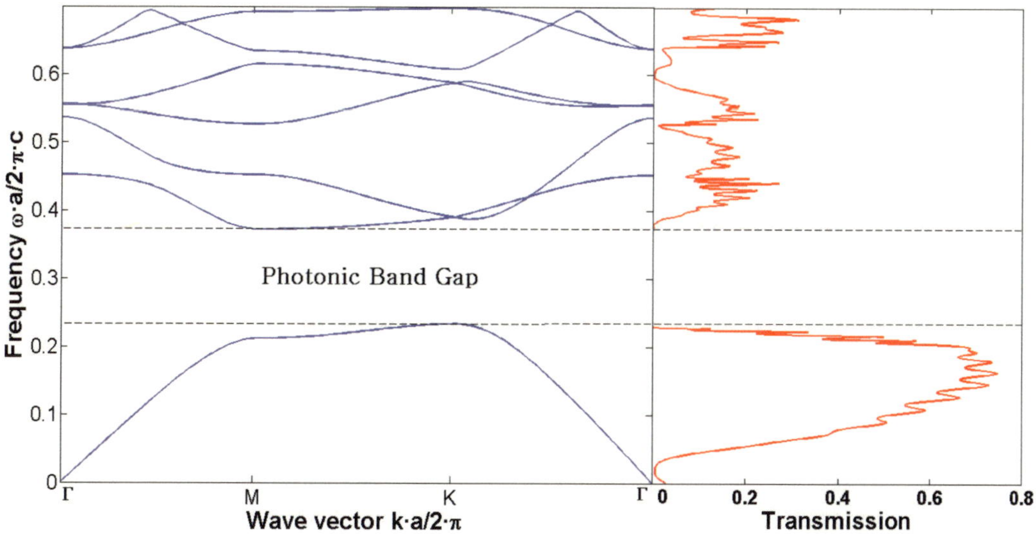

Fig. 8. Band structure and transmission diagram of a 2D photonic crystal made of air holes on silicon. The lattice period is *a* , the dielectric constant is ε=11,9716 and the hole radii r=0.36·a

3.2 Slow light photonic crystal waveguides and cavities

Unmodified photonic crystal structures, as the ones in the previous section, present a series of applications originating from the existence of a photonic band gap, as wavelength selective mirrors or stop-band filters. However, to fully exploit photonic crystal capacities for modelling electromagnetic propagation, one has to resort to the creation of defects within the otherwise perfectly periodic structure. Punctual and linear defects within the photonic crystal allow for tight confinement of light, at band gap frequencies, inside cavities and waveguides.

3.2.1 Photonic crystal slow-light waveguides

The band diagram of a photonic crystal waveguide on a two dimensional photonic crystal is shown in Fig.9. The periodicity has been broken in the y direction, due to the introduction of the linear defect, and thence only k_x is conserved. Therefore, the modes are not represented anymore over the irreducible Brillouin zone. Their projection on k_x is depicted instead. One can create a photonic crystal waveguide by simply removing a single row of holes. Such a waveguide, normally referred as W1, results in the appearance of a number of defect modes within the photonic band gap. As monomode behaviour is desirable, narrowing the waveguide is normally required. By reducing the width of W1 by a factor of 0.7, one gets a single defect mode within the band gap, the red-coloured band in Fig.9. Lateral confinement of the defect mode within the waveguide is ensured by the photonic bandgap, as the electric field density simulation in Fig.10 proves. Vertical confinement of light within the slab is achieved by total internal reflection at the interface between the high-dielectric constant slab and the surrounding air. The blue region in Fig. 9 represents the so called light cone, i.e. those extended modes propagating in air and not confined within the slab.

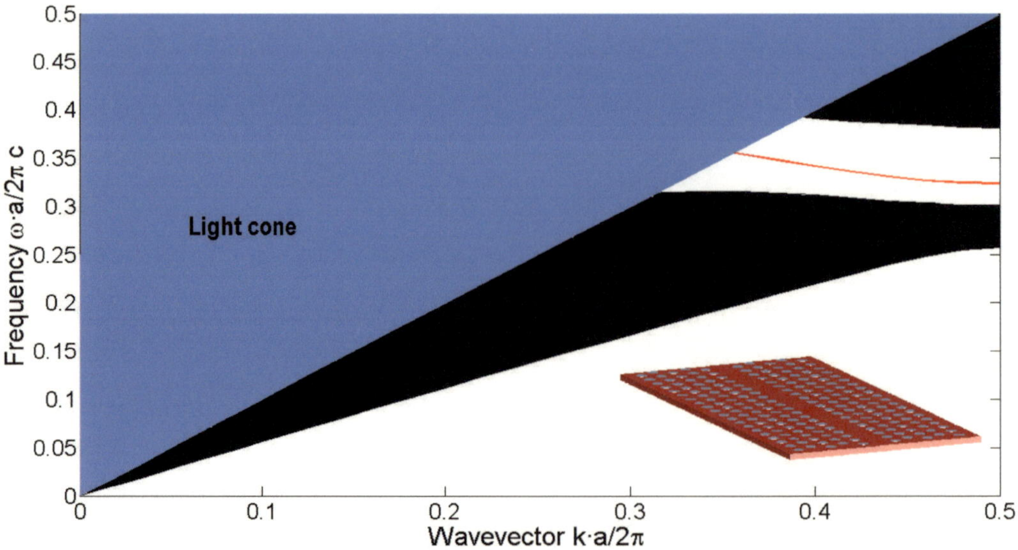

Fig. 9. Band diagram of a photonic crystal waveguide on a high-dielectric constant slab.

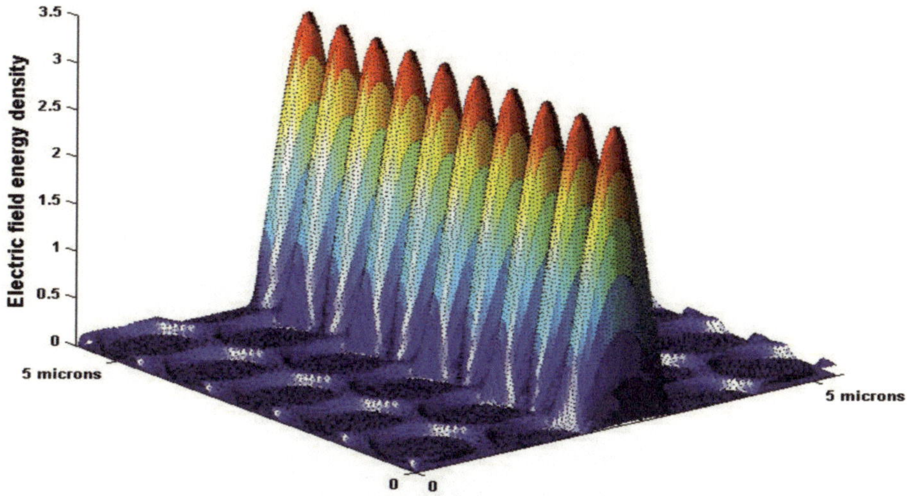

Fig. 10. Electric field energy density confined within the photonic crystal waveguide.

Group velocity is given by Eq. 1, i.e. by the slope of the modes appearing in the dispersion diagram. Computed group velocity for light pulses coupled to the defect mode is depicted in Fig. 11. The top graphic represents *vg* as a function of the longitudinal wave vector *kx* for the guided mode. It is noticeable that the group velocity is zero at the edges of the band and for certain wavelengths. Unfortunately, these working regions are not desirable in practice. At the band edges, any fluctuation in the structure, due to fabrication imperfections, causes oscillations between guided and not guided states. Moreover, the operational bandwidth for this ultra-low velocity is very small due to group velocity dispersion and higher-order dispersion. Special designs have been proposed to minimize dispersion and enable higher bandwidths (Baba, 2008; O'Faolain, 2009). The bottom graphic at Fig. 11 shows guided mode *vg* as a function of wavelength. By setting the lattice period *a* equal to 500nm the band gap is located around the third communications window. Since we are interested in guided modes and not in modes extended in air, we must only consider those frequencies out of the Light cone. These correspond to wavelengths superior to 1.4 µm.

Our group is working on optimized waveguide designs trying to achieve a balance between reduced group velocity and bandwidth. The waveguide depicted in Fig.12 has been created by diminishing the radii of a row of holes and filling it with a material of ε=7. Furthermore, to achieve monomode behaviour the waveguide width has been reduced by a factor of 0.64 (Andonegui et al., 2011). Group velocities of c/100 are achieved for a 33% of the k-vector space, achieving a good balance between low information velocity and bandwidth.

3.2.2 Photonic crystal cavities

The group velocity-bandwidth trade-off of photonic crystal waveguides can also be addressed by coupling a series of punctual defects (cavities) within the photonic band gap material. High quality factor (Q) photonic crystal cavities are capable to store photons for a relative long time in an extremely small volume. A high-Q photonic crystal cavity is the basic block of the so called coupled-resonator optical waveguides (CROWs). Remarkable achievements have been done in this field, as in (Notomi et al., 2008), where more than 100 high-Q cavities were coupled, achieving v_g of c/170 in pulse propagation experiments and notable storage capacity.

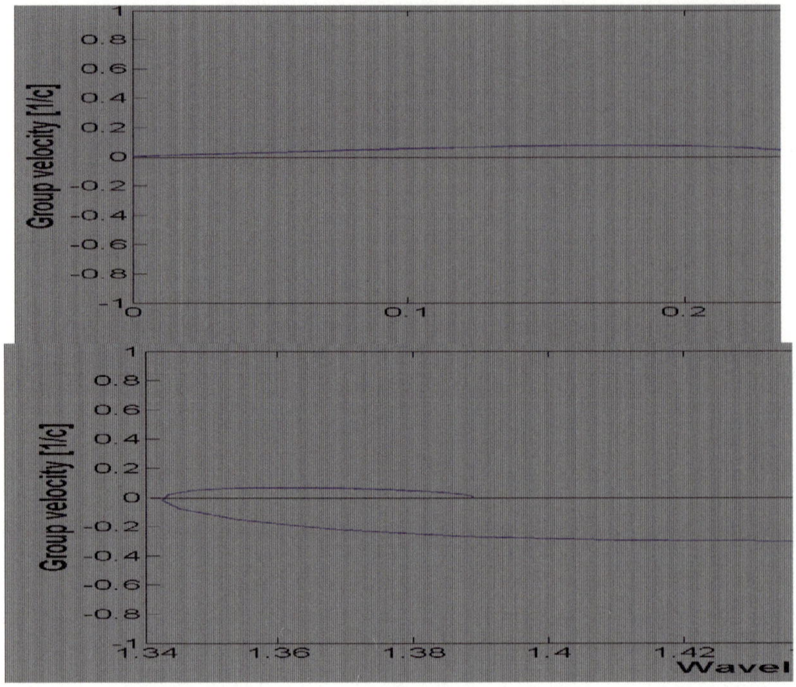

Fig. 11. Group velocity of light pulses coupled to the defect mode confined within the waveguide as a function of wave vector k_x (top) and as a function of wavelength (bottom)

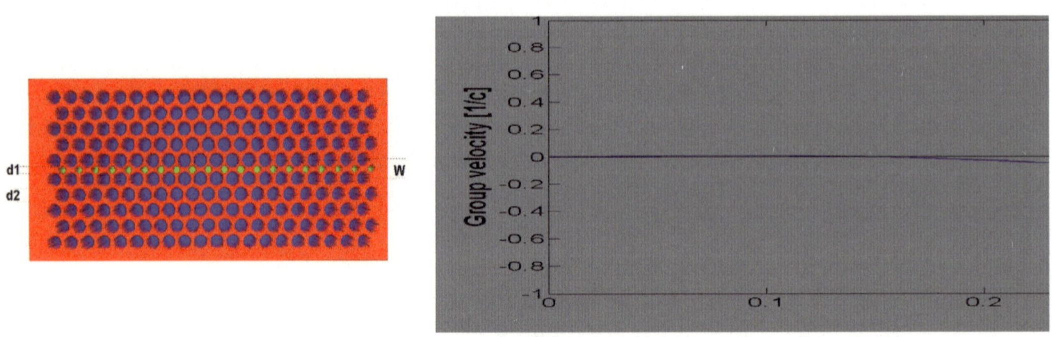

Fig. 12. Optimized photonic crystal waveguide on a triangular lattice of air holes on silicon with d1=0.51·a, d2=0.76·a, W=0.64·W (left) and group velocity of the defect mode (right)

By introducing a punctual defect the periodicity in both x and y direction is broken and therefore k vector is not conserved in any direction. Consequently, the band structure of a cavity is naïve and it is not usually represented. Nevertheless, it is very instructive to visualize the shape of the defect mode confined in the cavity, as in Fig. 13. Notice that the defect mode is parallel to the k-axis, giving useful information: a single resonant frequency remains confined into the cavity with zero group velocity, as illustrated in Fig.14.

A more sophisticated cavity consisting on a missing hole and a gradual change of surrounding holes radii is shown in the design of Fig.15. The cavity is adjacent to a waveguide so that the light at certain resonant frequency can be coupled from the waveguide to the cavity. This light will be coupled back from the waveguide to the cavity after a time, $t_{storage}$, proportional to Q.

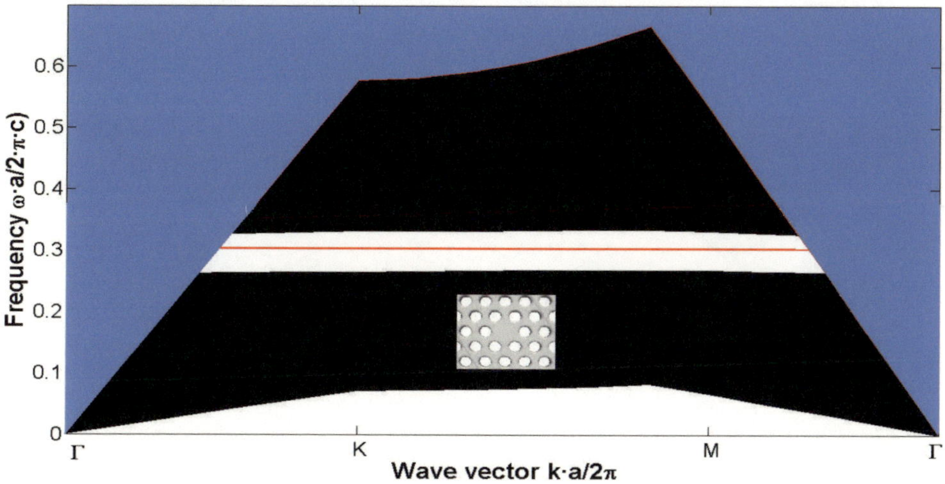

Fig. 13. Band diagram of a photonic crystal cavity on a high-dielectric constant slab.

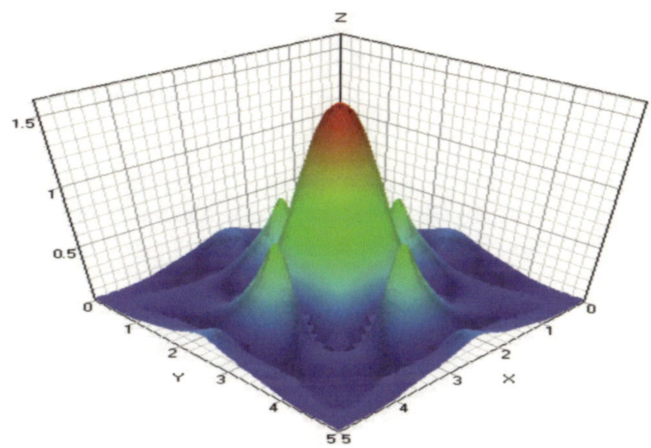

Fig. 14. Amplitude of the H_y component of the optical field confined within the cavity.

3.2.3 Tunability of photonic crystal slow light structures

Fast and fine device tunability is a requirement for many applications of slow light, e.g. for optical buffer memories. Fast reconfiguration of the photonic crystal device can be achieved by a variety of effects: by using thermo-optical effect, electro-optic effect, or carrier injection among others. Along this section, we show how photonic crystal waveguides and cavities can be fast and efficiently tuned by exploiting Pockels effect.

Lithium niobate ($LiNbO_3$) is an anisotropic crystalline material, i.e. its refractive index depends on the crystal axis direction. Consequently its response to the electro-optic effect is given by a matrix of coefficients giving the electro-optic response for each direction. $LiNbO_3$ index response to an applied field along the z axis, depicted in Fig.16, is given by the expression

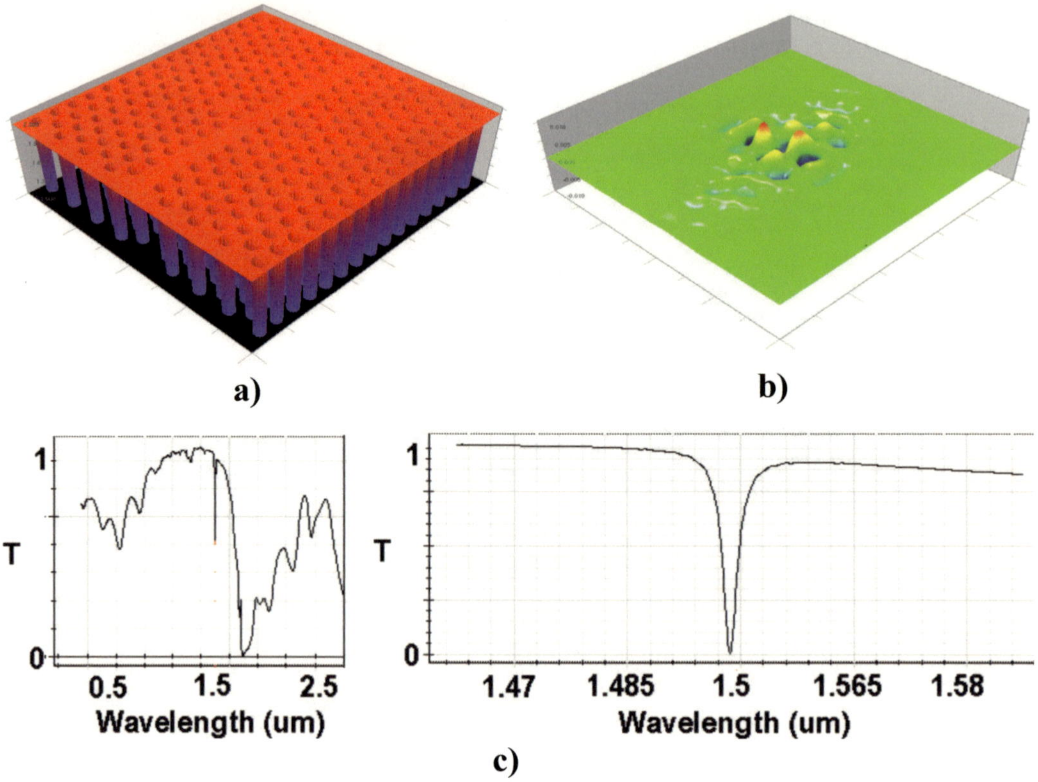

Fig. 15. a) Dielectric constant structure of a cavity coupled to a waveguide on a 2D triangular lattice of holes in lithium niobate; b) DFT of H⊥; c) Transmission diagram over wavelength and detail of the resonance and surrounding wavelengths.

$$\Delta n = \left(\frac{n^3}{2}\right) \cdot r_{33} \left(\frac{V}{d}\right) \tag{10}$$

where r_{33} is one of the matrix component and d is the electrode width. A refractive index change of $\Delta n = 1 \times 10^{-3}$is given by an electric field of $\Delta E_0 = -6.45$ V/µm. Taking into account that electrode width will be of the order of half a micron, small electric fields will be needed to achieve tunability.

We subsequently consider the appliance of an incremental electric field, by steps of $|\Delta E_0| = 64.5$ mV/µm, to a $LiNbO_3$ slab patterned with a 2D photonic crystal. This generates refractive index changes from 10^{-5} to 10^{-1}, later we will see how only a range of these index changes is achievable in practice. This is achieved by a means of a voltage supplied to electrodes placed on the surface of the photonic crystal device, as in the inset of Fig.15.

First, we pay attention to waveguides reconfigurability. The red line in the band diagram of Fig.9 represented the frequencies of the defect mode guided within the waveguide as a function of the longitudinal wave vector. The group velocity of the guided mode was also computed and represented in Fig.11. Now, we explore how the group velocity of the signals coupled to this mode can be dynamically switched. The graphic in Fig.17 is evidence for the defect mode frequency shift as a response to a refractive index change originating from the supplied voltage. When light at a proper wavelength proceeding from a narrow-width optical source (laser) is launched into the slow light device, it couples to the guided mode.

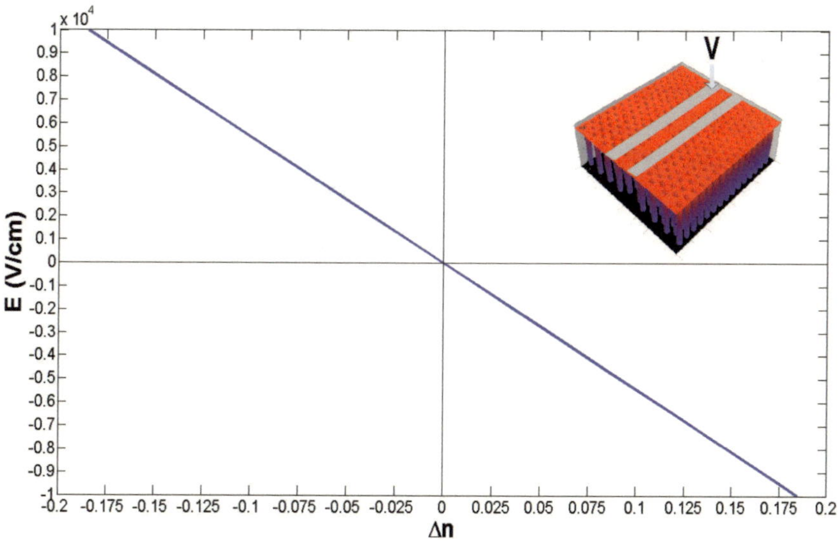

Fig. 16. Lithium niobate refractive index response to an electric field applied due to Pockels effect. The inset shows how to exploit this effect by applying a voltage to the electrodes placed on the lateral sides of the waveguide.

Depending on the source frequency (or wavelength) the signal will propagate with a given group velocity as given by the graphics in Fig.11. At this point, if a voltage is applied to the device electrodes, the mode shifts in frequency but the injected frequency remains the same. In consequence, the signal is switched to another velocity regime or even from a guided- to a non-guided regime. This simple mechanism enables a plurality of device applications from tunable delay lines to transistors, modulators or tunable filters.

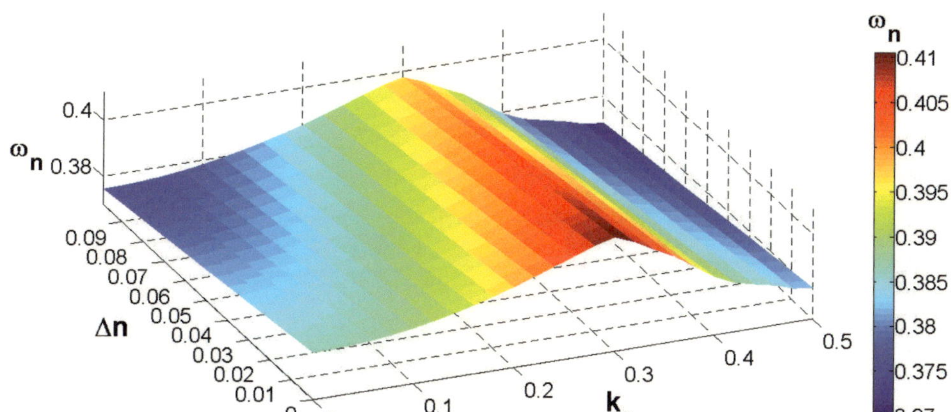

Fig. 17. Frequency shift of a photonic crystal waveguide guided mode as a function of refractive index change stimulated by Pockels effect.

The reconfiguration possibilities of waveguides are somehow limited when compared with that of the cavities. The reconfiguration capabilities of the photonic crystal cavity proposed in Fig. 15 via electro-optic effects are presented next. Recall that the cavity had a resonant frequency at around 1500nm, precisely at 1499.1nm. We alter the refractive index of the photonic crystal material from 0 to 0.1 in steps of 10^{-5} by exploiting Pockels. The overlap of transmission spectra for different refractive index values proves the resonant wavelength red-shift. This diagram resulting from our FDTD simulations is presented in Fig. 18. Subsequently we have focused in index changes achievable for reasonable values of applied electric field in normal applications, i.e. from 0.1 to 10MV/ μm. This reduces the range of achievable index changes to values up to $1.55 \cdot 10^{-3}$. In Fig.19 we represent the computed cavity resonant wavelengths as a function of the index change generated by Pockels effect using realistic electric field values. This simulation results are the proof of the concept of fast (switching time bellow 1 ns) and efficient (every $|\Delta E_0| = 64.5$ mV/μm implies $\Delta\lambda = 6$pm) reconfigurability.

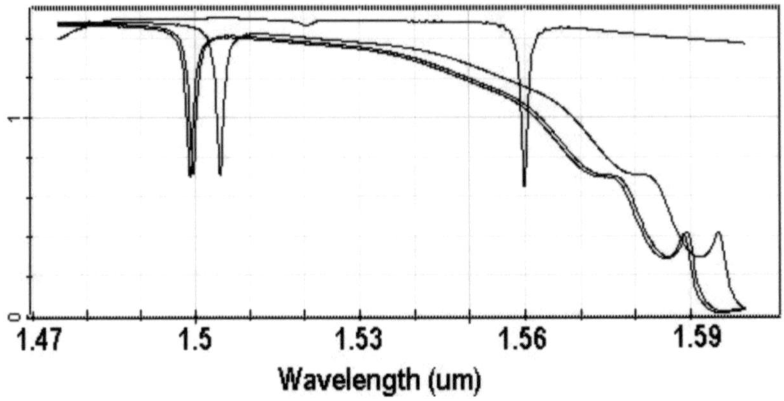

Fig. 18. Overlapped transmission spectrum for refractive index increments of Δn=(0, 0.001, 0.01, 0.1)

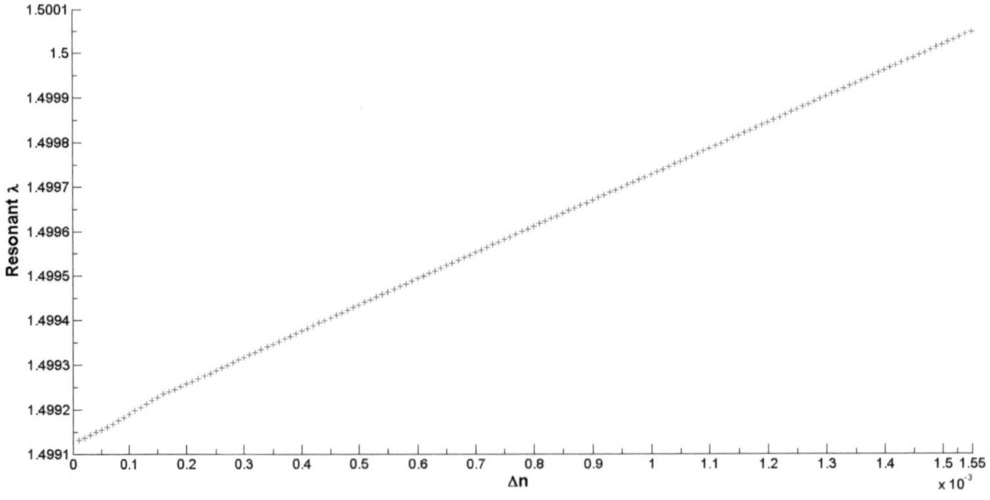

Fig. 19. Resonant wavelength versus refractive index change for the photonic crystal cavity shown in Fig.15.

3.3 Computational methods and fabrication techniques

Substantial work has been done to provide numerical solution of Maxwell equations. In this subsection a rough idea on the different computational methods to solve photonic crystal problems is given, for a broader notion we refer the reader to (Joannopoulus et al., 2008).

As a first approach, one can divide computational methods in frequency domain methods and time-domain methods, each of them useful for solving different problems typologies. Frequency domain methods are used to solve problems such as the computation of band diagrams and stationary mode profiles. On the other side, time-domain methods are better suited to perform computations involving time evolution of fields, such as transmission and reflection spectra or resonant cavities decay in time. Numerical methods can be alternatively classified on the basis of the used discretization schemes in: finite differences, finite elements, boundary-element and spectral methods.

Several commercial and open-source software packages implementing different numerical methods are available for computational photonics. Just to cite a few of these free-software products we will mention MPB (using plane wave expansion frequency-domain method) and Meep (implementing finite differences in time domain) MIT's packages and CAMFR (based in eigenmode expansion and advanced boundary conditions like perfectly matched layers).

Next, a coarse notion on the fabrication techniques to synthesize photonic crystals is given. A good set of references about 3D and 2D photonic crystal fabrication techniques is given in (Skorobogatiy & Yang, 2009). It has to be noticed that, in spite of its outstanding potential, photonic crystals mostly remain at research stage and this is mostly due to current technological limitations of fabrication techniques. Focused Ion Beam (FIB) and electron beam (e-beam) lithography combined with reactive ion-etching (RIE) are two the methods used in laboratories for high accuracy and high-resolution fabrication of planar 2D photonic crystals. However, it is necessary to start moving photonic crystal technology out of the laboratory and onto the production floor for building photonic devices for practical applications. Recent advances in nano-imprint lithography are fulfilling this goal (Kreindl et al., 2010).

4. Conclusion

Photonic band gap materials are a powerful tool for tailoring light propagation properties. This Chapter has put the emphasis on the control over the signal group velocity given the wide range of applications enabled by slow light. All slow light techniques reviewed rely on slowing down the information or the energy transported by light signals, more than on slowing the photons themselves.

Among the plurality of foreseen slow-light applications we have mainly highlighted two due to its high technological and societal expected impact. The development of optical buffer memories is of utmost importance for the deployment of all-optical networks. Slow light is a promising approach for fast, scalable and low-power consuming on-chip optical buffers. Given the state of the art of the technology, nowadays, such devices cannot replace electronic RAMs in their current functions; however major breakthroughs are still expected in this field. Concerning slow light application to biosensing, not so bandwidth demanding and less affected by losses, near future commercial prospects are encouraging.

Slow light engineered structures perform better than material-based methods for high bandwidths. On the other hand, their feasibility into devices for practical applications is higher, due to the materials employed and their operation conditions. In particular, photonic crystals are proving to be a suitable technology to take slow light into practice. Small operating power, compact footprint, and the possibility of monolithic integration with electronics and CMOS fabrication are some of their strong points. Along this Chapter the capabilities of fast and efficient switching using electro-optic effects have been justified on the basis of computing and simulations.

So far our work in this field has been focused on high-index contrast photonic crystal structures, exploiting the use of materials as silicon and lithium niobate. Our future research lines will explore the use of *intelligent* materials such as hydrogel polymers and chalcogenide glass (Eggleton et al.,2011), in order to achieve added functionalities to the photonic crystal devices.

In spite of all these promising and almost magic properties of photonic band gap materials, they still remain at research stage. Some difficulties such as high propagation losses in the slow light regime, dispersive effects, coupling inefficiencies and fabrication roughness or inaccuracies are somehow hampering their evolution to commercial devices. Recent advances in dispersion engineering and loss reduction-oriented design approaches, together with the continuous improvement of fabrication processes, are taking slow light in photonic crystal closer to practical applications. In addition to this, the refinement of nanoimprint lithography, as an alternative for accurate mass production, herald a brighter future for real slow-light photonic crystal devices.

5. Acknowledgements

This work has been partially supported by the Spanish Administration organism CDTI, under project CENIT-VISION 2007-1007.

6. References

Andonegi, I., Blanco, A., Garcia-Adeva, A. (2011). Characterization of Slow Light Regime in 2D Photonic Crystal Waveguides, *Proceedings of PPM 2011 Photonics, Plasmonics and Magnetooptics - ImagineNano 2011*, Bilbao, Spain, April 12 - 17, 2011

Baba, T. (2008). Slow Light in Photonic Crystals. *Nature Photonics,* Vol.2, (August 2008), pp. 465-473, ISSN 1749-4885

Biallo, D. et al. (2007). High Sensitivity Photonic Crystal Pressure Sensor. *Journal of the European Optical Society*, Vol. 2, (May 2007), pp. 1-5, ISSN 1990-2573

Bigelow, M. S., Lepeshkin, N. N. & Boyd, R. W. (2003). Observation of Ultraslow Light Propagation in a Ruby Crystal at Room Temperature. *Physical Review Letters*, Vol. 90, No. 11, (March 2003), pp. 1139031-4, ISSN 1079-7114.

Blanco, A., Beltrán, P., & Zubía, J. (2009). Slow Light Buffers for Future All-Optical Packet Switched Networks, *Proceedings of ICUMT 2009 International Conference on Ultra Modern Telecommunications & Workshops*, pp. 1-6, ISBN 978-1-4244-3942-3, Saint Petersburg, Russia, October 12-14, 2009

Blanco, A., Areizaga, E., Zubia, J. (2009). Slow Light for Microwave Photonics Applications, *Proceedings of MMS 2009 9th Mediterranean Microwave Symposium,* pp. 1-5, ISBN: 978-1-4244-4664-3, Tangiers, Morocco, November 15-17, 2009

Boyd, R. W., Narum, P., Slow- and Fast-light: Fundamental Limitations. *Journal of Modern Optics,* Vol.54, Nos. 16-17, (November 2007), pp. 2403-2411, ISSN 0950-0340.

Capmany, J., Novak, D. (2007) Microwave Photonics Combines Two Worlds. *Nature Photonics,* Vol. 1, (June 2007), pp.319-330, ISSN 1749-4885

Eggleton, B. J., Luther-Davies, B. & Richardson, K. (2011). Chalcogenide Photonics. *Nature Photonics,* Vol. 5, (February 2011), pp. 141–148, ISSN 1749-4885

El Daif, O. et al. (2010). Absorbing one-dimensional planar photonic crystal for amorphous silicon solar cell. *Optics Express,* Vol. 18, No. 103, (September 2010), pp. 293-299, ISSN 1094-4087

Dutton, Z., Ginsberg, N. S., Slowe, C. & Hau, L. V. (2004) The Art of Taming Light: Ultra-slow and Stopped Light. *Europhysic News,* Vol. 35 (March/April 2004), pp. 33 - 39, ISSN 1432-1092

Gonzalez-Herraez, M, Song, K. Y. & Thevenaz, T. (2005) Optically-controlled slow and fast light in optical fibers by means of stimulated Brillouin scattering. *Applied Physics Letters,* Vol. 87, Issue 8., (August 2005) p. 081113, ISSN 0003-6951

Hau, L.V., Harris, S. E., Dutton, Z. Light (1999) Speed Reduction to 17 Metres per Second in an Ultracold Atomic Gas. *Nature,* Vol. 397, (February 1999), p. 594, 1999, ISSN 0028-0836

Heitmann, W. & Nimtz, G. (1994). On Causality Proofs of Superluminal Barrier Transversal of Frequency band Limited Wave Packets. *Physics Letters A,* Vol. 196, (December 1994), pp. 154-158, ISSN 0375-9301

Joannopoulos, J., Johnson, S., Winn, J. & Meade R. (2008). *Photonic Crystals: Molding the Flow of Light,* Princeton University Press, ISBN 978-0-691-12456-8, Princeton, New Jersey, USA

Khurgin, J. & R. Tucker (2009). *Slow Light: Science and Applications,* CRC Press - Taylor and Francis Group, ISBN 978-1-4200-6151-2, Boca Raton, USA

Krauss, T. F. (2008). Why Do We Need Slow Light. *Nature Photonics,* Vol.2, No. 8, (August 2008), pp- 448-450, ISSN 1749-4885

Kreindl, G., Glinsner, T & Miller, R. (2010), Next-generation Lithography: Making a Good Impression. *Nature Photonics Technology Focus, Vol. 4,* (January 2010), pp.27-28, ISSN 1749-4885

McMillan, J. F., Yang, X., Panoiu, N. C., Osgood, R. M. & Wong, C. W. (2006). Enhanced Stimulated Raman Scattering in Slow-Light Photonic Crystal Waveguides. *Optics Letters,* Vol 31, (January 2006), pp. 1235–1237, ISSN 1539-4794

Melloni, A., Canciamilla, A., Ferrari, C., Morichetti, F, O'Faolain, L., Krauss, F., De La Rue, R., Samarelli, A. & Sorel, M. (2010) Tunable Delay Lines in Silicon Photonics: Coupled resonators and Photonic Crystals, a Comparison. *IEEE Photonics Journal,* Vol. 2, No. 2, (April 2010), ISSN 1943-0655

Notomi, M, Kuramochi, E. & Tanabe, T. (2008) Large-Scale Arrays of Ultrahigh-Q Coupled Nanocavities. *Nature Photonics,* Vol. 2, (December 2008), pp. 741-747, ISSN 1749-4885

O'Faolain, L, Li, J., Gomez-Iglesias, A., & Krauss, T. (2009). Low Loss Dispersion Engineered Photonic Crystal Waveguides for Optical Delay Lines, *Proceedings of 6th IEEE*

International Conference on Group IV Photonics, pp. 40 – 42, ISBN: 978-1-4244-4402-1, San Francisco, CA, USA, September 9-11, 2009

Pedersen, J., Xiao, S. & Mortensen, N.A. (2008). Slow-Light Enhanced Absorption for Bio-chemical Sensing Applications: Potential of Low-Contrast Lossy Materials. *Journal of the European Optical Society*, 08007 Vol. 3, Issue 3, (February 2008), pp. 1-4, ISSN 0963-9659

Sakoda, K. (2005). *Optical Properties of Photonic Crystals*, Springer, ISBN 3-540-20682-5, Berlin, Germany

Søndergaard, T. & Bozhevolnyi, S. I. (2007) Slow-plasmon resonant nanostructures: scattering and field enhancements, Physical Review B, Vol. 75, (February 2007), pp. 0734021-4, ISSN 1098-0121.

Skorobogatiy, M. & Yang, J. (2009) *Fundamentals of Photonic Crystal Guiding*, Cambridge University Press, ISBN 978-0-521-51328-9, New York, USA

Steinberg, A.M., Kwiat, P.G., Chiao, R.Y. (1993). Measurement of Single-Photon, Tunneling Time. *Physics Review Letters*, Vol.71, No. 5, (August 1993), pp. 708-711, ISSN 0031-9007

Tsakmakidis, K.L., Boardman, A. & Hess, O. (2007). 'Trapped Rainbow' Storage of Light in Metamaterials. *Nature*, Vol. 450, (November 2007), pp. 397-401, ISSN 0028-0836.

Tucker, R. S. (2009) Slow Light Buffers for Packet Switching, In: *Slow Light: Science and Applications*, Khurgin, J., pp. 347-365, CRC Press, ISBN 978-1-4200-6151-2, Boca Raton, USA

Willner, A.E., Zhang, B., & Zhang, L. (2009) Reconfigurable Signal Processing Using Slow-Light-Based Tunable Optical Delay Lines, In: *Slow Light: Science and Applications*, Khurgin, J., pp. 321-346, CRC Press, ISBN 978-1-4200-6151-2, Boca Raton, USA

4

Manipulation of Photonic Orbital Angular Momentum for Quantum Information Processing

Eleonora Nagali and Fabio Sciarrino
Dipartimento di Fisica, Sapienza Universitá di Roma
Italy

1. Introduction

More than a century ago, pioneering works carried out by Poynting (1909) and other physicists have provided the evidence of the validity of Marxwell equations for an electromagnetic field, showing how a beam of light carries energy and momentum, both in the linear and angular components. In particular the angular momentum of light, related to the generator of rotations in quantum mechanics, has been typically associated with its polarization, and more specifically with its circular polarization components. An optical beam traveling in the positive direction of the z axis that is circularly polarized, carries a z-component angular momentum content $\sigma = \pm\hbar$ per photon, which is positive if the circular polarization is left-handed and negative if it is right-handed. This angular momentum content is not just a formal property, but a very concrete one that can have significant mechanical effects as for example an observable induced rotation by absorption to a material particle Beth (1936); Friese et al. (1998)

At the same time, calculations of angular momentum for a free electromagnetic beam gave rise to a *second* contribution not related to photon spin, to which was attributed the name of *orbital angular momentum* (OAM). Unlike the spinorial angular momentum (SAM), considered as the intrinsic part of angular momentum since it does not depend on the specific reference frame, the orbital component is associated to the transverse spatial structure of the wavefront. More precisely, this angular momentum appears when the beam wavefront acquires a helicoidal structure, or equivalently, its field spatial dependence contains a helical phase factor having the form $e^{im\varphi}$, where φ is the azimuthal phase of the position vector r around the beam axis z and m is any integer, positive or negative, providing the direction and the "velocity" of the phase spiraling along the beam direction. It can be shown that in this case the optical beam carries an angular momentum along its axis z equal to $m\hbar$ per photon, in addition to the polarization one σ. When m is nonzero, the helical phase factor imposes the existence of an optical vortex at the center of the beam where the light intensity vanishes. Although the fundamental concept of orbital angular momentum associated with a light field was already known since the early forties, the research on the orbital angular momentum of light has begun only in 1992, with the appearance of a seminal publication carried out by Allen et al. (1992). In this work, Allen et al. demonstrated experimentally that a particular set of solutions of Helmoltz equation in paraxial approximation, the Laguerre-Gauss modes (LG), carry a fixed amount of orbital angular momentum. Moreover, such beam could be generated experimentally in the laboratory by manipulating gaussian beams emerging from a laser cavity. Interestingly, even though it was largely diffused to indicate the "single-photon contribution" to the whole value of OAM carried by the beam, the first experimental test on the OAM as an individual property of single photons has been carried out only in 2001 by Mair et al. (2001)

For all the reasons listed above, the orbital angular momentum is considered a recently discovered photonic degree of freedom. In general, only the global angular momentum is associated to an observable quantity, however in the paraxial approximation, both SAM and OAM can be manipulated and measured separately. Indeed the OAM of light can be exchanged with matter, thus opening new perspectives in several fields of classical and quantum physics as well as in biology. In contrast to SAM, which couples only with the material local anisotropy (birefringence), OAM couples mainly with material inhomogeneities characterized by a rotational asymmetry around the beam axis. This coupling may be considered a negative feature when OAM is considered for communication purposes, as it makes it very sensitive to turbulence or other sources of noise Paterson (2005), but it becomes an useful property when OAM is adopted as a tool for sensitively probing the properties of a given medium as considered in several recent works (see Molina-Terriza, Rebane, Torres, Torner & Carrasco (2007); Torner et al. (2005)). The use of OAM for probing can lead to microscopic imaging with a spatial resolution that is higher than the Rayleigh limit Tamburini et al. (2006) and, when OAM fields are used in combination with suitable fluorescence methods (e.g., the stimulated emission depletion), they enable new methods of far-field microscopy with theoretically unlimited resolution Harke et al. (2006); Hell (2009). Optimal spatially structured light beams have also been considered as tools to cage/uncage specific molecules for accurate and rapid biological imaging Shoham et al. (2005). Some of these approaches may have relevant applications in the imaging of biological tissues, e.g. for diagnostic or research purposes. Finally the characteristic doughnut profile of the intensity pattern of a LG mode allows an efficient ion and atom trapping with low scattering and hence heating of the atom, useful for atom optics and BECs purposes Andersen et al. (2006).

Beyond all these applications, it has been recognized that the orbital angular momentum has a great potential for quantum photonics, in particular regarding quantum information protocols implemented through quantum optics techniques. Quantum information (QI) is based on the combination of classical information theory and quantum mechanics. In the last few decades, the development of this new field has opened far-reaching prospects both for fundamental physics, such as the capability of a full coherent control of quantum systems, as well as in technological applications, most significantly in the communication field. In particular, quantum optics has enabled the implementation of a variety of quantum information protocols. The fundamental unit of information in QI theory is a two-level system, the quantum bit or qubit. Exploiting the features of quantum states, however, it has been proven that qubits allows the transfer of more information than the one encoded in a classical boolean alphabet and, at the same time, the quantumness of qubit systems ensure high level of security in communication processing. In this context, the information encoding based on two-dimensional system can be experimentally implemented by exploiting degrees of freedom of single photons as, for example, the polarization. Up to now, several quantum information protocols have been successfully implemented, thanks to a notable control on the polarization degree of freedom achieved through different efficient devices. However there may be a significant advantage in introducing the use of higher dimensional systems for encoding and manipulating the quantum information. Such *d*-level quantum systems, or *qudits*, provide a natural extension of qubits that has been shown to be suitable for prospective applications such as quantum cryptography and computation Cerf et al. (2002); Lanyon et al. (2009). In this framework the orbital angular momentum, defined in an infinitely dimensional Hilbert space, provides a natural choice for implementing *qudits* encoded in a single photon state (see Franke-Arnold et al. (2008); Molina-Terriza, Torres & Torner (2007)). This can be an important practical advantage, as it allows increasing the information content per photon, and this, in turn, may cut down substantially the noise and losses arising from the imperfect generation and detection efficiency, by reducing the total number of photons needed in a given process. Moreover, the combined use of different degrees of freedom of a photon, such as OAM and spin, enables the implementation of entirely new quantum tasks, as shown by Aolita & Walborn (2007); Barreiro et al. (2005; 2008)

Since the seminal paper of 1992 a large effort has been spent to develop optical tools able to manipulate and control efficaciously the orbital angular momentum degree of freedom. Up to now, by observing the transfer of OAM to matter, some devices have been adopted in order to generate/analyse LG modes (computer generated holograms, spatial light modulators), or manipulate the OAM analogously to what is commonly carried out through waveplates on polarization (cylindrical lenses mode converters). Despite these successes, the optical tools for controlling the OAM quantum states remain rather of limited use: a wider and more practical control of the OAM resource somehow analogous to that currently possible for the polarization degree of freedom is by the way under progress.

Here we present a brief introduction on orbital angular momentum in quantum optics, describing the main devices adopted in order to achieve an efficient manipulation. Furthermore, we describe some experiments that have been carried out by adopting an optical device, the *q-plate*, able to couple the spinorial and orbital contributions of the angular momentum of light by exploiting the properties of liquid crystals.

2. The orbital angular momentum of light

Light beams carry energy and momentum, the latter both in its linear and angular components. Thus, when an electromagnetic field interacts with matter, an exchange of energy and momentum occurs, manifested in interesting mechanical effects or in changing of the beam properties. Here we will focus on light angular momentum, composed by a spinorial and orbital component. The manipulation of spinorial angular momentum (SAM), commonly associated to the polarization of light, is largely diffused in all the different field of physics. On the other hand, the orbital angular momentum (OAM) of light did not go through the same diffusion for many years, since after seminal works of Poynting (1909) and Beth (1936), only in 1992 Allen et al. (1992) demonstrated that particular solutions of the Helmoltz equation in the paraxial regime exhibit an azimuthal phase structure typical of beams possessing OAM. Such beams, as for example the Laguerre-Gauss beams, can be experimentally manipulated and thus offer a valuable resource in quantum information, where the possibility to exploit the infinite-dimensionality of OAM opens interesting perspectives. Here we present theoretically the main elements that characterize, both in classical and quantum regimes, the orbital angular momentum of light, deriving the solutions of Helmoltz equation and presenting the main ones, namely the Hermite-Gauss and Laguerre-Gauss beams.

3. Classical orbital angular momentum of light

In classical physics the total angular momentum $\vec{J}$ of an electromagnetic field reads:

$$\vec{J} = \epsilon_0 \int d\vec{r}\vec{r} \times (\vec{E} \times \vec{B})$$

where $\vec{E}$ and $\vec{B}$ are the electric and magnetic fields, respectively. The angular momentum is a conserved quantity for a free field, due to the invariance under arbitrary rotations of free Marxwell equations Enk & Nienhuis (1994). Moreover, as the electric field can be separated in longitudinal and transverse component $\vec{E} = \vec{E}_{||} + \vec{E}_{\perp}$, it is possible to single out the contribution from the transverse part to the angular momentum related to the radiation field $\vec{J}_{rad}$ whose components reads:

$$\vec{J}_{rad} = \vec{L} + \vec{S} \tag{1}$$

where $\vec{S}$ is the *spin angular momentum* and $\vec{L}$ represent the *orbital angular momentum* contribution. The spin angular momentum does not depend on the specific coordinate system and hence it is often called as the intrinsic angular momentum. Moreover, since for a massless particle, as a photon, it is not possible to define a rest frame and hence a spin vector, it is possible only to define the projection of the spin along the propagation direction $\hat{z}$, whose values can be $\sigma = \pm s$ with s both integer or half-integer values Enk & Nienhuis (1994). For the photon, having $s = 1$, the helicity σ can takes only values $\sigma = \pm 1$ (in $\hbar$ units). As a form of angular momentum, spin angular momentum is physically associated to the rotation of the particle around its own axis, and for this reason physically is associated to the polarization of single photons. In particular, right-circular polarization is associated to $\sigma = -\hbar$ and left-circular polarization to $\sigma = +\hbar$. On the other hand as shown in eq.(1) light beams carry also orbital angular momentum, considered as the extrinsic component of angular momentum since it depends on the chosen coordinates system. It has been demonstrated that OAM is associated to the azimuthal phase structure of light beams, as will be described in section 3.2 (Allen et al. (1992)). Differently from SAM, OAM causes a rotation of the particle around the beam axis. The experimental evidence of the angular momentum and its components can be achieved only by studying the interaction between light and matter, where it is possible to observe a transfer of angular momentum. In particular, differences between SAM and OAM can be appreciated in the paraxial regime.

3.1 The paraxial approximation

In classical optics a beam that propagates along the $\hat{z}$ direction is described by the Helmoltz equation:

$$\nabla^2 U(\vec{r}) + k^2 U(\vec{r}) = 0 \tag{2}$$

where k is the wavevector $k = \omega c^{-1}$. For beams that are characterized by small divergence angle respect to the propagation direction, eq.(2) can be investigated within the *paraxial approximation*. Such approximation provides a proper description for laser beams, whose beam variation along the transverse plane $(\hat{x}\hat{y})$ are slow compared to the ones along the direction of propagation. In this case the beam can be represented as $U(\vec{r}) = A(\vec{r})e^{-ikz}$, where the exponential component is the plane wave contribution and $A(\vec{r}) \in \mathbb{C}$ is the complex amplitude whose variations are slow along the beam propagation and thus the following relation holds:

$$\frac{\partial^2 A(\vec{r})}{\partial z^2} << \frac{\partial^2 A(\vec{r})}{\partial^2}, \frac{\partial^2 A(\vec{r})}{\partial y^2} \tag{3}$$

Replacing in eq.(2) the expression of $U(\vec{r})$ and considering the paraxial approximation in eq.(3), we obtain the *paraxial Helmoltz equation*:

$$\frac{\partial^2 A(\vec{r})}{\partial x^2} + \frac{\partial^2 A(\vec{r})}{\partial y^2} - 2ik\frac{\partial A(\vec{r})}{\partial z} = 0 \tag{4}$$

This equation admits different families of solutions depending on the coordinates system adopted. For example adopting Cartesian coordinates, the solutions are the Gaussian beams, widely used in optics to describe laser beams propagation. However if the amplitude $A(\vec{r})$ is modulated by another function slowly varying on direction $\hat{z}$, other valid solutions of eq.(4) can be obtained. Among them, a common one is represented by the Hermite-Gauss (HG) modes, where the Gaussian envelope is multiplied by Hermite polynomials. Interestingly, if instead of a beam described by $U(x, y, z)$ the Helmoltz equation is solved in cylindrical coordinates that is, the function $U(r, \varphi, z)$ is adopted, we obtain another important family of solutions, known as Laguerre-Gauss (LG) modes. As it will be shown in the next section, LG modes are characterized by a dependence of the phase term to the azimuthal angle and hence carry a well-determinate value of orbital angular momentum. Finally, it is worth notice that other classes of solution can be found by adopting polar coordinates (Hypergeometric beams HG Karimi et al. (2007)), or elliptic coordinates (Ince-Gaussian beams), or modifying properly the envelopes (Bessel and Mathieu beams).

3.2 Laguerre-Gauss modes

When the paraxial equation is solved in cylindrical coordinates ($x = rcos\varphi, y = rsin\varphi, z = z$) the solution is represented by a function family known as Laguerre-Gauss (LG) modes, where the gaussian envelope is multiplied by Laguerre polynomials $L_p^m(\tau)$. The amplitude distribution of these modes reads:

$$u_{p,m}^{LG}(r,\varphi,z) = \frac{C}{\sqrt{1+\frac{z^2}{z_R^2}}} \left(\frac{r\sqrt{2}}{w(z)}\right)^m L_p^m\left(\frac{2r^2}{w^2(z)}\right) e^{-\frac{r^2}{w^2(z)}} e^{-i\frac{kr^2 z}{2(z^2+z_R^2)}} e^{-im\varphi} e^{i\varsigma(z)} \tag{5}$$

where C is a normalization constant, z_R is the Rayleigh range, $w(z) = w_0\sqrt{1+\left(\frac{z^2}{z_R^2}\right)}$ is the beam size along $\hat{z}$ starting from the beam waist w_0, and $\varsigma(z) = (2p+m+1)arctan\frac{z}{z_R}$ is the Gouy phase term. Here (p, m) are the radial and azimuthal index, respectively, and determine the order of the mode $N = 2p + |m|$. In particular (p, m) are integer numbers with $p \geq 0$ while m can assume both positive or negative values. The lowest order of these modes $p = m = 0$ is the Gaussian beam (TEM_{00} mode).

Let us focus on the phase term $e^{-im\varphi}$. This term leads to a phase front that *varies* with the azimuthal angle along the propagation direction: points on the phase front diametrically opposite are indeed dephased by a factor π. This means that the phase front follows an helicoidal pattern along $\hat{z}$ so that in the center of the transverse plane the contact between all different phases causes a *phase singularity*. The torque shown along the directional axis thus leads to an optical vortex, that is LG beams are characterized by a singularity in the center where the field intensity vanishes. For each optical vortex it is possible to associate a number, known as *topological charge*, that refers to the number of twist over a distance of the wavelength λ. Such number q is always an integer, and can assume both positive or negative values, depending on the direction of the twist. The larger is the number of torsions, greater is the twisting velocity along the beam direction. It is exactly this twisting that leads to a non-vanishing contribution of orbital angular momentum

Fig.(1) reports some examples of LG modes intensities and phases for different values of indices p, m. For $p = 0$ the intensity in the transverse plane is distributed among an annular pattern with a zero-intensity spot in the center, so that LG modes are known also as "doughnut modes". In the next section we will show that Laguerre-Gauss modes constitute a complete orthogonal basis of orbital angular momentum operator L_z eigenvectors, however here we present a notable result of Allen et al. (1992) where for the first time was demonstrated that beam characterized by a helical behavior of the phase front, as for LG modes, carry a well-defined amount of OAM per photon fixed by the value of the azimuthal mode index m

Let us first consider a vector potential written in Cartesian coordinates as:

$$\vec{A}(x,y,z) = u(x,y,z)e^{ikz} \tag{6}$$

Adopting the Lorentz gauge, the relation between magnetic $\vec{B}$ and electric $\vec{E}$ fields and potential vector are known, so that is possible to evaluate the mean value of the Poynting vector $\vec{P}$ that determines the energy carried by the beam:

$$p = \epsilon_0 < \vec{E} \times \vec{B} >= i\omega \frac{\epsilon_0}{2}(u\nabla u^* - u^*\nabla u) + wk\epsilon_0 |u|^2$$

This result can be extended to the case of a vector potential expressed in cylindrical coordinates, that is where the beam is characterized by the azimuthal angular dependence $e^{-im\varphi}$. Now we are interested in the estimation of the angular momentum carried by the beam along the $\hat{z}$ direction. The angular momentum reads:

$$\vec{j} = \vec{r} \times \vec{p}$$

where the Poynting vector provides three component $\vec{p} = (p_r, p_\varphi, p_z)$. Thus the component of angular momentum along the propagation direction of the beam j_z can be calculated as $j_z = rp_\varphi$ where:

$$p_\varphi = \epsilon_0 \frac{wm|u|^2}{r} \tag{7}$$

Just by replacing this result in the j_z expression, we get:

$$j_z = \epsilon_0 wm|u|^2 \tag{8}$$

The value of angular momentum carried by each photon can be obtained considering the ratio between angular momentum density and the energy one, and then integrating over all

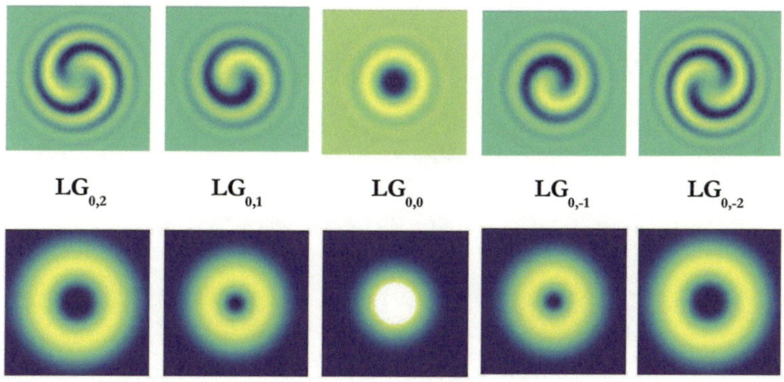

Fig. 1. Examples of LG modes. Top: phase profiles for different LG modes and Bottom: intensity profiles of the same modes.

the transverse plane:

$$\frac{J_z}{W} = \frac{\int\int r dr d\varphi (\vec{r} \times < \vec{E} \times \vec{B} >)_z}{c \int\int r dr d\varphi < \vec{E} \times \vec{B} >_z} = \frac{m}{\omega} \tag{9}$$

Equation (9) shows that each photon that constitutes the beam carries an amount of angular momentum equal to $m\hbar$ Allen et al. (1992; 1999). It is possible to demonstrate that this value refers to the orbital component of angular momentum and is related to the azimuthal phase factor. Indeed if we consider a circularly polarized beam, i.e. a beam with no vanishing contribution of spin angular momentum, its vector potential is:

$$\vec{A} = (\alpha\hat{x} + \beta\hat{y})u(x, y, z)e^{ikz}$$

and going through the same procedure carried before, including the transformation to cylindrical coordinates, the ratio between angular momentum and energy density leads to:

$$\frac{J_z}{W} = \frac{\int\int r dr d\varphi (\vec{r} \times < \vec{E} \times \vec{B} >)_z}{c \int\int r dr d\varphi < \vec{E} \times \vec{B} >_z} = \frac{m}{\omega} + \frac{\sigma}{\omega} \tag{10}$$

We observe from eq.(10) that each photon contributes to the total angular momentum J_z with two components. The σ/ω one belongs to the spinorial contribution, and is due to the polarization of the beam, since $\sigma = \pm 1$ depending on the right or left circular polarization. Analogously, we observe a further term purely related to the phase dependence $e^{-im\varphi}$ that does not belongs to the polarization. Such contribution is thus related to the orbital angular momentum, showing that, depending on the twisting velocity of the phase along the propagation direction, each photon of the beam carries $m\hbar$ OAM.

3.3 Hermite-Gauss modes

The transverse field distribution of laser cavities is typically well described in terms of Hermite-Gauss (HG) modes. These one form a complete orthogonal set that solve the paraxial equation in Cartesian coordinates and their amplitude is given, apart for a phase term, by the product of a gaussian function and Hermite polynomials of indexes $(\tilde{n}, \tilde{m})$ Walborn et al.

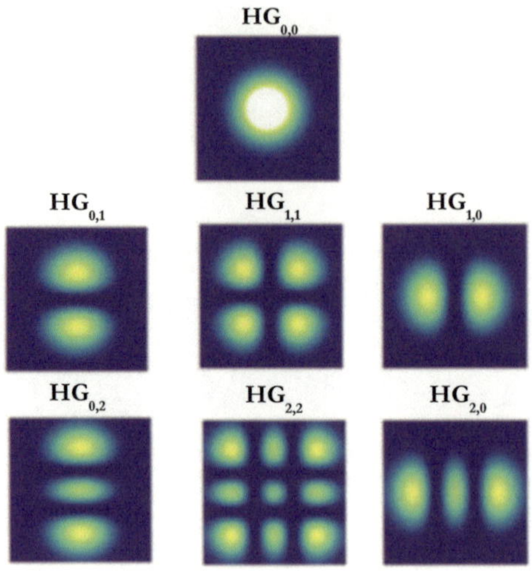

Fig. 2. Intensity profiles for different HG modes. As can be observed, the zero order of HG coincides with the one of LG modes that is, the gaussian mode.

(2005):

$$u^{HG}_{\tilde{n},\tilde{m}} = \sqrt{\frac{2}{2^{(\tilde{n}+\tilde{m})}\pi\tilde{n}!\tilde{m}!}}\frac{1}{w(z)}H_{\tilde{n}}\left(\frac{\sqrt{2}x}{w(z)}\right)H_{\tilde{m}}\left(\frac{\sqrt{2}y}{w(z)}\right)e^{-\frac{x^2+y^2}{w^2(z)}}e^{-ik\frac{(x^2+y^2)z}{2(z^2+z_R^2)}}e^{-i(\tilde{n}+\tilde{m}+1)\varsigma(z)} \quad (11)$$

where $\varsigma(z) = actan\frac{z}{z_r}$ and the last exponential term is the Gouy phase. This solution of the paraxial Helmoltz equation represents modes structurally stables, so that their intensity remain the same along the propagation Allen et al. (1999). The mode indexes $\tilde{n}, \tilde{m}$ define the shape of the beam profile along the $\hat{x}$ and $\hat{y}$ direction, respectively (Saleh & Teich (1991)). Some examples of intensity distribution of such modes are reported in Fig.(2). We observe that the intensity $I_{\tilde{n},\tilde{m}} = |u^{HG}_{\tilde{n},\tilde{m}}|^2$ has $\tilde{n}$ nodes on the horizontal direction, and $\tilde{m}$ along the vertical, while for $\tilde{n} = \tilde{m} = 0$ the shape is the one of a Gaussian beam (TEM_{00} mode). The order of the mode is defined as $N = \tilde{n} + \tilde{m}$. As can be observed from eq.(11), in HG modes each component of the transverse plane travel in phase, as for a plane wave. Indeed this means that surfaces that include all the points with same phase are planes separated by a distance equal to λ. For this reason there is a nil component of linear momentum along the axial direction, which means that such modes do not carry a well defined amount of OAM, as deducible by the lack of the phase term $e^{im\varphi}$. Hermite-Gauss modes form a complete basis set so that any arbitrary field distribution can be written as a superposition of HG modes with indexes $\tilde{n}, \tilde{m}$. However also Laguerre-Gauss modes form a complete basis set, hence a HG mode can be rewritten as a linear superposition of LG modes and *viceversa* through the relations:

$$u^{LG}_{m,p}(x,y,z) = \sum_{k=0}^{N} i^k b(m,p,k) u^{HG}_{N-k,k}(x,y,z) \quad (12)$$

where

$$b(m,p,k) = \frac{(N-k!k!)^{\frac{1}{2}}}{2^N m!p!}\frac{1}{k!}\frac{d^k}{dt^k}[(1-t)^m(1+t)^p]_t = 0 \quad (13)$$

Different indexes that characterize HG and LG are related one set to the other as:

$$m = \tilde{n} - \tilde{m} \tag{14}$$
$$p = min(\tilde{n}, \tilde{m}) \tag{15}$$

Coefficients given in eq.(13) are of crucial importance as they define the matrix element needed for the conversion between HG and LG modes (or viceversa): O'Neil & Courtial (2000). Moreover, the correspondence between HG and LG modes is fixed by the mode order N, defined as:

$$N = \tilde{n} + \tilde{m} = 2p + |m|$$

Inverting relations given in eq.(16) it is possible to estimate the indexes of HG modes in terms of the mode order N and the azimuthal index m:

$$\tilde{n} = \frac{N+m}{2}$$
$$\tilde{m} = \frac{N-m}{2}$$

All these relations can be exploited to generate Laguerre-Gauss modes starting from Hermite-Gauss modes emitted by lasers. Experimentally such mode converters based on eq.(12) are implemented through cylindrical lenses, as explained in section 4.3.

3.4 The orbital angular momentum in quantum mechanics

In 1992 Allen et al. demonstrated that Laguerre-Gauss modes carry a well defined amount of orbital angular momentum through classical mechanics approach, as shown in section 3.2. On the other hand, it is possible to demonstrate this property also following a quantum mechanical approach, showing that LG modes are eigenvectors of the orbital angular momentum operator L_z (Bransden & Joachain (2003); Sakurai (1994)).

The classical orbital angular momentum $\vec{L} = \vec{r} \times \vec{p}$ finds its counterpart in quantum mechanics using the definition of operator momentum $\hat{p} = -i\hbar\nabla$, so that

$$\hat{L} = -i\hbar(r \times \nabla)$$

Thus the cartesian components of the orbital angular momentum operator reads:

$$\begin{aligned} L_x &= -i\hbar\left(y\frac{\partial}{\partial z} - z\frac{\partial}{\partial y}\right) \\ L_y &= -i\hbar\left(z\frac{\partial}{\partial x} - x\frac{\partial}{\partial z}\right) \\ L_z &= -i\hbar\left(x\frac{\partial}{\partial y} - y\frac{\partial}{\partial x}\right) \end{aligned} \tag{16}$$

(17)

where the following commutation rules apply:

$$[L_i, L_j] = i\hbar\epsilon_{ijk}L_k \tag{18}$$

being ϵ_{ijk} the Levi-Civita tensor. It is convenient to study the orbital angular momentum in polar coordinates (r, θ, φ) so that the orbital angular momentum operator L_z can be written as:

$$L_z = -i\hbar \frac{\partial}{\partial \varphi}$$

The eigenvectors equation $L_z \Psi(\varphi) = E\Psi(\varphi)$ with $E = m\hbar$ leads to:

$$\Psi_m(\varphi) = \frac{1}{\sqrt{2\pi}} e^{im\varphi} \tag{19}$$

Moreover the imposition on the eigenvectors to be single valued leads to a constraint on the values of m, that is restricted to positive or negative integers or zero. It emerges from eq.(19) that in paraxial approximation LG modes represent indeed the eigenvectors of the orbital angular momentum operator L_z with eigenvalue $m\hbar$. Generally speaking, any mode with a phase dependence as in eq.(19) can be considered eigenvector of L_z and thus carry a well defined amount of OAM. Laguerre-Gauss modes form a complete Hilbert basis so that can properly represent the quantum photon states in paraxial regime, where spin and orbital contribution can be considered separately. Indeed photon represented by a single LG mode are in a quantum state related to a well-defined value of OAM. It is worth notice that as the eigenvalues of L_z can assume all integer values both positive or negative, the OAM is defined in a infinite-dimensional Hilbert space, thus opening exciting perspectives in the quantum information field.

4. Manipulation of orbital angular momentum

The experimental investigation of orbital angular momentum started in the '90s, thus allowing the idea of OAM as a relatively "'young"' degree of freedom of light. While many devices have been developed for the efficient manipulation of polarization states through birefringent media, as waveplates and polarizing beam splitter, the optical tools for generating and controlling the OAM photon states are rather limited. In last years many efforts have been made in order to implement a device able to generate and manipulate with high efficiency LG modes. In this section we will present the most common and reliable optical devices that exploiting the properties of beams with azimuthal phase structure $e^{im\varphi}$, have allowed the implementation of several quantum optics experiments.

4.1 Computer generated holograms

Computer generated holograms (CGH) are typical tools of diffractive optics, and are probably one of the most common device adopted for the generation and analysis of OAM states. CGHs can be described as diffractive gratings impressed on a film depending on the specific computer-calculated interference pattern formed by a reference plane wave and the desired beam that has to be generated, for example an optical vortex, propagating over a small angle respect to the reference beam. Indeed if a plane wave impinges on the hologram, we would expect to generate on the first diffracted mode an optical vortex equal to the former one. For example, if the reference beam is a LG with OAM m, its interference pattern with a plane wave to be impressed on the hologram is made by a diffraction grating with m lines forming a multi-pronged fork. Depending on the optical vortex that has to be generated/analyzed, a different hologram is then needed. Each hologram is then characterized by the transmittance function, related to the pattern that has to be impressed on the film. Generally, if we consider as reference optical vortex the beam $E(\varphi, z) = E_0 e^{im\varphi} e^{-ikz}$, and the oblique-propagation plane wave $U(x,z) = e^i(k_x x - k_z z)$, the interference pattern at $z = 0$ reads:

$$I = |E(\varphi, z) + U(x,z)|^2 = 1 + I_0 + 2\sqrt{I_0} cos(m\varphi + kx) \tag{20}$$

Taking the Fourier transform of the interference pattern, it is possible to obtain the generic transmittance function for a CGH in polar coordinates:

$$T(r, \varphi) = exp\{i\alpha cos(m\varphi + \frac{2\pi}{\Lambda} r cos\varphi)\} \tag{21}$$

where Λ is the period of the hologram grating, and α the amplitude of phase modulation. In order to illustrate how a hologram actually works, let us consider a standard single-fork hologram ($\Delta m = 1$), characterized by the transmittivity function:

$$T = cos^4\{\frac{1}{2}(\varphi(x,y) + \frac{2\pi}{\Lambda} x)\} \tag{22}$$

where $\varphi(x,y) = arctg\left(\frac{x}{y}\right)$. When a generic wave $E(\rho, \varphi) = E(\rho) e^{im\varphi}$ impinges on the hologram, the output field is determined by $\sqrt{T} E(\rho, \varphi)$, that gives rise to:

$$\frac{E(\rho)}{2}[1 + cos(\varphi(x,y) + \frac{2\pi}{\Lambda} x)] e^{im\varphi} = \frac{E(\rho)}{2} e^{im\varphi} + \frac{E(\rho)}{4} [e^{i(m+1)\varphi} e^{\frac{2\pi}{\Lambda} x} + e^{i(m-1)\varphi} e^{\frac{-2\pi}{\Lambda} x}] \tag{23}$$

Looking at the last term in eq.(23), it emerges how after the hologram the main part of the field is undiffracted and keeps the same value of the initial OAM. On the two first-order diffracted

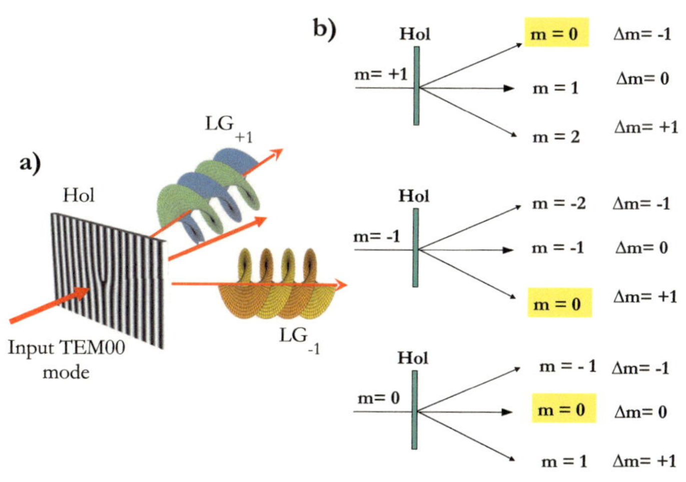

Fig. 3. **a)** Schematic representation of a hologram and its action on an optical field.**b)**Action of a single-fork hologram on different input states. In the yellow box have been enlightened the modes that can be coupled to single mode fiber for analysis.

mode, the beam carries a different value of OAM, equal to $m \pm 1$. A schematic representation of the hologram dynamic is reported in Fig.(3).

Analogously to polarizers, the holograms are used in two ways: (i) for generating a given input quantum state of OAM; (ii) for analyzing a given OAM component of an arbitrary input quantum state. When using the holograms for generating one of the above OAM states, a TEM_{00} input mode is sent into the hologram and the first-order diffracted mode is used for output. The input beam must be precisely centered on the hologram pattern center. When using the holograms for analysis, the input mode, having unknown OAM quantum state, is sent through the hologram (with proper centering). The first-order diffracted output is then coupled to a single-mode fiber, which filters only the $m = 0$ state, before detection. One of the main advantage of holograms device is the possibility to generate and analyze different OAM states, not only LG modes but also their superposition. Unfortunately the efficiency of this device is still low, reaching $\sim 30\%$ for phase holograms (higher transparency), and furthermore any measurement of OAM states in different basis implies a changing of the hologram itself, thus complicating experimental alignment.

4.2 Spiral phase-plates

The spiral phase plate (SPP) principle of function is to directly impose a phase shift on the incident light (Beijersbergen et al. (1994)). They are made by a transparent medium with index of refraction n, whose thickness d gradually varies with the azimuthal angle φ. A schematic representation of such device is reported in Fig.(4). The spiraling increasing thickness let the surface of the SPP look like a single period of helix. As the light beam crosses a dense medium, it is slowed down and hence with the increasing thickness of the plate, there will be a longer optical path corresponding to a wider phase shift δ. In particular each SPP introduces a phase

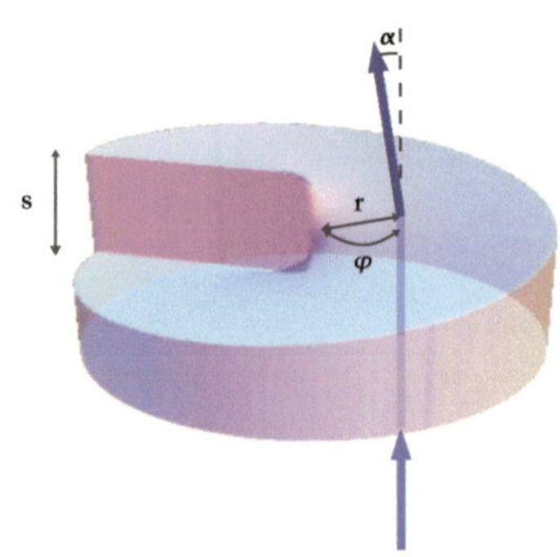

Fig. 4. Schematic representation of a spiral phase plate.

shift δ that is function of the working wavelength λ and of the height of the radial step s:

$$\delta = \frac{(n-1)s}{\lambda}\varphi \tag{24}$$

where we have considered a light beam that travels in the air as surrounding medium of the SPP. The phase shift introduced, due to its dependence to the azimuthal angle, induces a helical structure to the output phase front of the beam that is, a beam carrying orbital angular momentum. In order to design a device able to generate beams with a fixed value of OAM equal to $m\hbar$, the global phase delay has to be an integer multiple of 2π, which means that the physical height of the step s has to be:

$$s = \frac{m\lambda}{n-1} \tag{25}$$

The main difficulty in the SPP preparation is to properly fix and design the height s. Let us now demonstrate how the SPP generates beams carrying OAM equal to $m\hbar$. A beam is injected on the SPP in a certain position at distance r from the optical axis. The local azimuthal slope of the spiral surface is given by:

$$tan\theta = \frac{s}{2\pi r} \tag{26}$$

Since the beam crosses a dense medium with index of refraction n, it will be deflected by an angle α, determinated by the Snell's law:

$$sin(\alpha + \theta) = nsin\theta \tag{27}$$

For small angles, both the sine as well as the tangent functions can be approximated by the angle itself, hence leading to $\alpha = \theta(n-1)$. After the refraction there will be a non-vanishing component of linear momentum transferred from the SPP in the azimuthal direction, equal to: $p_\varphi \approx \frac{h\alpha}{\lambda}$ Thus the interaction between the SPP and the beam leads to a transfer of a component of orbital angular momentum given by:

$$L_z = rp_\varphi \approx \frac{rh}{\lambda}\alpha = \hbar\frac{s(n-1)}{\lambda} = m\hbar \tag{28}$$

which demonstrates the generation of a beam carrying a well defined value of OAM

Recently *adjustable* spiral phase-plates have been developed, which allow to work for multiple wavelengths in the wavelength region where the material transmits (Rotschild et al. (2004)). In addition, various values of m can be achieved with the same plate by tilting it. They are created by twisting a piece of cracked Plexiglas and orienting the device so that one tab of the phase plate is perpendicular to the incident light. A beam injected at the end of the crack will then produce an optical vortex because of the azimuthally varying tilt around the center of the phase plate.

4.3 Cylindrical lens

Most of the devices usually adopted in quantum optics experiments for the manipulation of OAM states, are not able, in general, to generate pure LG modes, but rather a superposition of modes with same azimuthal mode index. However in order to achieve a pure LG mode it is possible to exploit the relation between LG and HG modes (section 1.1.4) and the possibility to convert one into each other. Such conversion is made possible through the cylindrical lenses: Beijersbergen et al. (1993); Courtial & Padgett (1999).

In section 3.3 we have shown that since both HG and LG modes constitute a complete orthogonal set that solve the paraxial equation, linear superpositions of one set can be used to describe the other (eq.12). A HG mode, whose axes is rotated by 45°, can be expanded into the same constituent set:

$$u^{HG}_{\tilde{n},\tilde{m}}(\frac{x+y}{\sqrt{2}},\frac{x-y}{\sqrt{2}},z) = \sum_{k=0}^{N} b(n,m,k)u^{HG}_{\tilde{n}+\tilde{m}-k,k}(x,y,z) \quad (29)$$

where all terms are in phase. A device that is able to introduce the phase factor i^k, a phase shift of $\pi/2$ between successive terms in eq.(29), will allow the experimental conversion of HG modes in LG modes. If after implementing this $\frac{\pi}{2}$-converter, a π phase shift is introduced, then it is possible to change the reverse the helicity that is, the OAM sign. In Fig.(5) we report some examples of the transformation between Laguerre-Gauss and Hermite-Gauss modes. It has

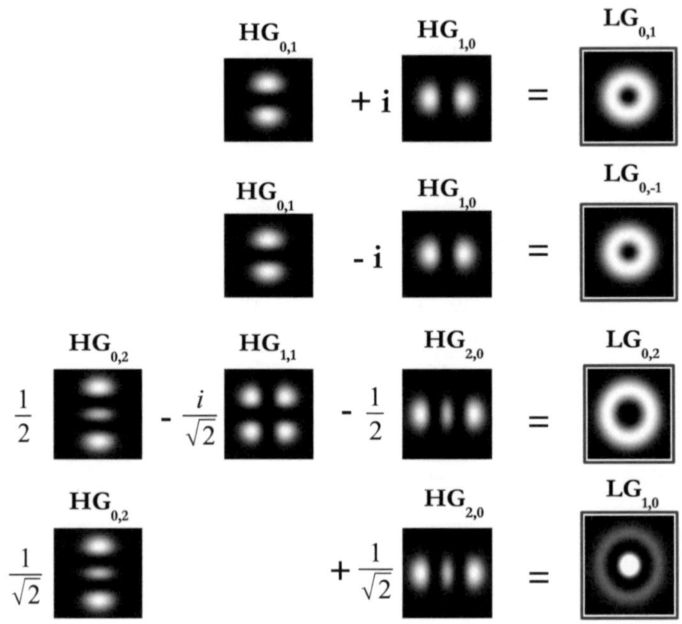

Fig. 5. Examples of transformations between HG and LG modes.

been experimentally demonstrated that a system of cylindrical lenses can implement both the $\frac{\pi}{2}$-converter and the π-converter: Alekseev et al. (1998); Beijersbergen et al. (1993); Courtial & Padgett (1999); Padgett & Allen (2002). Indeed it is possible to exploit the difference in Gouy phase shift between two HG astigmatic modes that take place in a HG mode focused in a cylindrical lens. When we deal with a non-astigmatic beam, the Gouy phase term is expressed by $(\tilde{n}+\tilde{m}+1)\varsigma(z)$. However for astigmatic beams the astigmatism can be characterized in terms of different Rayleigh ranges on planes xz (z_{R_x}) and yz (z_{R_y}), thus leading to a Gouy phase:

$$(\tilde{n}+\frac{1}{2})arctan(\frac{z}{z_{R_x}}) + (\tilde{m}+\frac{1}{2})arctan(\frac{z}{z_{R_y}})$$

It has been demonstrated by Beijersbergen et al. (1993) that a system of two identical cylindrical lenses at distance $\pm d$ from the waist of the diagonal HG input beam introduces a phase difference between two successive terms equal to:

$$\theta = 2[arctan\frac{1}{\beta} - arctan\beta] \quad , \ \beta = \sqrt{\frac{1-\frac{d}{f}}{1+\frac{d}{f}}} \tag{30}$$

with f focal length of the lenses. The condition $\theta = \pi/2$ is fulfilled for $p = -1 + \sqrt{2}$ that is, for lens with distance $d = f/\sqrt{2}$. The imposition of the mode-matching condition in order to restrict the astigmatism only in the region between the two lenses leads to $z_{R_x} = pd = f - d$ and $z_{R_y} = d/p = f + d$, as Rayleigh range for the input beam. As shown in Fig.(6-**a**), these conditions allow to implement experimentally the $\pi/2$-converter, which can be adopted also as converter from LG to HG modes simply injecting a LG beam and removing the i^k factor. Analogously to the action of a quarter waveplate (QWP) on polarization states, converting linear polarized states in circular polarized ones by introducing a $\pi/2$ phase difference between the linearly polarized components, the $\pi/2$-converter can be considered as the corresponding device for OAM states, converting a HG to LG by introducing a $\pi/2$ phase difference between the HG components. In order to implement the π-converter, it

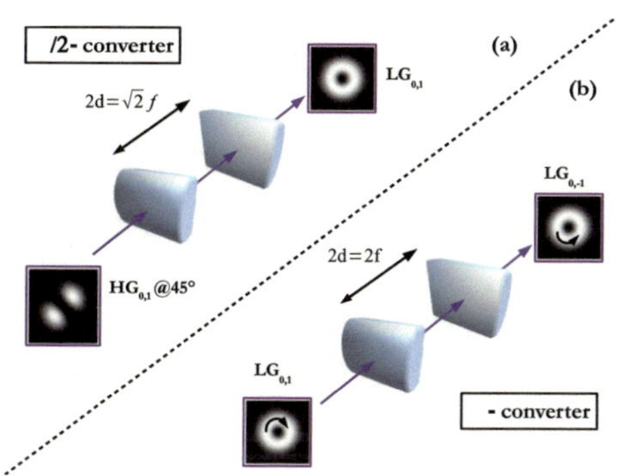

Fig. 6. Schematic representation of **a**: $\pi/2$-converter and **b**: π-converter adopting cylindrical lenses.

is possible to set $\theta = \pi$ which leads to $p = 0$ and thus $d = f$ and a collimated beam ($z_{R_y} \to \infty$). The representation of this converter, that allows to change the sign of m, is reported in Fig.(6-**b**). Once again, a full analogy with a polarization device, the half waveplate (HWP), can be drawn. Indeed the π-converter is similar to the HWP which converts right-circular polarization to left one by introducing a π shift between components.

4.4 Spatial light modulators

The Spatial Light Modulator (SLM) is a device, based on liquid crystals, able to modulate the phase of a light beam or both the intensity and the phase simultaneously (Yoshida et al. (1996)). Besides applications for optical manipulation, SLM are extensively adopted in holographic data storage, and in display technology. A scratch of the SLM structure is reported in Fig.(7). The pure phase SLM adopted for quantum optics experiments are reflective devices in which liquid crystal (LC) is controlled by a direct and accurate voltage, and can modulate a wavefront of light beam (liquid crystal on silicon technology - LCOS). Tha main structure of a SLM is composed by a transparent glass coated with a transparent electrode, then a LC

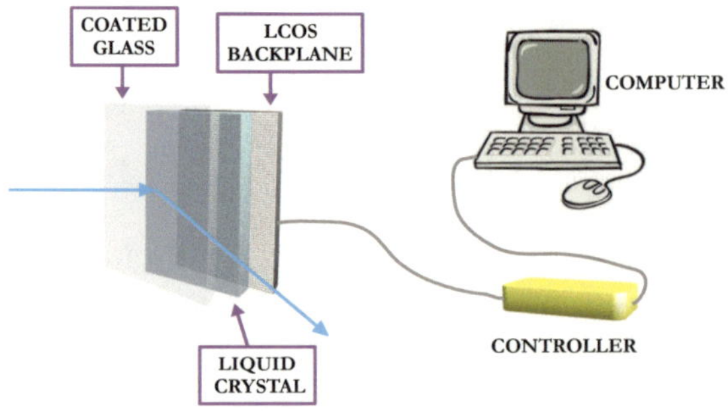

Fig. 7. Schematic representation of a spatial light modulator.

layer and a LCOS backplane, connected to a controller driven by the computer. The LCOS backplane is composed by an array of aluminum pixels, which serve as both reflective mirrors and electrodes. Each electrode is an independently controllable pixel, so that a high-resolution phase modulating array is obtained. Optical modulation is achieved by applying a voltage across the LC layer from the backplane pixels to the transparent electrode on the cover glass. Switching on/off the different pixels composing the backplane, it is then possible to change the refractive index of the LC over them. In this way is possible to create different patterns, like the fork hologram, that can be dynamically modified on demand. The amount of the phase shift depends primarily on three factors: the extraordinary index of the LC material, the thickness of the LC layer, and the wavelength of the input light. As an electric field is applied to a nematic LC layer, there is a corresponding reduction in the extraordinary index of the LC material, and a reduction in the phase shift induced. The main advantage adopting the SLM is then the possibility to work with a dynamic hologram on the same setup. On the other hand, the main disadvantage is the low contrast actually achievable, which is reflected in the efficiency of the device, typically around $10\% - 50\%$.

4.5 The q-plate

When a light beam interacts with matter, a transfer of angular momentum $\vec{J}$ can take place, obeying to the conservation of the global angular momentum of the system. In particular a photon absorbed by a medium can transfer only the spinorial component of angular momentum in *anisotropic* media, while the orbital component can be transferred in *inhomogeneous isotropic* transparent media: Beijersbergen et al. (1993; 1994); Beth (1936). A simultaneous exchange of both spinorial and orbital component of angular momentum is then expected to take place in a medium which is at the same time anisotropic and inhomogeneous. This is precisely the property of liquid crystals. Their structure is determined by the director axis, along witch all the molecules tend to be oriented. By applying electric or magnetic field, or by varying the temperature, it is possible to change the order parameter of the system, and hence change the properties of the LC.

A q-plate (QP) is a birefringent slab having a suitably patterned transverse optical axis, with a topological singularity at its center developed in Naples by Marrucci et al. (2006a). The device is composed by two thin transparent glasses, where in between are inserted some drops of a nematic liquid crystal. The inner surface of the glasses is coated with a polymide for planar alignment, so that it is possible to draw a desired pattern on the glass by simply pressing with a piece of fabric. Once the q-plate is assembled, the molecules of the LC get oriented following

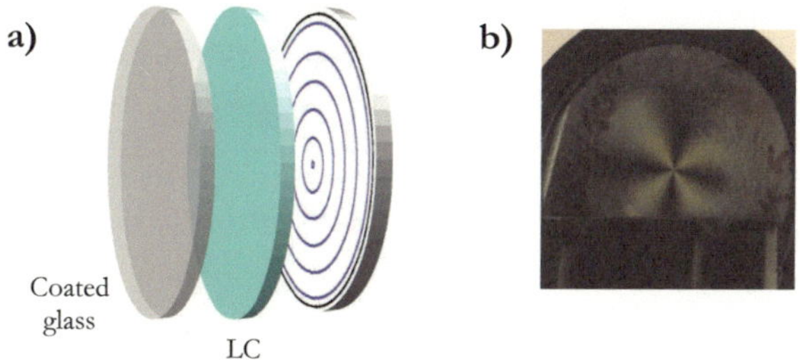

Fig. 8. **a)**Structure of a q-plate device and **b)** Photography of the singularity in a real q-plate device. The picture has been taken by inserting the q-plate between two polaroid films. (Marrucci et al. (2011))

the pattern drawn on the glass. In this way the QP represents an optical device where for each point there is an optical axis in a different position. The specific pattern drawn on the q-plate defines the "charge" q of the singularity that characterize the q-plate. The value of q could be an integer or either a half-integer.

Assuming the normal incidence for the beam of light that crosses the QP, the angle α that defines the local optical axis respect to the singularity of the q-plate is a linear function of the azimuthal angle φ:

$$\alpha(r,\varphi) = q\varphi + \alpha_0 \tag{31}$$

where q is the topological charge, and α_0 a constant. The local direction of the optical axis expressed in eq.(31) enlightens the presence of a topological defect at $r = 0$ that is, at the plane origin.

The working principle of the q-plate is based on the coupling between spin and orbital angular momenta in the NLC in order to exploit an effect similar to the "Pancharatnam-Berry phase" (PBP), already known for the polarization degree of freedom. Such effect is observed when a beam of light undergoes a continuous sequence of transformation on the Poincaré sphere following a closed path. As result, the final wave acquires a phase shift, known as Pancharatnam-Berry phase, that depends on the geometry of the closed path. It has been demonstrated that an analogous effect can be observed in the transverse plane when a wave undergoes a sequence of inhomogeneous polarization transformations having an initial and final homogeneous polarization state. As result, the final beam acquires an inhomogeneous geometrical phase reflected in an overall wave front reshaping, as shown by Marrucci et al. (2006b).

In a single-photon quantum formalism, the QP implements the following quantum transformations on the single photon state:

$$\begin{aligned} |L\rangle_\pi |m\rangle_o &\xrightarrow{QP} |R\rangle_\pi |m+2\rangle_o \\ |R\rangle_\pi |m\rangle_o &\xrightarrow{QP} |L\rangle_\pi |m-2\rangle_o \end{aligned} \tag{32}$$

where $|\cdot\rangle_\pi$ and $|\cdot\rangle_o$ stand for the photon quantum state 'kets' in the polarization and OAM degrees of freedom, and L and R denote the left and right circular polarization states, respectively. In the following, whenever there is no risk of ambiguity, the subscripts π and o will be omitted for brevity. In Fig.(9) is reported a schematic representation of the QP action. Any coherent superposition of the two input states given in Eq. (32) is expected

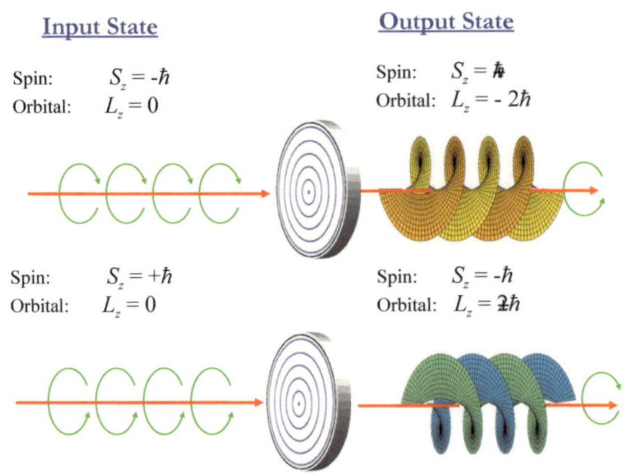

Fig. 9. Schematic representation of the action of a q-plate device with $\delta = \pi$ and $q = 1$, like the ones adopted in our experiments (Marrucci et al. (2011)).

to be preserved by the QP transformation, leading to the equivalent superposition of the corresponding output states, see Nagali et al. (2009). Explicitly, we have

$$\alpha|L\rangle_\pi|m\rangle_o + \beta|R\rangle_\pi|m\rangle_o \xrightarrow{QP} \alpha|R\rangle_\pi|m+2\rangle_o + \beta|L\rangle_\pi|m-2\rangle_o \tag{33}$$

These equations completely define the ideal behavior of the QP on the OAM and polarization subspaces of the photon.

Let us observe that the efficiency of the q-plate device, close to 85%, is related to the birefringent retardation δ introduced by the q-plate itself. An ideal q-plate should have δ uniform across the device and equal to $\delta = \pi$, in order to act on the polarization degree of freedom as a perfect half-waveplate. As have been shown, once a liquid crystal QP is assembled, the birefringent retardation can be tuned either by mechanical compression (exploiting the elasticity of the spacers that fix the thickness of the liquid crystal cell) or by temperature control Karimi et al. (2009), in order to reach the desired value of δ. From eq.(33) it is possible to conclude that for $\delta = \pi$, a QP modifies the OAM state m of a light beam crossing it, imposing a variation $\Delta m = \pm 2q$ whose sign depends on the input polarization, positive for left-circular and negative for right-circular. The handedness of the output circular polarization is also inverted, i.e. the optical spin is flipped (Calvo & Picón (2007)). In the experiments that will be presented in the following, we have adopted only QPs with charge $q = 1$ and $\delta \simeq \pi$. Hence, an input TEM_{00} mode (having $m = 0$) is converted into a beam with $m = \pm 2$.

5. Manipulation of polarization and OAM of single photons

Here we briefly review the experimental results on the adoption of the q-plate device in the single-photon regime.The spin-orbit coupling has been exploited in order to demonstrate the possibility of adopting the QP as an interface between polarization and OAM of single photon states.

5.1 Single-photon entanglement

The single photon transformations applied by the QP, expressed by eq. (33), describe the coupling of the OAM m and the polarization π degrees of freedom. Interestingly, this property can be exploited to generate single-particle entanglement of π and m degrees of freedom. Indeed when an input photon in a TEM_{00} mode and linear polarization is injected on a q-plate, the output state reads:

$$\left.\begin{array}{l} |H\rangle_\pi\,|0\rangle_m \\ |V\rangle_\pi\,|0\rangle_m \end{array}\right\} \overset{QP}{\to} \frac{1}{\sqrt{2}}(|L\rangle_\pi|-2\rangle_m \pm |R\rangle_\pi|+2\rangle_m) \tag{34}$$

This is an entangled state between two qubits encoded in different degrees of freedom. In particular $\{|+2\rangle_m, |-2\rangle_m\}$ is the basis for the OAM qubit which lies in the $|m|=2$ subspace of the infinite dimensional Hilbert space of orbital angular momentum. The property of the q-plate presented in eq.(34) has been experimentally verified through the reconstruction of the density matrix of the output state emerging from the QP. Such reconstruction is possible by exploiting the quantum state tomography technique (James et al. (2001)), whose peculiarity is to determinate the different elements of the density matrix by analyzing the state in different basis. Hence for the two-qubit quantum state reported in eq.(34), we have performed measurements both in π and m degrees of freedom. Besides the normal $\{|+2\rangle_m, |-2\rangle_m\}$ OAM basis, we had to carry out measurements also in the two superposition bases $\{|a\rangle_m, |d\rangle_m\}$ and $\{|h\rangle_m, |v\rangle_m\}$ by means of different computer-generated holograms Langford et al. (2004). We have considered as incoming states on the QP (**a**) $|H\rangle_\pi|0\rangle_m$,and (**b**) $|V\rangle_\pi|0\rangle_m$. As predicted by the transformation introduced by the QP, when the state $|H\rangle_\pi|0\rangle_m$ is injected, the output state in the basis $\{|L,+2\rangle, |R,-2\rangle\}$ reads:

$$\hat{\rho} = \frac{1}{2}\begin{pmatrix} 0&0&0&0 \\ 0&1&1&0 \\ 0&1&1&0 \\ 0&0&0&0 \end{pmatrix}$$

The experimental reconstruction is reported inf Fig.(10-**a**), with a concurrence of the state equal to $C = (0.95 \pm 0.02)$. Indeed the average experimental concurrence is $C = (0.96 \pm 0.02)$, while the average purity of the states is $P = Tr\rho^2 = (0.94 \pm 0.02)$ (Nagali et al. (2009)).

5.2 Quantum transferrers

Due to its peculiarities, the q-plate provides a convenient way to "interface" the photon OAM with the more easily manipulated spin degree of freedom. Hence as next step we have shown that such interface can be considered as a quantum "transferrer" device, which allows to transfer coherently the quantum information from the polarization π to the OAM m degree of freedom, and *vice versa*. In this Section, we present a complete description of two optical schemes, which have been shown in Nagali et al. (2009), that enable a qubit of quantum information to be transferred from the polarization to the OAM (*transferrer* $\pi \to o_2$), from OAM to polarization (*transferrer* $o_2 \to \pi$). Moreover, we tested also the combination of these

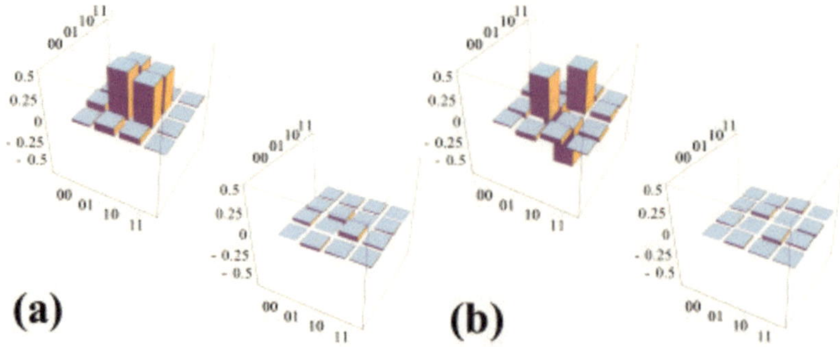

Fig. 10. Experimental density matrices (real and imaginary parts) for the single photon entangled state (Nagali et al. (2009)). The computational values $\{0,1\}$ are associated to the $\{|R\rangle, |L\rangle\}$ polarization states, and to $\{|+2\rangle, |-2\rangle\}$ for the orbital angular momentum m for the first and the second qubit, respectively. The incoming state on the QP is **(a)** $|H\rangle_\pi |0\rangle_m$,and **(b)** $|V\rangle_\pi |0\rangle_m$.

two schemes, thus realizing the *bidirectional transfer* polarization-OAM-polarization ($\pi \rightarrow o_2 \rightarrow \pi$). The latter demonstration is equivalent to demonstrate quantum communication using OAM for encoding the message. In other words, the qubit is initially prepared in the polarization space, then passed to OAM in a transmitting unit (Alice), sent to a receiving unit (Bob), where it is transferred back to polarization for further processing or detection.

5.2.1 Quantum transferrer $\pi \rightarrow o_2$

Let us consider as initial state the polarization-encoded qubit

$$|\Psi\rangle_{in} = |\varphi\rangle_\pi |0\rangle_o = (\alpha |H\rangle_\pi + \beta |V\rangle_\pi)|0\rangle_o \tag{35}$$

where $|0\rangle_o$ indicates the TEM_{00} mode. By passing it through a pair of suitably oriented quarter-waveplates (one with the optical axis parallel to the horizontal direction and the other at 45°), the photon state is rotated into the L, R basis:

$$(\alpha |L\rangle_\pi + \beta |R\rangle_\pi)|0\rangle_o \tag{36}$$

After the QP the quantum state of the photon is then turned into the following:

$$\alpha |R\rangle |+2\rangle + \beta |L\rangle |-2\rangle. \tag{37}$$

If a polarizer along the horizontal direction is used, we then obtain the state

$$|\Psi\rangle_{out} = |H\rangle_\pi (\alpha |+2\rangle_{o_2} + \beta |-2\rangle_{o_2}) = |H\rangle_\pi |\varphi\rangle_{o_2} \tag{38}$$

which completes the conversion. We note that such conversion process is probabilistic, since the state $|\Psi\rangle_{out}$ is obtained with a probability $p = 50\%$, owing to the final polarizing step. Moreover, since we are using the $\{|H\rangle, |V\rangle\}$ basis for the polarization encoding and the $o_2 = \{|+2\rangle, |-2\rangle\}$ for the OAM one, the transfer is associated also with a rotation of the Poincaré sphere.

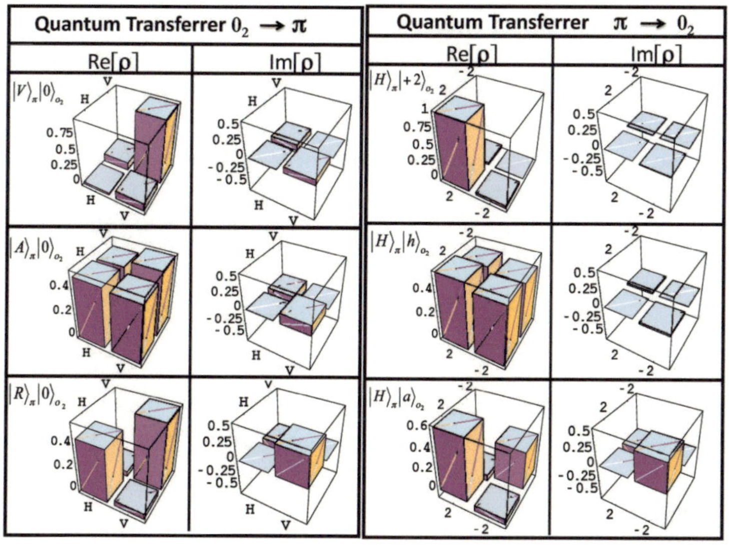

Fig. 11. **Right Side -** Experimental density matrices ρ (the left column shows the real part and right column the imaginary part) measured for the output of the $\pi \rightarrow o_2$ qubit transfer, for each of the three different predicted output states shown in the upper left corner of each row.**Left Side -** Experimental density matrices ρ (the left column shows the real part and right column the imaginary part) measured for the output of the $o_2 \rightarrow \pi$ qubit transfer, for each of the three different predicted output states shown in the upper left corner of each row. (Nagali et al. (2009))

The input arbitrary qubit is written in the polarization using two waveplates, as discussed previously. The experimental results for three specific choices of the input state are shown in Fig. (11). We find a good agreement with theory as demonstrated by the fidelity parameter, with an average fidelity value between the experimental states and the theoretical predictions equal to $F = (97.7 \pm 0.2)\%$.

Thus, we have demonstrated experimentally that the initial information encoded in an input TEM_{00} state can be coherently transferred to the OAM degree of freedom, thanks to the $\pi \rightarrow o_2$ converter, giving rise to the preparation of a qubit in the orbital angular momentum. As the initial information has been stored in the orbital part of the qubit wave-function, new information can be stored in the polarization degree of freedom, allowing the transportation in a single photon of a higher amount, at least two qubits, of information.

5.2.2 Quantum transferrer $o_2 \rightarrow \pi$

Let us now show that the reverse process can be realized as well, by transferring a qubit initially encoded in the OAM subspace o_2 into the polarization space. We therefore consider as initial quantum state of the photon the following one:

$$|\Psi\rangle_{in} = |H\rangle_\pi |\varphi\rangle_{o_2} = |H\rangle(\alpha| + 2\rangle + \beta| - 2\rangle) \tag{39}$$

By injecting the state $|\Psi\rangle_{in}$ in the q-plate device, and then rotating the output state by means of a pair of waveplates, we obtain the following state:

$$\frac{1}{2}\{\alpha|V\rangle|+4\rangle + \alpha|H\rangle|0\rangle + \beta|V\rangle|0\rangle + \beta|H\rangle|-4\rangle\} \tag{40}$$

Now, by coupling the beam to a single mode fiber, only the states with $m = 0$ that is, the TEM_{00} modes, will be efficiently transmitted. Of course, this implies that a probabilistic process is obtained again, since we discard all the contributions with $m \neq 0$ (ideally, again $p = 50\%$). After the fiber, the output state reads:

$$|\Psi\rangle_{out} = (\alpha|H\rangle + \beta|V\rangle)|0\rangle = |\varphi\rangle_{\pi}|0\rangle_{o} \tag{41}$$

which demonstrates the successful conversion from the OAM degree of freedom to the polarization one. The experimental results for three cases are shown in Fig.(11). We find again a good agreement with theory, with an average fidelity $F = (97.3 \pm 0.2)\%$.

6. Hybrid entanglement

Hybrid entangled states exhibit entanglement between different degrees of freedom of a particle pair. The generation of such states can be useful for asymmetric optical quantum network where the different communication channels adopted for transmitting quantum information exhibit different properties. In such a way one could adopt the suitable degree of freedom with larger robustness along the channel. From a fundamental point of view, the observation of non-locality with hybrid systems proves the fundamental independence of entanglement from the physical realization of the adopted Hilbert space. Very recently the hybrid entanglement of photon pairs between the path (linear momentum) of one photon and the polarization of the other photon has been reported by two different techniques (Ma et al. (2009); Neves et al. (2009)). Nevertheless, the capability of generating hybrid-entangled state encoded in the polarization and OAM of single photons could be advantageous since it could allow the engineering of qubit-qudit entangled states, related to the different Hilbert space dimensionality of the two degrees of freedom. It has been pointed out that such states are desiderable for quantum information and communication protocols, as quantum teleportation, and for the possibility to send quantum information through an optical quantum network composed by optical fiber channels and free-space (Chen & She (2009); Neves et al. (2009)).

In this section we review the realization of hybrid polarization-OAM entangled states, by adopting the deterministic polarization-OAM transferrer described in the previous Section. Polarization entangled photon pairs are created by spontaneous parametric down conversion, the spatial profile of the twin photons is filtered through single mode fibers and finally the polarization is coherently transferred to OAM state for one photon. A complete characterization of the hybrid entangled quantum states has been carried out by adopting the quantum state tomography technique. This result, together with the achieved generation rate, the easiness of alignment and the high quality of the generated state, can make this optical source a powerful tool for advanced quantum information tasks and has been presented in Nagali & Sciarrino (2005).

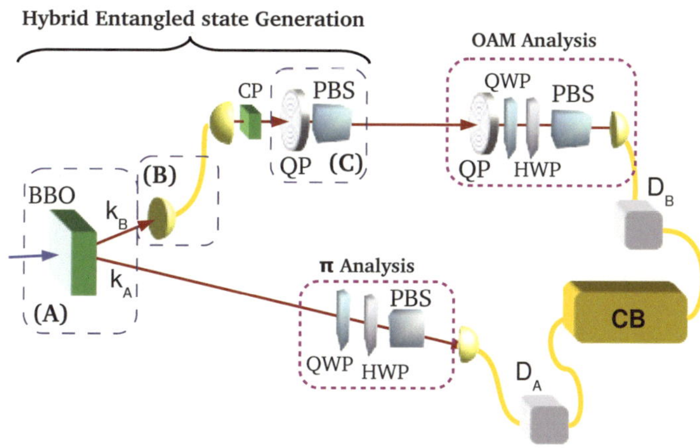

Fig. 12. Experimental setup adopted for the generation and characterization of hybrid π-OAM entangled states (Nagali & Sciarrino (2005)).**(A)** Generation of polarization entangled photons on modes k_A and k_B.**(B)** Projection on the OAM state with $m = 0$ through the coupling on a single mode fiber (SMF).**(C)**Encoding of the state in the OAM subspace o_2 through the $\pi \rightarrow o_2$ transferrer.

6.1 Experimental apparatus and generation of hybrid states

Let us now describe the experimental layout shown in Fig.(12). A 1.5mm thick β-barium borate crystal (BBO) cut for type-II phase matching Kwiat et al. (1995), is pumped by the second harmonic of a Ti:Sa mode-locked laser beam, and generates via spontaneous parametric fluorescence polarization entangled photon pairs on modes k_A and k_B with wavelength $\lambda = 795$ nm, and pulse bandwidth $\Delta\lambda = 4.5$ nm, as determined by two interference filters (IF). The spatial and temporal walk-off is compensated by inserting a $\frac{\lambda}{2}$ waveplate and a 0.75 mm thick BBO crystal on each output mode k_A and k_B Kwiat et al. (1995). Thus the source generates photon pair in the singlet entangled state encoded in the polarization, i.e. $\frac{1}{\sqrt{2}}(|H\rangle^A|V\rangle^B - |V\rangle^A|H\rangle^B)$. The photon generated on mode k_A is sent through a standard polarization analysis setup and then coupled to a single mode fiber connected to the single-photon counter modules (SPCM) D_A. The photon generated on mode k_B is coupled to a single mode fiber, in order to collapse its transverse spatial mode into a pure TEM_{00}, corresponding to OAM $m = 0$. After the fiber output, two waveplates compensate (CP) the polarization rotation introduced by the fiber. To transform the polarization entangled pairs into an hybrid entangled state the photon B is sent through the quantum transferrer $\pi \rightarrow o_2$, which transfers the polarization quantum states in the OAM degree of freedom. After the transferrer operation the polarization entangled state is transformed into the hybrid entangled state:

$$\frac{1}{\sqrt{2}}(|H\rangle^A_\pi|+2\rangle^B_{o_2} - |V\rangle^A_\pi|-2\rangle^B_{o_2})|0\rangle^A_o|H\rangle^B_\pi \tag{42}$$

In order to analyze with high efficiency the OAM degree of freedom, we exploited the $o_2 \rightarrow \pi$ transferrer. By this approach any measurement on the OAM state is achieved by measuring the polarization after the transferrer device, as shown in Fig.12. Finally the photon has been coupled to a single mode fiber and then detected by D_B connected to the coincidence box (CB), which records the coincidence counts between $[D_A, D_B]$. We observed a final coincidence

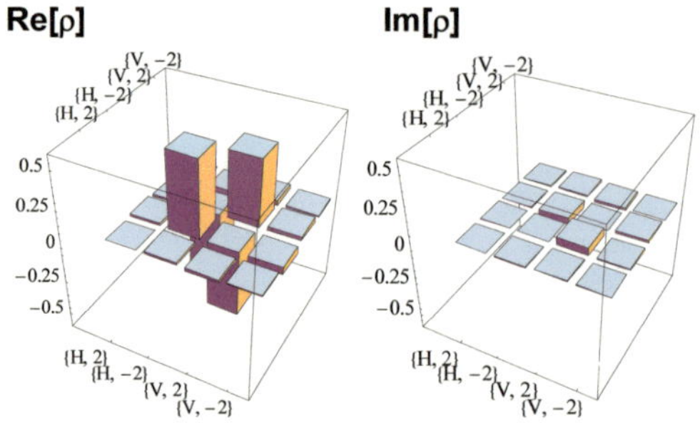

Fig. 13. Experimental density matrix of the hybrid entangled state generated after the transferrer transformation on photons on k_B mode. Each measurement setting lasted 15s (Nagali & Sciarrino (2005)).

rate equal to $C = 100Hz$ within a coincidence window of 3 ns. This experimental data is in agreement with the expected value, determined from $C_{source} = 6kHz$ after taking into account two main loss factors: hybrid state preparation probability p_{prep}, and detection probability p_{det}. p_{prep} depends on the conversion efficiency of the q-plate (0.80 ± 0.05) and on the probabilistic efficiency of the quantum transferrer $\pi \rightarrow o_2$ (0.5), thus leading to $p_{prep} = 0.40 \pm 0.03$. The detection efficiency includes the q-plate conversion efficiency (0.8), the transferrer $o_2 \rightarrow \pi$ (0.5), and the single mode fiber coupling (0.2). Hence $p_{det} = 0.08$.

6.2 Characterization of the state

To completely characterize the state in Eq. (42) we reconstructed the density matrix of the quantum state. The tomography reconstruction requires the estimation of 16 operators James et al. (2001) through 36 separable measurements on the polarization-OAM subspaces. We carried out the reconstruction of the density matrix $\rho^{A,B}_{\pi,o_2}$ after the polarization-OAM conversion. The experimental results are reported in Fig.13, with the elements of the density matrices expressed in the polarization and OAM basis $\{|H,+2\rangle, |H,-2\rangle, |V,+2\rangle, |V,-2\rangle\}$. The fidelity with the singlet states $|\Psi^-\rangle$ has been evaluated to be $F(|\Psi^-\rangle, \rho^{A,B}_{\pi,o_2}) = (0.957 \pm 0.009)$, while the experimental linear entropy of the state reads $S_L = (0.012 \pm 0.002)$. A more quantitative parameter associated to the generated polarization-entangled states is given by the concurrence $C = (0.957 \pm 0.002)$. These values demonstrate the high degree of hybrid entanglement generation.

To further characterize the hybrid quantum states, the violation of Bell's inequalities with the two photon system have been addressed. First, we measured the photon coincidence rate as a function of the orientation of the half-wave plate on Alice arm for two different OAM basis analysis, namely $\{|+2\rangle_{o_2}, |-2\rangle_{o_2}\}$ and $\{|h\rangle_{o_2}, |v\rangle_{o_2}\}$. The variation of the number of coincidences $N(\theta)$ with the angle θ is in agreement with the one expected for entangled states such as $N(\theta) = N_0(1 + cos\theta)$: Fig.14. The coincidence fringe visibility reaches the values $V = (0.966 \pm 0.001)$ and $V = (0.930 \pm 0.007)$. Hence, a non-locality test, the CHSH one (Clauser et al. (1969)), has been carried out. Each of two partners, A (Alice) and B (Bob) measures a dichotomic observable among two possible ones, i.e. Alice randomly measures

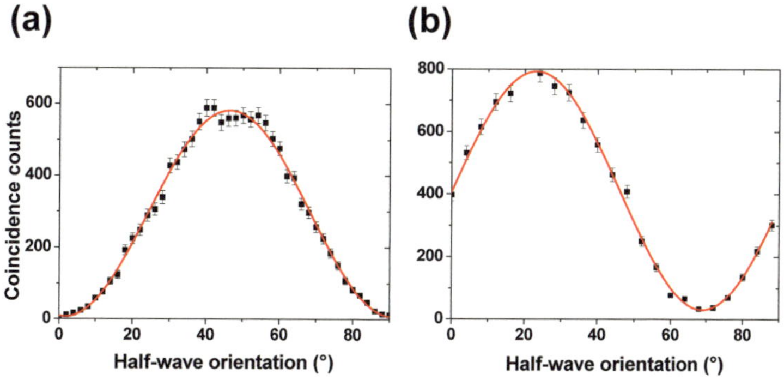

Fig. 14. Coincidence rate $[D_A, D_B]$ measured as a function of the angle θ of the half wave plate on the arm k_A for OAM detected state **(a)** $|+2\rangle$ and **(b)** $|h\rangle_{o_2}$ (Nagali & Sciarrino (2005)).

either **a** or **a′** while Bob measures **b** or **b′**, where the outcomes of each measurement are either $+1$ or -1. For any couple of measured observables $(A = \{\mathbf{a}, \mathbf{a'}\}, B = \{\mathbf{b}, \mathbf{b'}\})$, we define the following correlation function $E(A, B) = \frac{N(+,+)+N(-,-)-N(+,-)-N(-,+)}{N(+,+)+N(-,-)+N(+,-)+N(-,+)}$ where $N(i, j)$ stands for the number of events in which the observables A and B have been found equal to the dichotomic outcomes i and j. Finally we define the parameter S which takes into account the correlations for the different observables

$$S = E(\mathbf{a}, \mathbf{b}) + E(\mathbf{a'}, \mathbf{b}) + E(\mathbf{a}, \mathbf{b'}) - E(\mathbf{a'}, \mathbf{b'}) \tag{43}$$

Assuming a local realistic theory, the relation $|S| \leq S_{CHSH} = 2$ holds. To carry out a non-locality test in the hybrid regime, we define the two sets of dichotomic observables for A and B. For Alice the basis **a** and **a′** correspond, respectively, to the linear polarization basis $\{|H\rangle_\pi, |V\rangle_\pi\}$ and $\{|+\rangle_\pi, |-\rangle_\pi\}$. For Bob the basis **b** and **b′** correspond, respectively, to the OAM basis $\{cos(\frac{\pi}{8})|+2\rangle - sin(\frac{\pi}{8})|-2\rangle, -sin(\frac{\pi}{8})|+2\rangle + cos(\frac{\pi}{8})|-2\rangle\}$ and $\{cos(\frac{\pi}{8})|+2\rangle + sin(\frac{\pi}{8})|-2\rangle, sin(\frac{\pi}{8})|+2\rangle - cos(\frac{\pi}{8})|-2\rangle\}$. Experimentally we obtained the following value by carrying out a measurement with a duration of 60s and an average statistics per setting equal to about 1500 events: $S = (2.51 \pm 0.02)$. Hence a violation by more than 25 standard deviation over the value $S_{CHSH} = 2$ is obtained. This experimental value is in good agreement with an experimental visibility of $V = (0.930 \pm 0.007)$ which should lead to $S = (2.57 \pm 0.02)$.

7. Conclusion

Among all degrees of freedom offered by single photons, the orbital angular momentum (OAM) has a great potential in the quantum information field, as it provides a natural choice for implementing single-photon qudits, the units of quantum information in a higher dimensional space. This can be an important practical advantage, as it enables higher security in quantum cryptographic protocols, as well as implications in fundamental quantum mechanics theory. Moreover, the combined use of different degrees of freedom of a photon, such as OAM and spin, enables the implementation of entirely new quantum tasks.

The authors acknowledge the Future and Emerging Technologies (FET) programme within the Seventh Framework Programme for Research of the European Commission, under FET-Open Grant No. 255914, PHORBITECH.

8. References

Alekseev, A., Alekseev, K., Borodavka, O., Volyar, A. & Fridman, Y. (1998). Conversion of Hermite-Gaussian and Laguerre-Gaussian beams in an astigmatic optical system. 1. Experiment, *Technical Physics Letters* 24(9): 694–696.

Allen, L., Beijersbergen, M. W., Spreeuw, R. J. C. & Woerdman, J. P. (1992). Orbital angular momentum of light and the transformation of laguerre-gaussian laser modes, *Phys. Rev. A* 45(11): 8185–8189.

Allen, L., Padgett, M. & Babiker, M. (1999). The orbital angular momentum of light, *Progress in Optics*, Vol. 39, Elsevier, pp. 291 – 372.
URL: *http://www.sciencedirect.com/science/article/B7W5D-4SFRFMR-8/2/f396ecd62e319c9d81f0234f203b4548*

Andersen, M., Ryu, C., Clade, P., Natarajan, V., Vaziri, A., Helmerson, K. & Phillips, W. (2006). Quantized rotation of atoms from photons with orbital angular momentum, *Phys. Rev. Lett.* 97(17): 170406.

Aolita, L. & Walborn, S. P. (2007). Quantum communication without alignment using multiple-qubit single-photon states, *Phys. Rev. Lett.* 98(10): 100501.

Barreiro, J. T., Langford, N. K., Peters, N. A. & Kwiat, P. G. (2005). Generation of hyperentangled photon pairs, *Phys. Rev. Lett.* 95(26): 260501.

Barreiro, J., Wei, T. & Kwiat, P. (2008). Beating the channel capacity limit for linear photonic superdense coding, *Nature Physics* 4: 282–286.

Beijersbergen, M., Allen, L., Van der Veen, H. & Woerdman, J. (1993). Astigmatic laser mode converters and transfer of orbital angular momentum, *Optics Communications* 96: 123–132.

Beijersbergen, M., Coerwinkel, R., Kristensen, M. & Woerdman, J. (1994). Helical-wavefront laser beams produced with a spiral phaseplate, *Optics Communications* 112(5-6): 321–327.

Beth, R. A. (1936). Mechanical detection and measurement of the angular momentum of light, *Phys. Rev.* 50(2): 115–125.

Bransden, B. & Joachain, C. (2003). *Physics of atoms and molecules*, Pearson Education.

Calvo, G. & Picón, A. (2007). Spin-induced angular momentum switching, *Opt. Lett.* 32(7): 838–840.

Cerf, N. J., Bourennane, M., Karlsson, A. & Gisin, N. (2002). Security of quantum key distribution using *d*-level systems, *Phys. Rev. Lett.* 88(12): 127902.

Chen, L. & She, W. (2009). Increasing Shannon dimensionality by hyperentanglement of spin and fractional orbital angular momentum, *Opt. Lett.* 34(12): 1855–1857.

Clauser, J., Horne, M., Shimony, A. & Holt, R. (1969). Proposed experiment to test local hidden-variable theories, *Phys. Rev. Lett.* 23(15): 880–884.

Courtial, J. & Padgett, M. (1999). Performance of a cylindrical lens mode converter for producing Laguerre-Gaussian laser modes, *Optics Communications* 159(1-3): 13–18.

Enk, S. & Nienhuis, G. (1994). Spin and orbital angular momentum of photons, *EPL (Europhysics Letters)* 25: 497.

Franke-Arnold, S., Allen, L. & Padgett, M. (2008). Advances in optical angular momentum, *Laser & Photonics Reviews* 2(4): 299–313.

Friese, M., Nieminen, T., Heckenberg, N. & Rubinsztein-Dunlop, H. (1998). Optical torque controlled by elliptical polarization, *Optics letters* 23(1): 1–3.

Harke, B., Keller, J., Ullal, C., Westphal, V., Sch
"onle, A. & Hell, S. (2006). Resolution scaling in STED microscopy, *microscopy (STORM)* 3: 793–796.

Hell, S. (2009). Far-field optical nanoscopy, *Single Molecule Spectroscopy in Chemistry, Physics and Biology* pp. 365–398.

James, D. F. V., Kwiat, P. G., Munro, W. J. & White, A. G. (2001). Measurement of qubits, *Phys. Rev. A* 64(5): 052312.

Karimi, E., Piccirillo, B., Nagali, E., Marrucci, L. & Santamato, E. (2009). Efficient generation and sorting of orbital angular momentum eigenmodes of light by thermally tuned q-plates, *Appl. Phys. Lett.* 94(23): 231124.

Karimi, E., Zito, G., Piccirillo, B., Marrucci, L. & Santamato, E. (2007). Hypergeometric-Gaussian modes, *Opt. Lett.* 32(21): 3053–3055.

Kwiat, P. G., Mattle, K., Weinfurter, H., Zeilinger, A., Sergienko, A. V. & Shih, Y. (1995). New high-intensity source of polarization-entangled photon pairs, *Phys. Rev. Lett.* 75(24): 4337–4341.

Langford, N. K., Dalton, R. B., Harvey, M. D., O'Brien, J. L., Pryde, G. J., Gilchrist, A., Bartlett, S. D. & White, A. G. (2004). Measuring entangled qutrits and their use for quantum bit commitment, *Phys. Rev. Lett.* 93(5): 053601.

Lanyon, B. P., Barbieri, M., Almeida, M. P., Jennewein, T., Ralph, T. C., Resch, K. J., Pryde, G. J., O'Brien, J. L., Gilchrist, A. & White, A. G. (2009). Simplifying quantum logic using higher-dimensional hilbert spaces, *Nature Phys.* (7): 134.

Ma, X.-s., Qarry, A., Kofler, J., Jennewein, T. & Zeilinger, A. (2009). Experimental violation of a bell inequality with two different degrees of freedom of entangled particle pairs, *Phys. Rev. A* 79(4): 042101.

Mair, A., Vaziri, A., Weihs, G. & Zeilinger, A. (2001). Entanglement of the orbital angular momentum states of photons, *Nature* 412(6844): 313–316.

Marrucci, L., Manzo, C. & Paparo, D. (2006a). Optical spin-to-orbital angular momentum conversion in inhomogeneous anisotropic media, *Phys. Rev. Lett.* 96(16): 163905.

Marrucci, L., Manzo, C. & Paparo, D. (2006b). Pancharatnam-Berry phase optical elements for wave front shaping in the visible domain: switchable helical mode generation, *Applied Physics Letters* 88: 221102.

Marrucci, L., Karimi, E. Slussarenko, S., Piccirillo, B., Santamato, E., Nagali, E., F. Sciarrino. Spin-to-orbital conversion of the angular momentum of light and its classical and quantum applications, *J. Opt.* 13: 064001 (2011).

Molina-Terriza, G., Rebane, L., Torres, J., Torner, L. & Carrasco, S. (2007). Probing canonical geometrical objects by digital spiral imaging, *J. Europ. Opt. Soc. Rap. Public.* 07014 Vol 2 (2007).

Molina-Terriza, G., Torres, J. & Torner, L. (2007). Twisted photons, *Nature Physics* 3: 305–310.

Nagali, E. & Sciarrino, F. (2005). Generation of hybrid polarization-orbital angular momentum entangled states, *Phys. Rev. Lett* 95: 260501.

Nagali, E., Sciarrino, F., De Martini, F., Marrucci, L., Piccirillo, B., Karimi, E. & Santamato, E. (2009). Quantum information transfer from spin to orbital angular momentum of photons, *Phys. Rev. Lett.* 103: 013601.

Nagali, E., Sciarrino, F., De Martini, F., Piccirillo, B., Karimi, E., Marrucci, L., & Santamato, E. (2009). Polarization control of single photon quantum orbital angular momentum states, *Opt. Expr.* 17:18745.

Neves, L., Lima, G., Delgado, A. & Saavedra, C. (2009). Hybrid photonic entanglement: Realization, characterization, and applications, *Phys. Rev. A* 80(4): 042322.

O'Neil, A. & Courtial, J. (2000). Mode transformations in terms of the constituent Hermite-Gaussian or Laguerre-Gaussian modes and the variable-phase mode converter, *Optics Communications* 181(1-3): 35–45.

Padgett, M. & Allen, L. (2002). Orbital angular momentum exchange in cylindrical-lens mode converters, *Journal of Optics B: Quantum and Semiclassical Optics* 4: S17.

Paterson, C. (2005). Atmospheric turbulence and orbital angular momentum of single photons for optical communication, *Phys. Rev. Lett.* 94(15): 153901.

Poynting, J. H. (1909). The wave motion of a revolving shaft, and a suggestion as to the angular momentum in a beam of circularly polarized light, *Proc. R. Soc. London* 82(557): 560.

Rotschild, C., Zommer, S., Moed, S., Hershcovitz, O. & Lipson, S. (2004). Adjustable spiral phase plate, *Applied optics* 43(12): 2397–2399.

Sakurai, J. (1994). *Modern Quantum Mechanics*, Addison-Wasley.

Saleh, B. E. A. & Teich, M. C. (1991). *Fundamentals of Photonics*, Wiley Interscience.

Shoham, S., O'Connor, D., Sarkisov, D. & Wang, S. (2005). Rapid neurotransmitter uncaging in spatially defined patterns, *Nature Methods* 2(11): 837–843.

Tamburini, F., Anzolin, G., Umbriaco, G., Bianchini, A. & Barbieri, C. (2006). Overcoming the Rayleigh criterion limit with optical vortices, *Phys. Rev. Lett.* 97(16): 163903.

Torner, L., Torres, J. & Carrasco, S. (2005). Digital spiral imaging, *Optics Express* 13(3): 873–881.

Walborn, S. P., Pádua, S. & Monken, C. H. (2005). Conservation and entanglement of hermite-gaussian modes in parametric down-conversion, *Phys. Rev. A* 71(5): 053812.

Yoshida, N., Toyoda, H., Igasaki, Y., Mukohzaka, N., Kobayashi, Y., &Hara, T. (1996). Nonpixellated electrically addressed spatial light modulator (SLM) combining an optically addressed SLM with a CRT, Proc. SPIE 2885: 132-136.

Part 2

Photonic Materials

5

New Nanoglassceramics Doped with Rare Earth Ions and Their Photonic Applications

Aseev Vladimir, Kolobkova Elena and Nikonorov Nikolay
St. Petersburg State University of Information Technologies Mechanics and Optics, St. Petersburg Russia

1. Introduction

1.1 Why glassceramics?

Glass-crystalline materials (or glassceramics) are heterophase composite materials, usually consisting of a glassy matrix (glass phase) and micro- or nano-sized dielectric or semiconductor crystals (crystalline phase), or metallic particles distributed in it. Glassceramics are formed by the growth of crystalline phase inside of a glass matrix. The crystal growth can occur as a result of spontaneous thermal crystallization of glass, as in case of heat treatment. For example, there are following typical representatives of the spontaneous crystallization: glasses doped with microcrystals of CdS, CdSe, CdTe, PbS, PbSe (non-linear media) [1-7], glasses doped with semiconductor microcrystals AgBr, AgCl, CuBr, CuCl (photochromic media) [8], glasses doped with dielectric microcrystals Li_2O-SiO_2 (photoetchable media – FOTURAN ™, FOTOFORM ™, PEG ™ [9-11]. The crystal growth can occur as a result of photo-thermo-induced crystallization caused by UV photoirradiation and subsequent heat treatment. In this case, UV radiation generates centers of nucleation and the thermal treatment results in the growth of microcrystals in irradiated area of the glass host. Glasses doped with complicated microcrystals of NaF-AgBr (polychromatic glasses [12] and photo-thermo-refractive - PTR glasses [13]) are typical representatives of the photo-thermo-induced crystallization.

1.2 Why nanostructured glassceramics?

One of the major drawbacks of glass ceramic materials is a high light scattering occurring at the boundary of crystalline phase and glass phase. That is why current research in the development of optical glass-crystalline materials is aimed at decreasing of light scattering by means of formation of nanosize (5-30 nm) crystals or nanoparticles in the glass matrix. Only the nanoscale nature of the crystalline phase can significantly reduce the light scattering in heterophase composites (where the extinction coefficient can reach less than 0.01 cm^{-1}) and classify these materials as optical. Fig.1 illustrates transition from millimeter-size crystals to micrometer-size crystals and finally nano-size crystals in the glass host. The transition to the nanoscale crystalline phase not only leads to changes in physical, chemical and optical properties of glassceramics, but is also a cause for fundamentally new and unique properties. For example, nano-size of crystalline phase results in quantum-size, resonance, and other effects. Such materials can possess unique properties, that can't be realized in traditional materials. Following new materials can be synthesized on the base of nanoglassceramics: plasmonic materials, photonic crystals, metha-materials.

Currently, optical transparent glassceramics are of the great interest for the modern element base of photonics, because they stay at intermediate state between crystalline materials and glasses. These glassceramics combine the best properties of crystals (high emission cross-section, quantum yield of luminescence, mechanical and thermal strength etc.) and glasses (possibilities of pressing and molding, spattering, pulling optical fibers, and carrying out ion exchange to fabricate waveguide structures).

In this paper a new type of glassceramics, namely the nanostructured lead oxifluoride glasseramics, which was developed by the authors have been discussed. In addition some examples of using the novel nanoglassceramics for photonics applications have been outlined.

a. b. c.

Fig. 1. Transition from millimeter-size crystals (a) to a micrometer-size crystals (b) and to nano-size crystals (c) in the glass host.

2. Structure and properties of nanostructured oxifluoride glasseramics

The story of the oxifluoride glassceramics production technology starts in 1970-es. In [14] was made an attempt to synthesize the oxifluoride glasses, containing Ln_2O_3 (Y, La, Gd, Lu), Yb_2O_3, PbF_2, M_nO_m (M = B, P, Te, Si, Ge), activated by Er_2O_3 or Tm_2O_3. This attempt has resulted in production of the nontransparent glassceramic materials, containing the microcrystals with diameter of about 10 μm. The efficiency of luminescence, revealed by these media, was in several times larger than that of the etalon luminophors LaF_3 : Yb : Er. Later on, in 1993, there was published the first paper, devoted to synthesis of the transparent glassceramics, containing the cubic fluoride phase, activated by erbium and ytterbium ions [15]. There were for the first time produced the materials, which combine all the advantages of the glass-like alumosilicate matrix and the optical features of the low-phonon fluoride crystals.

Since recently the transparent fluorine-containing glassceramics matrices, containing the rare-earth ions, included into the fluorite-like nanocrystalline phases, are drawing the attention do to the series of spectroscopy advantages. It is obvious that from the point of view of laser active media development optimal are the materials, which are characterized by the low-frequency phonon spectrum and by the low content of the OH-groups, because in this case one can reduce the excitation energy losses due to the multi-phonon quenching process. For a long time there was a common opinion that only the fluorine-containing materials (like fluoride glasses and crystals) are optimal for the said problem solution. However, since recently the synthesis of the glassceramics materials like the oxifluoride silicate glasses has become the priority direction of studies [16-24]. Such composite materials combine the optical parameters of the low-phonon fluoride crystals and the good mechanical and chemical features of the silicate glasses. It was also revealed that some of the oxifluoride glass-like materials have a feature of forming the fluoride nanocrystals, doped by the rare-earth ions, during the process of heat treatment of the raw primary glass. Hence such a materials combine all the positive features of the fluoride nanocrystals, which control the optical properties of the rare earth ions, with that of oxide glasses like easy production technology and excellent macroscopy features like chemical and mechanical strength and the high optical quality. It is well known that in the case of production of the optically transparent glassceramics, including those, which are used for optical waveguide fabrication, it is very important to minimize the light losses for absorption and scattering. The Rayleigh scattering by the micro-inhomogeneities with the size about the radiation wavelength is a factor, limiting such materials use. It imposes the strict limitations over the size of the separated crystalline phase. According to the Rayleigh theory for the visible spectral range, the radius of the crystals, dispersed around the glass, has to be not more than 15 nm. The refraction indexes of the crystalline phase and of the amorphous matrix are to differ in not more than 0.1. Later on these limitations were somewhat softened. In [16] on the case of a model it was shown that one can produce the transparent glassceramics with the size of nanocrystals up to 30 nm and with the refraction indexes difference in not more than 0.3.

2.1 The glassy matrix for glassceramics production

The separation of the crystalline phase in a silicate glass is a traditional way of the opalescent glass production. It is well known that insertion of the fluorides into the glass, whose content is similar to the window glass, leads to forming of a large number of microcrystals in the glass volume. It leads to a drastic increase of a light scattering and to the so called opalescence effect. Hence, the fluorine content in a glass leads to intense phase separation; very often the second phase is represented by the introduced fluorides or by their derivatives. This fact has became the basis for the numerous studies, devoted to fabrication and investigation of the silicate nanoglassceramics, based on the fluorite-like nanocrystals like $Ba(Sr,Ca,)F_2$ or of the hexagonal LaF_3, activated by small concentration or rare-earth or of transient elements.

The separate group is represented by the glassceramics, fabricated on the basis of the glass-like systems with a big amount of fluorides. One can fill into this type the alkali-less germanate and silicate systems like GeO_2-PbO-PbF_2, SiO_2-PbO-10PbF_2 SiO_2 -Al_2O_3- CdF_2 -PbF_2- ZnF_2:(YF_3). In such systems the fluoride concentration can be as high as 60-70 mol.%. The process of crystalline phase formation in this case is not so obvious and needs a more detailed study.

Production of the starting glass and fabrication of the glassceramics on its basis

The glasses of the system of $0.3SiO_2$-$0.15AlO_{3/2}$-$0.29CdF_2$-$0.18PbF_2$-$0.05ZnF_2$-0.03 (Ln,Y) F_3 (the content of the glasses is given in molecular percents before synthesis), where Ln=La, Pr, Dy, Nd, Tb, Eu, Er, Sm, Tm, Ho, have been synthesized. The synthesis was carried out in platinum or corundum crucibles in an air or argon atmosphere, making it possible to produce glasses with a high transparency in a visible spectral range. With the purpose to prevent the spontaneous crystallization, the glass was produced in a gap between two cool glass-carbon plates. Hence the thickness of the produced glass was not more than 2 mm.

Heat treatment with the purpose of evaluation of the crystallization temperature and of the glassceramics production was carried out at the temperature of the crystallization start. It was determined from the thermal curve, measured by means of the differential scanning calorimeter (DSC), see Fig.1. This curve has two separate crystallization peaks, which is the obligatory condition for the transparent glassceramic production. The observed exo-peaks are produced by the bulk (522°C) and by the surface (607°C) crystallizations. The choice of the heat treatment temperature at the start of the first peak makes it possible to prevent completely the surface crystallization, which can lead to the uncontrolled growth of the large crystals. Basing on the DSC data (Fig.2), the specimens of glasses with all contents were exposed to temperature of 480-500°C for 0.5-10 hours.

In the Fig.2 are shown the thermal curves for the starting specimen (a) and for the glassceramics (b). The elimination of the first exo-peak, which is observed, is an evidence of the complete separation of the volume crystalline phase after heat treatment of the glass during two hours.

Hence on the base of DSC data we have determined the regimes of heat treatment of the starting glass, necessary for the nanocrystalline phase production: temperature 500°C, duration 2 hours.

2.2 The content of the crystalline phase and its growth kinetics

One has to note that, disregarding a large number of papers, considering the studies of the lead fluoride glassceramics, the chemical content of the nanocrystalline phases is still discussed. In [17] on the base of spectroscopy and X-ray phase data was concluded that the erbium-containing crystalline phase is a solid solution of erbium fluoride in β-PbF_2. In [18] there was investigated the separation of the crystalline phase during heat treatment of the glasses $50GeO_2$-$40PbO$-$10PbF_2$. Their heat treatment has resulted in separation of the fluorite-like phase. Change of the processing temperature from 350 to 395°C has resulted in crystals size growth from 11 to 16 nm. The size of the elementary cell was within the range of 5.82-5.83 Å, i.e. it was preserved constant within the experimental error limits of ±0.005 Å. It was supposed that the separated crystalline phase is comprised by β-PbF_2, doped by ions of Er^{3+}. In a search for confirmation there were synthesized the monocrystals of β-PbF_2 with various content of ErF_3. For the admixture of 20% of ErF_3 the size of the elementary cell was reduced from 5.94 to 5.816 Å. In the oxyfluoride crystals [19] the heat treatment of the glass with the content $32(SiO_2)$-$9(AlO_{1.5})$-$31.5(CdF_2)$-$18.5(PbF_2)$-$5.5(ZnF_2)$:$3.5(ErF_3)$ has resulted in separation of the crystalline phase with the fluorite structure and cell constant a= 5.72 Å.

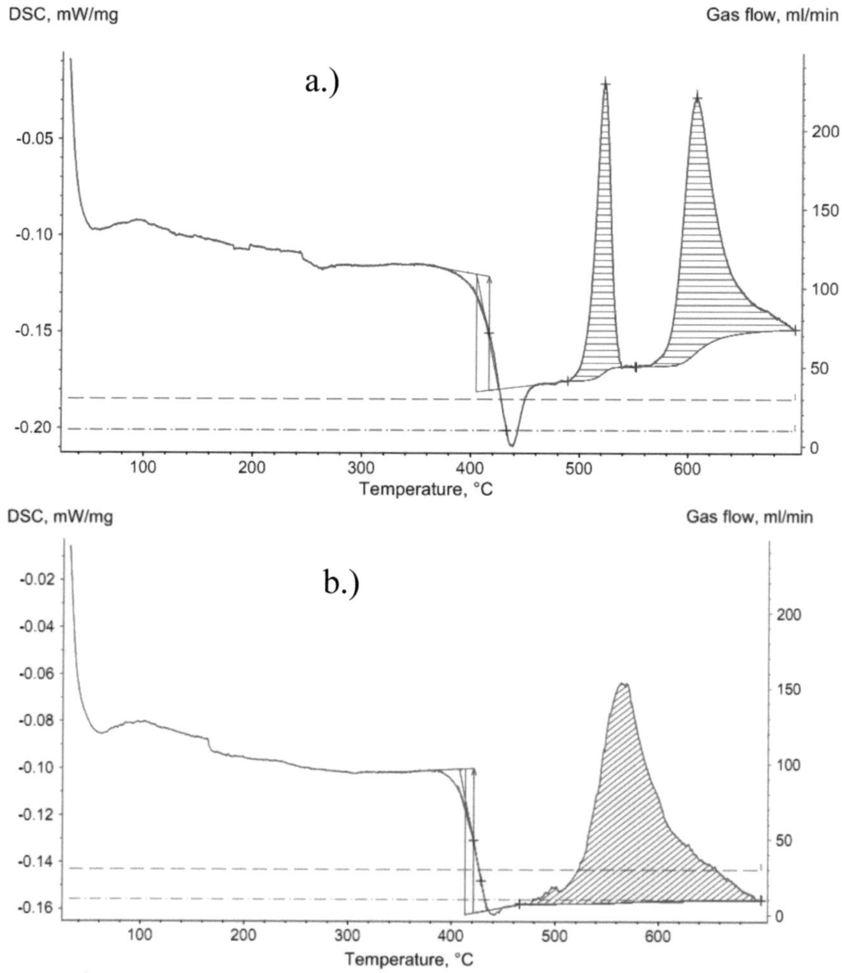

Fig. 2. Thermal curves of the starting glass (a) and of the glass, which was thermally processed for 2 hours (b).

Thus the precipitation of the fluorite-like crystalline phase of β-PbF_2 doped with ErF_3 has been earlier observed in various glass hosts. However the sizes of the elementary cell of the crystalline phase separated in glasses were differ from ones of the solid solution models. In all cases the sizes of elementary cell of the crystalline phase in glass host were much smaller than that observed in the solid solution models or in the crystals of β- PbF_2 doped with ErF_3. In [20, 21] there was revealed the dependence of the cell constant of the nanocrystals (a =5,72 - 5,81 Å) in the glass of system of $30SiO_2$ -$5ZnF_2$-(29-x)CdF_2-(18+x)PbF_2-7,5Al_2O_3-3 ErF_3 from the ratio of lead and cadmium fluorides concentration. There was made an assumption about formation of the mixed fluoride crystals with partial replacement of Pb^{2+} on Cd^{2+}.

In the present work a complex analysis of the precipitated crystalline phase has been carried out. The precipitation of the crystalline phase during heat treatment of glasses with the content of $30SiO_2$-$29CdF_2$-$18PbF_2$-$5ZnF_2$-7,5Al_2O_3, containing the fluorides of yttrium and of the lanthanum group elements, including the case of simultaneous addition of yttrium and of lanthanum group elements has been studied. The phase content and the cell constants have been determined with the use of XRD analysis. It was found out that thermal procession of the oxyfluoride matrix glass, lacking yttrium or lanthanum group elements, which was carried out for 2-3 hours under the temperature of 480-520°C, did not lead to crystalline phase separation. Only the glasses, containing the yttrium fluoride or lanthanum group fluoride (or their arbitrary concentration) has revealed separation of the crystalline fluorite-like phase (Fig.3).

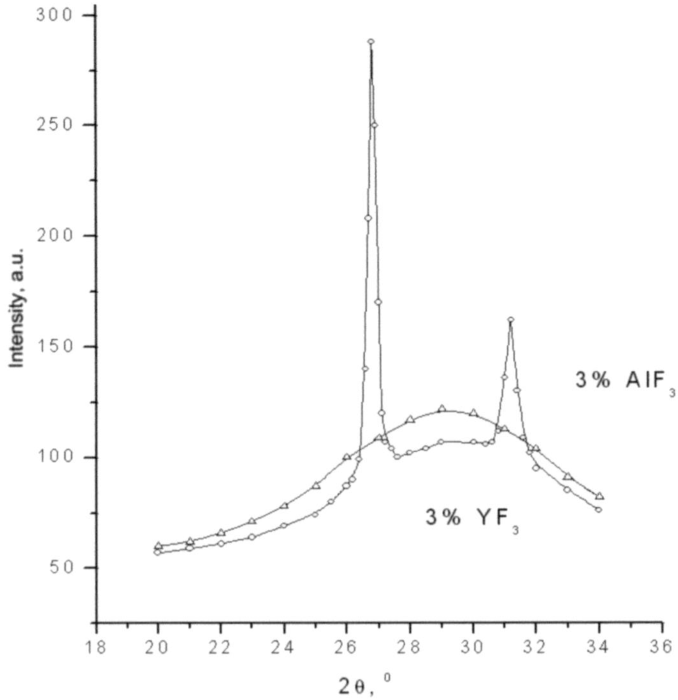

Fig. 3. X-ray diffraction curves for the angle range 2θ from 20 to 35°, 1 – thermally processed glass without YF_3 и lanthanum group fluorides, 2 – thermally processed glass, containing 3% of YF_3.

The presence of REI (rare-earth ions) in the crystalline phase is also confirmed by the absorption spectrum data. It is obvious, that redistribution of REI between the glass-like and crystalline phases has to result in variation of REI absorption and luminescence spectra. In the Fig.4 are shown as an example the absorption spectra of the starting glass, activated by erbium, and after its thermal processing.

The size of the crystals was evaluated by means of X-ray scattering to small angles. It was determined from the diffraction reflection width with the accuracy ±5%. After heat treatment of the glass at T= 480°C at 2-3 hours the size of the crystals has reached 250-300 Å. In such a regime the phase volume was close to the equilibrium one, i.e. it was not increasing in the case or increasing the heating duration or temperature. The further heat treatment has resulted in crystal size increase, while the integral intensity of diffraction reflection was practically constant. I.e., in this case mainly the re-crystallization processes took place; it is also indirectly confirmed by the spectroscopy data. In the Fig.5 is shown the kinetics of the crystals growth during heat treatment.

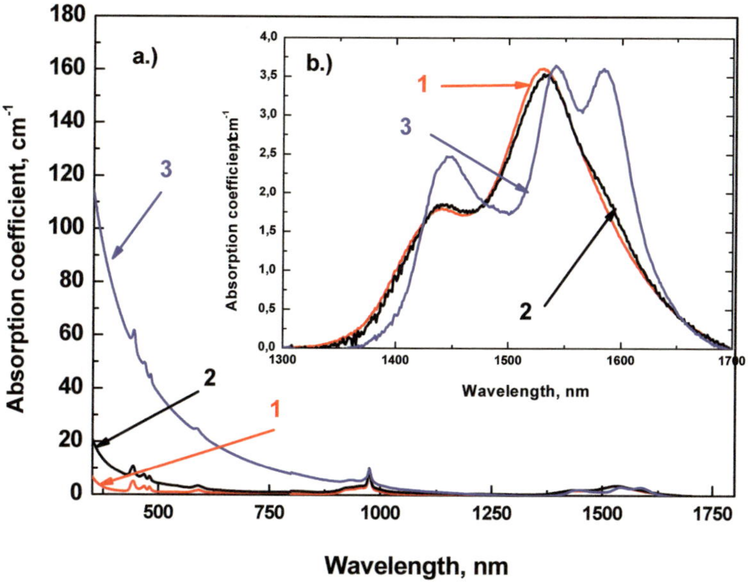

Fig. 4. Attenuation spectra of the (*1*) initial glass containing 3 mol % PrF_3 and (*2, 3*) nano-glassceramics obtained after heat treatment at T = 475°C for 4 (*2*) and (*3*) 32 h and (b) a fragment of the absorption spectra for the $^3H_4 \rightarrow {}^1G_4$ transition (normalized spectra).

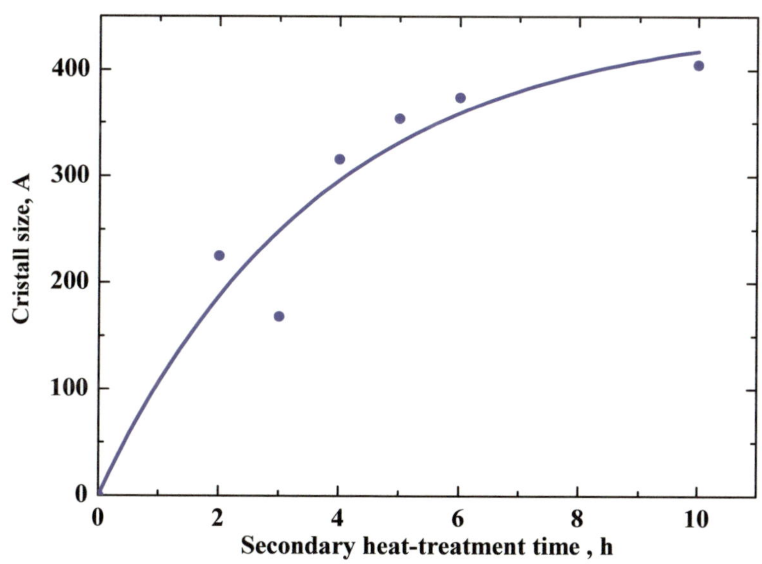

Fig. 5. Dependence of the crystal size on the heat treatment time at T= 480°C. ErF_3 concentration is 1.5 mol %.

There was observed the definite correlation between the concentration of the introduced fluorides of the lanthanum group elements and the integral intensity of the fluorite-like phase diffraction reflection. Hence one can conclude that the presence of fluorides of rare-earth elements and of yttrium totally determines the crystalline phase separation, and their concentration in the starting glass completely determines the volume of the extracted phase. One can treat it as an evidence of one and the same stoichiometry of the separated crystalline phases.

All the produced nanocrystalline phases have the fluorite-like structure - cubic, edge-centered, spatial group Fm3m. The lattice constants depend upon the REI radius (Table 1). The size of the elementary cell of the phase, produced by heat treatment of the glasses with YF_3, was equal to 5.74 Å. It differs significantly from the size of the elementary cell of the yttrium oxyfluoride of lead $PbYOF_3$, which is equal to 5.792 Å. The reduction of the size of the elementary cell of the phase separated in the studied glasses with YF_3 can be explained by presence of Cd. Hence the significant variation of the elementary cell size is connected with the formation of the solid solution

$$Pb_{1-x}Cd_xYOF_3.$$

	Lanthanum group element	R, Å	a, Å
1	Pr^{3+}	1	5.83
2	Nd^{3+}	0.99	5.82
3	Sm^{3+}	0.97	5.81
4	Eu^{3+}	0.96	5.8
5	Tb^{3+}	0.89	5.765
6	Dy^{3+}	0.88	5.75
7	Ho^{3+}	0.86	5.735
8	Er^{3+}	0.85	5.725
9	Tm^{3+}	0.84	5.715
10	Yb^{3+}	0.81	5.7
11	Y^{3+}	0.97	5.74

Table 1. Size of the elementary cell of the fluorite-like crystalline phases, separated in the investigated glasses, containing the fluorides of various lanthanum group elements and the ion radiuses of such elements and yttrium, R, according to Goldschmidt.

One the base of the investigations, which were carried out, one can make some conclusions and outline some recommendations for the synthesis of the lead oxifluoride glassceramics with the required spectral and luminescence properties:

- the presence of the fluorides of rare-earth elements and yttrium results in the precipitation of the crystalline phase in glass host, and the volume fraction of the precipitated nanocrystals is completely determined by their concentration in the starting glass;
- heat treatment of the starting glasses at 480-500°C for 2-3 hours results in the growth of crystalline phase in size of 250-300 Å. For such regimes of heat treatment the volume

fraction of the crystalline phase approaches equilibrium, i.e. it does not change after consequent heat treatment or its temperature increase;

- the crystalline phase, precipitated in the staring glass doped with YF_3, is the modified yttrium oxyfluoride of lead (Pb_{1-x} Cd_x YOF_3). The crystalline phases, precipitated in glass, which contains fluorides of various lanthanum group elements or such fluorides together with yttrium fluoride, are lanthanum or yttrium-lanthanum oxyfluorides of lead like Pb_{1-x} Cd_x $Y_yLn_{1-y}OF_3$. The crystalline phase has a fluorite-like structure (cubic, edge-centered, spatial group Fm3m). The value of "x" may be different for various elements of lanthanum group;
- the synthesis of the mixed yttrium and lanthanum group element nanocrystall can provide different required concentrations of RE ions in the crystal like Pb_{1-x} Cd_x $Y_yLn_{1-y}OF$. It can be achieved by simple variation of the raw material content.

2.3 The luminescence and spectral properties of the lead oxyfluoride nanoglassceramics

The conclusions of the Section 2.2 are confirmed by the luminescence and spectral studies. If the rare earth ion environment has changed - for instance, when it has moved to the crystalline phase - the energy gaps between its levels and manifolds change due to changing of the electric and magnetic field tension around the ion. These changes, obviously, are to reveal themselves in the shape and position of the absorption and luminescence spectra, and are also to modify the radiation probabilities.

In the novel lead oxyfluoride nanoglassceramics the rare earth ion shifts from the glass-like phase to a fluoride crystal one, i.e. its environment changes drastically. Let us consider these changes on the example of the europium, erbium and neodymium ions.

In the Fig.6 is shown the modification of the europium ions luminescence after the isothermal heat treatment (480°C) in its dependence upon the time of thermal treatment. The europium ion is traditionally used as the sounding ion for interpretation of its environment modification both in crystals [21] and in glasses [22], because this ion transitions reveal intense and well theoretically explained dependence upon the ligands fields. The analysis was carried out on the basis of modification of three bands in the three-valence europium spectrum, corresponding to the transitions from manifold $^5D_0$ to $^7F_0$ (extremely sensitive), $^7F_1$ (magnetic dipole) and $^7F_2$ (electric dipole).

The starting glass is characterized by the spectrum, where the most intense line corresponds to the degenerated transition $^5D_0 \rightarrow ^7F_2$. The second intensity is revealed by the transition $^5D_0 \rightarrow ^7F_1$. It is a triplet line, making thus the basement for the conclusion that Eu^{3+} is positioned in the low symmetry environment, usual for the fluoride glass-like media. Heat treatment for 30 minutes results in some decrease of intensity of transition $^5D_0 \rightarrow ^7F_2$ (620 nm) in comparison with $^5D_0 \rightarrow ^7F_1$ (590 nm). Heat treatment for 60 minutes leads to drastic changes. The triplet $^5D_0 \rightarrow ^7F_1$ is replaced by the doublet with the maxima at 16875 and 17150 cm^{-1}. This can be an evidence of improvement of environment symmetry of rare earth ions and of general transfer of europium to a crystalline phase [22]. Treatment for 90 minutes results in yet larger modification of the europium ion luminescence spectrum and reveals the complete transfer of REI to the nanocrystal. The analysis of the "extremely sensitive" transition line confirms the assumptions, which were done earlier. The frequency shift in the process of the glassceramics formation is practically absent, confirming thus the assumption that the europium ion is in a completely fluoride environment. The half width of the transition $^5D_0 \rightarrow ^7F_0$ is a measure of the rate of inhomogeneity of environment across the ensemble of the activating ions. Hence its reduce in nearly 2 times in comparison with the starting value after treatment for 90 minutes is an evidence of a nearly complete transfer of europium from glass to a crystalline phase. The modification of the ratio of the characteristic bands is also an evidence of a radical re-building of the europium ion environment, accompanied by its symmetry improvement.

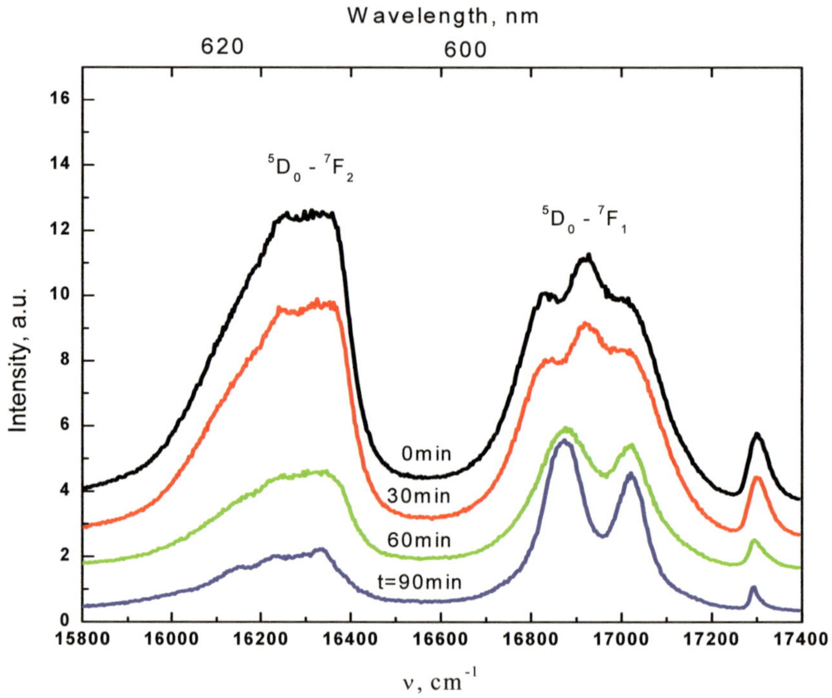

Fig. 6. Luminescence spectra of the glass, activated by 3 mol.% of EuF_3 before and after isothermal heat treatment (T=480 °C)

Similar modifications are observed in the absorption (Fig.7,a) and luminescence (Fig.7,b) of erbium ions. One can see that the heat treatment results in a strong deformation of the erbium ion absorption and luminescence contours. Transition from glass to nanoglassceramics definitely reveals the Stark structure of erbium, which is usual for the crystalline media. Such a behavior of the absorption and luminescence contours is an evidence of modification of the activation ion environment during transfer from glass phase to a crystalline one.

In addition the rare earth ion transfer to a crystalline phase results in modification of the radiation probabilities from different levels. It reveals itself in intensity growth of some bands in the luminescence spectrum and in reduces of some other bands. Let us consider this effect on the example of the glassceramics, activated by erbium (Fig.8).

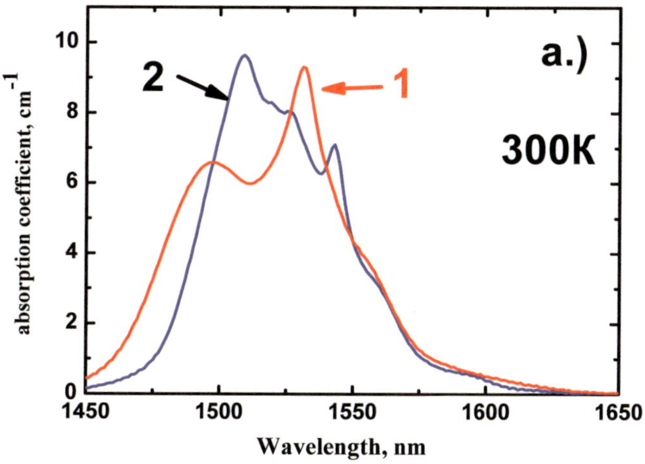

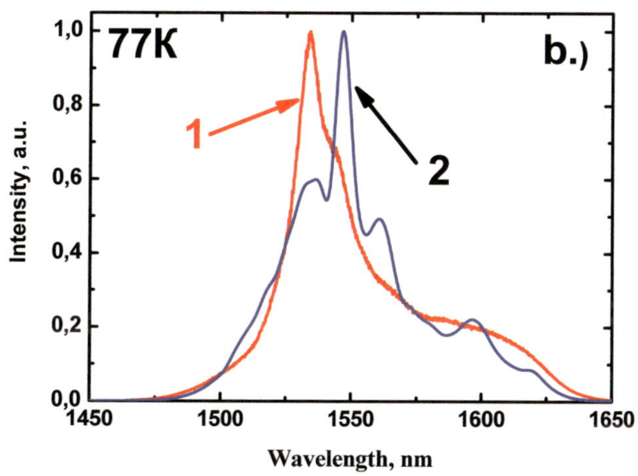

Fig. 7. Absorption (a) and luminescence (b) spectra for the transition $^4I_{15/2} \rightarrow {^4I_{13/2}}$ in the glass, containing 0.4 mol.% of ErF_3 (curve 1) and of nanoglassceramics, obtained after heat treatment (curve 2).

One can see that after the heat treatment the relative intensities of luminescence transitions have changed. For instance, the intensities of transitions with the maxima at 478, 530 and 540 nm have reduced in comparison with 660 nm, while a new transitions in an UV band (365 nm) have arose. In addition, the intensity of the transition in a blue range (405 nm) has grown up. The arise of additional bands became possible due to increase of probability of radiative transitions from these levels.

The heat treatment results in a significant modification of the spectral and luminescence features of the rare earth ions. The spectral and luminescence studies provide the independent confirmation of the X-ray phase analysis conclusion that the rare earth ions shift to the crystalline phase.

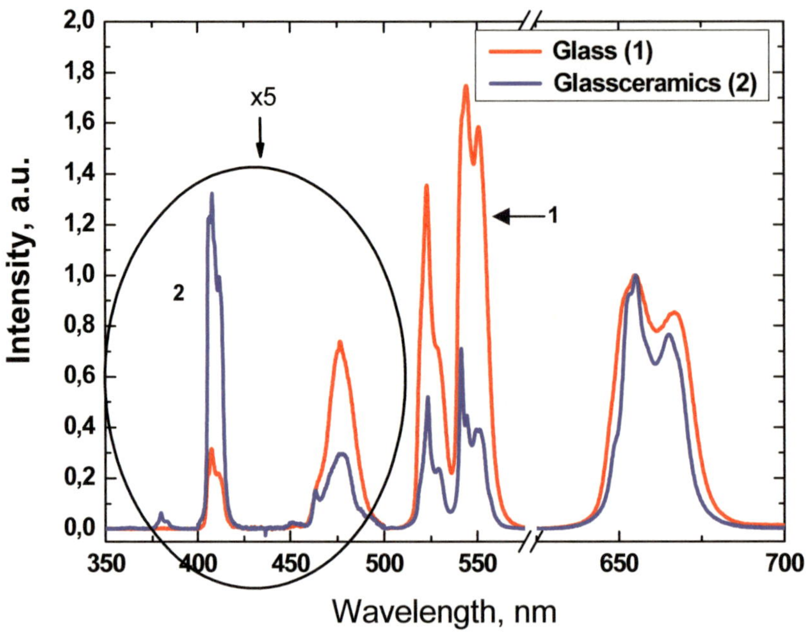

Fig. 8. Luminescence spectra of the (*1*) initial glass containing 0,2 mol % ErF_3 and (*2*) nanoglassceramics samples obtained over 10 hours at heat treatment. The spectra are normalized at peak of 660 nm

3. Application of new lead-oxifluoride nanoglassceramics for lasers and amplifiers

Various laser media are used for various types and applications of lasers and optical amplifiers. During the design the luminescence and spectral properties of the medium are chosen so as to provide the best fit to the solved problem. One has also to mention that the use of some rare earth ions imposes its own limitations over the medium use. For instance, the praseodymium ion practically does not luminescence in the oxygen glasses, and that is why the oxide-less fluoride glasses like ZBLAN are used in a praseodymium amplifiers. One has also to mention that the important feature of laser medium is its applicability for design of either super compact mini- or micro-chip lasers or amplifiers, when one has to increase the activator concentration, or of the extended fiber-optical lasers or amplifiers with the length of several meters. Hence the definite requirements are imposed onto the laser material. For instance, the material (glass, glassceramics etc.) has to be possible to accept a high concentration of the activator, while the factors, leading to the luminescence quenching (the presence of OH-groups or the nonlinear up-conversion) are to be minimized. In other case of fiber-optical lasers and amplifiers the material is to be possible to be used as the pre-form for the activated fiber production. The lead fluoride nanoglassceramics, activated by the rare earth ions, which was developed, can be used as an active medium for the miniature lasers of for the fiber-optical lasers and amplifiers. Let us discuss the laser features of the lead-fluoride nanoglassceramics on an erbium ion example and let us compare it with the other well known glass analogs.

Erbium ion activated glasses are widely used for creation of the mini-microchip and fiber lasers, emitting at 1.5 and 3 μm. The lasers and amplifiers at 1.5 μm are used for data transfer via fiber-optical communication lines and also in range finders, because their radiation fills into the so-called third window of the fused silica transparency and to the eye-safe spectral range. In the telecommunication tasks a very important feature of the fiber-optical amplifying medium is the width of its amplification spectrum, because the wider amplification range makes it possible to fill into and to provide simultaneous amplification of the big number of spectral channels. Unfortunately, disregarding a very good technology (from the point of view of fiber production), optical and luminescence spectral features of the erbium doped silicate and phosphate glasses, the halfwidth of their amplification spectrum is not large. It is equal to $\Delta\lambda$=20-30 nm, thus limiting significantly the number of spectral channels in the amplifier. The fluoride glasses ZBLAN, doped by erbium, have a wide amplification spectrum of $\Delta\lambda$ =50-80 nm. However, they meet some problems and limitations - first of all from the point of view of the fluoride fiber production and exploitation. One can overcome the said disadvantages and limitations in a new lead-fluoride nanoglassceramics, where erbium is in the crystalline fluoride phase and the glass matrix contains oxygen.

For instance, Fig.9 illustrates the spectra of erbium ions luminescence in a novel lead-fluoride nanoglassceramics and in the commercially available silicate glass.

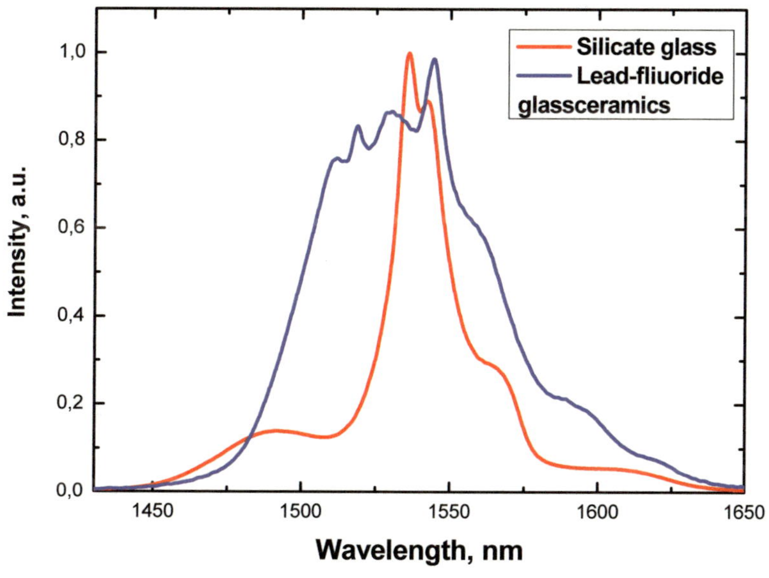

Fig. 9. Luminescence spectra of erbium ions in silicate glass and lead-oxifluoride nanoglassceramics.

One can see that for the nanoglassceramics the spectrum halfwidth is equal to $\Delta\lambda$= 66 nm, while that for the silicate glass it is equal to $\Delta\lambda$=20 nm. One has also to note that the new nanoglassceramics reveal the high (>80%) quantum yield of luminescence for the transition $^4I_{13/2} \rightarrow {}^4I_{15/2}$.

In the Fig.10 are shown the amplification cross-sections for the new nanoglassceramics in comparison with the starting glass (before treatment) and with the commercially available silicate glass.

One can see that the transfer from the starting glass to the nanoglassceramics by heat treatment results in increase of the amplification range from 48 to 64 nm. It is accompanied by the increase of the maximal amplification gain for the same pumping level. For instance, for the starting glass for the pumping level 70% the gain is equal to g=0,35 cm^{-1}, while for the glassceramics it is equal to g=0,42 cm^{-1}. One can also see that the amplification spectrum for the commercial silicate glass is much worsens than that of both of the untreated fluorine-containing glass and of the nanoglassceramics on its basis.

Let us briefly consider the medium for lasers, emitting at 3 μm, which are used first of all in medicine. Today the oxygen-less crystals like $LiYF_4$ or garnet crystals like YAG are used as the matrices for erbium ions. However, it is impossible to fabricate the optical fiber and fiber-optical amplifier on the base of these crystals. Hence the search of new media and realization of fiber-optical lasers, emitting at 3 μm, are today a very important task. The new lead-oxyfluoride nanoglassceramics, activated by erbium, can become an interest object from the point of view of realization of a fiber three-micron lasers.

In the Fig.11 is shown the luminescence spectrum of the erbium activated nanoglassceramics for the transition $^4I_{11/2} \rightarrow {}^4I_{13/2}$.

One can see from the Fig.11 that the heat treatment leads to broadening of the luminescence spectrum. It, for instance, makes it possible to tune the laser wavelength within the range 3-3.15 μm. The life time for transition of the $^4I_{11/2} \rightarrow {}^4I_{13/2}$ is a very important parameter for lasers operated at 3 μm. The life time for new nanoglassceramics achieves 5 ms. In comparison, for crystal of Er:YAG (that very often used as a laser media at 3 μm) the life time decay achieves 1 ms. Thus, the new nanoglassceramics is a very attractive candidate for 3 μm-fiber lasers.

Hence a novel lead-fluoride nanoglassceramics, activated by erbium, reveals the luminescence spectral and laser features, which are not worse than that of the well-known commercial glasses, and exceeds them for such parameters, as the amplification spectrum width and the lifetime of the metastable manifold. At the same time the structure and the content of the nanoglassceramics makes it possible to subject it to a traditional technology the optical fiber production. So one can treat the novel lead-fluoride nanoglassceramics, activated by erbium, is the prospective medium for fiber-optical lasers and amplifiers.

4. Application of new lead-oxifluoride nanoglassceramics for thermal sensors

The chapter demonstrates application of the new nanoglassceramics for luminescent fiber thermal sensors. Also the characteristics of the sensors are compared with traditional ones.

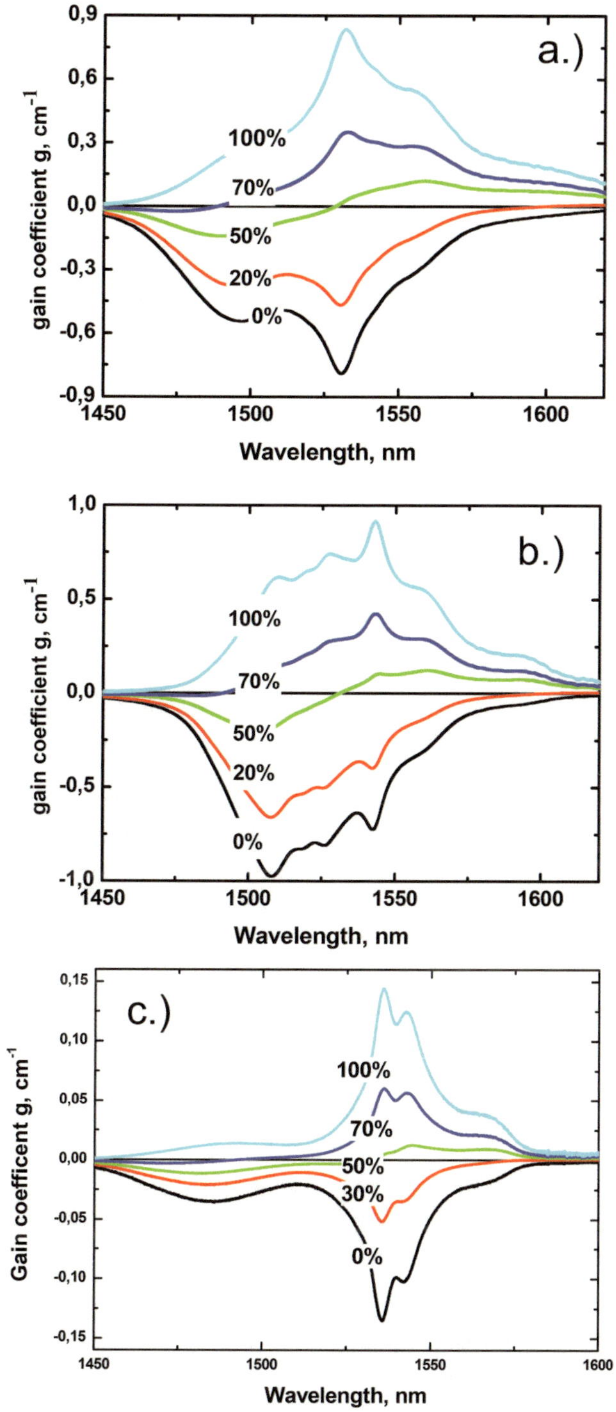

Fig. 10. Gain/loss spectra of initial glass (a), glassceramics (b) and commercial silicate glass (c), doped with erbium, for various pumping ratio N_2/N_{Er}. Pump wavelength 980 nm.

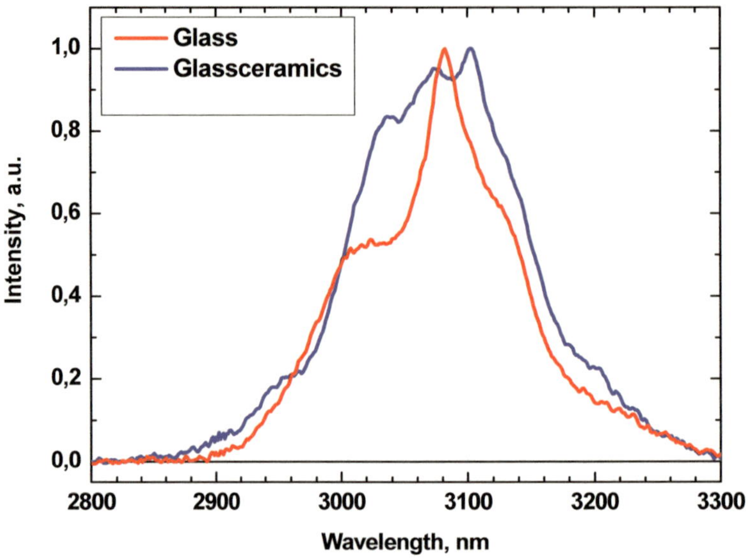

Fig. 11. Luminescence spectrum of the erbium ion in the starting glass and the glassceramics.

In today industry the most widely used (~ 60% of the overall measurements) are the temperature measurements. Wide range of measured temperatures, a big variety of measurement conditions and the requirements to the measuring devices, provide, on one hand, the big variety of the temperature measurement tools, but on the other hand require the development of the novel types of sensors and primary converters, which meet the constantly growing requirements to the accuracy, response and noise protection.

One can separate the big variety of devices for temperature measurement into two big categories - the electrical ones and optical. The optical sensors are used, for instance, for measuring the temperature at the remote object and in the intense electro-magnetic fields. One cannot provide it by means of the electric thermal converters due to high probability of "electric breakdown".

There exist three main types of the optical temperature sensors: the Fabry-Perot optical fiber interferometers; pyrometers - the contact-less sensors, implementing the radiation from the heated body; and the luminescent ones based on luminescence variation during heating.

The use of color pyrometers is limited by the fact that the body, heated to approximately 400 K, is radiating mainly in the mid-IR spectral range. Due to low sensitivity of majority of sensors in this spectral range, data collection is thus tantalized.

The use of fiber-optical interference sensors, based on the Fabry-Perot interferometer, comprised by the fused silica fiber is limited due to a comparatively high cost and to a complicated design, because the shift of the resonant wavelength is relatively small and requires the precise measurements. In addition, the interferometer quality degrades with time, especially under the high temperature action.

The luminescent sensors are the most prospective ones, because they combine the simple design and high accuracy (± 0,05 K). The majority of such sensors are based upon the following principles:

- the temperature shift of the luminescence band maximum across the spectrum;
- the temperature re-distribution of the energy across the excited levels, leading to redistribution of radiation intensity in the neighboring luminescence bands with temperature;
- changes in the luminescence quenching with temperature.

These effects are observed for the rare earth ions like erbium, neodymium, and dysprosium, and for transient metals like chromium, which are contained in the crystalline or glass-like state.

The sensors on the base of temperature shift of the luminescence band maximum across spectrum have not very high accuracy of temperature measurement (± 5K), in addition, they have rather large inertia (several seconds). Hence today they are not widely used and were replaced by the sensors of other kinds.

The advantage of the sensors on the base of temperature redistribution of energy across the excited levels is their high accuracy of temperature measurement up to ± 0.05K. Such an accuracy is achieved by the use of optical materials with the low phonon (<900 cm^{-1}) spectrum like fluozirconate or telluride glasses. It increases the intensity of radiation of the temperature-tied levels due to the up-conversion filling of the upper excited manifolds and thus characterized by high radiative probabilities.

The disadvantage of such devices is their small temperature range T = 7-500 K. For instance, the fluoride glasses like ZBLAN have a low temperature of softening and melting, while the telluride glasses reveal the high tendency to crystallization with the temperature growth. One can also treat as a disadvantage the high refraction index (n> 2.0), which tantalizes significantly the connection with the standard fused silica fiber (n = 1.47). In addition, these materials have high cost due to the use of the oxygen-less atmosphere (as in the case of the fluoride glasses) or to the toxic nature of raw materials (in the case of telluride glasses), which require some protection measures during the synthesis.

The advantage of the temperature sensors, based on the variation of the luminescence quenching time with temperature is their small inertia. So one can measure the temperature for the short period of time (several ms) and in the wide spectral range (T = 77-1000 K). The sensors have simple design and low cost. Disadvantage of such sensors is a low accuracy of temperature measurement (> ± 1 K). Such sensors employ the crystals of YAG:Cr or YAG:Dy as an active medium. The use of crystalline media provides additional complexity to the design, because such sensors are to combine the optical fibers for delivery of the exciting radiation and for collection of the registered signal, and of the active crystal, used as a sensitive element.

In such a design the thermally sensitive active crystal is connected with an active crystal by gluing. The length of active crystal is limited by the value of several millimeters, and thus its connection with the optical fiber (glued zone) is usually positioned in the zone of high temperature or of the high temperature gradient. This makes an effect onto the sensors reliability during their exploitation under the high temperatures (more than 200°C). It is impossible to replace such a design by a "purely fiber-optical one", containing only of the active crystalline fiber, because the technology of fibers production from the said crystals are yet not developed.

The active medium, which is used in such sensors, is to meet the following requirements: the luminescence has to be excited easily; they are to be easily connected with the fused silica fibers; the radiation is to fill into the spectral range, which is convenient for registration etc.

Erbium ions are often used as the luminescence ion, because they has two thermally tied manifolds $^2H_{11/2}$ и $^4S_{3/2}$. The energy gap between these two manifolds is relatively small, and the temperature growth can redistribute the energy between these levels (see Fig.12). One can excite erbium ions via ytterbium ions, which are the efficient sensitizer, and its absorption bands correspond to the spectral range of high power laser diodes (900-1000 nm).

The environment of the erbium ions strongly influences its luminescence properties, and hence the proper choice of the matrix makes it possible to increase the intensity of luminescence from the levels $^2H_{11/2}$ and $^4S_{3/2}$, which will increase the sensitivity of the sensors on this base.

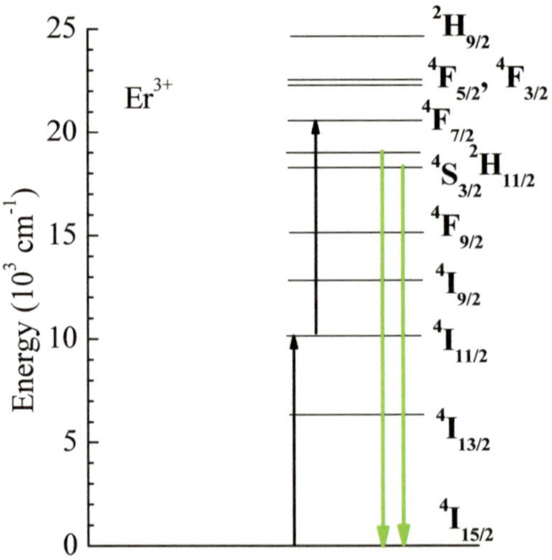

Fig. 12. The scheme of energy transfer from the Er^{3+} ion manifolds

The best intensity of luminescence from these levels is provided by erbium ions, placed in the oxygen-free environment – the fluoride, for instance. In such matrices the quantum yield of luminescence s increased, and the efficiency of up-conversion processes is also increased.

One can treat the glassceramics, which contain in their volume the fluoride nanocrystals – like CaF_2, PbF_2 etc. – as the prospective media for the luminescent temperature sensors. In present paper it has been shown the possibility to use the transparent glassceramics on the base of lead-fluoride nanocrystals, containing the rare-earth ions as the active medium for the new generation of the luminescent temperature sensors. In such glassceramics the rare-earth ions playing the role of nucleation centers for the crystalline phase, i.e. they are included into the crystal. In addition, the activated glassceramics can be used for the optical fiber production, solving thus the connection problem. It can be solved, for instance, by the way of welding of the activated and nonactivated fibers, improving thus the sensor stability to fast changes of temperature and to aggressive environment.

Change of the temperature results in changing the population of the erbium manifolds $^2H_{11/2}$ and $^4S_{3/2}$; in other words, takes place re-distribution of the intensity of luminescence bands with maxima at 522 and 547 nm. So, if the luminescence spectra for definite temperatures are known, one can make a conclusion about the object temperature. The temperature growth results in increase of excitation migrations; as a consequence, the number of the radiation-less transitions is growing and grows the probability of luminescence quenching at admixtures. It explains the gradual decrease of radiation intensity at 547 nm and accompanying it increase of intensity at 522 nm (Fig.13).

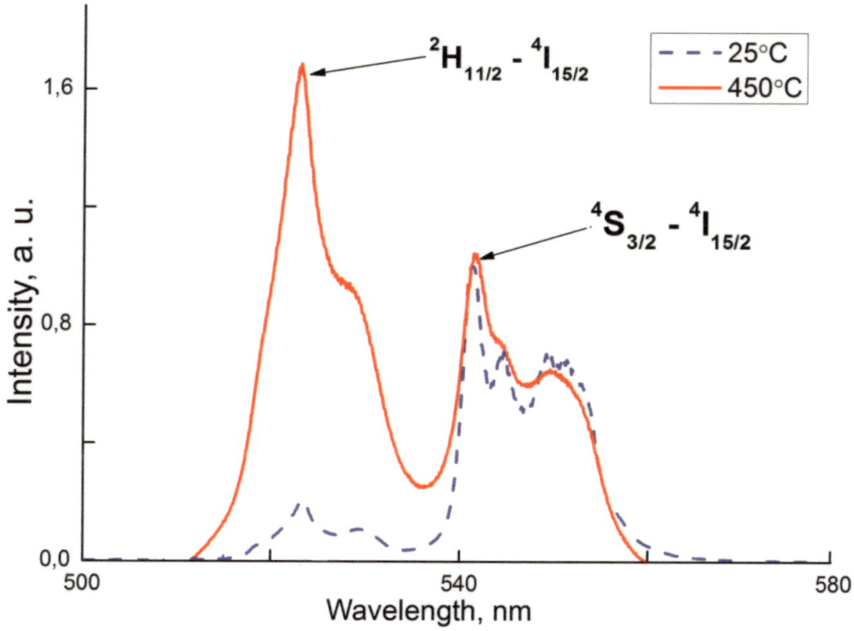

Fig. 13. Temperature dependence of the luminescence spectrum shape of the erbium-doped nanoglassceramics.

In the Fig.14 is shown the evolution of the luminescence spectrum of the erbium activated specimen of lead-fluoride nanoglassceramics with temperature. The shape of the luminescence spectra is explained by the Stark splitting of the erbium manifolds $^4S_{3/2}$ (547 nm) and $^2H_{11/2}$ (522 nm), caused by erbium transfer to the crystalline phase after the secondary heat treatment. If the temperature is increased, the maxima of the peak intensities are shifting one with respect to another, i.e. heating results in thermal redistribution of excitation between the transitions $^4S_{3/2} \rightarrow {}^4I_{15/2}$ and $^2H_{11/2} \rightarrow {}^4I_{15/2}$.

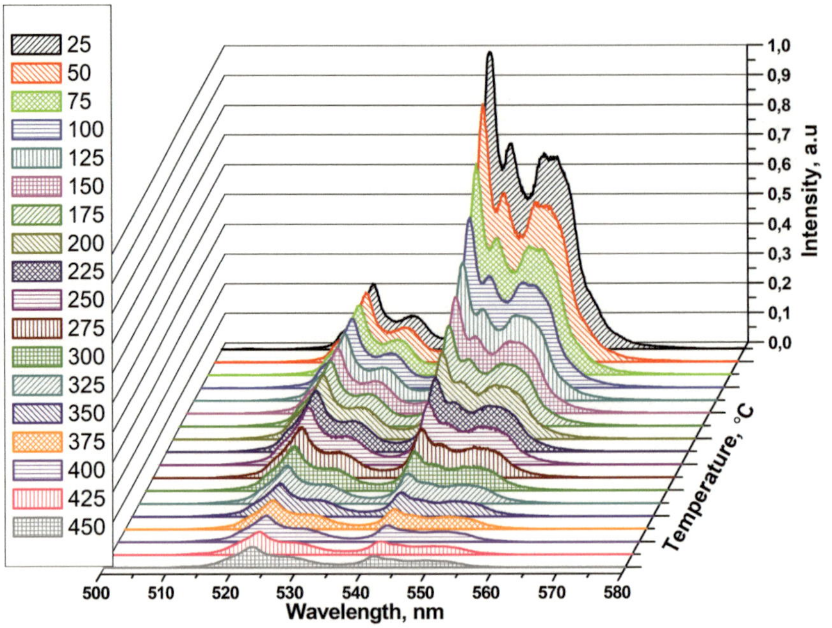

Fig. 14. Evolution of green erbium luminescence from the lead-fluoride nanoglassceramics specimen with temperature.

Let us consider problem of temperature sensitivity of the activated material and hence of the temperature sensor. Speaking of the temperature sensitivity we mean the variation of ratio of luminescence peaks at the selected wavelengths for temperature variation in 1 K. The larger is this variation, the larger is the sensitivity. It more evident from the Fig.15: here the specimen 2 is more sensitive to the temperature variation, than the specimen 1, because $\Delta_2 > \Delta_1$, i.e. the same temperature variation results in larger variation of the signal.

The increase of erbium ions concentration in glass and in nanoglassceramics leads to growth of the luminescence intensity across the overall observed temperature range (Fig.16). The luminescence intensity also grows up across the overall range of measured temperatures with the increase of the time of heat treatment of the erbium activated lead-fluoride nanoglassceramics (Fig.17).

Hence, varying the volume percentage of the crystalline phase by means of variation the time of heat treatment and also of erbium concentration one can control the curve slope, i.e. to control the temperature sensitivity of glassceramics. The dependencies, which were measured, prevent their exponential character across the overall temperature range, i.e. they are predictable and thus they can be used as a graduation curves for temperature evaluation. Hence the knowledge of ratio of the intensities of two erbium transitions makes it possible to determine the material temperature.

The temperature dependencies for the peak ratio for the starting glass and for the glassceramics are the good evidence that the lead-fluoride nanoglassceramics is the most sensitive to the temperature changes. One can treat this material as the prospective medium to be used in the fiber-optical luminescent temperature sensors. In comparison with known luminescent sensors the temperature sensor, made from the novel lead-oxyfluoride nanoglassceramics, will have a much shorter response time (tens of milliseconds), higher accuracy (±0.5K) and the rather wide range of measured temperatures (77-1000 K).

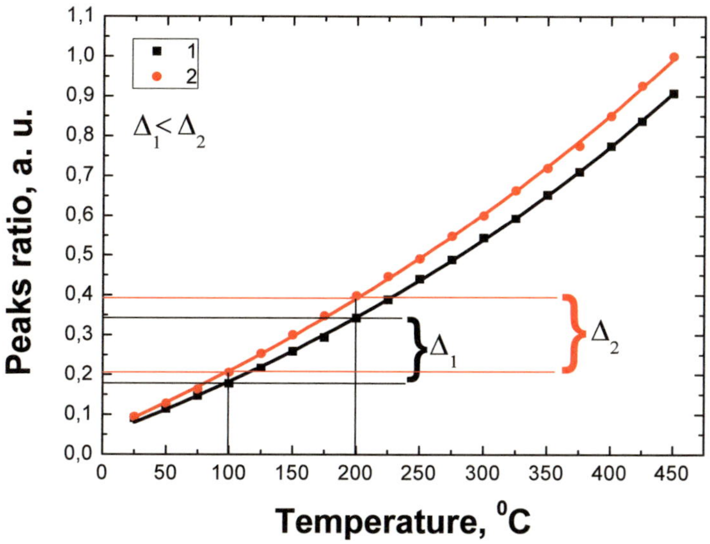

Fig. 15. Determination of the temperature sensitivity of the material.

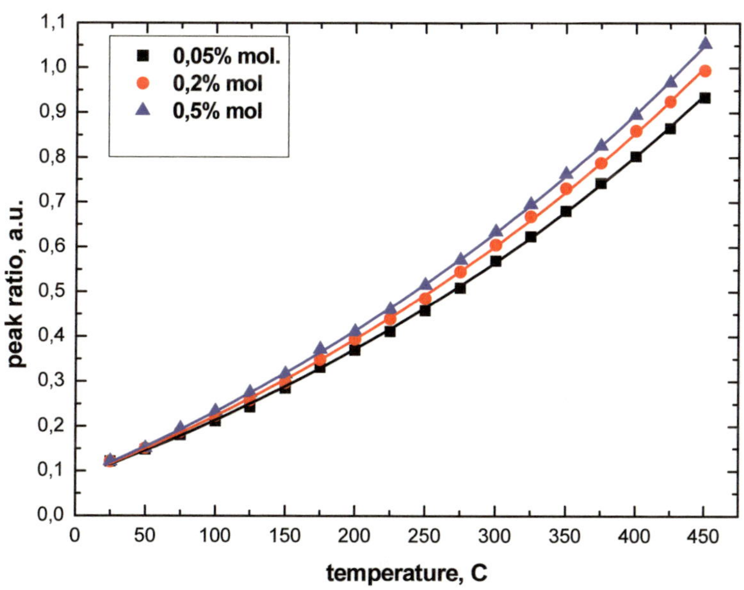

Fig. 16. Temperature dependence of ratio of luminescence peaks in the lead-fluoride nanoglassceramics for various erbium ions concentration and for the same duration of heat treatment (t = 10 h).

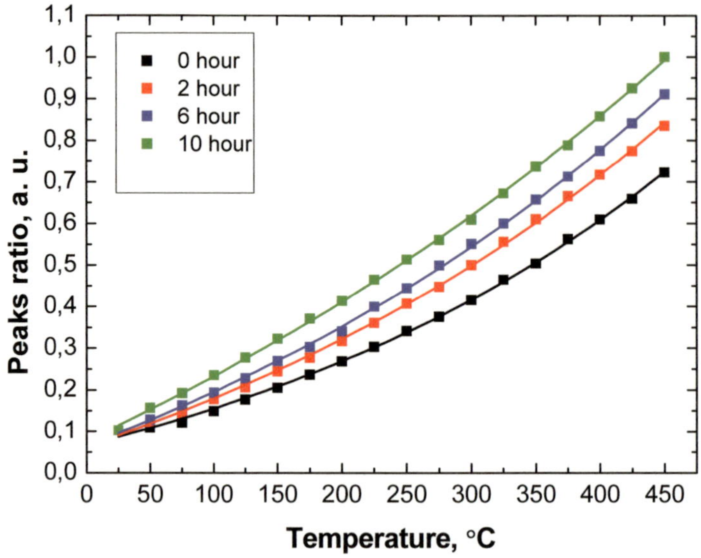

Fig. 17. Temperature dependence of the ratio of erbium luminescence peaks (concentration 0.2 mol %) of the starting glass and of the lead-fluoride nanoglassceramics with the different duration of heat treatment (t = 2, 6, 10 h).

5. Application of new lead-oxifluoride nanoglassceramics for luminophors of LEDs

The chapter demonstrates application of the new nanoglassceramics as a luminophor for light emitted diodes (LED). Also, some characteristics of the luminophor are compared with traditional ones.

Today there exist four main approaches to a white light diodes realization:

1. Fabrication of light emitting diodes (LEDs) on the base of ZnSe semiconductor. The structure comprises the blue ZnSe LEDs, "grown up" on the ZnS substratum. In this case the active region of conductor emits the blue light, while the substratum emits yellow. The white light ZnSe LEDs have some advantages. They are working at the voltage 2.7 V and are very stable with respect to static discharges. They can emit the light for the wide range of color temperatures (3500-8500 K). One can thus fabricate the devices with the "warmer" emission. They also have some disadvantages: they have a short lifetime, high electric resistance and still have not found the wide commercial application.
2. RGB mixing of colors. Blue, red and green LEDs are positioned tightly on one and the same matrix. Their radiation is mixed by some optical system – for instance, by a lens. As a result a white light is obtained. The main problem in this case is realization of

green LEDs as well as relatively shorter lifetime of blue diodes. Additional problems arise with technique of mixing the beams, increasing thus the cost of produced LEDs. RGB matrices are widely used in the light-dynamic systems. In addition the large number of LEDs in matrix provides high integral light flow and the high axial light intensity. However, due to the optical system aberrations the light spot has different colors in its center and at the edges. In addition, due to the non-uniform heat sink from the matrix edge and from its center. The LEDs are subjected to different heating. As a result, their color changes due to aging in a different manner and thus the integral color temperature and color are "drifting" during exploitation. Compensation of this unpleasant effect is rather complicated and expensive. The technology of light mixing makes it possible not only to obtain the white light, but also to move across the color diagram by means of changing the current through various LEDs. Special feature of RGB LEDs is the possibility to obtain not only the white light, but also a big variety of light colors with the use of addressing control; one can also obtain various color effects.

3. One can depose three luminophors, emitting red, green and blue light, onto the UV LED surface. Such a LED has a disadvantage - a technology problems with deposition of luminophor layers, as well as luminophor degradation under the UV radiation action. The phosphor-converted LEDs are much cheaper than the RGB LED matrices (per light flow unit) and provide the good white light. Other disadvantages of this type are: (i) the low light efficiency (less than that of RGB matrices) due to light conversion in the luminophor layer; (ii) the problems with homogeneity of luminophores deposition in technology process, and (iii) the degradation of luminophors with time.
4. One can use the yellow-green, green and red luminophors, deposited onto the blue LED. As a result, two or three radiations are mixed, providing the white or nearly white light. Today most usually are used InGaN based blue LEDs with luminophor, comprised by the nanopowder of yttrium-aluminum garnet, activated by cerium, emitting in a yellow-green spectral range. The luminophor - the powder of YAG:Ce^{3+}, deposited onto the diode surface, is covered by the protective polymer. The advantages of the polymer medium are the simple technology and the low cost. However, the polymer technology has its own limitation in the case of white LEDs. The polymer has low thermal conductivity, and thus the high-power diodes with the structure "YAG:Ce^{3+}-powder luminophor plus polymer coating" meet the heat sink problems. In addition, the polymers degrade during the long-term thermal and light action.. Hence in the case of high power white LEDs there exists the problem of replacement of the unstable polymer matrices by stronger ones.

The blue LEDs with YAG:Ce^{3+} powder as luminophor emit the so-called "cold" white light, because their radiation does not cover the overall visible range. This "cold" white light is not always comfortable for eyes, and this is the second disadvantage of such material as luminophor type. One can shift the color temperature range towards the "warm" white light by means of introducing the additional red band into the radiation spectrum. Thus one has to add to the LED of the "cold" white light - InGaN, emitting with the spectral maximum at 450 nm, accompanied by the yellow-green YAG:Ce^{3+} luminophor with the spectral maximum at 570 nm - some new component, introducing into the spectrum the red component with the maximum at 600-650 nm. The most prospective for this task are, for instance, the materials with the europium ion admixture. However, this type of LEDs has a significant disadvantage - the rather narrow bands of absorption by cerium ions in the nanocrystals of the yttrium-aluminum garnet. In the case of mass production there exists the variation of the luminescence band maximum due to the technology reasons. In this case is possible only partial overlapping or even mismatch of the bands of LED emission and cerium absorption, providing the significant influence onto the white LED emission intensity.

One can use this problem by use of other luminophor activators - for instance, terbium ions with the emission band maximum at 550 nm and ions of three-valence europium, which introduces into the blue LED emission the red component. As a result, the mixing of blue, green and red light takes place; such an approach is similar to RGB technology.

In present work we have tried to solve the mentioned problems by means of the novel lead-oxyfluoride nanoglassceramics, activated europium and terbium.

In the Fig.18 are shown the luminescence spectra of the lead-oxyfluoride nanoglassceramics, activated by three valence europium ions with various concentration under the excitation by the commercial blue LED with the additional YAG:Ce^{3+} luminophor. In this case the nanoglassceramics plays the role of the additional luminophor.

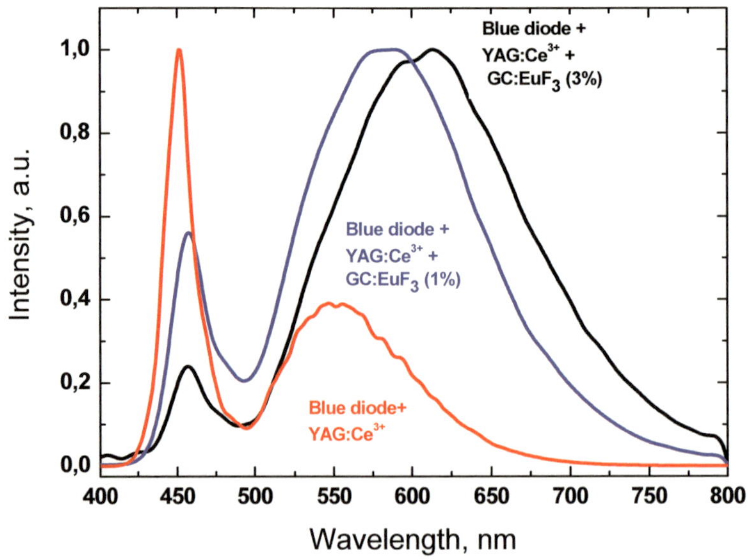

Fig. 18. Luminescence of Ce^{3+} in YAG powder and joint luminescence of Eu^{3+} ions with different concentrations in the glassceramics (GC) and of Ce^{3+} ions in YAG powder under excitation by the commercial blue diode.

The use of europium ions results in arises of the additional bands in the red spectral range. As a result there is observed the wide integral band of Ce^{3+} and Eu^{3+} luminescence with the maximum at ~580 nm. The increase of Eu^{3+} concentration leads to additional shift of the integral luminescence band maximum to 620 nm. Hence the color temperature is shifting towards higher values (6500 K), i.e. to transfer from the "cold" white light to a "warm" one.

The use of lead oxyfluoride nanoglassceramics, activated by other ions, for instance by terbium ions (green luminescence) and europium ions (red luminescence) makes it possible to use only one luminophore and to prevent the problems, caused by the use of polymer coating and of cerium ions, which have the narrow absorption bands (Fig.19).

The luminescence spectrum now reveals the peaks, produced by luminescence of the Tb^{3+} ions with the maximum at the wavelength 546 nm (transition $^5D_4 \rightarrow ^7F_5$) and of the Eu^{3+} ions with the maximums at 595, 615 and 700 nm (corresponding transitions $^5D_0 \rightarrow ^7F_0$; $^5D_0 \rightarrow ^7F_2$; $^5D_0 \rightarrow ^7F_4$). In a combination with the blue LED without YAG:Ce^{3+} luminophor these peaks provide the "warm" white light. One has to note specially that the nanoglassceramics, activated by europium and terbium, reveal high thermal and optical strength, and their use as a single luminophor makes it possible to solve the heat sink problems, usual for the high power LEDs. So one can treat the novel oxyfluoride nanoglassceramics, activated by the rare earth ions, as the prospective medium to be used as the luminophor in the high power white light LEDs.

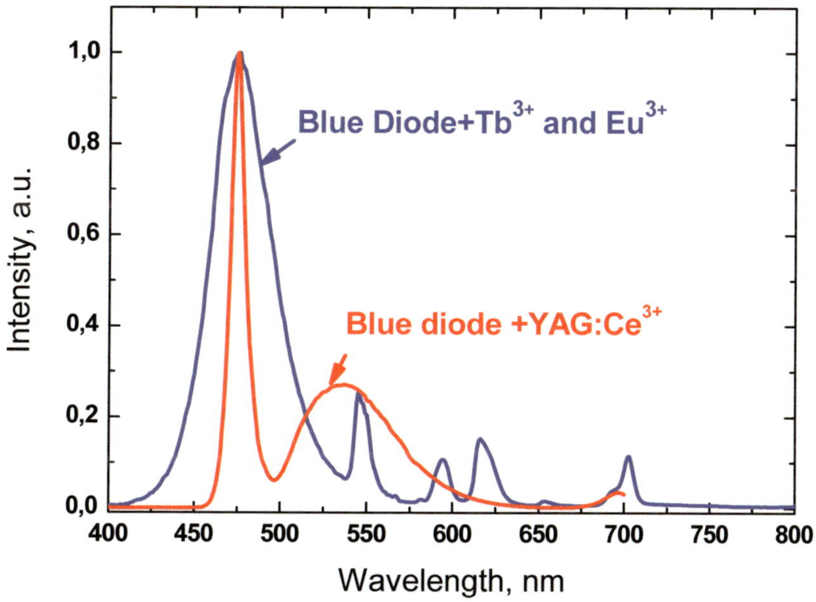

Fig. 19. Luminescence of the nanoglassceramics (GC), activated ions of Tb^{3+} and Eu^{3+} and of the YAG:Ce^{3+} powder, excited by the commercial blue diode.

6. Conclusion

Today the novel optical materials are determining the progress in photonics. In this paper it has been shown that one of prospective directions in the field of optical material science for photonics is the development of the nanostructurized glassceramic materials. One example of realization of a novel optical material - the transparent lead-oxyfluoride nanoglassceramics, doped with the rare-earth ions - has been presented. It was shown that due to spontaneous crystallization in a glass are growing the new crystalline phases – the yttrium-oxyfluoride of lead, the lanthanide oxyfluoride of lead and yttrium-lanthanide oxyfluoride of lead. The size of crystalline phase is 15-40 nm, providing thus transparency of nanoglassceramics in a visible and near-IR spectral ranges and thus putting it into the class of optical materials. One can also treat the novel optical nanoglassceramics as a multi-functional material, because it can be used in various fields of photonics. Several examples of the novel material applications in photonics have been demonstrated. Moreover, the features of the novel material with the known analogs have been compared. For instance, it was shown that the novel material can compete with traditional materials in production of fiber-optical lasers and amplifiers, working at 1.5 and 3 μm. It was shown that the novel material can be successfully used for the temperature sensors, including fiber-optical ones. It has been also demonstrated that the novel material can be successfully used as a luminophor for white light LEDs.

7. References

[1] T.P. Guerreiro, N.F.Borelli, J.Butty, N. Peygbhambarian, PbS quantum-dot doped glasses as saturable absorbers for mode locking of Cr:forsterite laser, Appl. Phys. Lett., 1997, V.71, № 12, P.1595-1607.

[2] N.F. Borelli, D.W. Smith, Quantum confinement of PbS microcrystals in glass, J. Non. Crystalline Solids, 1994, V.180, P.25-31.

[3] Ekimov, Growth and optical properties of semiconductor crystals in glass matrix, J. of Luminescence, 1996, V.70. P.1-20.

[4] E. V. Kolobkova, A. A. Lipovskii, N. V. Nikonorov, Nonlinear properties of phosphate glasses doped with CdS, CdSe and CdS_xSe_{1-x} microcrystals, Optics and Spectroscopy, 1997, V. 82, №3, P.390-392.

[5] K. V. Yumashev, A. M. Malyarevich, N. N. Posnov, V. P. Mikhailov, A. A. Lipovskii, E. V. Kolobkova, V. D. Petrikov, Nonlinear spectroscopy of phosphate glasses containing cadmium selenide nanoparticles, J. of Quantum Electron, 1998, V.28, №8, P.715–718.

[6] E. V. Kolobkova, A. A. Lipovskii, V. D. Petrikov, and V. G. Melekhin. Fluorophosphate glasses with quantum dots based on lead sulfide, Glass Physics and Chemistry, 2002, V. 28, № 4, P. 251–255.

[7] A.A. Lipovskii, V.D. Petrikov, V.G. Melehin, P. Lavalard, C. Laermans, Study of growth kinetics of CdTe nanocrystals in germanate glass by optical spectroscopy, Phys. Low -Dim. Structure, 2002, V.3, № 4, P. 51-60.

[8] A.V. Dotsenko, L.B.Glebov and V.A. Tsekhomsky, Physics and Chemistry of Photochromic Glasses, 1998, CRS Press, New York, 190 p.

[9] T. R. Dietrich et al. / Microelectronic Engineering, 1996, V.30, P.497-504.

[10] C.J. Anthony, P.T. Docker, P.D. Prewettand, K. Jiang, Focused ion beam microfabrication in Foturan™ photosensitive glass, J. Micromech. Microeng. 2007, V.17, P.115–119.

[11] See:http://www.mikroglas.com/foturane.htm.

[12] G. H. Beall, Polychromatic glasses and method, US Patent 4309217, 1982.

[13] V.A. Aseev, N.V. Nikonorov, Spectroluminescence properties of photo-thermo-refractive nanoglassceramics doped with ytterbium and erbium ions, J. Optical Technologies, 2008, V.75, №12, P.81-85.

[14] F. Auzel, D. Pecile, D. Morin, Er^{3+} doped ultra-transparent oxy-fluoride glass-ceramics for application in the 1.54 μm telecommunication window, J. Electrochem. Soc. 1975. V.122, P. 101-108.

[15] Y.H. Wang, J. Ohsaki, New transparent vitroceramics colored with Er^{3+} and Yb^{3+} for efficient frequency up-conversion, J. Physics letters, 1993, V.63, №24, P.3268-3270.

[16] V.K. Tikhomirov, V.D. Rodriguez, J. Mendez-Ramos, P. Nunez, A.B. Seddon, Comparative spectroscopy of $(ErF_3)(PbF_2)$ alloys and Er^{3+} - doped oxyfluoriode glass-ceramics, Optical Material, 2004, V. 27, P.543-547.

[17] G. Dantelle, M. Mortier,G. Patriarche,D. Vivien, Comparison between nanocrystals in glass-ceramics and bulk single crystals Er^{3+} - doped PbF_2, J. of Solid State Chemistry, 2006, V. 179, P. 2003-2006.

[18] M. Beggiora, I.M. Reaney, A.B. Seddon, D. Furniss, S.A. Tikhomirova, Phase evolution in oxy-fluoride glass ceramics, J. Non-Crystalline Solids, 2003, V. 326-327, P.476-483.

[19] M. Mortier, P. Goldner, C. Chateau, M. Genotelle, Erbium doped glass–ceramics: concentration effect on crystal structure and energy transfer between active ions, Journal of Alloys and Compounds, 2001, V 323-324, P. 245-249.

[20] V.A. Aseev, V.V. Golubkov, A.V. Klementeva, E.V. Kolobkova, N.V. Nikonorov, Spectral luminescence properties of transparent lead fluoride nanoglassceramics doped with erbium ions, Optics and Spectroscopy, 2009, V.106, № 5, P.691–696.

[21] E. V. Kolobkova, V. G. Melekhin and A. N. Penigin, Optical glass-ceramics based on fluorine-containing silicate glasses doped with rare-earth ions, Glass Physics and Chemistry, 2007, V.33, № 1, P.8–13.

[22] E.V. Kolobkova, N.O. Tagil'tseva, P.A. Lesnikov, Specific features of the formation of oxyfluoride glass-ceramics of the SiO_2-PbF_2-CdF_2-ZnF_2-Al_2O_3-Er(Eu,Yb)F_3 system, Glass Physics and Chemistry, 2010, V.36, №3, P.317-324.

[23] V.A. Aseev, V.V. Golubkov, E.V. Kolobkova, N.V. Nikonorov, Lead oxifluoride lanthanides in glass-like matrix, Glass Physics and Chemistry, 2012, V.38, №1, P11-18.

[24] R.W. Hopper, Stochastic theory of scattering from idealized spinodal structures: II. Scattering in general and for the basic late stage model, J. Non-Cryst. Solids, 1985, V.70, №1, P.111-117.

6

Fianite in Photonics

Alexander Buzynin
A. M. Prokhorov General Physics Institute, Russian Academy of Sciences
Russia

1. Introduction

The further progress in photonics, as well as in many other technological fields is connected with application of new materials. Fianite is the material of such kind. Fianites are single crystals of zirconia- or hafnia-based cubic solid solutions with yttrium, calcium, magnesium or lanthanides (from gadolinium to lutetium) stabilizing oxides (ZrO_2 (HfO_2)$\cdot R_2O_3$, where R - Y, Gd ... Lu). Industrial technology of synthesis of fianite has been for the first time developed in Russia in the Lebedev Physical Institute of the Russian Academy of Sciences (FIAN in Russian), as has entitled crystals[1, 2]. Serial production of the crystals has been already started in the early seventies of XX century [3-5]. Currently, fianite crystals are in the second position by the volume of worldwide production following silicon. Fianite single crystals – zirconia-based solid solutions (or "yttrium stabilized zirconia" - YSZ) are widely known worldwide as jewelry material (fig. 1).

Fig. 1. Great color variety of the crystals combined with unique optical properties makes fianite single crystals a promising material for jewellery, arts and Crafts (left); fianite substrates 3" in diameter (right).

Recently, in the countries with the developed microelectronics a significant growth of interest to various aspects of fianite application in semiconductor technologies has been observed. Fianite is an extremely promising multipurpose material for new optoelectronics technologies due to its unique combination of physical and chemical properties. It can be used in virtually all of the main technological stages of the production of micro-, opto- and SHF-electronics: as a bulk dielectric substrate, a material for buffer layers in heteroepitaxy; a material for insulating, antireflection, and protective layers in the devices and as a gate dielectric [6-22].

The use of fianite, as well as ZrO_2 and HfO_2 oxides instead of SiO_2 as gate dielectrics in CMOC technology, which can be considered for microelectronics as a basic one, is of peculiar interest [14, 15]. That is associated with the increase of leakage currents by the increase of the integration level when conventional SiO_2 is used. Therefore, a change of SiO_2 over dielectrics with higher values of dielectric constant (high-k-materials) is required. Due to higher value of dielectric constant (25÷30 for fianite [4, 14, 15] instead of 12 for SiO_2) it is possible to provide the same electric capacity using much more thick layers of the gate oxide.

A number of modern aspects of the application of fianite in photonics are analyzed in this chapter.

2. Techniques for the synthesis of fianite crystals Fianite substrates

Peculiarities of the synthesis, the investigation techniques and properties of the crystals have been considered in details in [3-5]. In this chapter only brief information concerning synthesis of the crystals and manufacturing the substrates is presented. A novel laser technique developed for instant monitoring of defects in the substrates and in bulky fianite and sapphire crystals is also considered

2.1 Crystal growth of fianites using installations with cold containers of 130-700 mm diameter

The growth technique of the crystals was elaborated and developed using following installations: "Crystal -407" (5.28 MHz frequency, 60 kW power, ØCC 130mm); "Crystal -403" (1.76 MHz, 160 kW, ØCC 400mm); "Crystal -403M" (0.4-0. 88 MHz, 600 kW, ØCC 700mm) (Fig.2).

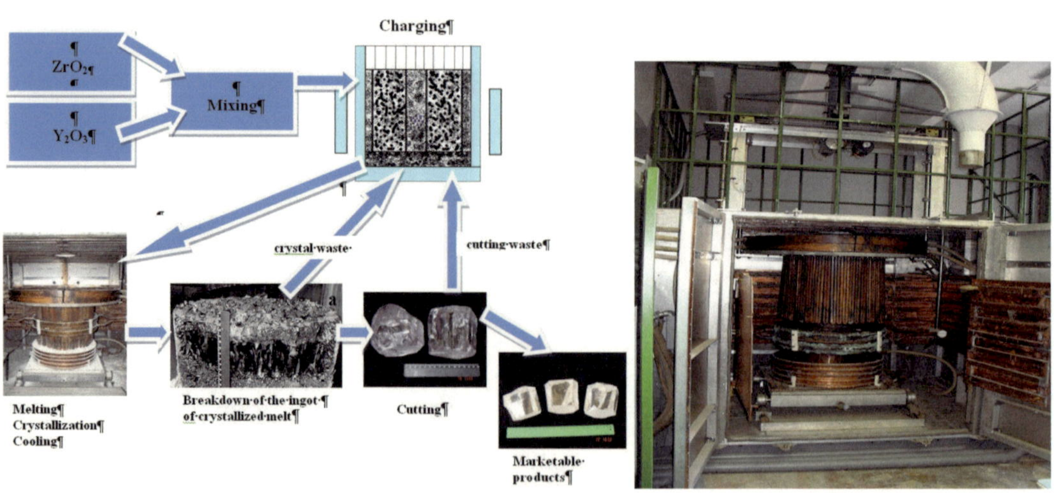

a b

Fig. 2. The scheme of manufacturing of zirconia-based crystals (a); Installations for direct RF melting of dielectric materials in a cold container (CC) "Crystal -403M" (0.4-0. 88 MHz frequency, 600 kW power, ØCC 700mm) (b).

2.2 Technology for mechanical machining of fianite crystals

Fianite lend itself to a machining considerably readily, similarly to sapphire crystals. Dislocation densities in ZrO_2 - (8-20) mol% Y_2O_3 crystals have been measured:

in central parts - 10^3 cm^{-2}
in periphery - 10^5cm^{-2}

Followed by annealing (2100^{0}C, vacuum) dislocation density decreased to 10^2 -10^3cm^{-2}.

Pre-epitaxial treatment of surface of the substrates. With the purpose to guarantee optimal physical-chemical state of fianite substrates various techniques and conditions of pre-epitaxial treatment have been studied. Treatment at 1000-1400°C temperatures in air during 1-4 h was used as one of such techniques. The high-temperature annealing provides a relief of stress occurred in the surface layer at mechanical treatment, removal of impurities from the surface and increasing of phase and structural perfection.

The effect of high-temperature annealing on surface quality of the substrates has been studied. In Fig.3a scratches occurred in course of polishing of the substrate by ACM 1/0 diamond paste are apparently observable. The following annealing (1250^0 C in air) did not result in smoothing of the relief as a whole but caused re-structuring of the surface layer (Fig. 3b) and flatten the relief in micro-locations at scratch residues and results in 2-3-fold decrease of high bump- valley drops (Fig. 3, right side)

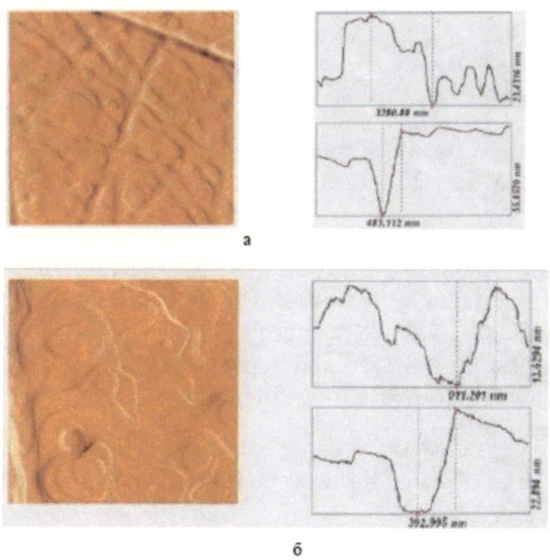

Fig. 3. AFM image of fianite substrate (left) and surface profile (right) after chemical-mechanical polishing (a) and subsequent high temperature annealing (b).

The studies of the effect of thermal treatment of the substrates on roughness of polished surface have shown that high-temperature annealing (1250-1400^0 C) conducted following chemo-mechanical treatment promoted an increase of structural and phase homogeneity of the surface of zirconia-based crystal substrates.

Half-width of the rocking curve (HWRC) is another parameter featuring quality of a substrate surface. The HWRC values significantly decreased due to the annealing.

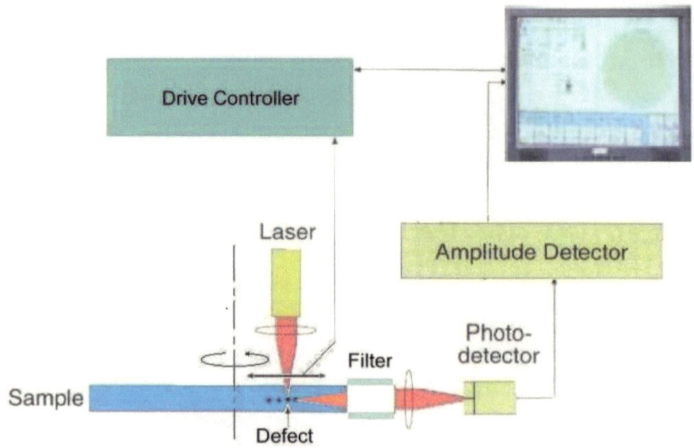

Fig. 4. Draft-scheme of the apparatus for laser control of sapphire and fianite wafers and bulky crystals

The technique for production of epi-ready substrates has been developed. The epi-ready substrates of 2″ and 3″ diameter have been manufactured from zirconia-based crystals. The polished surface was mirror-flat, scratches, etching pits, fractures and other defects are absent. The profilograms showed no noticeable deviations from R_2 - 5 nm height. Typical roughness σ values for the epi-ready polished surface of fianite are less than 0.5 nm.

2.3 New laser technique for express-monitoring of the defects in a volume of wafers and crystals - Development of the technique and equipment

The Laser technique is based on a laser emission at the wavelength, which coincides with the region of transparency of sapphire and fianite, for example, radiation of CO_2-laser in the middle IR range ($\lambda \sim 5.4$ μm).

Such laser radiation readily penetrates through the flawless regions of fianite and sapphire, and scatters on impurity inclusions within fianite volume and micro-bubbles within sapphire volume. The scattered part of radiation passes through the filter and is registered by a photodetector arranged perpendicularly to the direction of the laser beam. Subsequent computer processing of the signal provides information on the defects.

Besides information on presence or absence of the defects in a volume of wafers and crystals, it is also possible to observe its two- or three-dimensional distribution over square of a plate and volume of a crystal.

This technique allows:

- monitoring of all defective wafers and separating really defective wafers with micro-bubbles (typical defects of sapphire) and impurity inclusions (typical defects of fianite) from tentatively defective ones with surface defects only, those which may be relieved by special treatment (thus increasing the yield);
- selecting defective segments in the bulk crystals for the most economic scribing.

The data obtained can be sorted and saved in a computer memory for the subsequent analysis of reasons of the defect formation in bulk crystals with the purpose of the technology improvement.

The laser technique has the following advantages. It is: contactless, nondestructive, express, rather easily automated, technological and easy in use. So it is convenient for the use in commercial production with hundred percent inspections of the products and for the solution of some scientific and technological problems.

Development of the laser control equipment. Draft-scheme of apparatus for laser control of sapphire and fianite wafers and bulky crystals is shown in Fig. 4.

The laser control apparatus has the following characteristics:

- Apparatus sizes are - 1м x 1м x 1.5 м;
- Wafer diameter - from 0 up to 2" (can be increased up to 6"and more);
- Wafer width 0.5 - 3 mm and more, bulk crystals height up to 20 cm and more;
- The surfaces can be polished, finely ground or roughly ground;
- Time for the test of one substrate (slice) with 2" diameter is less than 10 seconds, 6"- about one minute.

Fast horizontal scanning of the whole wafer is provided by its rotation and the radial movement of the light from motionless laser (the optical system is used). At tomographic investigations of bulky crystals a periodical crystal movement along vertical axis with a given step (for example 0.5 or 1 cm) is added.

Operational characteristics of the apparatus can be further improved [23, 24].

3. Fianite as a substrate and buffer layer for Si, Si-Ge and $A^{III}B^{V}$ compounds epitaxy

Fianite has a number of advantages over other dielectric materials as a substrate and buffer layer for the epitaxy of Si and $A^{III}B^{V}$ compounds [6-12, 16-18].

A wide spectral range (260–7500 nm) of the fianite transparency completely covering absorption and emission ranges of $A^{III}B^{V}$ compounds and its solid solutions makes "semiconductor-on-fianite" structures promising for the development of various optoelectronic devices with advanced characteristics, including photodiodes with Schottky barrier, photoresistors, emitting and laser diodes, avalanche photodetectors.

Thin films of fianite and related solid solutions such as $Zr(Ce)O_2$ can be used as **insulating layers** (alternative to SiO_2, SiC, Si_3N_4) in development of Si-, Ge- and GaAs-based "semiconductor-dielectric" multilayer structures. Fianite is also good **gate dielectric** for Si- as well as for $A^{III}B^{V}$ -based devices (including GaN-based) due to its high dielectric constant value (25…29.7). Thin fianite films are a barrier for diffusion of impurities and provide significant (up to 1000-fold and even more) decrease of current loss in highly-integrated devices [14, 15].Due to high chemical inertness fianite films can also be used as **protective coatings.**

The first epitaxial Si films on YSZ were grown in [6]. The first successful results on epitaxial MOCVD growth of various $A^{III}B^{V}$ compounds (GaAs, InAs, InGaAs, AlGaAs, GaAsN and GaN) on YSZ are presented in a number of studies [10, 16, 17], InN on YSZ - in [21, 22]. In [17, 18] «capillary epitaxy technique» - the new effective way of heteroepitaxy - was developed. It has been shown that the use of capillary forces in the method positively influences both on the mechanism of epitaxial growth, and on quality of $A_{III}B_V$ epitaxial films, and also reduces the minimum thickness of a continuous layer [17, 18].

An application of fianite as either monolithic substrate or buffer layer in "semiconductor-on-dielectric" technology is of peculiar importance for micro- and opto-electronics. The technology allows improving such characteristics of integrated circuits as operation speed, critical operational temperature and radiation resistance. Due to a decrease of loss of current and stray capacitance energy consumption of the devices is decreasing. Moreover, the devices based on "semiconductor-on-dielectric" structures are more reliable, especially under extreme operational conditions. Currently, "silicon-on-insulator" structures are one of the most dynamically developing directions in the field of semiconductor material science. However, electrophysical and operational parameters of the devices, as well as its radiation resistance and reliability significantly suffer because of structural imperfection of silicon layers. In case of "silicon-on-sapphire" structures the imperfection is determined, in particular, by a difference in crystallographic structure of silicon and sapphire, as well as by autodoping of a silicon film by aluminum penetrating from the sapphire substrate in concentrations up to 10^{18}–10^{20}cm^{-3}. Considering crystal-chemical and physical characteristics of fianite, the material is more preferential for the epitaxy of Si as an alternative substrate in comparison with sapphire.

In comparison with the other dielectrics, there are the following merits of fianite in application as **a substrate material and buffer layer** for Si and $A^{III}B^V$ compounds epitaxy:

- High resistivity - >10^{12} Ohm•cm at 300 K;
- Similarly to Si, Ge and $A^{III}B^V$ compounds it is of cubic structure (in contrast to hexagonal of sapphire) and has low mismatch by its lattice parameters with these compounds. In particular, the mismatching of fianite (15% Y_2O_3) with silicon is ~5,3 % (Fig. 5);
- it is possible to alter fianite cubic lattice constant in solid solutions by varying ratio of the main (zirconium or hafnium dioxide) and stabilizing oxides (yttria, rare earth oxides from gadolinium to lutetium and alkaline-earth oxides) that allows an optimum matching between substrate and cubic lattice of semiconductor films thus improving its structural perfection. For example, the values of lattice parameter mismatching between Si and fianite crystals of $(ZrO_2)_{100-x} \times (Y_2O_3)_x$ compositions are 5.7, 5.3% and 4.4% at x = 9, 15 and 21, respectively;

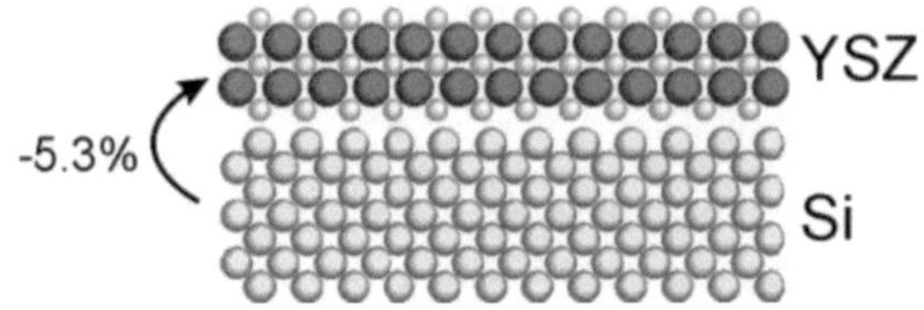

Fig. 5. Correlation of the lattice parameters of Si and fianite $(ZrO_2)_{0,85}\cdot(Y_2O_3)_{0,15}$.

- Fianite is characterized by low cation diffusion up to 1000–1200°C temperatures that reduces interdiffusion of substrate and film impurities and uncontrollable self-doping of a film (typical for sapphire) hindering the film parameters;
- Due to its excellent stability at elevated temperatures, the upper limit of the corresponding structure operational temperatures depends on physical properties of a semiconductor only. Elevated temperature is not critical for the substrate.

- Fianite is very promising material for the development of semiconductor-on-fianite structures for various optoelectronic devices with enhanced characteristics. It has broad band of optical transparency (260 to 7500 nm), which completely overlaps the absorption and emission bands of Si and $A^{III}B^{V}$ compounds and their solid solutions
- Application of thin layers of fianite on Si and GaAs instead of its monolithic substrates allows avoiding spatial limitations of the structures and decreasing the net cost. At the same time, the structures on "fianite/Si" and "fianite/GaAs" episubstrates have better heat conductivity in comparison with the structures on monolithic substrates.

3.1 Silicon-on-fianite epitaxial structures

The first studies on silicon epitaxy on fianite single crystal substrates have been carried out in France and USA [7, 8]. Silicon films on fianite substrate were deposited by chloride and hydride epitaxy at 900…1100°C. The films obtained were of polycrystalline structure and, consequently, featured with poor electrophysical parameters. However, at the same time it was shown that silicon-on-fianite structures sustaining actually all advantages of silicon-on-sapphire are free from its principal drawbacks.

At the epitaxy of Si on fianite a formation of SiO_2 intermediate layer between the film and the substrate was observed [7, 8]. Subsequent annealing of the structure leaded to the increase of SiO_2 layer thickness. It was demonstrated [8] that the layer can improve properties of silicon-on-fianite epitaxial structure because its formation:

- Removes mechanical stress in the layer-substrate interface;
- Smoothens over negative effect occurring due to a difference of linear expansion coefficients between fianite and silicon;
- Improves insulation of the integrated circuit elements (ICE) based on Si;
- Acts as a barrier for metal impurities diffusing from the substrate and forming deep levels in silicon;

The formation of SiO_2 intermediate layer at high-temperature epitaxy is associated with peculiar properties of fianite. In contrast to the other dielectrics, fianite features with a unique peculiarity as a solid electrolyte: starting from 650°C it becomes actually oxygen-transparent due to high mobility of oxygen. The reason for significant mobility of oxygen in fianite crystals is an occurrence of oxygen vacancies due to Zr^{+4} to Y^{+3} cation substitution at formation of the solid solution. High mobility of oxygen in fianite crystals is determined by an occurrence of oxygen vacancies at $ZrO_2(HfO_2)$ – R_2O_3 (here: R - Y, Gd…Yb) solid solutions formation due to $Zr^{+4}(Hf^{+4})$ to R^{+3} cation substitution. The process results in the oxygen non-stoichiometric ZrO_2 (HfO_2) based phase [4]. Because of high mobility of oxygen at high-temperature epitaxy (900…1000°C), which was used in [6-8] the formation of ether SiO_2 continuous layer or its islets between the substrate and the film was shown to be inevitable.

The phenomenon occurs even at the epitaxy initial stages when a continuous epitaxial film is forming. It was shown [9] that the formation of SiO_2 layer or isles at the initial stage of molecular-beam epitaxy on fianite results in 3-dimensional mechanism of growth, formation of structural defects and hindered the synthesis of Si films of single crystal structure. The occurrence of oxide SiO_2 isles at the initial epitaxy stages and polycentric growth of Si layers were shown possible to avoid only by using a set of techniques, those which prevent diffusion of oxygen from the substrate to the film at initial stage of the process. In particular, high structural perfection of the Si-on-fianite films was achieved by using low-temperature (T<650°C) molecular-beam epitaxy [9].

3.2 Ge and GeSi films on fianite substrates

Growth of Ge and Ge-Si heterostructures on fianite substrates was carried out using the installation shown in Fig. 6.

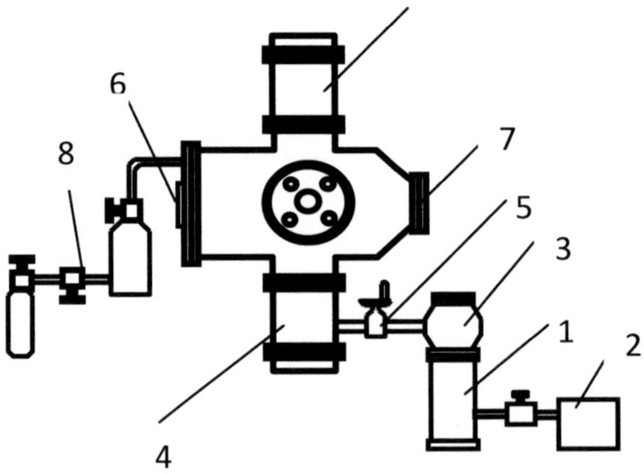

Fig. 6. Draft-scheme of ***HWCVD*** installation for $Si_{1-x}Ge_x$ growth : 1 - diffusion pump, 2 - forepump, 3 - liquid nitrogen trap, 4 - getter-ion pump, 5 – high-vacuum shuttle, 6 – charging flange for sources, 7- charging flange for substrate, 8 – system for supply of GeH_4

The growth cell comprised a stainless steel cylinder of 290 mm inner diameter and 360 mm length. There were two charging flanges in the cylinder faces, one for 3 pairs of current leads and the other for current leads for a substrate.

Base pressure in the chamber $\sim 1{\cdot}10^{-8}$ torr was maintained by pump-down using two hetero-ionic pumps. High-vacuum gate was used for isolation of the growth cell and the pumps from other parts of the vacuum system. Forepumping of the chamber was performed using diffusion pump. The diffusion pump allowed to exhaust any gas (including GeH_4) both in atomic and molecular state. FM-1 oil with low vapor pressure was used as a pressure fluid. There was a nitrogen trap above the diffusion pump preventing reverse diffusion of the oil from preevacuation and diffusion pumps into the growth cell. The (100) and (111) oriented fianite single crystal plates were used as substrates. Silicon atomic beam was maintained by sublimation of the element single crystal (high-resistance) in form of 4x4x90 mm ingot sections. The sources were mounted on the cooled current leads. There was Ta plate of 80×5×0,5 mm size istalled in one of the sources position.

Before the epitaxial growth the sources and substrates were subjected to 10 min annealing at 1350 and 1250°C, respectively, then temperature of the source was increased to 1380°C, as the substrate temperature was decreased to assigned values (600-700°C) and the buffer layer was grown. The pressure in the cell corresponded to basic one.

In order to grow Ge layers the cell was filled with GeH_4 up to $1{\cdot}10^{-3}$ – $5{\cdot}10^{-6}$ torr, the pressure was maintained constant by a system of the gas feeding. Simultaneously the Ta plate situated in vicinity of the substrate was heated to T = 1200°C. With the purpose to avoid destruction of germane on evaporators (Ti) following pre-epitaxial annealing of the sources and substrates the sublimating pumps were switched off and the growth was carried out at pumping-down using only diffusion- and for-pumps. It is worth to note that the gas filling up to such high pressure (~10^{-3} torr) is impossible in MBE installations with electron-beam heating. The pressure in the cell was tentatively assigned by ionization vacuum gage indications. Nevertheless, this peculiarity in GeH_4 pressure measurement did not impede the controlled growth of Ge films at 700-750°C temperature of the substrate. Ge films were continuous and homogeneous. Solid solution GeSi with up to 80% Si content on fianite substrates (111) and (100) also was obtained. Vacuum annealing at 1250 C during 10 min was used as pre-epitaxy treatment. The growth was carried out under $5 \cdot 10^{-4}$ torr germane pressure and at600° C substrate temperature. Simultaneously, Ta plate positioned in vicinity of the substrate was heated to 1200°C. At fig. 7 X-ray diffraction pattern of Ge film (0.3 μm thickness) on fianite substrate (111) is shown, Ge(111) 27.3° and YSZ(111) 30.0°peaks are apparent. Heteroepitaxial Ge films obtained show high structural perfection. The half-width of the X-ray curve for Ge film of 0.3 μm thickness was 0.31° (fig. 7).

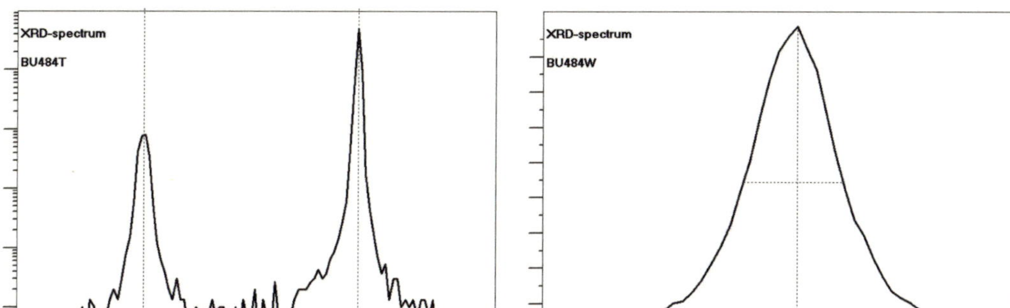

Fig. 7. Spectra of θ/2θ- scanning (a) and rocking curve (b) of Ge layer on (111) fianite.

The surface morphology of the Ge epitaxial layers grown on (100) and (111) fianite substrates (fig. 8a) as well as the peaks of Raman scattering near 300 cm^{-1} (fig. 8b) are identical to those of bulk Ge Therefore, it is possible to conclude that there are no stains in the Ge/fianite layer.

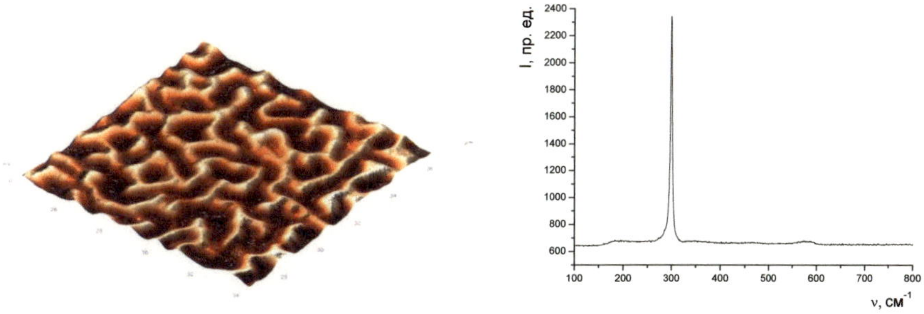

a b

Fig. 8. Surface morphology (a) and Raman spectrum (b) of Ge film on fianite; T_s=700°C, t=60 min (AFM).

3.3 Epitaxial films of $A^{III}B^{V}$ on-fianite

Crystallochemical and physical properties of fianite are favorable not only for silicon but also for $A^{III}B^{V}$ compounds epitaxy (Table 1).

Crystal	Lattice		T_m,°C (melting point)	Therm.Exp.Coeff. 10^{-6} deg^{-1}	E_g, eV
	type	a, Å			
$(ZrO_2)_{100-x}$ $(Y_2O_3)_x$	Cubic (fluorite)	5.141(x=10) 5.157(x=15) 5.198(x=21)	2800	11.4 (15–1000°C)	
GaAs	Cubic (sphalerite)	5.65	1283	5.4	1.43
GaP	Cubic (sphalerite)	5. 445	1467	4.7	2.26
GaN	Hexagonal (wurtzite)	a=3.186; c=5.178	1700	5.6 ; 7.8	3.4
GaN	Cubic (sphalerite)	4.52	1700	3.9	3.2
InN	Hexagonal (wurtzite)	a=3.54 c=5.70	1200	12.7	0.7
InN	Cubic (sphalerite)	4.98	1200	4.4	0.67

Table 1. Some properties of fianite crystals and $A^{III}B^{V}$ compounds.

First successful results on growth of $A^{III}B^{V}$ compound epitaxial films on fianite substrates were presented in [10, 17]. GaAs, InAs, GaN and other $A^{III}B^{V}$ semiconductor compound films have been grown on fianite, as well as on silicon and gallium arsenide with fianite buffer layer substrates by means of metal-organic Chemical Vapor Deposition (MOCVD). A new efficient epitaxy technique – "capillary epitaxy" has been suggested. The technique allowed synthesizing of $A^{III}B^{V}$ compounds films by MOCVD on fianite substrates. Samples of structurally perfect sub-micron (up to 0.1 μ) epitaxial films of $A^{III}B^{V}$ compounds have been obtained using this technique. The samples demonstrated high electrophysical parameters [16, 17, 19, 25, 26]. In [20] GaN epitaxial films have been grown on fianite substrates by MOVP technique. It was observed that the epitaxial growth of GaN on fianite significantly depends on conditions of initial stage of the process.

In [12, 21, 22] fianite substrates were successfully tested for growth of InN heteroepitaxial films. InN films of cubic structure have been grown on (001) fianite substrates by plasma-stimulated molecular-beam epitaxy (RF-MBE) at 400–490 °C temperature. The lattice mismatch of InN and fianite at (001) plane is very low (less than 2.3%), in contrast to 17% for InN – sapphire and more than 10% for InN – GaAs. Due to this fact, InN films grown on (001) fianite substrate were superior InN films grown on sapphire [12] and (001) GaAs substrates by its crystallographic perfection [22].

Therefore fianite is apparently in advance as a substrate for InN epitaxy as compared to sapphire. A new effective method of heteroepitaxy, capillary epitaxy, was proposed in [17]. It allows us in particular to obtain films of $A^{III}B^{V}$ compounds on fianite using the MOCVD approach.

3.3.1 GaAs on fianite films - MOCVD capillary epitaxy of III-V on fianite

The investigations showed that continuous GaAs layers on fianite can be obtained only in a very narrow range of epitaxial conditions. In particular, temperature range of 550-600°C is necessary. The minimum thickness of a continuous layer was 1.5-2.0 μm. The epitaxial films had polycrystalline structure and rough surface. Structural and electrical properties of GaAs films could be improved using capillary epitaxy. The essence of this method is that a thin (less than 50 nm) film of an III-group element is initially deposited on the fianite surface and then saturated with a V-group component with the formation of a thin continuous epitaxial III-V layer. After this procedure, the film growth continues to obtain the necessary thickness under conventional epitaxial conditions.

The use of capillary forces in the first (heteroepitaxial) stage of GaAs film formation led to improvement of epitaxial quality. Electron microscopy of the GaAs films at the initial growth stages showed that the transition from the standard MOCVD growth to capillary epitaxy leads to a change in the growth mechanism. Three-dimensional island mechanism changes to the two-dimensional one with propagation of the growth steps (Fig. 9, A). This process is similar to graphoepitaxy [27, 28] from aqueous solutions with addition of surfactants, where an increase in the substrate wettability also significantly improves the quality of graphoepitaxial layers [27] (Fig. 9, B).

In both cases, the height of the crystallization medium (melt or solution) decreases in the initial stage due to the capillary forces. This effect impedes growth of epitaxial nuclei in the direction normal to the substrate surface and facilitates their growth in the tangential direction. As a result, the substrate orienting role increases and a transition to the layer-by-layer growth mechanism occurs with a decrease in the growth step height. As a result, the minimum height of the continuous layer decreases and the film structural quality is improved. It has been shown that the use of capillary force in this method has a positive influence on both the mechanism of epitaxial growth and the quality of $A^{III}B^{V}$ epitaxial films. It also reduces the minimum thickness of a continuous layer [17, 19]. Virtually the same approach has now begun to be used with success in the works of other authors in order to obtain $A^{III}N$ films on various substrates [29].

The use of capillary epitaxy made it possible to decrease minimum thickness of a continuous GaAs/fianite film to 25 nm and to improve its structural quality and surface morphology. The technique was also efficient for growing other $A^{III}B^{V}$ compounds on fianite.

3.3.2 Study of impurities content in GaAs-on-fianite films using mass-spectrometry analysis

Mass-spectrometry analysis using single crystal GaAs standard curve has shown concentration of the impurities in GaAs-on-fianite films grown using the capillary epitaxy technique to be in the range of $5{\cdot}10^{16}$–$5{\cdot}10^{17}$cm^{-3} (Tab. 2). Layerwise mass-spectrometry analysis of the GaAs/fianite structures has shown uniform distribution of the impurities in GaAs film. Somewhat increase of Ca, Na and Cr concentrations in the film-substrate interface seems to be associated with a formation of oxides in the interface.

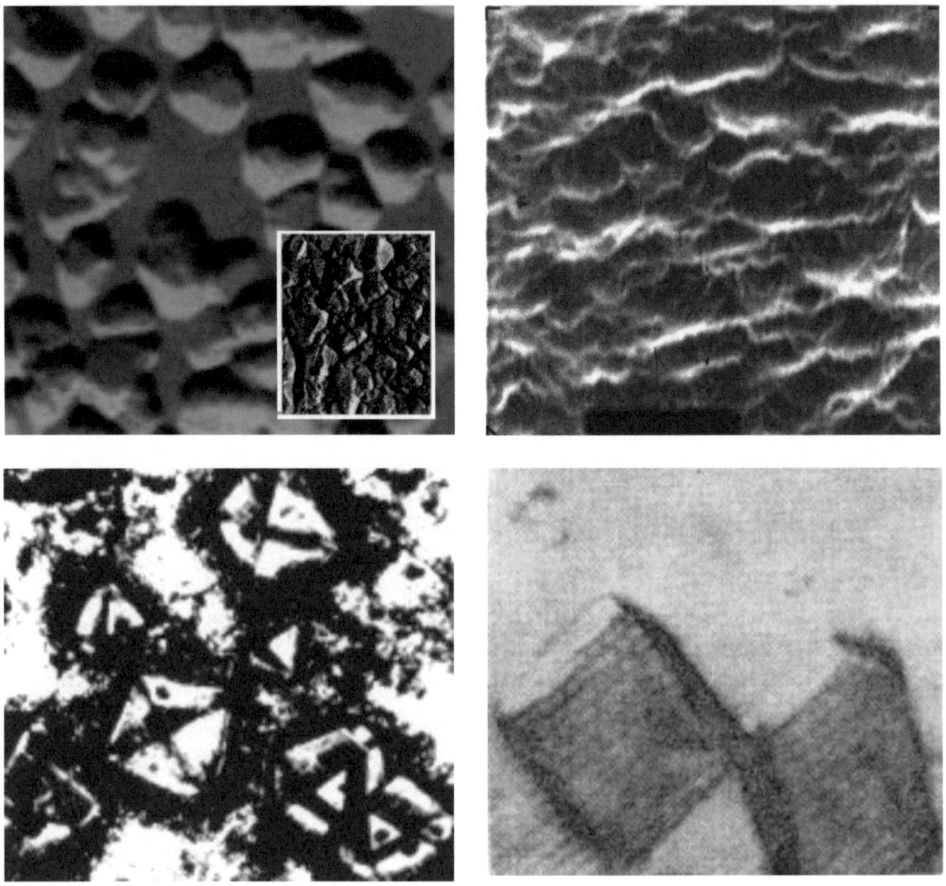

Fig. 9. Analogy between capillary epitaxy and graphoepitaxy: A - Electron microscopy image of GaAs on YSZ initial stage of growth (20000^X): Conventional MOCVD, height of the islets is up to 3000 nm, left; Capillary epitaxy technique, minimal layer thickness is 50 nm, layers growth is visible, right [17]; B - *Optical microscopy image of NH_4 J on amorphous Al graphoepitaxy growth:* without (left) and with (right) the use of surface-active substances, magnification 100^X [27].

3.3.3 The deposition of GaAs, GaSb, GaAs: Sb films and GaSb/InA supper-lattice on fianite substrates by means of laser sputtering

Our experiments have shown that the conventional "direct" growth of heteroepitaxial films InGaAs on fianite substrates resulted in the films with rough surface. So the buffer layers were elaborated to improve the results. The buffer layer must have very high structural perfection and mirror-homogeneous surface. The multiple experiments were conducted for growth of GaAs, GaSb, GaAs: Sb buffer layers on fianite (100) and (111) substrates as well as well as GaSb/InAs superlattice by using laser spraying. This superlattice is working as a filter which prevents the defects penetration into InGaAs film and first of all, formation of growing dislocation. Furthermore Sb is an effective surfactant [52] which significantly improves the layer morphology.

Impurity	Fianite crystal, mass%	Fianite substrate, mass%	GaAs-on-fianite film, atoms cm^{-3}
Al	0.0004	0.001	$5x10^{17}$
Ca	0.001	0.003	$5x10^{17}$
Mg	0.0005	0.0005	
Na	0.0001	0.003	$2x10^{17}$
K	0.0005	0.001	$5x10^{16}$
Si	0.001	0.015	$1x10^{17}$
Cu	0.0005	0.0005	
Fe	0.0004	0.0004	$5x10^{16}$
Mn	0.0001	0.001	
La	0.0006	0.006	
Cr			$1x10^{16}$
C			$1x10^{17}$

Table 2. Concentrations of the impurities in the crystals, fianite substrates and GaAs films

The studies have shown that it was complicated to obtain thin and homogeneous layers of $A^{III}B^{V}$ compounds on fianite substrates. It may be related to rather high mismatching of the lattice parameters of fianite and $A^{III}B^{V}$ compounds leading to growth according Volmer-Weber mechanism. Formation of the continuous layer occurred through 3-dimensional nuclei, their subsequent growth and joining. Low nuclei density results in formation of highly inhomogeneous rough surface that hinders subsequent formation of a flat layer. Laser sputtering technique is considered to maintain high nuclei density, so, before joining the nuclei are of sufficiently small size that promotes formation flat continuous layer.

In order to obtain flat layers laser sputtering technique was used in the study.

The Q-switched Nd laser and single crystal GaAs and InAs targets were used. The superlattices were grown by optical switching of the layer beam between the targets. Mirror-flat GaSb, GaAs: Sb layers, as well as penta-periodic InAs/GaSb supperlattices of 0.15 μm total thickness were deposited using this technique.

The X-ray diffraction investigations of GaAs:Sb (111) layers on fianite (111) showed their single-crystal structure (fig. 10a). It was shown that the spectral dependence of photoconductivity of GaSb layers on fianite substrates (fig. 10b) have a maximum of photoconductivity at the edge of fundamental absorption. This effect may be due to high velocity of the surface recombination.

The width of the rocking curve for these layers as FWHMω [GaSb (111)] = 0.23^{o}. The image of the surface of GaAs:Sb (0.2 μm fickness) on fianite is shown in fig. 11a. It is apparent, that the surface of the layer is mirror-flat and sufficiently homogeneous. The microrelief of the layer surface is shown in fig. 11b. According to our estimations roughness of the layer is less than 4 nm (Sq = 0.003778 μm).

In the penta-periodic InAs/GaSb supperlattices of 0.15 μm total thickness grown on (111) fianite substrates electron mobility approaches to 580 cm^2/ V×s.

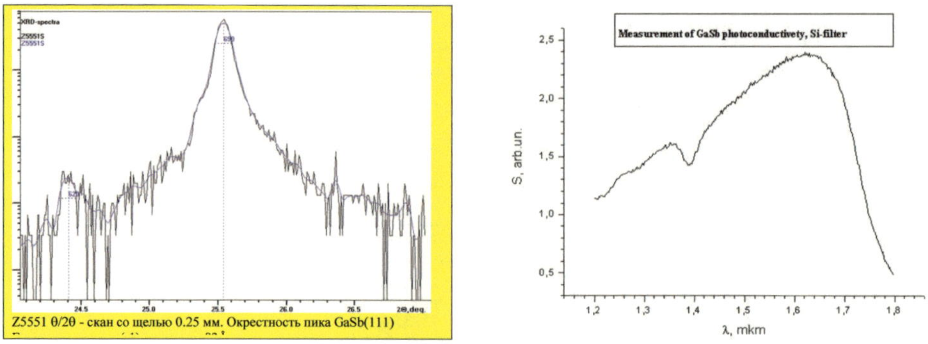

a b

Fig. 10. X-ray rocking curve of layer GaSb(111)/fianite(111) (a); rocking curve width FWHMω GaSb(111) = 0.23°; photoconductivity of GaSb on fianite substrate (b).

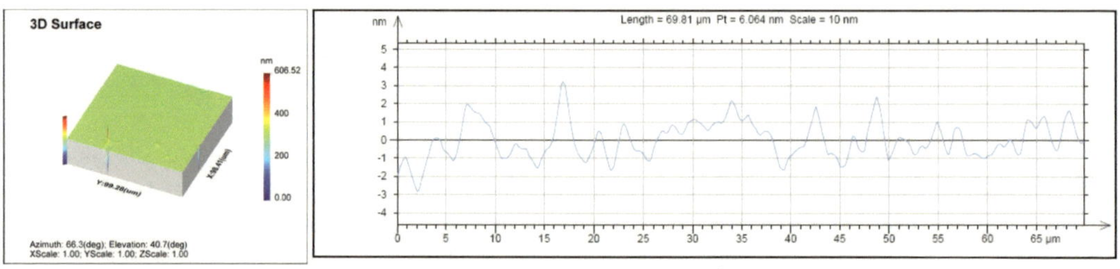

a b

Fig. 11. Interference microscope images of surface (a) and the surface relief (b) of buffer layer GaAs:Sb on fianite (Interference microscope Talysurf).

The GaSb layers, as well as InAs/GaSb short-period supperlattices are suitable for the development of IR detectors operating in 2-3 μm range. In our studies they were used as buffers for $A^{III}N$ growth on fianite substrates.

3.3.4 Deposition of GaAs, AlGaAs, InGaAs – Based multilayer structures on fianite

The results on epitaxial growth of $A^{III}B^{V}$ compounds films obtained in the studies described above were used for obtaining of AlGaAs/InGaAs/GaAs multi-layer heterostructures on fianite. These structures were used in FET. Sequential growth $A^{III}B^{V}$ heteroepitaxial layers on the fianite substrates was conducted according to topologic scheme of PHEMT (Pseudomorphic High Electron Mobility Transistor) for microwave frequency FET operating in 10-40 GHz range (Tab. 3) using «Aixtron AIX 200RF»installation. Capillary epitaxy MOCVD technique in 550–600°C temperature range was used.

Grown by «capillary epitaxy» techniques series of GaSb and GaAs:Sb buffer layers on (111) and (100) fianite substrates were developed to decrease the surface roughness of the PHEMT heterostructure. The buffer layers had a uniform mirror-smooth surface with about 5 nm roughness. Application of the developed buffers made it possible to obtain an AlGaAs/InGaAs/GaAs heterostructures with uniform mirror-smooth surface on fianite substrates and to decrease its roughness by a factor of 10 (to 25 nm). As a result, sufficiently homogeneous AlGaAs/InGaAs/GaAs multi-layer heterostructures with smooth slightly bloom surface were grown on (001) fianite substrates of 50 mm diameter. Roughness of the heterostructure surface measured using Talysurf interference microscope (3-dimensional topography) was 0.25 μm. This structure was grown using «AIXTRON» installation on (100) fianite ellipsoidal substrate of 2 inch major diameter. The surface of multilayer structure is rather uniform but its roughness reaches the value of 25 nm.

n^+ GaAs:Si	n_{Si} ~$6x10^{18}$ cm^{-3}	40 nm
i-$Al_xGa_{1-x}As$	x~0.24 (>0.23)	25 nm
i-GaAs		~0.6 nm
δ-Si	n_{Si} ~$4,5x10^{12}$ cm^{-2}	
i-GaAs		~0.6 nm
i-$Al_xGa_{1-x}As$	x~0.24	4 nm
i-GaAs		1 nm
i-$In_yGa_{1-y}As$	y~0.18 (<0.2)	11 nm
i-GaAs		30 nm
i-$Al_xGa_{1-x}As$	x~0.24	50 nm
i-GaAs	n< $8x10^{14}$ cm^{-3}	0,5-0,8 μm
CP AlAs/GaAs		(1 nm/ 2 nm) x 5
GaAs: Sb		100 nm
	Fianite substrate	400 μm

Table 3. PHEMT heterostructure for FET operating in 10-40 GHz range

Structural perfection of AlGaAs/InGaAs/GaAs multi-layer heterostructures on fianite was investigated by means of XRD. DRON-4 device (Ge(004) monochromator, CuKα1 radiation) was used. Θ/2Θ - spectra were recorded at symmetric reflection mode by scanning with 0.1 steps of texture maxima rocking. X-ray diffraction Θ/2Θ - spectrum of GaAs (001) / fianite (001) is shown in Fig. 12. The peaks of $(Zr,Y)O_2$ (004), 2θ = 73.4 substrate and of GaAs(004), 2θ = 66.05° buffer layer were recorded. The width of the layer rocking curve $FWHM_{\omega} = 1$ that is the evidence of a mosaic structure of GaAs layer. The grain-boundary angle was ~ 1° (Fig. 12a). Preliminary conditions of the growth of AlGaAs/InGaAs/GaAs heterostructures on fianite has shown that the use of (111) fianite substrate with GaAs:Sb buffer layer allowed reaching of mirror-flat homogeneous surface and 10-fold decrease of its roughness up to 0.025 μm value.

Layer-by-layer SIMS analysis of the heterostructures on fianite was carried out using «Shipovnik 3» and «TOF SIMS-5» devices. These devices provide detailed information on elemental and molecular composition in thin sub-surface layers, as well as 3-dimensional analysis. Sputtering was carried out by Cs^+,2 keV, raster 250×250 μm, negative ion detection mode, the probe beam Bi^+, 25 keV, depth resolution DZ > 7 nm. The analysis of the AlGaAs/InGaAs/GaAs heterostructures obtained on fianite (Fig. 12b) has shown that its inner topology was in conformity with the assigned scheme (Tab. 3) of the PHEMT-structure.

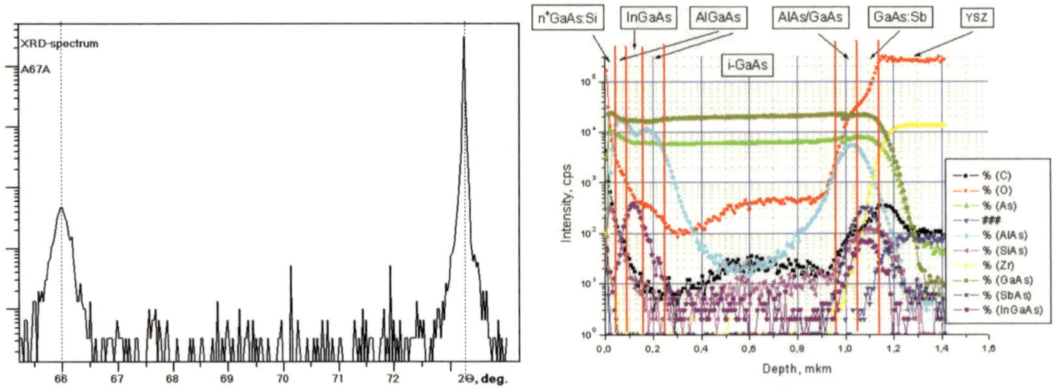

a b

Fig. 12. X-ray diffraction pattern (a) and layer-by-layer secondary ion mass-spectrometry (b) of the multilayer heterostructure AlGaAs/InGaAs/ GaAs (001) / fianite (001).

3.4 $A^{III}N$ films on fianite substrates and buffer layers

Principal difficulty of growth of perfect heteroepitaxial GaN layers is an absence of suitable substrates having good matching with the heteroepitaxial layer. Currently, for the growth of GaN layers Al_2O_3, ZnO, MgO, SiC, Si, GaAs substrates are in use. Usually, a material with wurtzite structure is grown on a hexagonal substrate, whereas sphalerite - on a cubic one. Fianite as a substrate material for cubic InGaN epitaxy has a number of advantages, such as favorite crystallochemical parameters and high chemical stability. Besides fianite, Si and GaAs substrates with fianite buffer layer were developed in scope of the work. Synthesis of the layer was carried out by laser deposition technique. The growth of fianite films on silicon substrates was conducted with the purpose to evaluate prospects of the use of less expensive large silicon substrates with fianite sublayer instead of monolithic fianite because maximum dimensions of the silicon-on-fianite structures are limited by size and quality of fianite crystals and the corresponding substrates (currently ~50 mm). Another purpose of the study was determination of suitability of fianite not only as a substrate material but also as a gate dielectric. Producing of such substrates will allow integrating GaN-based optoelectronics with a well-developed silicon electronics and gallium arsenide electronics and optoelectronics.

3.4.1 GaN films on fianite substrates

Growth of the films on (111) and (100) oriented fianite substrates was carried out using nucleus layers. 3 types of the nucleus layers were used:

1. Low-temperature GaN nucleus layer with annealing in hydrogen-ammonia atmosphere;
2. Low-temperature AlN nucleus layer with annealing in hydrogen-ammonia atmosphere;
3. High-temperature AlN nucleus layer.

At the use of all of the types of the nucleus layers fianite substrates were annealed in pure hydrogen at ~1070^{0}C before deposition of the nucleus layers.

Hydrogen is conventional carrier gas in MOGPE of III-V materials because it is rather readily can be purified. Similarly, in MOGPE of nitrides of III group hydrogen for the first time was used as a carrier gas. However, later it was demonstrated that in contrast to classic III-V semiconductors, GaN and InN are unstable under hydrogen atmosphere and undergo destruction (etching) at the temperatures used for growth of these crystals. This is an evidence that hydrogen as a carrier gas at the epitaxy of nitrides of III group elements actively participates in the process occurring on the surface of growing layer, in contrast to GaAs. Therefore, in most cases for growth of nitrides of III group by MOGPE ammonia is used as nitrogen source and supplied into reactor in large quantities. For a long time ammonium was because that it inhibits the destruction of a growing film and makes the effect of hydrogen negligible. However, it appears that it is far from the case and hydrogen significantly influenced on the process of the nitrides growth.

The studies have shown that at annealing of LT-GaN nucleus layer, the latter undergo etching in H_2-NH_3 flow hindering growth of a high-quality GaN layer. Application of the low-temperature AlN nucleus layer with annealing in hydrogen-ammonia atmosphere, as well as the high-temperature AlN nucleus layer on (111) and (100) oriented fianite substrates resulted in formation of hexagonal GaN layers comprising a textured polycrystal of hexagonal modification. Scattering angles of the texture for the GaN layers grown on the (111) and (100) oriented substrates were 10° and 15°, respectively.

It has been shown that the high-temperature annealing of LT-GaN buffer layer at 1000-1100C promotes improvement of structural perfection GaN heteroepitaxial layer. The GaN layers on fianite substrates exhibited an intense photoluminescence with maximum at 365 nm.

The conditions of growth of single-crystal GaN layers on (111) and (100) fianite substrates by MOCVD without buffer layer at 850°C substrate temperature has been determined. The spectra of $\theta/2\theta$ scanning were obtained with monochromator Ge(400) (fig. 13).

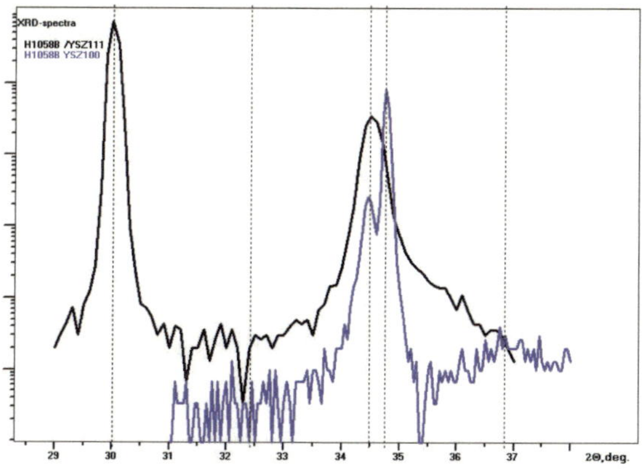

Fig. 13. XRD spectra of GaN on (111) and (100) fianite substrate.

Two peaks of the substrate were observed at 30° YSZ(111) and 34.8^{o} YSZ(200). The layer provides a single GaN(0002) peak at 34.5°. Since GaN (0002) peak is close to YSZ(200) a narrow slit in front of the detector was inserted with the purpose to increase the resolution. GaN_{hex} (0001) was detected on the both substrates at $FWHM\omega < 1°$ that corresponds to epitaxial growth. Traces of the polycrystalline phase at 32.4° (suggested 0.1-1.0 intensity units) were not detected.

3.4.2 AlN films on fianite substrates

The AlN films on fianite substrates were grown using MOGPE technique. The $Al_xGa_{1-x}N$ direct gap semiconductors are very useful in the development of UV photodetectors. By altering Al content in GaN-based solid solutions, it is possible to obtain the material with a forbidden band ranging in 3.43-6.2 eV thus covering 200-365 nm spectral band. This spectral band is of practical importance in UV astronomy, ozone layer monitoring, combustion and water sensors. These films are both of original interest, as well as are useful as nucleating and buffer layers in GaN epitaxy.

Growth of the films was started from thin 20-50 nm nucleating layer. Two growth modes were used: at 650 C with subsequent annealing in ammonia-hydrogen media at 1100 C during 30 min followed by growing-up of the basic layer and high-temperature growth of AlN at the same temperature. Before the deposition of AlN films the fianite substrates were annealed in pure hydrogen at ~1070°C. Mirror-flat homogeneous AlN films with the roughness not exceeding 0.6 nm (Fig .14) were deposited on (100) and (111) fianite substrates.

Layer-by-layer analysis of AlN nucleating layer on the fianite substrates was carried out by SIMS using TOF SIMS-5 device (sputtering by Cs+, 2 keV, 250 x 250 raster, negative recording mode, Bi+ probe beam 25 keV).

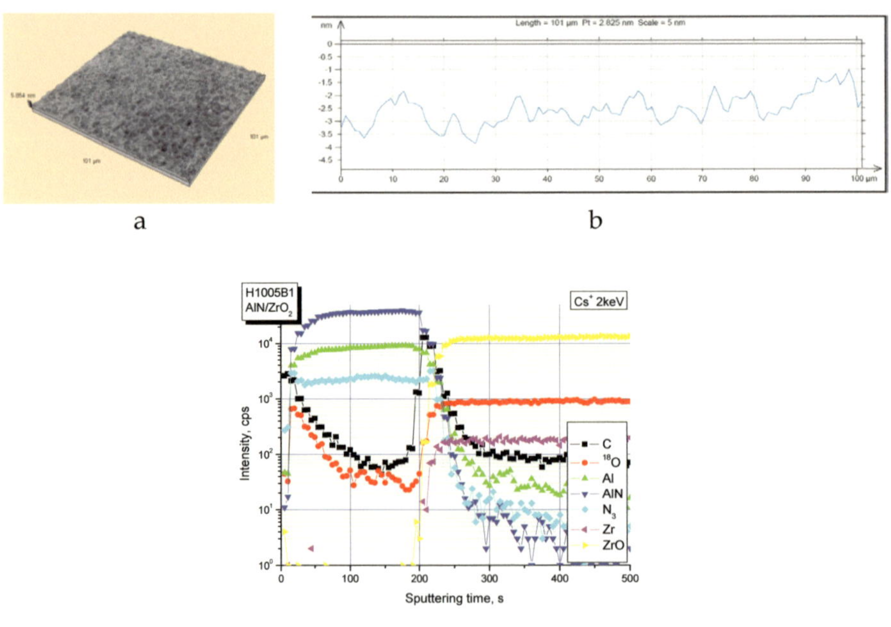

Fig. 14. Interference microscope image (Interference microscope "Talysurf") of surface (a) and surface relief (b); and layer-by-layer secondary ion mass-spectrometry (c) of low temperature AlN seeding layer on fianite substrate.

The study has shown that the layers had uniform distribution of its constituents, the concentration profile of Zr atoms at the hetero-interface being very sharp (Fig. 14c). The use of AlN nucleating layers on the fianite buffering layers allows deposition of continuous and homogeneous GaN layers of hexagonal modification.

3.4.3 Electrically active defects in GaN films on GaAs substrates with fianite buffer layers

Comparative study of density and electric activity of structural defects in the GaN epitaxial films grown on GaAs substrates with various buffer layers were carried out by **Induced bias technique**. Induced bias (IBT) technique has been developed rather recently [30, 31]. It is contact-free similarity of induced current technique (EBIC-mode). IBT is nondestructive contact-free diagnostic technique of semiconducting materials and microelectronic devices. IBT is based on detecting of voltage (or charge) generated by an electron probe of scanning electron microscope (SEM). Draft-scheme is shown in Fig. 15 a.

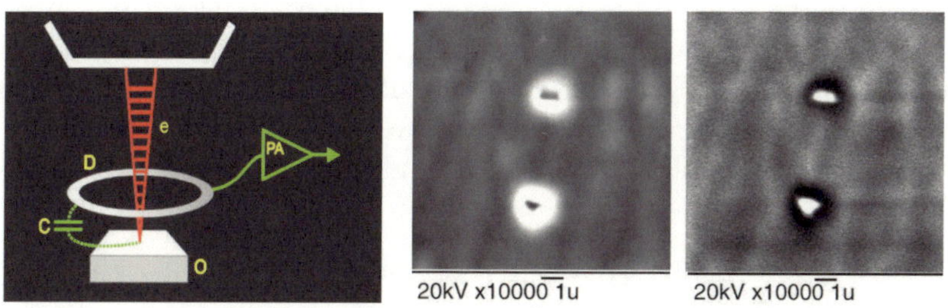

a b c

Fig. 15. Outline of induced potential method (a) and scanning electron microscope images of electrically active polygonal defects in GaAs films: secondary-electron emission mode (b); b - induced potential mode (c).

The electron probe (e) scans the surface of a crystal under the study (O). Metal ring (D), in which surface charge generated by electrons through capacitive coupling is induced, is a detector of the signal. The signal from the ring electrode is monitored in the SEM display (or by other measurement equipment) through charge-sensitive amplifier (PA) (Fig. 15 a). The technique allows qualitative monitoring of semiconductor plates, structures and devices identifying electric active inhomogeneities such as dislocations, stacking faults, micro-fractures, extent of doping by various dopants, all *p-n* junctions and Schottky barriers, etc (see for example Fig. 15 b,c). Quantitative measurements of local fundamental characteristics of semiconductors are also possible (diffusion distance, nonequilibrium carrier lifetime, its surface recombination rate, diffusion barrier height).

The studies have shown that the use of GaAs substrates with porous GaAs layer resulted in a decrease of the electric activity of structural defects in the GaN films and in an increase of its electrical uniformity as compared to GaN films grown on monolithic GaAs substrates. The use of GaAs substrates with double buffer layer (fianite on porous GaAs) allows additionally decreasing concentration of the electrically active defects in the GaN films to more than an order of magnitude (Fig. 16).

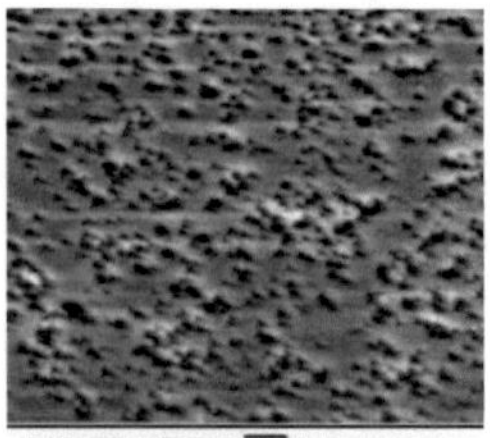

a

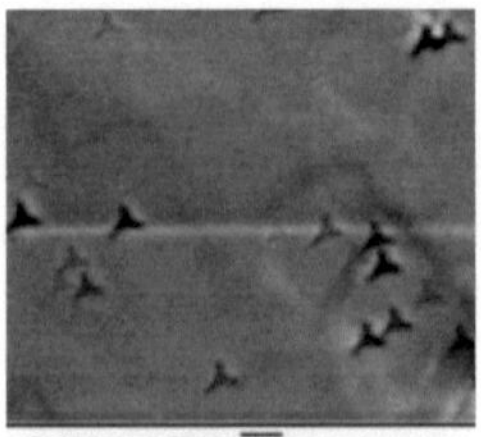

b

Fig. 16. Electrically active defects in GaN film on GaAs substrate with buffer layers:
a – single buffer (fianite); b – double buffer (fianite on porous GaAs).

3.4.4 GaN films on Si and GaAs substrates with fianite buffer layers

Silicon and gallium arsenide are promising substrates for GaN and other $A^{III}N$ epitaxy due to its high quality, large dimensions and a low net-cost, as well as possibility to integrate GaN-based devices with high-developed silicon and gallium arsenide electronics and opto electronics. However, there are three considerable problems occurring at GaN epitaxy: first, a significant parameter mismatch of GaN layer and Si or GaAs substrates, second, difference of thermal expansion coefficients of the layer and substrates and third, insufficient chemical and thermal stability of the substrates at the epitaxy temperature. Application of various buffer layers, in particular, fianite-based, can be an efficient method for solution of the above problems.

GaN epitaxial films were grown by MOGPE technique using capillary epitaxy on Si and GaAs substrates with various buffer layers. Tri-methylgallium (TMG), arsin (AsH_3) and ammonia (NH_3) were used as Ga, As and N sources, respectively. **Single** (fianite, layer of porous Si or GaAs material) and **double** (fianite on porous Si and GaAs) were tested. The first "prominent" porous buffer layer was suggested to allow decreasing thermoelastic strains in the second heteroepitaxial buffer thus improving its morphology and structure. The upper buffer layer, being chemically stable in the growth medium, provides fine matching with the working heteroepitaxial layer.

The epitaxial structures grown were studied using various techniques: photoluminescence (PL), scanning electron microscopy under induced current (IC) and induced bias conditions and secondary-ion mass spectroscopy (SIMS).

It was established that the use of fianite buffer layer on Si substrate prevents formation of amorphous silicon nitride. The GaN films grown on Si substrates with fianite buffer layer were of hexagonal modification (α-GaN) and had mosaic single crystal structure. It was established that the use of porous Si in the complex fianite/Si buffer allows improving of adhesion of GaN layer and its uniformity by phase composition and thickness.

Layerwise SIMS analysis of the GaN films grown on Si and GaAs substrates with fianite buffer layers has shown that the fianite layer serve as a barrier for diffusion of silicon and arsenium into GaN film from Si and GaAs substrates, respectively (Fig. 17). Good insulating properties of ZrO_2 in double buffer give some possibility to use the technology "Semiconductor on dielectric" which is promising to improve the integration level.

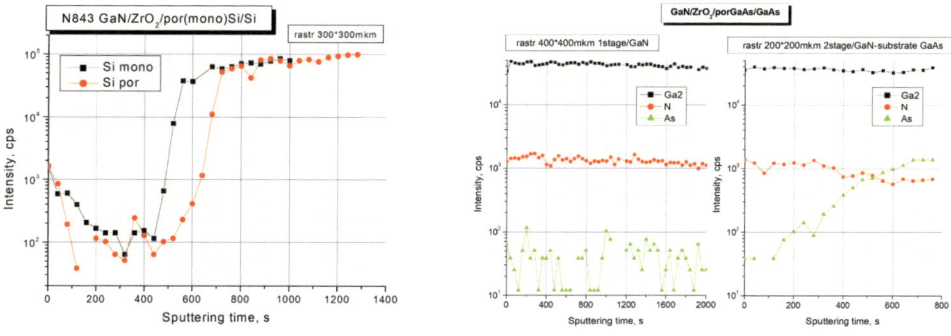

a b

Fig. 17. Layer-by-layer secondary ion mass-spectrometry of GaN/fianite/por(mono)Si/Si (a) and GaN/fianite/porGaAs/GaAs (b) structures.

Comparative studies of PL spectra recorded at 300ºK of GaN films grown on a monolithic GaAs substrate and GaAs substrates with various kinds of buffer layers have been carried out (Fig. 18): 1 - single buffer "porous GaAs"; 2 - double buffer "fianite on porous GaAs"

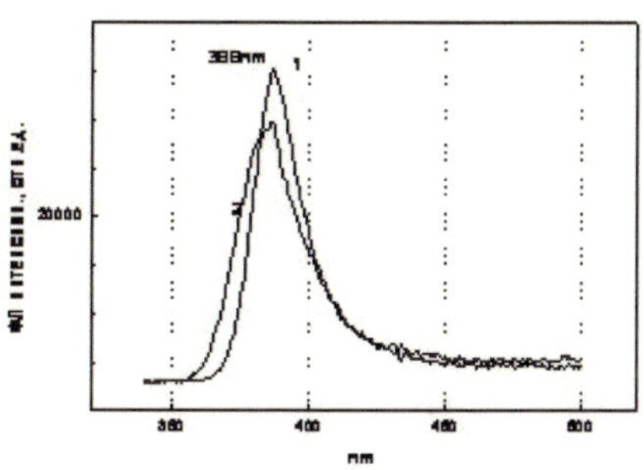

Fig. 18. Photoluminescence spectra of GaN films (300ºK) on GaAs substrate with buffer layers: porous GaAs) (1) and double buffer -fianite on porous GaAs (2).

The position of PL peaks in the spectra corresponded to characteristic peak of cubic GaN. Consequently, the use of the single buffer layer of porous GaAs, as well as double buffer layer (fianite on porous GaAs) allows growing GaN films of cubic modification. The growth of GaN film grown on monolithic GaAs substrate in contrast resulted to formation of hexagonal modification.

4. Functional fianite films on Si, Ge and GaAs substrates

4.1 Techniques for deposition of fianite films on Si and GaAs substrates

In recent years a considerable attention was drawn to fianite films on silicon due to its electric and optic device applications, such as isolating layers in SOI (silicon-on-insulator) devices [32], gate dielectric in Si- [33, 34], SiGe- [35] and $A^{III}B^{V}$ -based [36] device structures, buffer layers for producing of optic coatings of films of various semiconductors [37-40], superconductors [41-43], ferroelectrics, etc.

Various techniques can be used for the producing of fianite films on silicon and other semiconductors, including magnetron [39, 40, 44-46], laser and electron-beam [47-49] sputtering, molecular-beam epitaxy (MBE), as well as gas-phase chemical deposition [50].
The choice of a specific technique is determined by further designation of a fianite film, possibility to produce the film of maximum structural perfection, as well as technologic potentialities of a technique. So, MBE technique is more suitable for deposition of the thinnest fianite film for the use as a gate dielectric. Magnetron and laser sputtering are more favorable for fianite layers used as buffer layers with subsequent growing semiconductor films, including $A^{III}B^{V}$ compounds. In [39] fianite films were deposited on Si and GaAs substrates using magnetron, laser and electron-beam sputtering techniques. The films obtained by magnetron sputtering were of the best structural perfection [39].

4.2 Growing of fianite layers on silicon and gallium arsenide

The growth of fianite films on silicon and gallium arsenide substrates was carried out with the purpose to evaluate the prospects of using less expansive and more large Si and GaAs substrates with fianite sublayer instead of monolithic fianite substrates because, currently, maximum size of the latter is ~50 mm. Another purpose was determination of an opportunity to use fianite not only as a substrate but also as insulating layers material alternative to SiO_2, SiC, Si_3N_4, protecting and insulating layers, as well as a gate dielectric for multi-layer "semiconductor-dielectric" structures. Producing of such substrates will allow integrating GaN-based optoelectronics with a well-developed silicon electronics and gallium arsenide electronics and optoelectronics. Magnetron and laser sputtering were used for deposition of fianite films on silicon and gallium arsenide fianite films on porous Si and GaAs.

With the purpose to improve quality of fianite films and its adhesion to Si and GaAs substrates opportunities of the use of porous layers of the material were studied.

The following results were obtained:

- appropriate regimes of deposition of porous GaAs layer on GaAs (111) substrates of n- and p-conductivity types were developed;
- appropriate regimes of deposition of the uniform mirror-flat fianite layer on GaAs (111) substrates of 18x18 mm size were established;
- it has been demonstrated that the use of the porous layers allowed an improvement of adhesion of fianite with GaAs layers;
- the samples of fianite/GaAs, fianite/Si epitaxial substrates have been obtained for subsequent growth of $A^{III}N$ films.

High mechanical and chemical stability of fianite and absence of pores confirmed the prospects of its application as protective and stabilizing coatings substrates.

4.2.1 Magnetron sputtering technique

Magnetron systems are related to diode-type sputtering systems. The sputtering occurred due to bombardment of a target surface by gas ions (usually Ar) forming in plasma of anomalous glow discharge. A material ions knocked out the target subjected to the bombardment are captured by magnetic field and maintained complex cycloidal movement by closed trajectory in vicinity of the target surface. High sputtering rate, which is a feature of magnetron systems, is achieved by an increase of the ion current density due to localization of plasma by means of high transverse magnetic field. The increase of sputtering at simultaneous decrease actuation gas pressure allows a significant decreasing contamination of the films by alien gas impurities. Fianite was grown up on Si and GaAs substrates using unbalanced magnetron system. Fianite crystals were used as a target. Si substrate subjected to the sputtering was heated by IR radiance. Preparation of the substrates included degreasing, removing of the oxide and passivating of the surface in ammonium-peroxide solution. Optimization of the conditions of the growth of fianite films on Si substrates was carried out by varying of the sputtering rate, temperature of the substrate and residual gas pressure.

Bombardment of the target leads to dissociation of zirconium and yttrium oxides to ZrO, Zr, YO, Y, O_2. That is why such parameters as sputtering rate and residual gas pressure considerably influence on stoichiometry of the resulting film. Energy of the evaporating particles is rather low (~0.5-10 eV), so for the epitaxial growth of fianite film a high temperature of Si substrate and optimal rate of the condensate supply are necessary.

4.2.2 Laser sputtering technique

Experimental installation for deposition of fianite films was a sputtering system composed by vacuum device and eximer laser. The system has been designed and manufactured in IPM RAS.

Operational oxygen pressure was maintained by vacuum system supplied with a mechanical pump and CHA-2 letting system. Evaporation of the target was performed by LPX200 eximer laser radiation working on KrF mixture. Wavelength of the radiation was 248 nm, pulse duration 27 ns, the pulse energy 350 MJ (pulse power 1.3×10^7W), repetition frequency 50 Hz. Optical system providing a focusing of the laser beam on the target surface consisted of qrtz prisms and 30 cm focal distance lens. The laser beam spot on the target surface was 1×4mm^2. The energy density on the target surface was ~10 J/cm^2. The distance between the target and substrate was 60 mm. Cylindrical targets of 15-20 mm diameter and 10-30 mm length were used in the installation. In order to prevent local overheating of a target and to provide uniform material drift rotation and axial movement of the target was used. Possibility of conducting pre- and post-growth annealing under oxygen atmosphere at 10 Pa - 100 kPa pressure and at up to 750°C temperature is a peculiarity of the installation.

Ceramic target of $(ZrO_2)_{1-x}(Y_2O_3)_x$ with x=0.1 composition was used for deposition of fianite films. The deposition was carried out on Si and GaAs substrates heated to 600-800°C temperature under oxygen atmosphere at approximately 10 Pa. The growth rate of YSZ films was about 0.02 nm per pulse. Contactless heater of substrates (heating by irradiance) was an original peculiarity of the sputtering system. The heater comprises vertically positioned quartz tube (of 30 mm inner diameter) supplied with refractory stainless steel heating coil on its outer surface with up to 1 kW power of the heater. Monitoring and maintenance of the assigned temperature (with ±5°C precision) were carried out using precise regulating device and Pt-Rh thermocouple positioned under the heating coil. A substrate was fitted in a holder and positioned inside of the quartz tube. Loading of substrates and oxygen supply was maintained through the upper end of the tube.

Technology of growth of dielectric fianite films using the laser sputtering consists of the following stages:

1. A substrate is loaded to the sputtering system and vacuum chamber is evacuated up to ~1 Pa residual pressure.
2. Letting-to-oxygen is done up to the pressure required.
3. Rotational movement drive of the target is switched on.
4. A substrate is heated up to deposition temperature.
5. The eximer laser (the pulse energy 350 MJ, repetition frequency 50 Hz) is switched on and the sputtering is started.
6. Followed by the achievement of assigned thickness of the film the laser is switched off.
7. Followed by the end of the film growth the chamber is filled with oxygen up to the pressure required.
8. The structure is annealed.

The substrate heater is switched off and the substrate is cooled to room temperature.

4.2.3 Initial stages of deposition and structure of fianite buffer layers on Si and GaAs substrates

The application of fianite as a buffer layer will allow a solution route to another very important problem - epitaxy of $A^{III}N$ compounds on Si and GaAs substrates having large dimensions, high quality and low net cost.

Single crystalline heteroepitaxial fianite layers of 1000 A thickness were grown on silicon substrates of up to 50 mm diameter in vacuum chamber at p ~$2 \cdot 10^2$ Pa pressure, sputtering rate V_s ~60 A/min and substrate temperature T_s ~800°C.

The studies have shown that the layer became continuous as from 100 A thickness.

X-ray structural studies of ZrO_2-Y_2O_3/Si structures have shown that the fianite film is single phased and consisted of two layers with different rocking curve values: 0,20° for the upper layer and 0,96° for the lower one. Epitaxial relation between the film and the substrate was (100) [100]Si//(100)[100]ZrO_2-Y_2O_3. The relation was established using diffraction measurements under following regimes: Θ/2Θ scanning (simultaneous rotation of the detector and sample over goniometer axis) and Ψ- scanning (rotation of the plate in a proper plane at fixed detector position). The former regime was used to determine orientation of the composition plane, the latter - mutual orientation of unit cells of the film and the substrate in the composition plane.

Spectra of the Ψ- scanning of (ZrO_2-Y_2O_3)/Si structure for the asymmetric (422) reflection of the film (b) and the substrate (a) are shown in Fig. 19.

The absence of additional peaks and high peak maximum-to-background ratio (~10^3) are the evidence for ZrO_2-Y_2O_3 layer is a perfect single crystal film. The fianite buffer layers grown on Si and GaAs were used for $A^{III}N$ compounds epitaxy.

4.2.4 Some difficulties in deposition of the fianite layers on silicon

Growth of fianite-on-silicon structures of high quality featuring with sharp interfaces is associated with significant difficulties because of a number of principal problems.

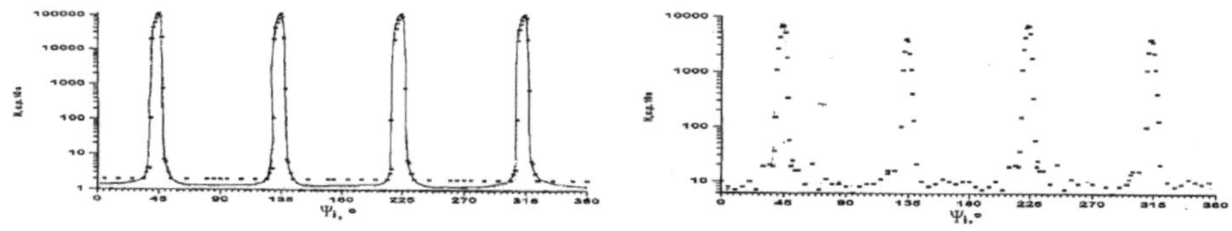

a b

Fig. 19. Ψ scanning spectra for (422)reflection of YSZ substrate (a) and the film (b)

First, silicon surface readily undergoes to transformation into SiO_2 amorphous layer due to either interaction with oxygen-containing fianite film, or oxidative atmosphere usually used at the fianite growth. As it has been shown by calculations, fianite should not react with silicon substrate to form SiO_2, which has low dielectric constant value, at a direct contact [51]. However, in practice, it is very difficult to avoid formation of this layer at the fianite deposition or subsequent high-thermal treatment [52,53]. Therefore, a development of special technological tools is necessary. One of the routes to solve the problem has been suggested by the authors [54]. Thin Zr or Y layer was deposited on Si substrate before fianite deposition. The metals absorb oxygen from SiO_2 layer because free energy of both fianite and Y_2O_3 formation is lower than of SiO_2 one [55]. That leads to a decrease of the layer thickness.

Second, oxygen from the fianite layer readily diffuses to a silicon substrate or reacts with silicon surface. Secondary phases occurring as a result of the reaction disturb silicon crystal lattice and hinder a perfect growth. Under these circumstances, the fianite layers on Si substrates are of amorphous or polycrystalline structure. At the development of gate dielectric technology these issues are of peculiar importance because thickness of the last layer is about some nanometers.

Therefore, the above data show that the problem of deposition of fianite layers on Si substrates is of great interest. The problem of improvement of quality of the layers seems to be very urgent because of a number of principal difficulties occurring due to peculiarities of physic-chemical properties of the materials considered resulting in reactions at the growth and subsequent thermal treatment stages. The synthesis of perfect fianite layers on Si requires a development of special methods to decrease the influence of amorphous SiO_2 layer at the substrate-layer interface.

4.3 Development of the techniques of fianite films etching

To choose the most appropriate method and conditions of fianite film etching, we have tried out the main methods of etching used in microelectronics technologies: liquid (wet), plasmachemical, and ion-beam methods of etching.

4.3.1 Liquid etching

For fianite film liquid etching (by analogy with ZrO_2) the following etchants were used:

- etchant $HCl:HF:H_2O$ (10:1:5). Fianite films were found to be resistant to this etchant;
- strong acids: H_2SO_4, HF, HCl, HNO_3;
- aqua-regia: $HNO_3:HCl$ (3:1).

The film was found to be chemically resistant to all of the above listed reagents. So we tried out all the most chemically active reagents that traditionally are used for dielectric film liquid etching in photosensitive devices production. On the one hand, this evidences resistivity of photosensitive devices with fianite protective layers to corrosive medium exposure. On the other hand, such properties cause technological difficulties. To settle the problem, we have searched for other methods of etching.

4.3.2 Plasmachemical etching

In case of plasmochemical etching, a mask of 1.3 μm thick FP91-20-1 photoresist hardened at 120°C during 20 minutes was used. A diode type plant was used. The discharge power was between 200 and 350 Wt, discharge frequency was 105 KHz. Temperature inside the reaction chamber was varied from 20°C to 90°C, etching time – from 10 to 35 minutes. Mixtures of CCL_2F_2+Ar, CCL_2F_2+He, and $CCL_2F_2+O_2$ of various percentages were used as reagent gases. The pressure in the processing chamber during samples etching varied from 0.2 to 0.85 mBar. The samples surface texture after etching depends upon the etchant, the process conditions, and the preliminary surface treatment. Prior to etching, the samples were treated with Ar and He ions. Loose texture surface was observed on the samples; such texture had been formed probably by precipitation of products of reaction on the sample surface. Water vapor and oxygen may be absorbed on the reactor surface and slow down etching till they completely react with the working gas. The period of etching slow down may be decreased by eliminating of the said factors with the help of a "loading lock". For this purpose, another construction of the plant was chosen that applied the reactive-ion technology of film etching with the loading lock. Use of such plant made it possible to combine glow discharge plasma and the chemical medium providing etching. The medium consists of charged particles, radicals and neutral particles participating in chemical reactions on the film surface. Volatile products are formed in the medium. Positive ions being accelerated in the interelectrode space bombard the surface of plates thus finishing material removing.

The following conditions of the process of fianite film etching were examined:

Working pressure in the reactor	0.03-0.08 mBar
Discharge power	320-800 Wt
Discharge frequency	13.56 MHz
Etching time	10-45 minutes

Adding of O_2 to chloride bearing plasma increases concentration of Cl and suppresses polymer film forming on the sample surface. Adding of inert gases stabilizes plasma. Stabilization may be achieved due to thermal properties of the discharge gas used, especially in case of helium adding.

Unfortunately, in this method the rate of film etching was found to be low; besides, the problem of mask selection occurred. That is why, the studied technologies of plasmachemical and reactive-ion etching for fianite films are rather inefficient. But they may be used for gold contacts etching.

Of cause, capabilities of the investigated technologies may be expanded by use of more active reagents, such as CCl_4BC_3. But, such reagents are referred to extremely hazardous substances. Any work with such substances requires availability of specific production or laboratory premises, equipping of which is allowed only in specific industrial areas and causes additional labour and financial costs. That is why further researches were devoted to fianite film etching using ion-beam method.

4.3.3 Ion-beam etching

This method is based on material scattering under ion bombardment. The sample (fianite-semiconductor structure) was fixed on a holder. The holder was cooled to avoid overheating. Ions were generated in a direct current discharge in a separate "ion" gun, were focused and accelerated towards the samples treated.

A fianite film of 1000 Å was etched through a photoresist mask with the help of ion-beam etching during 35 minutes.

4.4 Characterization of the fianite films

The fianite films were studied by means of scanning electron microscopy, ellipsometry and CV-parameters measurement techniques. The films parameters were found as follows:

- optic refractive index $n_{\text{ок}} \sim 2{,}1{\div}2{,}2$;
- dielectric constant $\varepsilon \sim 25$;
- absence of defects of porosity type (in 30 mm diameter sample).

4.4.1 The capacity-voltage characteristic measurements of fianite-on-silicon" structures

The capacity-voltage (CV) characteristics of the structures supplied with fianite films deposited on *p*-Si and *n*-Si substrates were measured.

Capacity measurements provide evaluation of dielectric properties of the films under the study: dielectric constant ε and dielectric loss $tg\delta$. The application of multifrequency device allows determination of frequency dependencies of dielectric constant and high-frequency loss in dielectric films. Since the dielectric film is deposited on semiconductor a MIS structure (metal-insulator-semiconductor) is formed, so the CV-measurement provides additional information concerning the semiconductor and the dielectric-semiconductor interface, namely, type of the semiconductor conductivity (n- or p) and concentration of the dopant, flat band barrier voltage V_{fb}, density of boundary states and a charge induced in the dielectric.

The device used for CV- measurements allowed determining of capacity and high-frequency conductivity of the structures, as well as its dependency on the applied voltage. The measurements were carried out at 500 KHz and 1 MHz frequencies. Direct potential bias range was ±40 V. Thermally sputtered Al of 1 mm surface diameter was used as the contacts. The results obtained are shown in Fig. 20.

"Al-fianite-Si" MIS structure parameters: flat band barrier voltage – 4 V for 180 nm film and 1.5 V for 20 nm film; density of boundary state charge~ $+10^{12}$ см$^{-2}$.

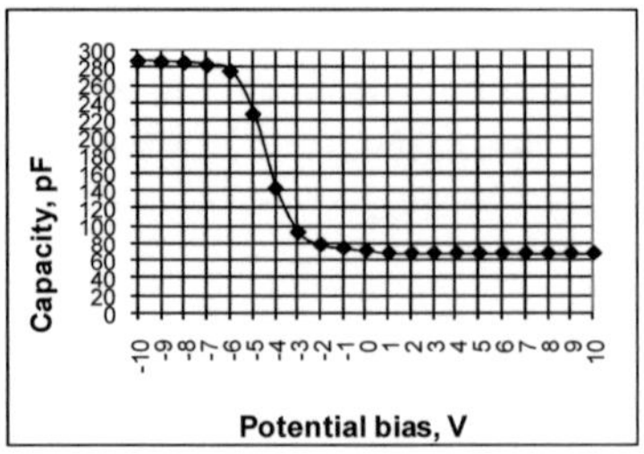

Fig. 20. CV- characteristics of fianite-on- *p*-Si sample

4.4.2 Investigation of ZrO_2 films on Si and Ge substrates by means of scanning electron microscopy

The ZrO_2 films were studied using scanning electron microscopy. All of the films studied were porous-free. Since square of the samples studied was 5-6 cm^2, it is possible to consider the porosity value at least not exceeding 0.15-0.2 cm^{-2}. For comparison, it is worth to mention that porosity of SiO_2 films is 4-8 cm^{-2} . Therefore, it is possible to consider ZrO_2 films as the protective layer for Ge devices actually superior SiO_2 films because its porosity decreased in 1.5-2 orders of magnitude.

The study of morphology of the films deposited by magnetron sputtering technique at high magnification has shown its satisfactory homogeneity. Some regions of the surface featured by a relief composed by quasi-spherical hills of 500-600 nm in diameter and exhibiting lateral periodicity. Analytical study of the films has shown an absence of inclusions of impurities.

An attempt to study mechanism of formation of the films with the purpose to optimize conditions of magnetron sputtering was done using electron microscopy (JSM JEOL 5910 LV). The particles were identified by means of electron probe. The film was removed by polishing using diamond paste with 2.5-4μm particle size. This abrasive size was chosen to minimize decreasing particle size of the film constituents at the polishing. The obtained material was flushed by ethanol (9-12 purity grades "for microelectronics") and the suspension was put in plastic syringes (1 ml). In order to disintegrate aggregates ultrasonic (US) treatment was carried out. The US dispersion was conducted using «Sapphire 3M-1.3» US device with 35 GHz operational frequency. The syringes were inserted to the device chamber filled with water. The chamber was thermostated at 27°C. Followed by 3 min of the US treatment the suspension was aspirated onto conductive (graphitized) ribbon for subsequent microscopy study. The study has shown that the largest constituents of the zirconia film were quasi-spherical particles of 50-100 nm size that explained X-ray amorphous nature of the film. It is possible to suggest that formation of larger elements of the relief occurred by enlargement of such particles. The reasons of local enlargement (formation of spherical hills) can be gradients of temperature and mass-transfer, as well as occurrence of impurities. The observations allowed refining the refine conditions of the sputtering of ZrO2 and fianite films in order to minimize surface roughness.

4.5 Fianite as a gate dielectric

Recently, a sharp surge of interest in the use of fianite as a gate dielectric in CMOS technology has been observed. It is associated with an increase of leakage current at the use of conventional SiO_2 by increasing of integration level. That requires a change of SiO_2 over dielectrics with higher dielectric constant (high-k materials) [33-35,51]. The resent studies have limited possible alternatives to fianite, HfO_2, ZrO_2 and its silicates. For example, ZrO_2 has high dielectric constant value, good dielectric properties (5.8 eV energy gap width) and rather good crystallochemical matching with Si [56] (see Fig. 5). Intel Corp. - one of the leaders of the world electronics, has demonstrated that the change of SiO_2 over HfO_2 as a gate dielectric in 45 nm technological process allows decreasing leakage currents (which became a serious problem for transistors) by more than two orders of magnitude [57].

Comparison of fianite and SiO_2 films [34] with electrical equivalent oxide thickness of about 1.46 nm has shown that the leakage current for fianite was four orders of magnitude lower than that of conventional SiO_2 gate oxides.

The hysteresis and interface state density in this film was measured to be less than 10 mV and $2.0 \times 10^{11} eV^{-1}cm^{-2}$. It demonstrated that crystalline oxide on semiconductor could be used for future generation of semiconductor-based devices.

It is worth to note that quality of the synthesized fianite, as well of the interfaces [85], is very important for integration of such a dielectric to the CMOS technology currently in use.

Synthesizing of fianite-on-silicon structures of high quality featuring with sharp interfaces is associated with significant difficulties because of a number of principal problems.

First, silicon surface readily undergoes to transformation into SiO_2 amorphous layer due to either interaction with oxygen-containing fianite film, or oxidative atmosphere usually used at the fianite growth. In practice, it is very difficult to avoid formation of this layer at the fianite deposition or subsequent high-thermal treatment. Therefore, a development of special technological tools is necessary.

Second, oxygen from the fianite layer readily diffuses to a silicon substrate or reacts with silicon surface resulting in SiO_2 formation having low dielectric constant value. At the development of the gate dielectric technology these issues are of peculiar importance because thickness of the last layer is about some nanometers.

One of the routes to solve this problem is in application of low-temperature growth and annealing regimes, as those, which were used in the series of experiments described below, Type of a substrate and the annealing media were also varied. Conditions of the synthesis of the fianite/Si structures are given in Tab. 4. XRD technique has shown that fianite layers obtained by laser deposition at room temperature were of amorphous structure.

Sample	T of growth, C	Annealing, 600 C, 10 min	Film thickness, nm	Substrate
z 1	room	without annealing	~20	Si <B>
z 2	room	vacuum	~20	Si <B>
z 3	room	oxygen	~20	Si <B>
z 4	room	oxygen	~20	Si <Sb>
z 5	600	oxygen	~20	Si <B>

Table 4. Parameters of growth and annealing of the fianite-on-Si films

Subsequent post-growth recrystallization annealing resulted in arising of a polycrystalline phase in the layer. At the same time, the layers sustained mirror-flat and uniform. Profile of the surface of z4 sample (Table 4) obtained using Talysurf interference microscope is shown in Fig. 22 a. Roughness of this ZrO_2 surface was estimated as Sq = 0.852 nm that is not practically differ from roughness of the Si substrate used for the fianite growth (Sq = 0.7877 nm).

Preliminary studies of gate properties of thin (10–15 nm) fianite films obtained by laser deposition on Si substrates have been carried out. The studies conducted on the test structures with deposited Al contacts have shown that thin fianite films featured with low values of loss currents, minimum values being 10^{-12} A/cm^2 at 1V voltage (Fig. 21 b, samples z 3 and z 4).

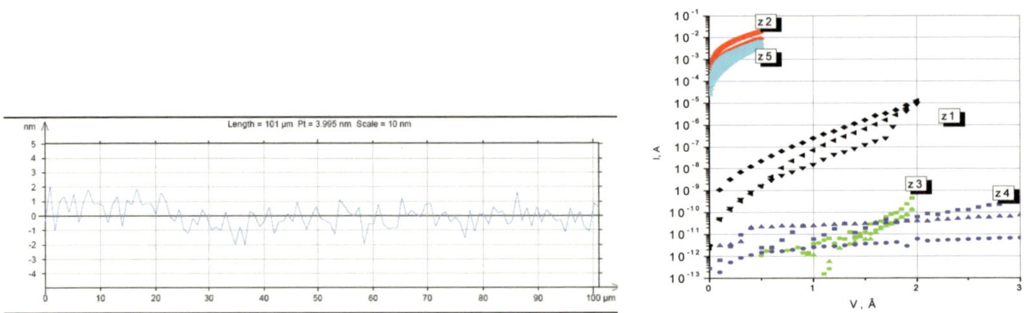

a b

Fig. 21. Surface roughness of fianite film on Si substrate, sample Z4 (a) and leakage current of Al/fianite/Si structure (b), samples z 1 - z 5 were prepared under different conditions.

4.6 Fianite and ZrO_2 as protective and stabilizing layers on Ge and Si substrates and multilayer structures

4.6.1 Deposition modes

For magnetron deposition of fianite and zirconium dioxide films, 2 types of vacuum evaporation Leybold Heraeus units were used with different target dimensions: 70 mm in diameter for fianite and 203 mm - for ZrO_2 (table 5).

HF magnetron and direct voltage sputtering techniques were tested. The latter technique did not provide sufficient film growth rate, that is why magnetron HF sputtering (13.56 MHz) was chosen. The optimal modes of fianite and ZrO_2 sputtering are also shown in table 5.

Target Material	Fianite	ZrO_2
Plant Z	Z-400	Z-550
Target diameter	∅ 70 mm	∅ 203 mm
Argon pressure	$5*10^{-3}$ mBar	$5*10^{-3}$ mBar
Power	~ 500 Wt	~ 400 Wt
Film growth rate	100 Å/min	~ 50 Å/min

Table 5. Optimal Modes of Fianite and ZrO_2 Sputtering

In case of low magnetron power, plasma is unstable ("blinking plasma"); in case of larger values of discharge power, the growth rate increases, but irregularity of substrate surface layers and growing film coarse-graining are possible. Fianite sputtering requires higher power than in case of ZrO_2; provided that the growth rate is twice as much than in case of ZrO_2.

The developed technique of magnetron sputtering made it possible to vary the fianite film thickness between 600 and 2000 Å. Ge and Si plates with fianite film thereon were made using this technology. Ge samples with fianite film were used to try out further operations of device structures making: photolithography and etching.

4.6.2 Protective and stabilizing properties of fianite films on Ge

Inorganic dielectric coatings are usually used for passivating and protection of p-n transition surface, as shielding and thermal compensation layer at ion implanting and for interference antireflecting protection. Passivation of the surface is the most important issue for manufacturing ot germanium photodiodes because natural GeO and GeO_2 oxides are unstable and, so, can not be considered as the only passivating coatings. It is one feature distinguishing Ge and Si devices (the latter have stable and effective coating of its own SiO_2 oxide). This oxide film deposited from a gas phase is of the most frequent use for photodiodes with p+ - n-structure. It has positive charge and by attracting electrons to the surface prevents growth of p-channels thus decreasing probability of generation in the layer. It is worth to note that for improved reliability and stability of characteristics of photodiodes it is necessary to maintain surface state density at 10^{11} cm^{-2} eV^{-1} level. However, this passivating technique is far from ideal because high porosity of SiO_2 films that decreases humidity resistance and reliability of the devices.

In order to improve dielectric properties of the protective coating fianite films deposited by magnetron sputtering were used. The opportunity of its application for maintaining high-quality practically porous-free protective coating has been confirmed earlier by the experiments.

It has been demonstrated that the use of the fianite protective layer in Ge-structures instead of SiO_2 eliminated pulse noise and thus considerably improved photoelectric and performance characteristics of these devices. It has been established that the improvement was related to more uniform nature of the fianite films, in particular, absence of pores, in comparison with SiO_2 films, which containing defects in form of pores.

4.6.3 Some properties of the device structures supplied with zirconium dioxide films

Photoelectric characteristics and noise of germanium photodiodes supplied with ZrO_2 and SiO_2 films described above have been investigated. Monochromatic sensitivity of these photodiodes is typical for germanium devices and equals to 0.5-0.6 A/W (at 1.06 and 1.55 μm wavelengths). The change SiO_2 over ZrO_2 resulted in somewhat decrease of a dark current (on average for 10%). Main improvement of the photodiodes quality achieved due to the application of ZrO_2 films revealed at the noise studies. Under the voltage exceeding operational one (that corresponds to accelerated reliability testing conditions) the check samples with SiO_2 films have shown pulse noise of telegraphic type in the oscillogram, which can be associated with processes of energizing- deenergizing of the surface conducting channels [59]. The defects occurring because of the presence of pores in SiO_2 films are a probable cause of arising of the channels. In the batch with ZrO_2 protective films only shot noise, which is in principle unavoidable, was observed. More detailed results of the device studies are presented in [60].

Thus, the studies performed on fianite and zirconium dioxide films, as well as on the device structures developed using these films have demonstrated the advantages of zirconia-based solid solutions in application to photosensitive apparatus technology.

4.7 Studies of optical properties of ZrO_2 films

Optical refraction of ZrO_2 equals to 1.98÷2.1, that is close to fianite one, therefore this material is also promising for antireflection coatings. Determination of the refraction constant n and monitoring of the film thickness d were carried out using ellipsometry technique. The experimentally determined values of d depended on duration of the films growth and varied within 600Å - 1100Å range.

The films obtained have shown rather high refraction: ~2÷2.1. These values were significantly higher than that of SiO_2 (1.45).

In theory, considering an incident beam from air (vacuum), it is possible to decrease the reflection to zero when the refraction constant of an antireflecting film corresponds the following equation:

$$n=\sqrt{n_{\mathrm{n}}},$$

where n_n – refraction constant of a semiconductor. In case of Si and GaAs $n_n \sim 3.5 \div 4$, thus $\sqrt{n_n} \sim 1.9 \div 2$. Therefore, the ZrO_2 films obtained actually satisfy perfect antireflection of Si and GaAs – based devices from the viewpoint of n. Moreover, the difference in n-values of SiO_2 and ZrO_2 films provides an opportunity for the antireflection over a broad spectral range due to application of binary SiO_2+ ZrO_2 antireflecting coatings. The dependency of the reflection constant on wavelength of silicon sample coated with ZrO_2 film of 1200Å thickness is presented in Fig. 22. Theoretical absorption minimum corresponds to λ = 4 nd=4·2.1·0.12 ≈ 1 μm.

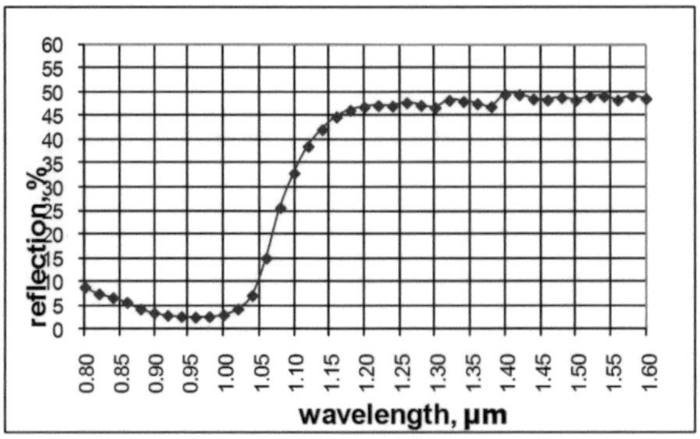

Fig. 22. The dependency of the reflection constant on wavelength of silicon sample coated with ZrO_2 film of 1200Å thickness.

As it is apparent from Fig. 23, the reflection minimum was approached at λ_{min} =0.97 μm. Thus, the experimental results are in conformity with the theory practically complete.

Therefore, the ZrO_2 film ensures high antireflection quality: at λ_{min} the reflection loss does not exceed 2-3 %. The data obtained confirm that ZrO_2 is an excellent material for antireflecting films, as well as fianite.

5. Silicon and III-V solar cells with fianite antireflecting layer

5.1 Anti-reflection properties of fianite film on Ge and Si

In theory, it is possible to eliminate the reflection completely (at the corresponding thickness of a film d) at $n_d = \sqrt{n_f}$, where n_f – optical refraction constant of a semiconductor. Since for Si and Ge the constants equal to 3.7 and 4, respectively, the reflection is completely eliminated at $n_d = \sqrt{n_п} \approx 2$. Therefore, a dielectric having its optical refraction constant $n_d = \sqrt{n} \sim 2$ (at $n = 3.7 \div 4$) can be considered as an optimal material for the antireflecting film for solar cells and the other photosensitive devices. Theoretically, it is the case at the film thickness, which is equal to a quarter of optical wavelength $W = \lambda / 4n_d$, such dielectric allows a complete elimination of the reflection loss (R=0).

The refraction constant of SiO_2 ($n = 1.47$) is considerably lower than that value. At this n value it is impossible to maintain the reflection loss lower than 10%. Refraction constants of fianite and ZrO_2 are within (2.15÷2.18) and (2,13÷2.2), respectively, - that is close to the above optimum value. Thus providing an evidence that fianite and ZrO_2 are very promising as antireflecting coatings for solar cells and the other photosensitive devices based on Ge, Si and $A^{III}B^{V}$ compounds.

Experimental dependencies of antireflection (as dependencies of the reflection on wavelength) of fianite films on Si and Ge have been plotted (Fig. 24).

The plots apparently demonstrate that the reflection drops to 0 – 1.5 % in the minima.

Experimental study of antireflective properties of fianite oxide applied to Ge was performed. By the reason that, germanium photodetectors are designed for detecting radiation generated by lasers with wavelengths λ= 1.06; 1.3; 1.54 μm, the thickness of the antireflective fianite film was chosen as W=1300 Å; such thickness provides for minimal reflection losses in the said wavelength range λ= 1.06-1.54 μm. Fig. 23 a shows the comparison of experimental (thin line) and theoretical (bold line) R(λ) curves. The theoretical R(λ) curve was calculating using the following formula:

$$R = 1 - \frac{4 n_{\Pi} n_{ок}^2}{n_{ок}^2 (n_{\Pi} + 1)^2 - (n_{\Pi}^2 - n_{ок}^2)(n_{ок}^2 - 1)\sin^2(2\pi n_{ок} W / \lambda)}.$$

According to the above formula, reflection may fall practically to zero at the optimal value of n_{ok} (note, that in case of SiO_2 anti-reflective film, for which n_{ok}=1.47, it is impossible to obtain reflection lower than 10%). The minimal reflection is achieved at the following wavelength λ_{min}:

$$\lambda_{\min} = 4 W n_{ок}.$$

As it is shown on Fig. 24a, in the range of fundamental absorption (for λ<1.65 μm) the experimental curve 1 coincides with the theoretical curve 2. Some discrepancy at higher wavelengths (λ>1.65 μm) appears due to deep penetration of such radiation and its reflection from the back surface. It is important that at the optimal wavelength (λ=1.12 μm), fianite film provides for ideal antireflective properties – the reflection is actually absent. In rather large range 0.88 - 1.55 μm, into which radiation wavelengths of most wide spread lasers fall, the losses for reflection do not exceed 10%.

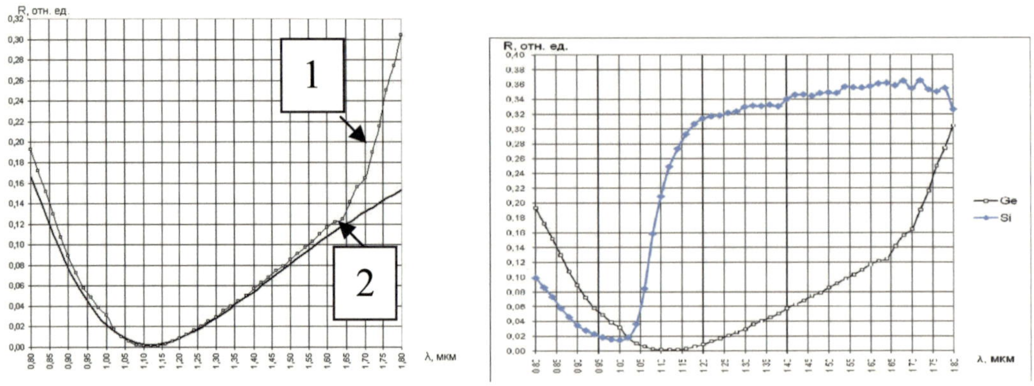

a b

Fig. 23. Experimental (1) and theoretical (2) dependencies of the reflection on wavelength in Ge-fianite antireflecting film system (1300 Å) (a); experimental dependencies of reflection of fianite film on Si and Ge (b).

The experimental dependences of enlightenment (the dependence of reflectance on the wavelength) of cubic zirconia films on Si and Ge, (Ge on their optical properties similar to GaAs) exhibit excellent antireflective properties of cubic zirconia (Fig. 23 b). As is evident from the graphs, the minimum reflection can drop to 0-1, 5%. Position of the minimum depends on the thickness of the film. When it gets thinned twice the minimum would be in the solar spectrum. Plateau in the curve shows the reflection from the back side of the substrate in the transmission range for Si. So the gain due to the use of antireflecting fianite film reaches 20-30%.

So, it was experimentally proved that in case of use of 1300 Å thick fianite film, reflection may actually fall to zero in the wavelength range λ= 1.06 – 1.54 μm.

A new, non-standard, fianite use as a reflecting film (in contrast to anti-reflective film!) was proposed. Such unexpected application may appear useful for screening of peripheral (non-photosensitive) photodetector areas. For standard screening of such areas, forming of proper photosensitive areas, metallic masks sputtered to SiO_2 have been used. But such solution causes notable spurious capacitance of the metal-oxide-semiconductor structure; provided that such capacities are inadmissible in a number of photodetectors, in particular – in high frequency photodetectors. In case of screening by the reflecting oxide (for this purpose the thickness should be chosen as $W=1/2\ \lambda n_{ok}$), no surface capacity is being formed, of cause; spurious capacitance is absent. In such case, fianite film may reflect about 60% of radiance from the surface.

5.2 Silicon solar cells with fianite antireflecting layer

Experimental dependencies of antireflection of fianite films deposited on commercial solar cells were recorded. The reflection spectra of fianite obtained on two such samples are shown in Fig. 25a. The plots (Fig. 24) demonstrate excellent antireflecting properties of the fianite films. The plots also apparently demonstrate that the reflection drops to 0 - 1.5 % in the minima. A position of the minimum depends on the film thickness. At the film thinning the minimum occurs in the solar spectrum. Therefore, energy gain due to the application of the antireflecting fianite films approaches to 20-30%.

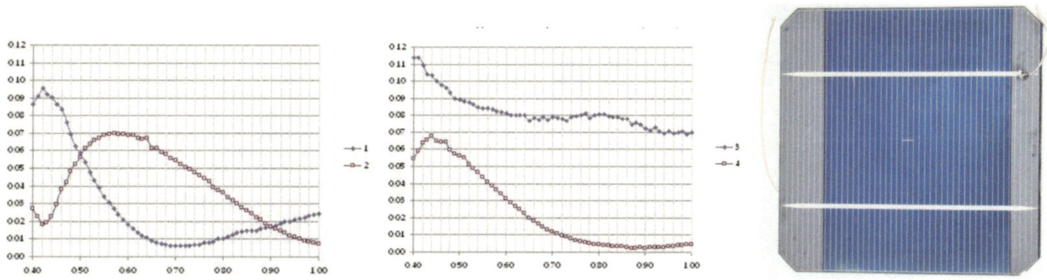

a b c

Fig. 24. Antireflecting properties of fianite films of 580 Å (a) and 1050 Å (b) thickness obtained on the industrial items (c) of Si solar cells of 4″ x 4″ size.

5.3 New ecologic technique of formation of *p-n* junctions in Si for solar sells

The majority of modern technologies of semiconductor devices are based on generation of different conductivity areas in the semiconductor and in particular of *p-n* junctions. For this purpose the crystals are doped, by means of three basic processes: diffusion, ion implantation, and irradiation. The donor and acceptor impurities coexist in real silicon crystals. Therefore there is an alternative possibility: to redistribute available impurities to fabricate the areas of different conductivity type. Traditional doping-based techniques (diffusion, ion implantation, a radiating doping) have the common drawbacks such as : (1) high temperatures of the processing. The diffusion doping is usually carried out at the temperature higher than 1100°C. After ion or radiating doping the subsequent high temperature annealing of the radiation defects should be carried out; (2) undesirable contamination of the crystal by new impurities could occur; (3) nearly all dopants are poisonous, leading to contamination of the environment. In the present work we consider a new technique of formation of p-n junctions in silicon, which provides facilitation and cost reduction of production process of semiconductor devices such as solar cells.

5.4.1 Experimental

The *p*-Si wafers were cut from boron-doped CZ crystals of various resistivities (with the boron concentration about 10^{15} cm^{-3}). The wafers were irradiated by 1-5 keV Ar ions in a gas discharge plasma. Then the wafers were cleaved, and the depth profile of the conductivity type was inspected by SEM-EBIC (scanning electron microscopy-electron beam induced current) technique. For this purpose, a high quality Schottky barrier was made by metal coating. This process resulted in formation of an *n-p* junction at some depth, which is clearly revealed as a sharp peak of EBIC signal (Fig. 25).

The formation of an n-type region below the irradiated surface was also confirmed by a conventional thermo-probe technique. Upon increasing the irradiation time, the depth of the n–p junction (denoted by X_d) increases (Fig. 26). Some n-type (phosphorus-doped) wafers were also irradiated and inspected; in this case no p–n junction was found [61, 62]. The junction depth X_d is a non-linear function of the irradiation time t. No junction was found in reference non-irradiated wafers. There is some "dead" time (1–15 min) in the junction propagation. After prolonged irradiation, the junction reaches some final position (Fig. 26 b) that can be quite close to the back (non-irradiated) surface of the wafer. In some range of duration (neither too short nor too long) the depth is roughly proportional to $t^{1/2}$ – a typical dependence for a diffusion process.

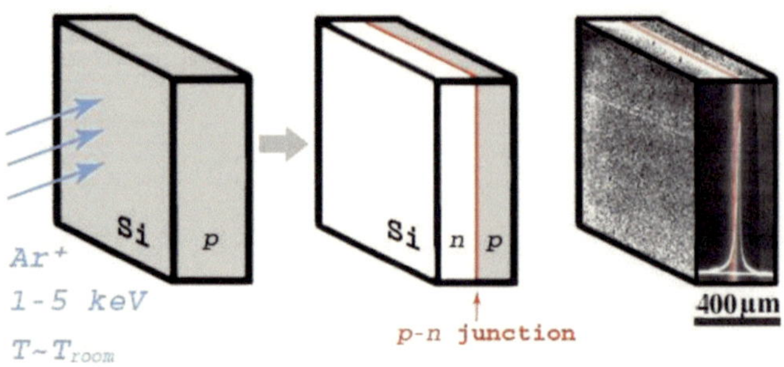

Fig. 25. Scheme of experiment

Peculiarities of n–p junction propagation. The n–p junction propagation was found to be sensitive to the state of the wafer surface. If the irradiated surface is bright polished, the junction moves faster, in comparison to the abrasion-polished surface. The surface defects, like scratches, cause a local distortion of the junction shape. The scratches at the backside 'attract' the junction. On the contrary, near the wafer edges, the junction propagation is retarded (Fig. 27). Striation non-uniformity of Si affects the shape of inversion p-n junction too (Fig. 28).

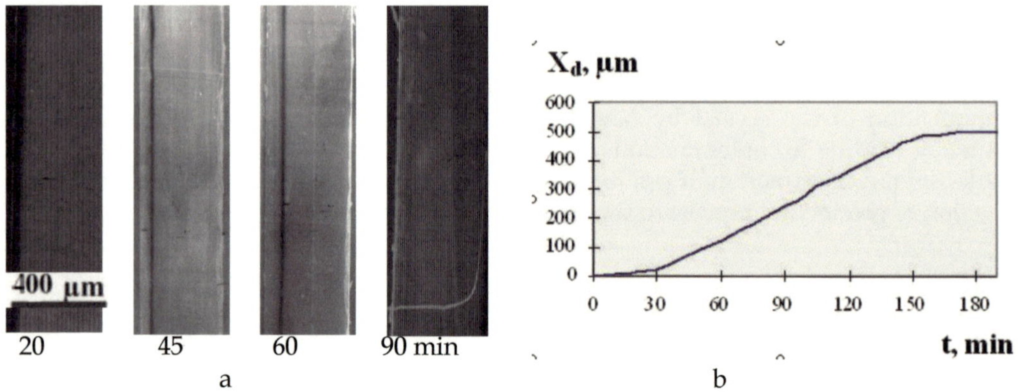

Fig. 26. SEM microphotograph made both in secondary emission and EBIC modes of inversion p-n junctions on a cleaved Si wafer (a). The wafers were irradiated at the left side. The dark vertical strip is the image of p-n junction. The p-n junction depth X_d in dependence of the time of exposure to Ar ions (b).

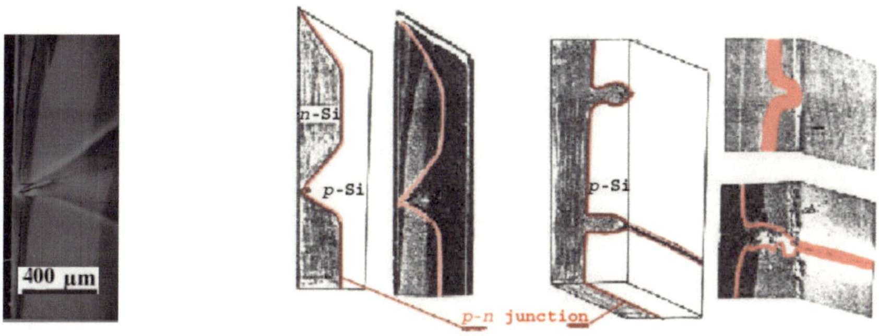

Fig. 27. Influence of surface damage in Si wafer on the shape of inversion *p-n* junction

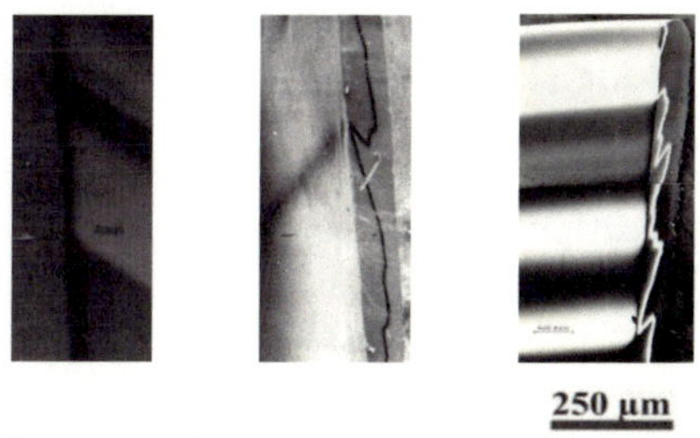

Fig. 28. Influence of non-uniformity of Si wafer on the form of inversion *p-n* junction

These results indicate to the role of the irradiation-induced self-interstitials: the local self-interstitial concentration is sensitive to the sinking ability of the sample surface. Particularly, the scratches at the backside may getter the surface impurities from the adjacent regions of the surface, thus improving the sinking ability of those regions.

One can argue that the *p–n* inversion is caused by some fast-diffusing donor impurity introduced during Ar irradiation. To check this possibility, we used the secondary-ion mass-spectrometry. An irradiated sample with a shallow *p–n* junction and a reference non-irradiated sample were inspected using layer-by-layer etching. No difference in the impurity content between the two samples was found which proves that the irradiation did not lead to any contamination of the sample near the surface.

It is therefore accepted that the *p–n* inversion is caused by in-diffusion of intrinsic point defects (self-interstitials) which leads to a loss of boron acceptors by kicking out the boron atoms B_s into the interstitial state B_i. The B_i atoms are known to be donors in *p*-Si [63]. Most likely, B_i will be paired to B_s, into neutral B_iB_s defects. The conductivity will then change to n-type, due to either isolated B_i or due to residual donors (phosphorus and grown-in thermal donors) that are present already in the initial state, before the irradiation.

The thermal donors are well known to be produced by a heat treatment around 450°C, and to be annihilated by annealing at $T > 600$°C. It was found that after several hours at 450°C, the *n–p* junction persisted. However, after one hour at 750°C the p–n junction disappeared. This result can be treated as an indication of role of the thermal donors. On the other hand, it can be attributed to conversion of B_i back into B_s by annealing at higher T.

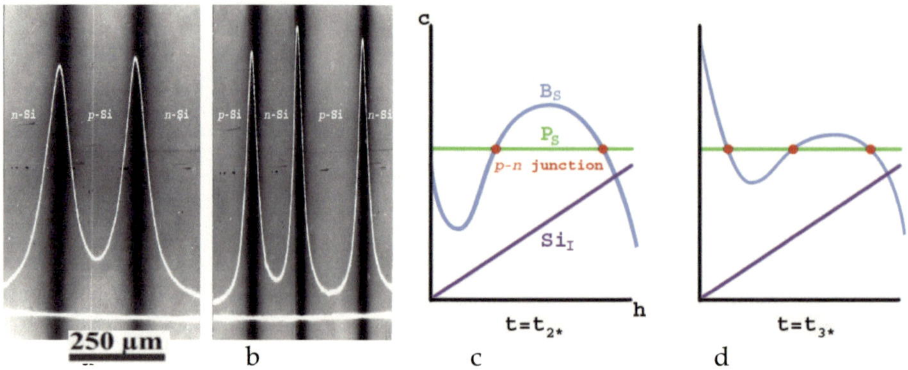

Fig. 29. Schematic profiles of self-interstitials and substitutional boron after irradiated at the front side (a) and back side (b), the horizontal line shows the zero level; model of the depth profile of the SEM–EBIC signal for a sample with two (c) and three (d) irradiation-induced p–n junctions.

By varying the irradiation conditions (for instance, using a two-side irradiation), multiple junctions can be produced. An example of a double and a triple junction is shown in Fig. 29.

5.4.2 Model

A proposed mechanism of this process consists mainly in the following [64]. The irradiation of the sample by inert ions generates a flux of silicon interstitial atoms Si_I directed from a surface in the bulk of the sample. Due to very high diffusivity of Si_I [65] (even at low temperatures), the steady non-uniform distribution of Si_I in a sample is formed (Fig. 30 a). Equilibrium concentration Si_I at low temperatures is very low, therefore a huge supersaturation of Si_I is created, which results in a sharp increase in the boron interstitial component, B_i. Reaction of kicking-out boron and the backward reaction ($B_i \rightarrow B_s + Si_I$) establish dynamically equilibrium ratio between B_i and Si_I. This ratio is proportional to the supersaturation of Si_I. Therefore the loss of boron acceptors will be more pronounced in the wafer part with a higher concentration Si_I. As a result the local inversion of conductivity occurs in this part. This model implies that the self-interstitials diffuse very fast at low T (below 100 $^{\circ}$C), and penetrate to the depth of at least 300 µm within 100 min (Fig. 30b). Accordingly the self-interstitial diffusivity, at the irradiation temperature, is at least as high as 10^{-7} cm^2/s.

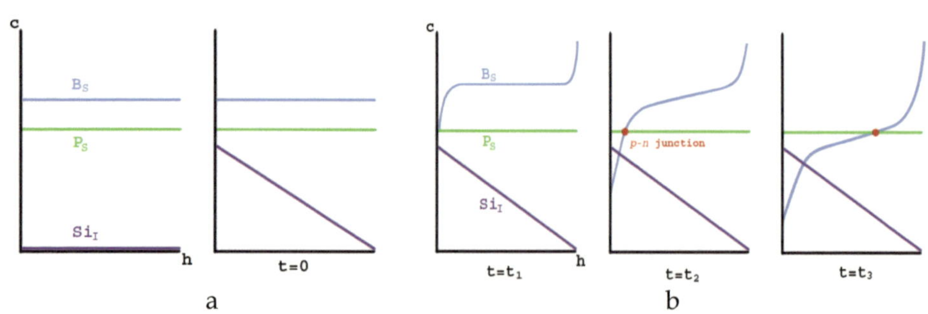

Fig. 30. Schematic profiles of self-interstitials and substitutional boron in the beginning of the process (a) and at successive time of intermediate stage of junction propagation (b).

In the subsequent discussion, we concentrate on the boron acceptor loss, assuming that the near-surface region contains some concentration of donors, N_d, which is less than the initial concentration of the boron acceptors, N_o.

A change in the substitutional boron concentration, due to the kick-out reaction (and due to the inverse reaction of kicking out the silicon lattice atoms by B_i), is described by a simple equation,

$$dN_s/dt = -\alpha\,(N_sC - KN_i) \tag{1}$$

where C is the local (depth-dependent) self-interstitial concentration, α is the kinetic constant of the direct kick-out reaction and K is the equilibrium constant in the mass-action law that relates the concentration for the case of equilibrium between the reacting species ($N_sC/N_i = K$). The highest self-interstitial concentration, C_f, is reached near the front surface; it is defined by the balance of the production rate (proportional to the Ar flux) and the consumption rate by local Ar-produced vacancies and by sinking of self-interstitials at the front surface. With specified C, the concentration ratio of the interstitial and substitutional boron species, N_i/N_s, tends to C/K due to the reaction (1). A strong loss of acceptors occurs if $C >> K$. It is therefore assumed that this inequality holds at least at the front surface: $C_f >> K$.

Initial stage of boron acceptor loss. At short irradiation time, the term KN_i in Eq. (1) is negligible, and the boron concentration near the front surface is lost exponentially,

$$N_s(t) = N_o \exp(-\alpha\, C_f\, t) \tag{2}$$

The *n–p* junction appears when N_s becomes less than N_d. This moment (t_d) lies experimentally, between 1 and 10 min. The product $\alpha\, C_f$ is estimated, from E_q. (2), to be in the range 0.01–0.001 s^{-1}.

Propagation of the n–p junction The near-surface region - where a large fraction of boron is already displaced into interstitial state Bi (and then paired into B_iB_s) –expands as more self-interstitials diffuse from the front surface into the bulk. The mass action law, $N_sC/N_i=K$, is valid at duration longer than the kick-out reaction time (10 min or less). The boron-depleted region corresponds, approximately, to the condition $C(x) > K$. At not too long duration, the self-interstitials penetrate to some limited depth (Fig. 30 b), and the n–p junction resides at some intermediate position within the sample. Finally, the C(x) profile approaches a steady-state linear shape: the interstitials generated at the front surface are consumed at the back surface (Fig. 30 b). Therefore, the *n–p* junction does not reach the back surface but stops at some final position, just like observed.

The self-interstitial flux into the sample bulk is, approximately, $DC_f/\ x_d$, where x_d is the size of the boron-depleted region (x_d is almost identical to position X_d of the n–p junction). The total amount of the remaining boron is equal to $C_o(L-x_d)$, where L is the wafer thickness. The boron loss rate, dQ/dt, is twice as large as the above self-interstitial flux (each consumed Si_I leads to a loss of two Bs: one by kick-out, and the other by pairing of B_i to B_s). The following equation provides a solution for the junction depth $x_d(t)$,

$$x_d = (4DC_ft/N_o)^{1/2} \tag{3}$$

The DC_f product is estimated to be $5x10^7$ $см^2с^{-1}$ from Eq. (3). Above, we estimated the product $\alpha\, C_f$ (where α is the kick-out kinetic coefficient). By these numbers, the $\alpha\,/\,D$ ratio is of the order of 10^{-10} cm. If the kick-out reaction were limited just by self-interstitial diffusion (which means that any 'encounter' of a self-interstitial with the boron atoms immediately leads to the boron displacement into the interstitial state), the coefficient a/D ratio would be equal to $4\pi r=4 \times 10^{-7}$ cm, where r is of the order of the interatomic distance. The difference between the two numbers indicates some kinetic barrier (roughly, 0.25 eV) for the kick-out reaction.

A possibility of long-range migration of interstitial boron. It was assumed in the above discussion that the boron atoms displaced into interstitial state do not diffuse much from the initial location. The alternative possibility is that the B_i species are of high mobility (comparable to the self-interstitial mobility), and therefore they can migrate to the distance comparable to the sample thickness. In this case, a considerable spatial redistribution of boron impurity would occur. The final profile of B_i would be smoothed by diffusion to some constant (depth-independent) concentration N_i. The mass-action law would then imply that the substitutional boron concentration, $N_s = N_iK/C$, is inversely proportional to C(x). Therefore, substitutional boron would accumulate near the back surface, where C(x) is at minimum. Such a profile of substitutional impurity (with a well-pronounced accumulation at the back surface) is typical during in-diffusion of Au and Pt impurities [65, 66].

A formation of triple junction (Fig. 29 b) can be accounted for by the long-distance migration of B_i. The boron profile after the first irradiation is of the type shown in Fig. 30 b ($t=t_2$), with just one junction. The second (back-side) irradiation creates an *n*-region near the back surface, and also results in the boron acceptor accumulation near the front side (now non-irradiated). Then a region adjacent to the front side becomes again of *p*-type conductivity. The resulting structure is *p–n–p–n* (Fig. 29b, d).

5.4.3 Application of solar and the other

Due to physical nature and peculiarities of generation of the inverse *p-n* junction its application is the easiest and the most efficient in those semiconductor technologies, which involve the development of two-dimensional (flat) *p-n* transitions over a considerable square [67]. Solar cells are related to the devices of such a type. Further development of these investigations will lead to the new technology of *p-n* junction formation in silicon, providing simplification and cost reduction of production process of solar cells and another various semiconductor devices.

5.5 The development of III-V heterostructures for solar cells

The following four types of heterostructures based on InGaAsP compounds have been developed for solar cells by means of MOGPE using Aixtron AIX 200RF installation and capillary epitaxy technique under improved conditions and by using the refined technologies of the epitaxial growth on GaAs substrates of *n*- and *p*-types of conductivity:

1. *p*-GaAs substrate, 2.7 μm absorption range, InGaP window - 0.2 μm, contacting layer GaAs – 0.2 μm;
2. *p*-GaAs substrate, InGaP barrier layer - 0.03 μm, 3 μm – GaAs absorption range, InGaP window - 0.06 μm, contacting layer – 0.6 μm;
3. **n**$^+$- GaAs substrate, barrier InGaP - 0.2 μm, absorption range - undoped GaAs 3 μm, emitter - *p*-GaAs - 0.4 μm, *p*-InGaP window- 0.02 μm, contacting layer p-GaAs - 0.4 μm;
4. n^+- GaAs substrate, buffer n^+-GaAs-0.6 μm; barrier n^+-InGaP-0.05 μm; absorption layer undoped GaAs-3.3 μm; emitter p^+-GaAs-0,07 μm; window p^+- InGaP0,07 μm; contacting layer p^{++}-GaAs-0.2 μm; antireflecting layer fianite-0.1 μm;.

Prototypes of the solar cells have been manufactured using the obtained samples. Fianite films were used as antireflecting and protective coatings [68, 69]. The films were deposited by means of magnetron sputtering (Fig. 31).

The study of characteristics of the prototypes of the heterostructure solar cell has shown 20-30 % efficiency gain due to the application of the fianite antireflecting films.

a b

Fig. 31. Prototype sample of the heterostructure solar cell of 40x40 mm size supplied with fianite antireflecting coating: functional side with Au contacting routs (a), reverse side with Au+Ti contacting layer (b)

6. Conclusions

The unique properties of fianite as monolithic substrute and buffer layer for Si, Ge and $A^{III}B^{V}$ compounds epitaxy; protecting, stabilizing and antireflecting coatings, as well as a gate dielectric in photosensitive opto-electronic devices have been demonstrated. The results obtained in this work have actually demonstrated advantages of fianite as novel multipurpose material for new optoelectronics technologies.

7. References

[1] V.I. Aleksandrov, V.V. Osiko, A.M. Prokhorov, V.M. Tatarintsev. New technique of the synthesis of refractory single crystals and molten ceramic materials. Vestnic AN SSSR №12 (1973) 29-39. (in Russian)

[2] V.I. Aleksandrov, V.V. Osiko, V.M. Tatarintsev, A.M. Prokhorov. Synthesis and crystal growth of refractory materials by RF melting in a cold container. Current Topics in Materials Science. 1 (1978) 421-480.

[3] E.E. Lomonova, V.V. Osiko: Growth of Zirconia Crystals by Skull-Melting Technique. In: Crystal Growth Technology. Ed. by H.J. Scheel and T. Fukuda (John Wiley @ Sons, Chichester, England 2003) pp. 461-484

[4] Yu.S. Kuz'minov, E.E. Lomonova, V.V. Osiko Cubic zirconia and skull melting Cambridge International Science Publishing Ltd., UK, 346 pp. 2009

[5] V.V.Osiko, M.A.Borik, E.E.Lomonova Synthesis of refractory materials by skull melting technique. Handbook of Crystal Growth, Springer, 2010, p.433-469.

[6] I. Golecki, H.M. Manasevit, L.A. Moudy, J.J. Yang, and J.E Mee. Heteroepitaxial Si films on yttria-stabilized, cubic zirconia substrates. // Appl. Phys. Lett., 1983, v. 42, No. 6, p. 501-503.

[7] D. Pribat, L.M. Mercandalli, J. Siejka, J. Perriere Interface oxidation of epitaxial silicon deposits on (100) yttria stabilized cubic zirconia, J.Appl.Phys. 58, 313-320 (1985)
[8] L.M. Mercandalli, D. Diemegand, M. Crose, Y. Sierka. Recent progress in epitaxial growth of semiconducting materials on stabilized zirconia single crystals, Proc.Soc.Photo-Opt. Istr.Eng. 623, 133-210 (1986)
[9] V.G. Shengurov, V.N. Shabanov, A.N. Buzynin, et al., Mikroelektronika, no. 6 (1996) 204.
[10] A.N. Buzynin, V.V. Osiko, E.E. Lomonova, Yu.N. Buzynin, A.S. Usikov. Epitaxial films of GaAs and GaN on fianite substrate. // Wide–Bandgap semiconductors for High Power, High Frequency and High Temperature. (Mat. Res. Soc. Simp. Proc. 1998). Vol.512, Pittsburg, PA, p. 205–210.
[11]A.N. Buzynin, V.V. Osiko, Yu.K. Voron'ko, et al. // Izv. Akad. Nauk, Ser. Fiz. 69 (2005) 211.
[12] P.A. Anderson, C.E. Kendrick, R.J. Kinsey, et. al. (111) and (100) YSZ as substrates for indium nitride growth. // Physica status solidi (c), 2005, v. 2, N.7, p. 2320 - 2323.
[13] Y.Z. Yao-Zhi Hu, S.P. Sing-Pin Tay, J. Vac. Sci. Technol. B 19 (2001) 1706.
[14] S.J. Wang, C.K. Ong, S.Y. Xu, et al. Electrical properties of crystalline YSZ films on silicon as alternative gate dielectrics, Semicond. Sci. Technol. 16, L13-L16 (2001)
[15] S.J. Wang, C.K. Ong. Rapid thermal annealing effect on the electrical properties of crystalline YSZ gate dielectrics. // Semiconductor Science and Technology, 2003, v. 18, p. 154-157.
[16] A.N. Buzynin, V.V. Osiko, E.E. Lomonova, at. el. Epitaxial films of III–V compound on fianite substrates. // Proc. of International Congress on Advanced Materials, their Processes and Applications, 25–28 September 2000, Munich, Germany, Article 672.
[17] A.N. Buzynin, V.V. Osiko, Yu.N. Buzynin, B.V. Pushnyi. Growth of GaN on fianite by MOCVD capillary epitaxy technique. // MRS Internet Journal of Nitride Semiconductor Research. 1998, Vol.4, Article 49. (http://nsr.mij.mrs.org/4/49/).
[18] A.N. Buzynin, V.V. Osiko, Yu.N. Buzynin et al. // Izv. Ross. Akad. Nauk, Ser. Fiz., 2002, v. 66, n. 9, p. 1345-1350.
[19] A.N. Buzynin, V.V. Osiko, E.E. Lomonova, Yu.N. Buzynin, A.S. Usikov. Epitaxial films of GaAs and GaN on fianite substrate. // Wide–Bandgap semiconductors for High Power, High Frequency and High Temperature. (Mat. Res. Soc. Simp. Proc. 1998). Vol.512, Pittsburg, PA, p. 205–210.
[20] R. Paszkiewicz, B. Paszkiewicz, R. Korbutowicz, et al, MOVPE GaN Grown on Alternative Substrates // Cryst. Res. Technol.2001, v.36, N 8-10, p. 971-977.
[21] P.D.C. King, T.D. Veal, S.A. Hatfield, et. al. X-ray photoemission spectroscopy determination of the InN/yttria stabilized cubic-zirconia valence band offset. // Appl. Phys. Let., 2007, v. 91, p. 112103-1 - 112103-3.
[22] T. Nakamura, Y. Tokumotoa, R. Katayamaa, et. al. RF–MBE growth and structural characterization of cubic InN films on yttria–stabilized zirconia (001) substrates. // J. of Crystal Growth, v. 301–302, April 2007, p. 508–512.
[23] A.N.Buzynin and V.P. Kalinushkin, Laser characterization of sapphire defectts. Proc. of Inter. Congress" Materials Week-2001": Advanced Materials, their Processes and Applications, 1-4 October 2001, Munich, Germany, Article 389.
[24] A.N.Buzynin and V.P. Kalinushkin New technique of laser characterization of sapphire and fianite crystals // Abstracts of the Laser Interaction with Matter International Symposium" (LIMIS2010),
Aug. 15th to 18th, 2010, Changchun, China.
[25] G.G. Shahidi. SOI technology for the GHz era. // IBM J. of Research and Development, 2002, v.46, No. 2/3, p. 121–132.

[26] A.N. Buzynin, V.V. Osiko, Yu.N. Buzynin, et al. // Bulletin of the Russian Academy of Sciences: Physics 74 (2010) 1027-1033
[27] N.N. Sheftal and A.N. Buzynin// Vestnik Moscow Univ. 3 (1972) 102; Nat. Trans. Cent. The Yohn Crenan Library, 35w, 33rd St., Chicago IL 60616 USA.
[28] Smith, H. I.; Flanders, D. C.// Appl. Phys. Lett. 1978, *32*, 349.
[29] Masatomo Sumiya, Youichi Kurumasa, Kohji Ohtsuka, et al., J. Cryst. Growth, vol. 237—239 (2002) 1060.
[30] E.I. Rau, A.N. Zhukov, E.B. Yakimov, // Solid-State Phenomena 1998, v. 327, p. 53–54.)
[31] A.N.Buzynin, Yu.N.Buzynin, A.V.Belyaev, A.E.Luk'yanov, E.I.Rau. Growth and defects of GaAs and InGaAs films on porous GaAs substrates. // Thin Solid Films 2007, v. 515, p. 4445–4449.
[32] G.G. Shahidi. SOI technology for the GHz era. // IBM J. of Research and Development, 2002, v.46, No. 2/3, p. 121–132.
[33] S.J. Wang, C.K. Ong, S.Y. Xu, P. Chen, W.C. Tjiu, J.W. Chai, C.H. Huan, W.J. Yoo, J.S. Lim, W. Feng, W.K. Choi Cristalline zirconia oxide on silicon as alternative gate dielectrics, Appl.Phys.Letters 78, 1604-1606 (2001)
[34] S.J. Wang, C.K. Ong, S.Y. Xu, P. Chen, W.C. Tjiu, A.C.H. Huan, W.J. Yoo, J.S. Lim, W. Feng, W.K. Choi Electrical properties of crystalline YSZ films on silicon as alternative gate dielectrics, Semicond. Sci. Technol. 16, L13-L16 (2001)
[35] T. Ngai, W.J. Qi, R. Sharma, J. Fretwell, X. Chen, J.C. Lee, S. Banerjee Electrical properties of ZrO_2 gate dielectric on SiGe, Applied Physics Letters 76, 502-504 (2000)
[36]S Abermann, G Pozzovivo, J Kuzmik et al MOCVD of HfO_2 and ZrO_2 high-*k* gate dielectrics for InAlN/AlN/GaN MOS-HEMTs. // Semicond. Sci. Technol., 2007, v. 22, p. 1272–1275
[37] H. Fukumoto, T. Imura and Y. Osaka. Heteroepitaxial growth of yttria–stabilized zirconia on silicon // Jpn. J. Appl. Phys., 1988, v.27, No.8, p. L1404–L1405
[38] T. Hata, K. Sasakia, Y. Ichikawa and K. Sasakia. Yttria-stabilized zirconia (YSZ) heteroepitaxially grown on Si substrates by reactive sputtering. // Vacuum, 2000, v.59, issues 2-3, p. 381-389
[39] A.N. Buzynin, V.V. Osiko, V.V. Voronov, Yu.N. Buzynin at al. Epitaxial Fianit films on Si and GaAs. // Bulletin of the Russian Academy of Sciences. Physics., 2003,v.67, № 4, p.586–587
[40] V.G. Beshenkov, A.G. Znamenskii, V.A., Marchenko at al. Widening temperature range of epitaxial growth of YSZ films on Si [100] Расширение области температур эпитаксиального роста пленок YSZ на Si [100] under magnetron sputtering. // Journal of technical physics, 2007, v.77, N5, p. 102–107
[41] J.M. Vargas, P. Brown, T. Khan et. al. Superconducting half-wave microwave resonator on YSZ buffered Si(100). // Applied Superconductivity, 2001, v. 11, N. 1, p.392 - 394
[42] C A. Copetti, H. Soltner, J. Schubert, et. al. High Quality Epitaxy of $YBa_2Cu_3O_{7.5}$ on Silicon- on-sapphire with the Multiple Buffer Layer YSZ/CeO_2. Applied Physics Letters, 1993, v. 63, N 10, p.1429-1431
[43] T. Khan, P. Brown, Yu. Vlasov et al. $YBa_2Cu_3O_7$ on Sputter Deposited ZrO_2 Buffered (100) Si - Processing and Characterization," Advances in Cryogenic Engineering (Materials), 2000, v. 46, p. 901-905
[44] J. Sanghun; T. Matsuda; U. Akira; W. Kiyotaka; I. Yoko; H. Hyunsang; Interfacial properties of a hetero-structure YSZ/p-(100)Si prepared by magnetron sputtering. Vacuum, 2002, vol. 65, nº1, p. 19 - 25
[45] N. Wakiya, T. Yamada, K. Shinozaki, and N. Mizutani: Heteroepitaxial growth of CeO_2 Thin Film on Si(001) with an Ultra Thin YSZ Buffer Layer. Thin Solid Films, 2000, v.371, p.211-217.

[46] S. Kaneko, K. Akiyama, T. Ito et. al. Single domain epitaxial growth of yttria-stabilized zirconia on Si (111) substrate. // Ceramics International, 2008, article in press.
[47] R. Lyonnet, A. Khodan, A. Barthe´le´my, et al. Pulsed laser deposition of $Zr_{1-x}Ce_xO_2$ and $Ce_{1-x}La_xO_{2-x/2}$ for buffer layers and insulating barrier in oxide heterostructures. // J. of Electroceramics, 2000, 4, No. 2/3, p. 369–377
[48] N. Pryds, B. Toftmann, J.B. Bilde-Sørensen et al. Thickness determination of large-area films of yttria-stabilized zirconia produced by pulsed laser deposition Applied Surface Science, 2006, v. 252, issue 13, p. 4882-4885
[49] A.P Caricato., G.A. Barucca, Di. Cristoforo et al. Excimer pulsed laser deposition and annealing of YSZ nanometric films on Si substrates // Applied Surface Science, 2005, v. 248, issues 1-4, p. 270 - 275
[50] Sang-Chul Hwang and Hyung-Shik Shin. Effect of Deposition Temperature on the Growth of Yttria-Stabilized Zirconia Thin Films on Si(111) by Chemical Vapor Deposition. // J. of the American Ceramic Society, 1999, v. 82, issue 10, p 2913-2915
[51] A. Osinsky, S. Gangopadhyay, J. W. Yang, et al. // Appl.Phys.Lett. 72 (1998) 561
[52] M. F. Wu, A. Vantomme and G. Langouche, *B.* S. Zhang and Hui Yang, Appl.Phys.Lett. 80 (2002) 4130
[53] Supratik Guha and Nestor A. Bojarczuk, Appl.Phys.Lett. 72 (1998) 415
[54] Lianshan Wang, Xianglin Liu, Yude Zan, Jun Wang, Du Wang, Da-cheng Lu, and Zhanauo, Appl.Phys.Lett. 72 (1998) 109
[55] N. P. Kobayashi,"' J. T. Kobayashi, P. D. Dapkus,"' W.-J. Choi, and A. E. Bond, Appl.Phys.Lett. 71 (1997) 3569
[56] A. Ohtani, K. S. Stevens, and R. Beresford, Appl.Phys.Lett. 65 (1994) 61
[57] A.J. Steckl, J. Devrajan, C. Tran and R. A. Stall, Appl.Phys.Lett. 69 (1996) 2264
[58] Y.Z. Yao-Zhi Hu, S.P. Sing-Pin Tay, J. Vac. Sci. Technol. B 19 (2001) 1706
[59] A.N. Buzynin, T.N. Grishina, T.V. Kiselyov, et al. Zirconia-based Solid Solutions – New Materials of Photoelectronics. // Optical Memory & Neural Networks (Information Optics). № 4, 2009, p.323-331
[60] A.N.Buzynin, E.E.Lomonova, T.N.Grishina, et al. Semiconducting photosensitive structures with passivating protective zirconia coating. // Applied physics. № 2. 2009, p. 105-109
[61] A.N. Buzynin, A.E. Luk'yanov, V.V. Osiko, V.V. Voronkov, Mater. Res. Soc. Proc. (Pittsburg, PA) 378 (1995) 653.
[62] A.N. Buzynin, A.E. Luk'yanov, V.V. Osiko, V.V. Voronkov, Mater. Res. Soc. Proc. (Pittsburg, PA) 510 (1998).
[63] R.D.Harris, J.L.Newton and G.D.Watkins, Phys.Rev. B 36, 1094 (1987).
[64] A.N.Buzynin, A.E.Luk'yanov, V.V.Osiko, V.V.Voronkov. Nuclear Instrum. and Methods in Physics Research, B 186 (2002), p.366-370.
[65] N.A. Stolwijk, J. Holzl, W. Frank, E.R. Weber, H. Mehrer, Appl. Phys. A 39 (1986) 37.
[66] H. Zimmermann, H. Ryssel, Appl. Phys. A 55 (1992) 121.
[67] Patent "Metod of p-n junction formation in silicon", № RU 2331136 C2., 10.08.2008.
[68] A.N. Buzynin, V.V. Osiko, Yu.N. Buzynin, et al Fianite: a multipurpose electronics material //Bulletin of the Russian Academy of Sciences: Physics, 2010, Vol. 74, No. 7, pp. 1027–1033.
[69] A.N. Buzynin, Yu.N. Buzynin, V.V. Osiko, et al Antireflection fianite and ZrO_2 coatings for solar cells //Bulletin of the Russian Academy of Sciences: Physics, 2011, Vol. 75, No. 9, 1213-1216.

7

Hybrid Polyfluorene-Based Optoelectronic Devices

Sylvain G. Cloutier
École de Technologie Supérieure
Canada

1. Introduction

It is now well established that controlled delocalization in π-conjugated chains can lead to unique optoelectronic properties in polymer materials (MacDiarmid, 2001). While still being in a relatively early developmental stage, conjugated-polymer systems with a myriad of optoelectronic properties can now be synthesized at relatively low costs. Albeit a very promising technology, there remains some key challenges to address before efficiently integrating these conjugated-polymer systems into large scale applications including displays, biomedical imaging & sensing, lab-on-a-chip, solid-state lighting and photovoltaic devices and architectures (Arias et al., 2001; Moons, 2002; Morteani et al., 2003).

Pending material issues still limit the functionality and the overall performances of these emerging material systems, while photo-chemical degradation can severely restrict their lifetimes. Since the main photo-chemical degradation process is usually a photo-oxidation reaction that truncates the conjugation length of the polymer chain to reduce the π-electron delocalization, this undesireable process can be significantly suppressed by simply permeating the structure with a transparent dielectric coating. Using such coating technologies, lifetimes of 20 years have now been demonstrated for commercial polymer-based photovoltaic devices and displays.

In the last few years, the hybrid integration of semiconductor nanocrystals within conjugated polymer host systems has grown into a very active research area as it provides a new pathway of (1) improving the conjugated polymers' optoelectronic properties and/or (2) providing added functionality to the conjugated polymer-based structures.

This chapter will present a general overview of hybrid polymer-nanocrystal material systems and their application as low-cost optoelectronic devices. Using a device-engineering perspective, we will focus our attention on the synthesis & processing, structural and optoelectronic properties of polyfluorene-based systems interfaced with lead-sulfosalt (PbS) semiconductor nanocrystals grown by hot-colloidal methods. Using this specific hybrid material system as our case study, we will begin by providing a general understanding of pure polymer-based type-II heterostructures and their limitations. Then, we will demonstrate how the incorporation of lead-chalcogenide quantum dots can be used to (1) add new functionality and (2) improve the performances of those all polyfluorene-based optoelectronic devices. Most importantly, we will also see that the hybrid integration of conjugated polymers and colloidal quantum dots also raises many important fundamental questions and crucial technical challenges to address before achieving low-cost hybrid optoelectronic devices with superior performances.

In the long-term, we strongly believe this emerging class of hybrid polymer-based heterostructures will potentially transform the field of opto-electronics by providing low-cost and high-performance semiconductor-based nanocomposite materials and devices for applications such as light sources, biomedical & lab-on-a-chip devices, flexible and/or high-performance optoelectronics platforms and photovoltaics.

2. Polyfluorene-based type-II heterostructures

Compared to more commonly-used π-conjugated polymer families such as the polyphenylenes (PPPs and PPVs), polythoiphenes (PTs), polypyrroles (PPYs) and polyanilines (PANIs), polyfluorenes tend to be easier to process and less sensitive to photo-chemical degradation while still offering very decent optoelectronic properties. This combination of facile processing and durability makes polyfluorenes an ideal case-study platform.

In particular, relatively efficient polymer light-emitting structures for the visible have been previously realized using polyfluorene-based type-II heterostructures fabricated using poly(9,9-dioctylfluorene-co-N-(4-butylphenyl)diphenylamine (or **TFB**) as the hole-transporting material and poly(9,9'-dioctylfluorene-co-benzothiadiazole (or **F8BT**) as the electron-transporting polymer (Moons, 2002).

Meanwhile, decent photovoltaic structures have also been realized using similar polyfluorene-based type-II heterostructures. However, the best performances reported so far for polyfluorene-based photovoltaics were also using F8BT as the electron-transporting polymer but replacing the hole-transporting TFB with poly (9,9'-dioctylfluorene-*co*-bis-*N,N'*-(4-butylphenyl)-bis-*N,N'*-phenyl-1,4-phenylenediamine) (or **PFB**) (McNeill et al., 2009).

Finally, it is important to mention that while hole-transporting π-conjugated polymers with conductivities over 1000 $\Omega^{-1}cm^{-1}$ have been reported, decent electron-transporting polymers are much more difficult to come by. As such, the electron-transporting material often limits the overall performances of polymer-based optoelectronic devices (Moons, 2002).

2.1 Blended all polyfluorene-based type-II heterostructures for light-emitting and photovoltaic device architectures

TFB-F8BT system provides a great material platform to understand the basic principles associated with conjugated polymer-based optoelectronic devices. For the typical light-emitting diode (LED) configuration shown in Figure 1(a), the hole-transporting TFB (labeled HTL) and electron-transporting F8BT (labeled ETL) are used to provide a type-II heterostructure. In this configuration, holes are injected from the transparent ITO anode while electrons are injected from the Aluminum cathode. To facilitate carrier injection and reduce exciton quenching at the electrode-polymer interface, optional hole- and electron-injection layers can be introduced in such device architectures. Here, a thin layer of poly(ethylenedioxythiophene):polystyrenesulphonate (or **PEDOT:PSS**) can be used to facilitate the hole injection and improve the structural quality of the TFB film by alleviating the surface roughness of the indium-tin oxide (ITO) substrate. In contrast, a thin layer of low work-function metal such as Calcium can be used to facilitate the electron-injection on the other side of the junction. In such a case, the Aluminum contact remains necessary to prevent oxidation of the low work-function metal. In this system, injected carriers bind into an exciton at the TFB-F8BT interface. Due to the band alignment, this exciton is much more likely to migrate to the electron-transporting F8BT until it recombines radiatively and generates the emission.

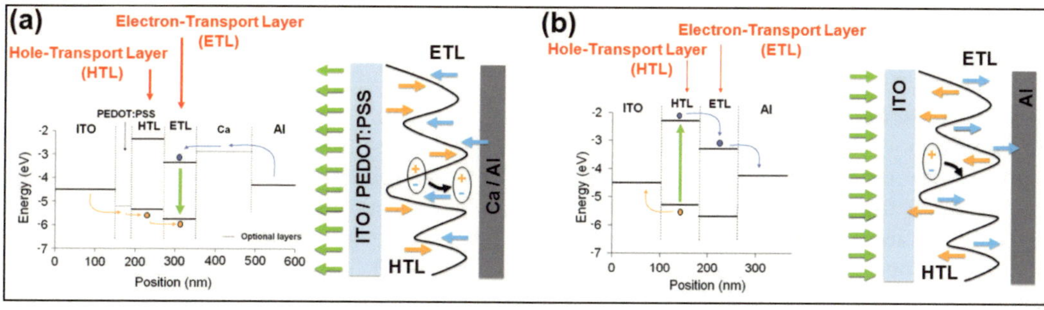

Fig. 1. Energy diagrams and schematics of TFB-F8BT polyfluorene-based type-II heterostructures. (a) For use as visible light-emitting diode. (b) For use as a photodetector or solar cell device structure.

In contrast, Figure 1(b) illustrates how a similar heterostructure can be used as a photovoltaic device. There, the exciton is photo-generated in the hole-transporting TFB and dissociates upon meeting the energy barrier at the TFB-F8BT interface to allow carrier extraction. Of course, this structure does not require the PEDOT:PSS and Ca layers previously used to facilitate carrier injection.

Due to the very low mobilities in conjugated polymers compared with conventional semiconductors, it is clear that the bulk of these devices' optoelectronic properties stem from the interface between the hole- and electron-transporting polymers. In the case of polymer-based LED structures, the exciton will usually recombine within tens of nanometers from the ETL-HTL interface. Meanwhile, any exciton generated more than tens of nanometers from the ETL-HTL interface in photovoltaic device structures will recombine radiatively before reaching the surface and those carriers will be lost. To enable an easy processing, these all polyfluorene-based type-II heterostructures are usually formed using a **blended** precursor solution containing both polymers dissolved in a given solvent. When this blend is deposited on the ITO substrate by spin- or dip-coating, phase-separation occurs as the solvent evaporates. This leads to the formation of HTL-rich and ETL-rich domains such as shown in Figure 2 (Moons, 2002).

2.2 The importance of the domain sizes and the crystalline phase in polyfluorene-based thin-film structures

Based on the previous discussion, we now understand that the optoelectronic properties of those conjugated polymer-based heterostructures will depend largely on the interface between the hole-transporting and electron-transporting polymers. As such, an intuitive

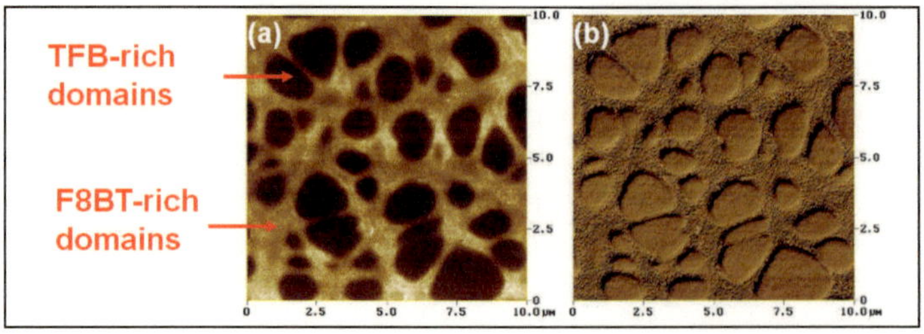

Fig. 2. Example of a large domain structure in a polyfluorene-based type-II heterostructures using xylene-based precursor measured using AFM. Adapted from (Moons, 2002) with permission.

way to controllably alter the optoelectronic properties of those structures would be to control the domain size. One relatively simple way of doing this is by changing the solvent used in the precursor solutions. Since the domains form by phase-separation, faster evaporations rates (using solvents such as chloroform, acetone or hexane) will generally yield significantly smaller domains compared with lower evaporation rates (using solvents such as toluene or xylene). As an example, Figure 3 shows a comparison of the photoluminescence from films of pure TFB, pure F8BT and 1:1 blends of TFB:F8BT obtained both from toluene- and chloroform-based precursors.

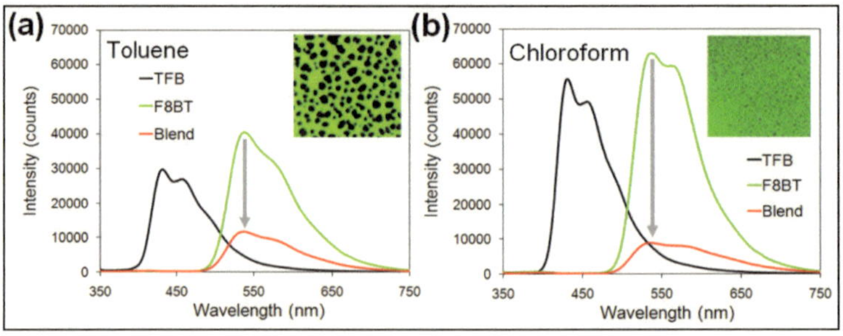

Fig. 3. Blended TFB-F8BT polyfluorene-based type-II heterostructures using different solvents. (a) Fluorescence emission from pure TFB, pure F8BT, and 1:1 ratio TFB:F8BT blended films obtained from toluene-based precursor solutions. (b) Same measurements obtained from chloroform-based precursors. The insets shows confocal fluorescence images of the domain structures in blended films. The insets are Adapted from (Moons, 2002) with permission.

In the toluene-based blend, the quenching of the F8BT emission by the TFB is much less pronounced compared with the chloroform-based blend. Indeed, the larger domains in the toluene-based blend provide more leasure for the photo-generated excitons to recombine radiatively before reaching another domain interface. In the chloroform-based blend, the probability for radiative-recombination is much lower since the photo-generated excitons are more likely to hit another domain interface and dissociate before they can recombine radiatively. This results precisely in the severely quenched F8BT fluorescence seen in Figure 3(b). Based on this result, it is obvious that larger domains will generally be preferable for light-emitting diode structures while smaller domains will be more desirable for photo-detector or photovoltaic device architectures.

For example, Figure 4 shows the optimal visible light-emitting diode architecture we obtain using a blended TFB:F8BT type-II heterostructure. Here, the structure is formed by spin-coating a blended 1:1 toluene-based precursor atop a thin PEDOT:PSS hole-injection layer previously spin-coated directly on a commercial ITO substrate. After annealing at 160°C, the top electrode was evaporated through a shadow mask.

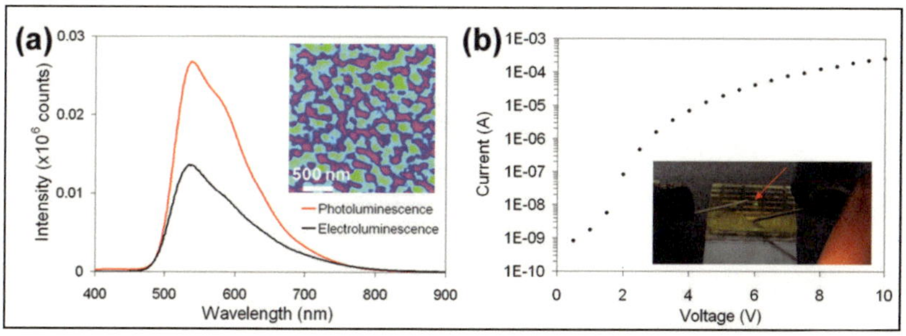

Fig. 4. Blended TFB-F8BT polyfluorene-based light-emitting diode. (a) Typical photoluminescence and electroluminescence from the TFB:F8BT blend at 19 V. The inset shows the typical domain structure obtained by phase separation between the hole-transporting TFB (pink) and the electron-transporting F8BT (green). (b) The current-voltage characteristics for this 520 nm LED structure. The inset shows a typical small-area device and the arrow points to the active device area.

3. Migrating the emission of polyfluorene-based LEDs towards the near infrared using lead-sulfosalt (PbS) colloidal quantum dots

Due to their relatively large HOMO-LUMO separations, conjugated polymer-based light-emitting diodes are perfectly suited for operation in the visible but their potential for near-infrared operation remains limited. As we mentioned previously, the hybrid integration of semiconductor nanocrystals and conjugated polymer material systems can provide an easy pathway for (1) improving the conjugated polymer-based devices' optoelectronic properties and/or (2) providing added functionality to the conjugated polymer-based device structures.

Indeed, semiconductor quantum dots have been recently used to controllably-alter the optoelectronic properties of a wide variety of host systems for biosensing, light-emitting or photovoltaic applications (Bakueva et al., 2003; Liu et al., 2009; McDonald et al., 2005; Steckel et al., 2003; X. Zhang et al., 2007).

Owing to low bandgaps, ultrafast recombination processes and large nonlinear coefficients, crystalline lead-salt chalcogenides (a sub-group of the IV-VI semiconductor family) have been one of the basic materials used in modern infrared light sources & lasers, photodetectors and high-performance thermoelectric for the last 50 years (Klann et al., 1995; Preier, 1979). Bulk lead sulfosalt (PbS) is well-suited for infrared optoelectronics, having a direct 0.41 eV bandgap and uncommonly large exciton binding energy (close to 300 meV).

Meanwhile, the first colloidal synthesis of chalcogenide semiconductor nanocrystals (CdS) in the mid-1980's (now referred-to as *colloidal quantum dots*) has provided a new pathway to producing low-cost optoelectronic materials with novel physical properties (Brus, 1984; Rossetti et al., 1983; Steigerwald et al., 1988). Then, it was only a matter of time before lead-salt nanocrystals were synthesized using the colloidal method (Hines & Scholes, 2003; I. Kang & Wise, 1997; Machol et al., 1993; Wang et al., 1987; Yang et al., 1996).

As shown in Figure 5, the TFB:F8BT conjugated polymer-based heterostructures also constitute an ideal host system for PbS semiconductor nanocrystals to provide low-cost and high-performance hybrid heterostructures for key applications such as lighting & displays, biomedical devices, lab-on-a-chip, flexible optoelectronics, night-vision and solar-energy harvesting device architectures.

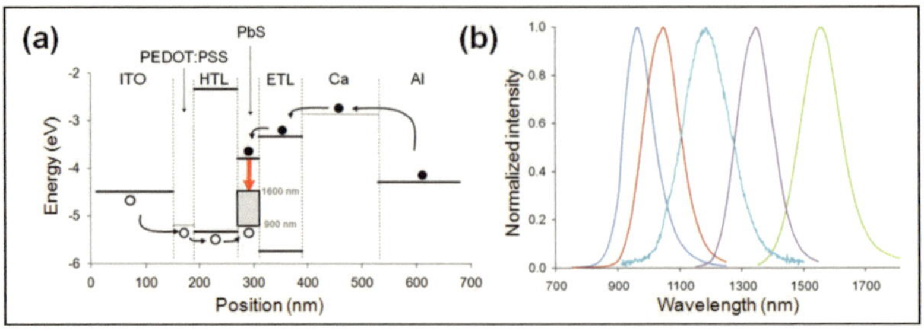

Fig. 5. Hybrid polyfluorene-based light-emitting heterostructures. (a) The TFB:F8BT system provides an ideal host system for PbS nanocrystal incorporation. (b) The incorporation of PbS nanocrystals can migrate their operation to the near-infrared (between 900 - 1600 nm), depending on the their size.

3.1 Colloidal synthesis of lead-sulfosalt nanocrystals with different shapes and sizes

Controlling the structural and optoelectronic properties of PbS nanostructures through improved hot-colloidal chemistry remains a very active field of research. As shown in Figure 6, the recent evolution of the early hot-colloidal lead-salt nanocrystal synthesis (Hines & Scholes, 2003) has lead to more complex nanostructures including nanorods & nanowires (Acharya et al., 2008; Dom et al., 2009; Ge et al., 2005; Yong et al., 2006; F. Zhang & Wong, 2009), star-shaped nanocrystals (Lee et al., 2002; Zhu et al., 2007), nanocubes (Zhao & Qi, 2006), octahedral nanocrystals (Cho et al., 2005; Koh et al., 2010) and core-shell nanocrystals (Petryga et al., 2008; Stouwdam et al., 2007; Swart et al., 2010; Warner & Cao, 2008). Most of these developments have occurred over the last five years.

Fig. 6. Colloidal synthesis of exotic PbS nanostructures. (a) Typical PbS quantum dot, (b) PbS/CdS core-shell nanocrystal, (c) PbS nanocubes, (d) star-shaped PbS nanocrystals and (e) PbS nanowires, all synthesized in our lab using variations of the hot-colloidal method. The inset in (e) shows the selective area electron diffraction (SAED) used to confirm the single-crystal structure of the nanowires.

3.2 Directed self-assembly of lead-salt nanocrystals

More recently, the self-assembly of lead-salt nanocrystals into more complex nanowire (1D) (Cho et al., 2005; Jang et al., 2010; Koh et al., 2010), monolayer (2D) (Anikeeva et al., 2007, 2008; Coe-Sullivan et al., 2003; Konstantatos et al., 2005; Steckel et al., 2003; X. Zhang et al., 2007) and nanocrystalline films and superlattices structures (3D) (Hanrath et al., 2009; Klem et al., 2007, 2008; Luther et al., 2008; Talapin et al., 2005) with a wide range of most promising optoelectronic properties has rapidly become a very active field of research, largely due to its facile solution-based processing.

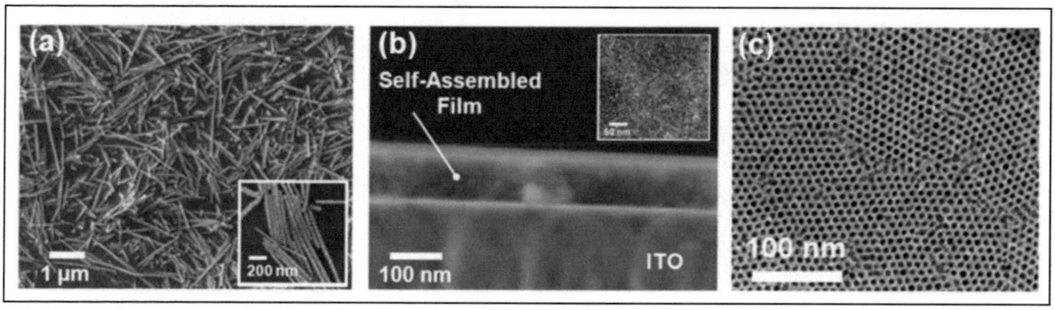

Fig. 7. (a) PbS nanowires formed by oriented-attachment of colloidal nanocrystals and (b) PbS nanocrystal films obtained by directed self-assembly. The inset shows the top-view of the self-assembled film. (c) This assembly process can be controlled down to reasonably-well organized monolayers.

Recently, exciting reports such as the observation of superb multiple-exciton generation efficiencies (Sargent, 2009; Sukhovatkin et al., 2009), highly-efficient hot-electron injection (Tisdale et al., 2010), and cold-exciton recycling (Klar et al., 2009), have propelled nanocrystalline lead-chalcogenide film structures to the forefront of cutting-edge research (M. S. Kang et al., 2009; W. Ma et al., 2009; Sambur et al., 2010; Steckel et al., 2003). Figure 7 shows typical examples of nanowires (1D), monolayers (2D) and films (3D) fabricated via the directed self-assembly of PbS nanocrystals synthesized by hot-colloidal method.

3.3 The incorporation of PbS nanocrystals in polymer-based host systems

Due to their band-structure alignment, we have shown that such PbS nanocrystals would be ideal for hybrid integration into the TFB:F8BT heterostructure to help migrate its operation towards the near-infrared. The most intuitively-obvious thing to do would be of course to simply mix colloidal quantum dots within the blended precursor prior to deposition. While this approach does work, threshold voltages and currents are generally very high, while quantum efficiencies and net output powers tend to be very low (Choudhury et al., 2010; Konstantatos et al., 2005). Indeed, this approach suffers from major fundamental drawbacks.

Assuming that the quantum dots are distributed homogeneously in the blended film, we also know that only the quantum dots located within tens of nanometers for the interfaces will be active. As such, this approach requires very large concentrations of quantum dots while most of them remain inactive. Moreover, it is likely that the incorporation of such large concentrations of quantum dots in the polymer host will have detrimental consequences on the performances of the polymer host itself.

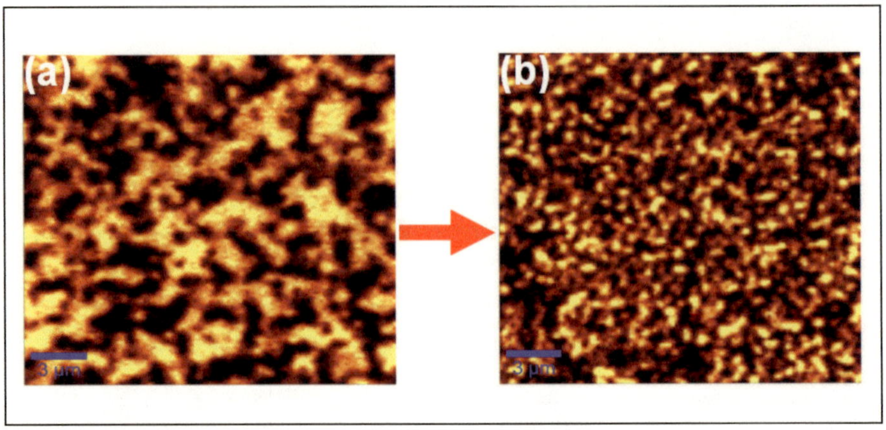

Fig. 8. The consequences of quantum-dot incorporation on blended polyfluorene-based film structures. Confocal fluorenscence mapping of the domain structure for the same toluene-blended TFB:F8BT (a) without PbS quantum dots and (b) impregnated with PbS quantum dots. The scale bars are 3 μm.

To verify the consequences of the PbS quantum dots incoporation, we used confocal fluorescence mapping of blended TFB:F8BT films with and without the colloidal quantum dots. The visible emission intensity maps (collected with a silicon detector) shown in Figure 8 reveal the domain structure for the blended films with and without quantum dots. There, the bright regions are F8BT-rich domains, while the dark regions represent TFB-rich domains. Without colloidal dots, Figure 8(a) shows significantly larger domains compared with the same blended film impregnated with near-infrared colloidal quantum dots (Fig. 8b).

Another factor to consider is the potential consequences of the nanoparticles incorporation on the crystalline phase of both polymers. Indeed, most polymers are known to consist of crystalline domains surrounded by amorphous chains. Indeed, the degree of crystallinity and the organization of those domains will also significantly impact the optoelectronic properties of the polyfluorenes (X. Ma et al., 2010). To study the consequences of nanoparticle incorporation on the structural organization of the polymers, we used electrospinning to pull polymer nanofibers (Sharma et al., 2010). Using a split collector during the electrospinning, it is possible to align those polymer fibers across the collector gap as shown in Figure 9(a). Using the 2D X-ray diffraction facility at the DND-CAT Synchrotron Research Center (AdvancedPhoton Source at the Argonne National Laboratory), we were able to observe the clear diffraction orders indicating that the crystallites in pure polymer fibers favor an alignment along the fibers such as shown in Figure 9(b). However, the incorporation of inert silica nanoparticles inside the polymer fibers disrupts this orientational ordering and yields a ring-like diffraction pattern shown in Figure 9(c), now suggesting a randomized polycrystalline structure such as shown in Figure 9(d). Indeed, by integrating for all azymuthal angles, we can confirm that the degree of crystallinity remains similar while the crystallites no longer show preferred alignment along the polymer fibers. This effect can also have very detrimental consequences of the optoelectronic properties of the polyfluorene-based device architectures incorporated with high concentrations of quantum dots.

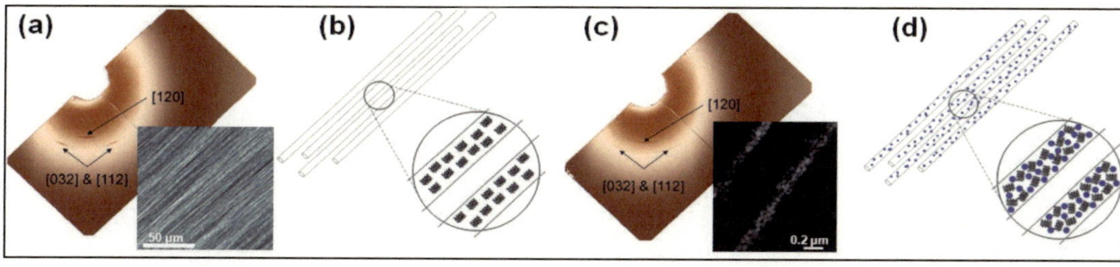

Fig. 9. (a) 2D X-ray diffraction pattern of pristine PEO polymer nanofibers (b) Schematic (not drawn to scale) showing oriented crystallite arrangement within the fibers. The white spaces between the polymer crystallites represent regions of amorphous polymer domains. (c,d) Results for PEO fibers incorporated with nanoparticles. Adapted from (Sharma et al., 2010) with permission. Copyright 2010 American Chemical Society.

3.4 The incorporation of PbS nanocrystals in polyfluorene-based bilayered type-II heterostructures

To alleviate those two fundamental problems, a better option explored more recently consists in depositing a thin self-assembled monolayer of quantum dots directly at the junction of the organic heterostructure between the hole- and electron-transporting organic materials (Steckel et al., 2003). However, this approach renders the whole fabrication sequence significantly more complex and usually involves the thermal evaporation of short-molecule organic semiconductors atop the quantum dots.

With long-molecule organics, this approach is even more delicate. One might suggest that simply spin-coating a pure F8BT layer atop a pure TFB layer or vice-versa would work. However, conjugated polymers from a same family tend to dissolve in similar solvents. As a consequence, the solvent of the second layer will significantly deteriorate the first layer. As shown in Figure 10, a viable approach to fabricating decent polyfluorene-based bilayered heterostructures consists in lifting-off a F8BT layer from one substrate and re-depositing directly on a TFB layer previously spin-coated on another substrate.

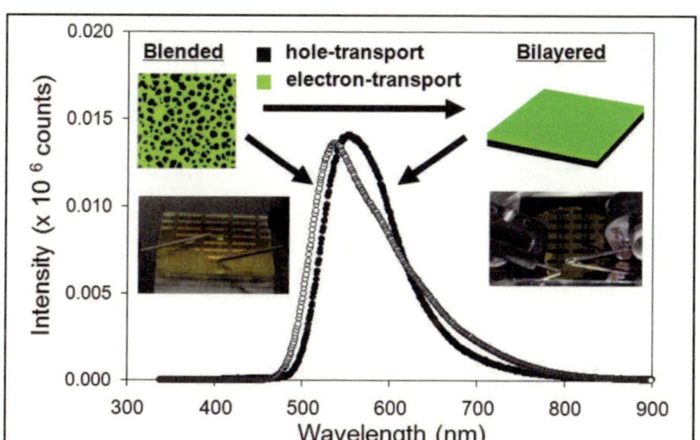

Fig. 10. Electroluminescence of blended (◦) and bilayered (•) TFB-F8BT light-emitting diodes under 19V forward bias.

As shown in Figure 11, this approach can allow the incorporation of a thin monolayer of PbS quantum dots directly at the junction between the TFB and F8BT. Still, their performances greatly suffer from poor injection efficiencies and from significant carrier losses into the organic layers (Konstantatos et al., 2005), most especially in the electron-transporting layer.

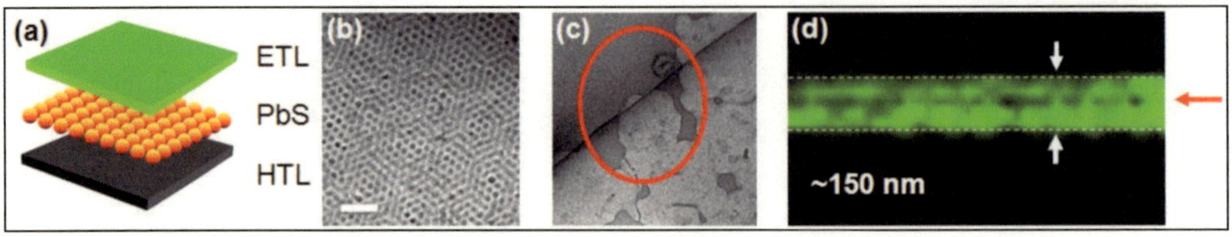

Fig. 11. Incorporation of PbS nanocrystals in polyfluorene-based bilayered type-II heterostructures. (a) Device schematics. (b) Self-assembled monolayer of PbS quantum dots. (c) Cross-sectional TEM of the structure after lift-off and deposition of the polymer layer atop the structure. (d) Cross-sectional confocal fluorescence intensity mapping showing the thin quantum-dot layer at the interface.

4. A new approach: The fabrication of high-performance hybrid polymer-nanocrystal heterostructures for near-infrared optoelectronics

As we have seen, PbS nanocrystals are typically synthesized using slightly modified versions of the widely-popular hot-colloidal method (Hines & Scholes, 2003). As such, the oleate capping group keeping colloidal nanocrystals stable in solution also severely limits charge transport between nanocrystals (S. Zhang et al., 2005). For this reason, previously-proposed near-infrared hybrid LED structures rely on colloidal quantum dots embedded within a polymer host matrix (Choudhury et al., 2010; Konstantatos et al., 2005), or use a monolayer of nanocrystals located directly at the junction of an organic heterostructure (Steckel et al., 2003). In both cases, we have seen that their performances greatly suffer from poor injection efficiencies and from significant carrier losses into the organic layers, thus providing output powers of a few micro-Watts (μW) at best (Konstantatos et al., 2005; X. Ma et al., 2010).

More recently, dithiol-based ligand exchange has been explored as a way of producing high-quality self-assembled PbS nanocrystalline film structures such as shown in Figure 12, with application in low-cost photovoltaic and photo-detector platforms (Klem et al., 2008). In this process, short dithiol linker molecules with strong thiolated bonds on both ends are used to exchange the long capping groups around colloidal nanocrystals, resulting in highly-conductive films of strongly-coupled cross-linked nanocrystals. As we will see, this breakthrough also provides new and exciting possibilities for novel near-infrared light-emitting device structures.

4.1 Conductivity and mobility of dithiol-treated nanocrystalline PbS film structures

For the first time, we investigated the optoelectronic properties of PbS nanocrystalline film structures cross-linked using carefully-controlled ethanedithiol (EDT) and benzenedithiol (BDT) ligand-exchange processes. To characterize the electronic properties of those self-assembled nanocrystalline film structures, we use two established methods. First, the charge extraction in linearly increasing voltage (CELIV) method described in Figure 13 can be used to measure both the conductivity and the majority carrier mobility. In view of the

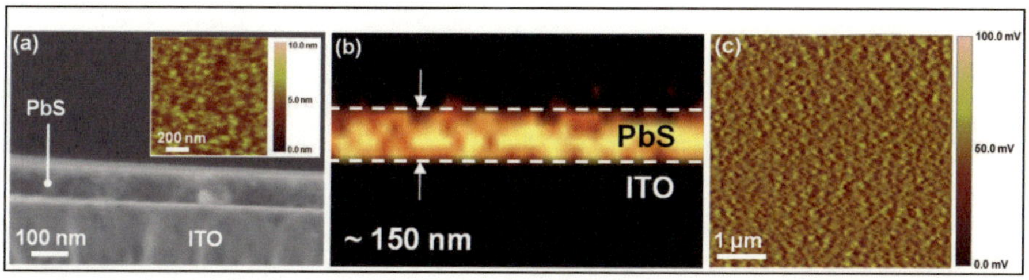

Fig. 12. Self-assembled PbS nanocrystalline film structure. (a) Cross-sectional SEM micrograph of a self-assembled film of PbS colloidal quantum dots formed by dithiol ligand-exchange chemistry. The inset shows a typical AFM mapping of the nanocrystalline PbS film surface revealing a 2 nm roughness parameter. (b) Cross-sectional map of the intensity of the 451 cm^{-1} Raman line associated with the 2LO-phonon vibration of the PbS film structure measured by confocal micro-Raman spectroscopy. (c) Conductive AFM (TUNA) mapping of the nanocrystalline film showing a good conductivity uniform across the surface.

p-type doping of the PbS nanocrystalline films, the CELIV measurement provides us with the hole-mobility for the dithiol-treated nanocrystalline films.

For CELIV measurements, the samples consist of a 250~350 nm-thick nanocrystalline films fabricated using the layer-by-layer spin coating method and sandwiched between ITO and Al contacts such as shown in Figure 13(a). As shown in Figure 13(b,c), a linearly-increasing bias is then applied across the sample while the transient currents are measured through the voltage drop across a 200 Ω load. The conductivity σ can then be obtained using (Juška et al., 2001):

$$\sigma = \frac{3}{2}\frac{d\Delta j}{t_{\max}A} \tag{1}$$

while the hole mobility μ_h is calculated using (Juška et al., 2001):

$$\mu_h = \frac{2d^2}{3At_{\max}^2\left(1 + 0.36\frac{\Delta j}{j_0}\right)} \tag{2}$$

The parameters t_{max}, j_0 and Δj in those equations can be obtained directly from the CELIV measurement, such as shown in Figure 13(b,c).

In contrast, the time-of-flight (TOF) method can be used to measure the minority carrier mobility. For TOF measurements, the samples have the same structure as for the CELIV measurements but require a much thicker nanocrystalline layer (1.0 ~ 1.5 µm) formed by depositing multiple layers of quantum dots. The 5 ns pulses of a Q-switched Nd:YAG laser are then used to excite the sample from the ITO side, while a reverse-bias is applied on the sample to extract the photo-generated minority carriers. In this case, the transient currents can also be recorded using the voltage drop across a 200 Ω load as shown in Figure 13(d). The electron mobility μ_e is then directly obtained as (Tiwari & Greenham, 2009):

$$\mu_e = \frac{d^2}{V\tau} \tag{3}$$

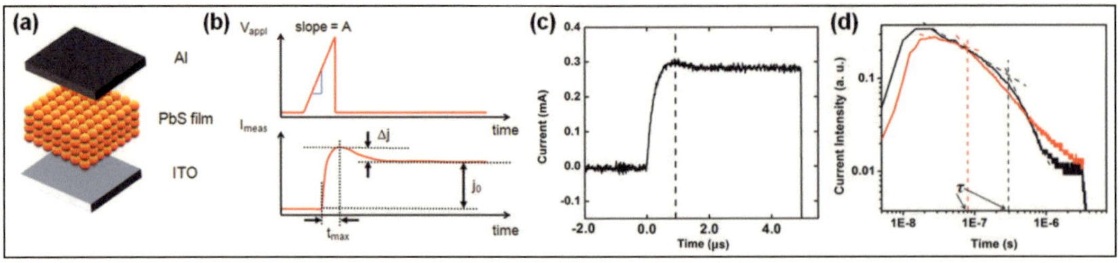

Fig. 13. CELIV and TOF measurements of self-assembled PbS nanocrystalline film structure. (a) Device schematics. (b) For CELIV measurements, a linearly-increasing bias is applied across the sample while the transient currents are measured through the voltage drop across a 200 Ω load. (c) Typical CELIV transient current measured for an EDT-treated PbS film of thickness d= 250 nm using a linearly-increasing voltage with slope A = 400 KV/s. (d) TOF measured for EDT-treated nanocrystalline film structure with d = 1.2 μm under V = 15 V bias (black) and for BDT-treated nanocrystalline film structure with d = 1.1 μm under V = 27 V bias (red).

where d is the sample thickness, V the applied bias across the sample and τ is the transit time obtained from the TOF measurement such as shown in Figure 13(d).

The results of the conductivity, hole- and electron-mobility measurements for both EDT- and BDT-treated nanocrystalline film structures are summarized in Figure 14.

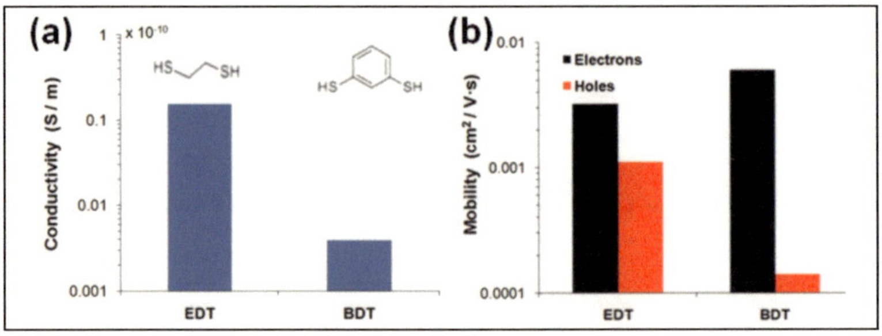

Fig. 14. Conductivity, hole- and electron-mobility measurements for EDT- and BDT-treated nanocrystalline film structures. (a) Conductivity measured using CELIV. (b) Electron-mobility measured using TOF (black) and hole-mobility measured using CELIV (red).

The smaller conductivity in the BDT-treated films simply suggests a significantly lower p-type doping concentration, most likely resulting from a better passivation of surface states compared with EDT. Moreover, the electron- and hole-mobilities are comparable for EDT-treated films. While we observe a modest increase in the electron mobility for the BDT-treated film despite using longer BDT molecules, it also leads to a significant drop in hole mobility. This is mostly because conjugated-dithiol molecular conductors (such as BDT) were previously shown not only to provide a physical linking between nanocrystals, but also a *conductive path* for electron transfer between nanocrystals through delocalization of the molecular electronic orbitals (Dadosh et al., 2005; Nitzan & Ratner, 2003). As such, the conjugated linker's LUMO and HOMO now provide additional energy barriers between nanocrystals for electrons and holes respectively. Since this energy barrier between nanocrystals is now significantly higher for holes compared with electrons, the carrier transport gets affected accordingly.

With BDT, the conjugation also provides substantial weight on the thiolated bonds that are directly coupled with the nanocrystals carriers wave-functions (Dadosh et al., 2005), thus directly increasing carrier transport for electrons. Indeed, previous experimental and theoretical models imply that the carrier transport through conjugated aryl dithiol molecules (such as BDT) occurs through the LUMO (electrons) level and not the HOMO (holes) (Nitzan & Ratner, 2003). This is consistent with both the increase of electron mobility and the large offset in electron and hole mobilities we observe for the BDT-treated nanocrystalline films.

4.2 Hybrid polymer-nanocrystal heterostructures for near-infrared optoelectronics

As we have seen with both the hybrid polymer-nanocrystal blends discussed in section 3.3 and the hybrid bilayered heterostructures discussed in section 3.4, their performances greatly suffer from poor injection efficiencies and from significant carrier losses into the organic layers (Choudhury et al., 2010; Konstantatos et al., 2005; Steckel et al., 2003). As we mentioned before, constraints on the electron-transporting material especially limit the overall performances of polymer-based optoelectronic devices (Moons, 2002).

As such, a new approach summarized in Figure 15 consists in substituting entirely the electron-transporting organic material, with a self-assembled film of cross-linked PbS nanocrystals deposited atop a hole-transporting polymer film structure. As shown in Figure 15(c), this new hybrid polymer-nanocrystal architecture allows direct electron injection without any additional injection layer and provides significantly better electron transport, thus leading to highly-improved LED devices.

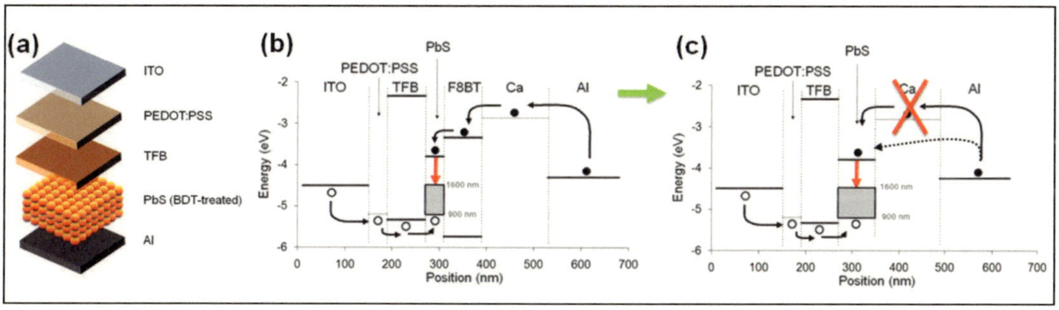

Fig. 15. Novel hybrid polymer-nanocrystal heterostructures for near-infrared LEDs. (a) Device schematics. (b) Energy diagram of the previously studied near-IR LED structures using nanocrystals embedded within a polymer host matrix, or a monolayer of quantum dots introduced at the junction of the organic heterostructure in-between the hole- and electron-transporting organic materials. (c) Energy diagram of the proposed near-IR LED architecture replacing the electron-transporting organic layer, with a self-assembled film of cross-linked PbS nanocrystals deposited atop a hole-transporting polymer film structure.

4.3 High-performance near-infrared LEDs using hybrid polymer-nanocrystal heterostructures

As shown in Figure 16, high-performance near-IR LEDs can be fabricated at extremely low cost using a very simple all solution-based processing approach.

To fabricate those structures, the ITO substrate is first patterned using standard photolithography and wet-etching processes before spin-coating of a very thin 30 nm-thick layer of PEDOT:PSS. As we mentioned before, this thin layer of PEDOT:PSS plays the dual role of facilitating the hole-injection while alleviating the detrimental effects of the surface roughness of the ITO film on the structural properties of the hole-transporting polymer. A toluene-based TFB solution is then spin-coated to provide a 120 nm-thick TFB film atop the PEDOT:PSS. Since the PEDOT:PSS is immune to toluene, there is no solvent-compatibility issues. Then, a solution of PbS quantum dots suspended in hexane is spin-coated atop the TFB, prior to performing the ligand-exchange process using the dithiol molecule diluted in acetonitrile. While the PEDOT:PSS would be affected by both the hexane and acetonitrile solutions, the TFB provides an efficient protection barrier for both solvents, thus preserving the structural integrity of the whole structure.

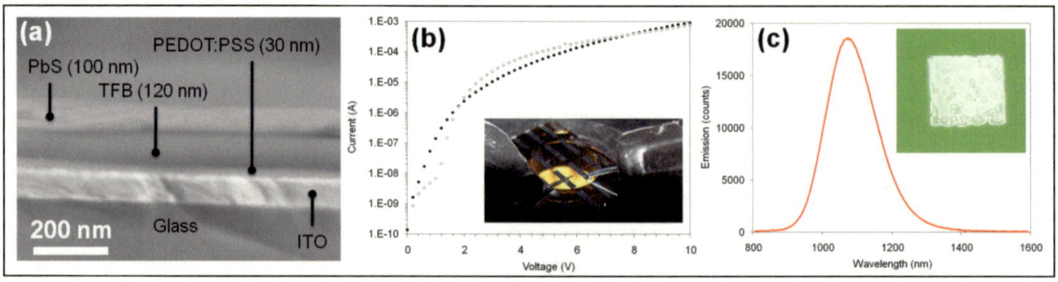

Fig. 16. High-Performance near-infrared LEDs using all solution-based processing. (a) Cross-sectional SEM micrograph showing the optimized structure of the device. (b) Device current-voltage characteristics for hybrid LED structure (•) and a polymer-only control device (◦). The inset shows the actual device atop the entry port of an integrating sphere while the near-IR electroluminescence is collected through the transparent substrate. (c) Emission spectra of a typical 1050 nm LED. The inset image shows the actual near-IR emission of a 1 mm^2 device measured using a near-IR camera coupled to a 2X objective.

As shown in Figure 16(b), the polymer-only control device reaches a clear single-carrier (hole) trap-limited regime around 1 Volt, before reaching a space-charge limited operation regime around 2 Volts. This is consistent with the large energy barrier at the TFB-aluminum interface. For the LED device with the BDT-treated nanocrytalline film structure, measurements indicate a much higher current density at low voltages originating from the efficient electron-injection at the metal-PbS interface. Here, the lower slope in the trap-limited region simply suggests different transport and trapping mechanisms in the nanocrystalline film compared with the TFB. These highly-efficient LED structures can operate anywhere between 1000 and 1600 nm depending on the nanocrystals used while providing external quantum efficiencies as high as 0.7% and ouput powers close to 80 μW.

While the conventional ethanedithiol (EDT)-based ligand-exchange treatment is known to work well for photovoltaic structures (Luther et al., 2008), it yields only relatively poor LED structures compared to the phenomenal results achieved using benzendithiol (BDT)-based treatment. This disparity can be readily explained now based on the conductivity and mobility results presented in Figure 14. Indeed, the EDT treatement provides higher conductivities due to higher p-type doping and comparable electron- and hole- mobilities. As such, both the hole current and nonradiative Auger recombination will be orders of magnitude larger than for BDT-treated films. Moreover, the large hole-current would for EDT-treated films would ideally require a hole-barrier at the metal-nanocrystal interface. While we tried to use a TiO_2 barrier to reduce the hole-current for such films, we observed that this barrier significantly impedes the electron injection.

With the BDT-treatment, the hole-mobility drops significantly. As such, there is no need to have a hole-blocking barrier at the metal-PbS interface since the holes don't make it to this interface anyway. Even better, this dramatic reduction in hole mobility is associated with a modest increase in electron mobility. As we know, everything happens at the junction of this hybrid polymer-nanocrystal heterostructure. Using the BDT-treated nanocrystalline films, electrons can be very efficiently injected from one side and holes from the other. Moreover, the hole-transporting polymer bilayer provides an efficient electron barrier while the BDT-treatment provides a good mobility-barrier for holes in the nanocrystalline film. As such, the carriers are efficiently delivered and confined close to the junction (active region). Due to the low hole-mobility, the excitons then bind and stay close to the junction, having plenty of time to recombine radiatively while avoiding metal quenching from the metal-PbS interface.

5. Conclusion

While π-conjugated polymer-based light-emitting diodes are perfectly suited for the visible, their potential for near-infrared operation remains limited. However, the hybrid integration of semiconductor nanocrystals and conjugated polymer material systems can provide an easy pathway for (1) improving the conjugated polymer-based devices optoelectronic properties and/or (2) providing added functionality to the conjugated polymer-based device structures. Because the oleate capping groups keeping colloidal lead-chalcogenide nanocrystals stable also inhibit carrier transport, previously-proposed hybrid near-infrared LED structures usually rely on nanocrystals embedded within a polymer host matrix, or use a self-assembled monolayer of colloidal quantum dots located at the junction of an organic heterostructure directly between hole- and electron-transporting organics. We have demonstrated why both these hybrid polymer-nanocrystal blends and the hybrid bilayered heterostructures greatly suffer from poor injection efficiencies and from significant carrier losses into the organic layers, while limited electron-transporting materials especially limit the overall performances of those polymer-based optoelectronic devices.

Here, we report an all solution-based method producing efficient hybrid polymer-nanocrystal multilayered heterostructures for light-emission in the near-infrared (1050-1600 nm). After optimization device structure, we obtain low-cost near-infrared light-emitting diodes with external quantum efficiency (EQE) as high as 0.7% and up to 80 μW output from devices entirely processed in ambient air and with no encapsulation. This approach relies on a carefully-controlled layer-by-layer benzenedithiol (BDT) ligand-exchange to achieve direct charge injection and better transport. In comparison with this BDT treatment, the conventional ethanedithiol (EDT)-based treatment provides poor LED structures. As we show, the high performances of our devices can be explained by the different doping levels and electron & hole mobilities resulting from the BDT versus EDT treatments.

In the future, this easy, robust, low-temperature and substrate-independent approach has the potential to become extremely useful for flexible and/or reconfigurable integrated optoelectronic platforms, biological imaging & sensing, lab-on-a-chip and thermoelectronic platforms. Moreover, this method could also be extended to other colloidal nanocrystals such as PbSe (for longer wavelengths) or CdSe (for the visible). Finally, the large refractive index of these self-assembled nanocrystalline film structures offers the possibility of incorporating these electroluminescent structures directly onto silicon substrates to work as light source or in more complex optoelectronic device architectures.

6. Acknowledgment

I am most thankful to my entire team and all my collaborators of the last 5 years for their contribution direct or indirect to this research. Most especially, I would like to thank Xin Ma and Fan Xu who have done tremendous work to advance this emerging field of research. Finally, most of this research work was kindly supported through the AFOSR (FA9550-10-1-0363), the DARPA-COMPASS and the DARPA-Young Faculty Award programs, to whom I am most thankful. The majority fraction of this work was conducted while the author worked at the University of Delaware.

7. References

Acharya, S., Gautam, U. K., Sasaki, T., Bando, Y., Golan, Y., & Ariga, K., Ultra Narrow PbS Nanorods with Intense Fluorescence, *Journal of the American Chemical Society*, Vol.130 (2008), pp. 4594-4595

Anikeeva, P. O., Halpert, J. E., Bawendi, M. G., & Bulovic, V., Electroluminescence from a Mixed Red-Green-Blue Colloidal Quantum Dot Monolayer, *Nano Letters*, Vol.7 (2007), pp. 2196-2200

Anikeeva, P. O., Madigan, C. F., Halpert, J. E., Bawendi M. G., & Bulovic, V., Electronic and excitonic processes in light-emitting devices based on organic materials and colloidal quantum dots, *Physical Review B*, Vol.78 (2008), pp. 085434-8

Arias, A. C. MacKenzie, J. D., Stevenson, R., Halls, J. J. M., Inbasekaran, M., Woo, E. P., Richards, D., & Friend, R. H., Photovoltaic Performance and Morphology of Polyfluorene Blends: A Combined Microscopic and Photovoltaic Investigation, *Macromolecules*, Vol.34 (2001), pp. 6005-6013

Bakueva, L., Musikhin, S., Hines, M. A., Chang, T.-W. F., Tzolov, M., Scholes, G. D., & Sargent, E. H., Size-tunable infrared (1000-1600 nm) electroluminescence from PbS quantum-dot nanocrystals in a semiconducting polymer, *Applied Physics Letters*, Vol.82 (2003), pp. 2895-2897

Brus, L. E., Electron-electron and electron-hole interactions in small semiconductor crystallites: The size dependence of the lowest excited electronic state, *Journal of Chemical Physics*, Vol.80 (1984), pp. 4403-4409

Cho, K.-S., Talapin, D. V., Gaschler, W., & Murray, C. B., Designing PbSe Nanowires and Nanorings through Oriented Attachment of Nanoparticles, *Journal of the American Chemical Society*, Vol.127 (2005), pp. 7140-7147

Choudhury, K. R., Song, D. W., & So, F., Efficient solution-processed hybrid polymer–nanocrystal near infrared light-emitting devices. *Organic Electronics*, Vol.11 (2010), pp. 23-28

Coe-Sullivan, S., Woo, W.-K., Steckel, J. S., Bawendi, M., & Bulovic, V., Tuning the performance of hybrid organic/inorganic quantum dot light-emitting devices, *Organic Electronics*, Vol.4 (2003), pp. 123-130

Dadosh, T., Gordin, Y., Krahne, R., Khivrich, I., Mahalu, D., Frydman, V., Sperling, J., Yacoby, A., & Bar-Joseph, I., Measurement of the conductance of single conjugated molecules, *Nature*, Vol.436 (2005), pp. 677-680

Dom, A., Wong C. R. & Bawendi, M. G., Electrically Controlled Catalytic Nanowire Growth from Solution, *Advanced Materials*, Vol.21 (2009), pp. 3479-3482

Ge, J.-P., Wang, J., Zhang, H.-X., Wang, X., Peng, Q., & Li, Y.-D., Orthogonal PbS Nanowire Arrays and Networks and Their Raman Scattering Behavior, *Chemistry - A European Journal*, Vol.11 (2005), pp. 1889-1894

Hanrath, T., Choi, J. J. & Smilgies,D.-M., Structure/Processing Relationships of Highly Ordered Lead Salt Nanocrystal Superlattices, *ACS Nano*, Vol.10 (2009), pp. 2975-2988

Hines, M. A., & Scholes, G. D., Colloidal PbS Nanocrystals with Size-Tunable Near-Infrared Emission: Observation of Post-Synthesis Self-Narrowing of the Particle Size Distribution, *Advanced Materials*, Vol.15 (2003), pp. 1844-1849

Jang, S. Y., Song, Y. M., Kim, H. S., Cho, Y. J., Seo, Y. S., Jung, G. B., Lee, C.-W., Park, J., Jung, M., Kim, J., Kim, B., Kim, J.-G., & Kim, Y.-J., Three Synthetic Routes to Single-Crystalline PbS Nanowires with Controlled Growth Direction and Their Electrical Transport Properties, *ACS Nano*, Vol.4 (2010), pp. 2391-2401

Juška, G., Viliūnas, M., Arlauskas, K., Nekrašas, N., Wyrsch, N., & Feitknecht, L., Hole drift mobility in μc-Si:H, *Journal of Applied Physics*, Vol.89 (2001), pp. 4971-4977

Kang I. &Wise, F. W., Electronic structure and optical properties of PbS and PbSe quantum dots, *Journal of the Optical Society of America B*, Vol.14 (1997), pp. 1632-1646

Kang, M. S., Lee, J., Norris, D. J., & Frisbie, C. D., High carrier densities achieved at low voltages in ambipolar PbSe nanocrystal thin-film transistors, *Nano Letters*, Vol.9 (2009), pp. 3848-3852

Klann, R. H., Hofer, T., Thomas Buhleier, R. E., & Thomas Tomm, J. W., Fast recombination processes in lead chalcogenide semiconductors studied via transient optical nonlinearities, *Journal of Applied Physics*, Vol.77 (1995), pp. 277-286

Klar, T. A., Franzl, T., Rogach, A. L., & Feldmann, J., Super-efficient exciton tunneling in layer-by-layer semiconductor nanocrystal structures, *Advanced Materials*, Vol.17 (2009), pp. 769-773

Klem, E. J. D., MacNeil, D. D., Cyr, P. W., Levina, L., & E. H. Sargent, E. H., Efficient solution-processed infrared photovoltaic cells: Planarized all-inorganic bulk heterojunction devices via inter-quantum-dot bridging during growth from solution, *Applied Physics Letters*, Vol.90 (2007), pp. 183113-3

Klem, E. J. D., Shukla, H., Hinds, S., MacNeil, D. D., Levina, L., & Sargent, E. H., Impact of dithiol treatment and air annealing on the conductivity, mobility, and hole density in PbS colloidal quantum dot solids, *Applied Physics Letters*, Vol.92 (2008), pp. 212105-3

Koh, W.-K., Bartnik, A. C., Wise F. W., & Murray, C. B., Synthesis of Monodisperse PbSe Nanorods: A Case for Oriented Attachment, *Journal of the American Chemical Society*, Vol.132 (2010), pp. 3909-3913

Konstantatos, G., Huang, C., Levina, L., Lu, Z., & Sargent, E. H., Efficient Infrared Electroluminescent Devices Using Solution-Processed Colloidal Quantum Dots, *Advanced Functional Materials*, Vol.15 (2005), pp. 1865-1869

Lee, S.-M., Jun, Y.-W., Cho, S.-N., & Cheon, J., Single-Crystalline Star-Shaped Nanocrystals and Their Evolution: Programming the Geometry of Nano-Building Blocks, *Journal of the American Chemical Society*, Vol.124 (2002), pp. 11244-11245

Liu, J., Wang, S., Bian, Z., Shan M., & Huang, C., Organic/inorganic hybrid solar cells with vertically oriented ZnO nanowires, *Applied Physics Letters*, Vol.94 (2009), pp. 173107-3

Luther, J. M., Law, M., Song, Q., Perkins, C. L., Beard, M. C., & Nozik, A. J., Structural, Optical, and Electrical Properties of Self-Assembled Films of PbSe Nanocrystals Treated with 1,2-Ethanedithiol, *ACS Nano*, Vol.2 (2008), pp. 271-280

Ma, W., Luther, J. M., Zheng, H., Wu, Y., & Alivisatos, A. P., Photovoltaic devices employing ternary PbS_xSe_{1-x} nanocrystals, *Nano Letters*, Vol.9 (2009), pp. 1699-1703

Ma, X., Xu F., & Cloutier, S. G., High-Performance 1550 nm Polymer-Based LEDs on Silicon using Hybrid Polyfluorene-Based Type-II Heterojunctions, *Proceedings fo the 7th Inter. Conference on Group-IV Photonics*, Beijing, China, October 2010

Machol, J. L., Wise, F. W., Patel, R. C., & Tanner, D. B., Vibronic quantum beats in PbS microcrystallites, *Physical Review B*, Vol.48 (1993), pp. 2819-2822

MacDiarmid, A. G., Synthetic metals: a novel role for organic polymers, *Current Applied Physics*, Vol.1 (2001), pp. 269-317

McDonald, S. A., Konstantatos, G., Zhang, S., Cyr, P. W., Klem, E. J. D., Levina L., & Sargent, E. H., Solution-processed PbS quantum dot infrared photodetectors and photovoltaics, *Nature Materials*, Vol.4 (2005), pp. 138-142

McNeill, C. R., Watts, B., Thomsen, L., Belcher, W. J., Greenham, N. C., Dastoor, P. C., & Ade, H., Evolution of Laterally Phase-Separated Polyfluorene Blend Morphology Studied by X-ray Spectromicroscopy, *Macromolecules*, Vol.42 (2009), pp. 3347-3352

Moons, E., Conjugated polymer blends: linking film morphology to performance of light emitting diodes and photodiodes, Journal of Physics: Condensed Matter, Vol.14 (2002), pp. 12235-12252

Morteani, A. C., Dhoot, A. S., Kim, J.-S., Silva, C., Greenham, N. C., Murphy, C., Moons, E., Cina, S., Burroughes, J. H., & Friend, R. H., Barrier-Free Electron-Hole Capture in Polymer Blend Heterojunction Light-Emitting Diodes, *Advanced Materials*, Vol.15 (2003), pp. 1708-1715

Nitzan, A., & Ratner, M. A., Electron transport in molecular wire junctions, *Science*, Vol.300 (2003), pp. 1384-1389

Pietryga, J. M., Werder, D. J., Williams, D. J., Casson, J. L., Schaller, R. D., Klimov, V. I., & Hollingsworth, J. A., Utilizing the Lability of Lead Selenide to Produce Heterostructured Nanocrystals with Bright, Stable Infrared Emission, *Journal of the American Chemical Society*, Vol.130 (2008), pp. 4879-4885

Preier, H., Recent Advances in Lead-Chalcogenide Diode Lasers, *Applied Physics*, Vol.20 (1979), pp. 189-206

Rossetti, R., Nakahara, S., & Brus, L. E., Quantum size effects in the redox potentials, resonance Raman spectra, and electronic spectra of CdS crystallites in aqueous solution, *Journal of Chemical Physics*, Vol.79 (1983), pp. 1086-1088

Sambur, J. B., Novet, T., & Parkinson, B. A., Multiple exciton collection in a sensitized photovoltaic system, *Science,* Vol.330 (2010), pp. 63-66

Sargent, E. H., Infrared photovoltaics made by solution processing, *Nature Photonics,* Vol.3 (2009), pp. 325-331

Sharma, N., McKeown, S. J., Ma, X., Pochan, D. J., & Cloutier, S. G., Structure-Property Correlations in Hybrid Polymer-Nanoparticle Electrospun Fibers and Plasmonic Control over their Dichroic Behavior, *ACS Nano,* Vol.4 (2010), pp. 5551-5555

Steckel, J. S., Coe-Sullivan, S., Bulović, V., & Bawendi, M. G., 1.3 μm to 1.55 μm Tunable Electroluminescence from PbSe Quantum Dots Embedded within an Organic Device, *Advanced Materials,* Vol.15 (2003), pp. 1862-1866

Steigerwald, M. L., Alivisatos, A. P., Gibson, J. M., Harris, T. D., Kortan, R., Muller, A. J., Thayer, A. M., Duncan, T. M., Douglas, D. C., & Brus, L. E., Surface Derivatization and Isolation of Semiconductor Cluster Molecules, *Journal of the American Chemical Society,* Vol.110 (1988), pp. 3046-3050

Stouwdam, J. W., Shan, J., van Veggel, F. C. J. M., Pattantyus-Abraham, A. C., Young J. F., & Raudsepp, M., Photostability of Colloidal PbSe and PbSe/PbS Core/Shell Nanocrystals in Solution and in the Solid State, *Journal of Physical Chemistry C,* Vol.111 (2007), pp. 1086-1092

Sukhovatkin, V., Hinds, S., Brzozowski, L., & Sargent, E. H., Colloidal quantum-Dot photodetectors exploiting multiexciton generation, *Science,* Vol.324 (2009), pp. 1542-1544

Swart, I., Sun, Z., Vanmaekelbergh D., Liljeroth, P., Hole-Induced Electron Transport through Core-Shell Quantum Dots: A Direct Measurement of the Electron-Hole Interaction, *Nano Letters,* Vol.10 (2010), pp. 1931-1935

Talapin, D. V., & Murray, C. B., PbSe Nanocrystal Solids for n- and p-Channel Thin Film Field-Effect Transistors, *Science,* Vol.310 (2005), pp. 86-91

Tisdale, W. A., Williams, K. J., Timp, B. A., Norris, D. J., Aydil, E. S., & Zhu, X.-Y., Hot-electron transfer from semiconductor nanocrystals, *Science,* Vol.328 (2010), pp. 1543-1547

Tiwari, S., & Greenham, N. C., Charge mobility measurement techniques in organic semiconductors, *Opt. Quant. Electron.,* Vol.41 (2009), pp. 69-89

Warner J. H., & Cao, H., Shape control of PbS nanocrystals using multiple surfactants, *Nanotechnology,* Vol.19 (2008), pp. 305605-5

Wang, Y., Suna, A., Mahler, W., & Kasowski, R., PbS in polymers: From molecules to bulk solids, *Journal of Chemical Physics,* Vol.87 (1987), pp. 7315-7322

Yang, J. P., Qadri, S. B., & Ratna, B. R., Structural and Morphological Characterization of PbS Nanocrystallites Synthesized in the Bicontinuous Cubic Phase of Lipid, *Journal of Physical Chemistry,* Vol.100 (1996), pp. 17255-17259

Yong, K.-T., Sahoo, Y., Choudhury, K. R., Swihart, M. T., Minter, J. R., & Prasad, P. N., Control of the Morphology and Size of PbS Nanowires Using Gold Nanoparticles, *Chemistry of Materials,* Vol.18 (2006), pp. 5965-5972

Zhang, F., & Wong, S. S., Controlled Synthesis of Semiconducting Metal Sulfide Nanowires, *Chemistry of Materials,* Vol.21 (2009), pp. 4541-4554

Zhang S., Cyr, P. W., McDonald, S. A., Konstantos, G., & Sargent, E. H., Enhanced infrared photovoltaic efficiency in PbS nanocrystal/semiconducting polymer composites:

600-fold increase in maximum power output via control of the ligand barrier, *Applied Physics Letters*, Vol.87 (2005), pp. 233101-3

Zhang, X., Dong, X., Liu, Y., Kai, G., Wang, Z., Li, L., Han X., & Li, Y., Near-infrared emission from PbS Quantom Dots in polymer matrix, *Optoelectronics Letters*, Vol.3 (2007), pp. 337-342

Zhao, N., & Qi, L., Low-Temperature Synthesis of Star-Shaped PbS Nanocrystals in Aqueous Solutions of Mixed Cationic/Anionic Surfactants, *Advanced Materials*, Vol.18 (2006), pp. 359-362

Zhu, J., Peng, H., Chan, C. K., Jarausch, K., Zhang, X. F., & Cui, Y., Hyperbranched Lead Selenide Nanowire Networks, *Nano Letters*, Vol.7 (2007), pp. 1095-1099

8

Plasmonics for Green Technologies: Toward High-Efficiency LEDs and Solar Cells

Koichi Okamoto
Institute for Materials Chemistry and Engineering, Kyushu University
Japan

1. Introduction

Nowadays, energy issues became very important problem for us. We spend a lot of energy for illumination at night, so developing high-efficiency light sources is very important to save our energy. Recently, solid-state light-emitting devices have been developed and expected as new-generation light sources because of their advantages such as small, light-weight, long lifetime, easy operation, and saving energy. Since 1993, InGaN quantum wells (QW)-based light-emitting diodes (LEDs) have been continuously improved and commercialized as light sources in the ultraviolet (UV) and visible spectral regions. In 1996, white light LEDs, in which a blue LED is combined with yellow phosphors, have been developed and offer a replacement for conventional incandescent and fluorescent light bulbs. However, these devices have not fulfilled their original promise as solid-state replacements for light bulbs as their light-emission efficiencies have been limited. The most important requirement for competitive LEDs for solid-state lighting is improvement of their quantum efficiencies of light emissions.

Making energy is also very important so much as saving energy. Renewable energies have attracted a great deal of attention as a new energy source instead of fossil resource which is going to be exhausted. The solar energy is one of the most important renewable energy resources and the photocurrent conversion efficiencies of several kind of solar cells have been rapidly developed. Especially, the crystalline solar cells with silicon or compound semiconductors were well developed and their efficiencies were almost reached to the theoretical limits. The drastic cost reduction is much important for such crystalline solar cells to use for much wider areas. For example, making ultra-thin device structures is required to save the materials. On the other hand, amorphous or organic solar cells are very cheap and easy to treat them but the efficiencies are still very low. The improvements of the efficiencies and device lifetime are most important for such solar cells.

A lot of effort and time have been used to improve the efficiencies of LEDs and solar cells, but still it has been very difficult to achieve dramatic improvements. Here I introduce the unique approach to increase these efficiencies based on "Plasmonics". These studies should bring the new application field of plasmonics for green technologies.

2. Fundamental and application of plasmonics

Conduction electron gas in a metal oscillates collectively and the quantum of this plasma oscillation is called plasmon. A special plasma oscillation mode called surface plasmon (SP) exists at an interface between a metal, which has a negative dielectric-constant, and a positive dielectric material (Raether, 1988). The charge fluctuation of the oscillation of the SP is accompanied by fluctuations of electromagnetic fields, which is called surface plasmon polariton (SPP). Schematic diagram of the SP and the SPP mode generated at metal/dielectric interface were shown in Fig. 1. The SPP can interact with light waves at the interface and it brings novel optical properties and functions to materials. The technique controlling and utilizing the SPP is called "plasmonics" and has attracted much attention with the recent rapid advance of nanotechnology (Barnes et. al., 2003; Atwater et al., 2007).

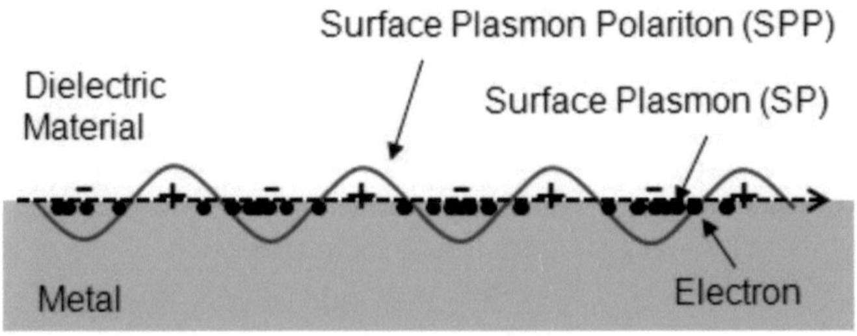

Fig. 1. Schematic diagram of the surface plasmon (SP) and the SP polariton (SPP) generated at the metal/dielectric interface.

The wave vector of the SPP (k_{SP}) parallel to the interface can be written with the following equation when the relative permittivity of the metal is $\varepsilon_1 = \varepsilon_1' + \varepsilon_1''i$ and that of the dielectric material is ε_2.

$$k_{SP}(\omega)=\frac{\omega}{c}\sqrt{\frac{\varepsilon_1'(\omega)\varepsilon_2(\omega)}{\varepsilon_1'(\omega)+\varepsilon_2(\omega)}}+\frac{\omega}{c}\left(\frac{\varepsilon_1'(\omega)\varepsilon_2(\omega)}{\varepsilon_1'(\omega)+\varepsilon_2(\omega)}\right)^{\frac{3}{2}}\frac{\varepsilon_1''(\omega)}{2\varepsilon_1'(\omega)^2}i \quad (1)$$

where, ω and C are the frequency of the SPP and the light velocity in vacuum, respectively. The first and second terms of this equation give the dispersion and the damping factor of the SPP. The k_{SP} values are much larger than the wave vector of the light wave propagated in the dielectric media. This fact suggests that the SPP can propagate into nano-spaces much smaller than the wavelength. This enables us to shrink the sizes of waveguides and optical circuits into nano-scale.

Wave vectors of the SPP perpendicular to the interface in a metal or a dielectric material must be an imaginary number because k_{SP} is larger than the light line. This suggests that the electromagnetic fields of the SPP are strongly localized at the interface and it makes giant fields at the interface. This huge field enhancement effect is also one of the most important features of the SPPs. It has been applied to high sensitive sensors using the surface enhanced Raman scattering (SERS), surface plasmon resonance (SPR), and so on.

One futuristic application of plasmonics is the development of high-efficiency LEDs. LEDs have been expected to eventually replace traditional fluorescent tubes as new illumination sources. For example, InGaN-based QWs provide bright light sources, however, their efficiencies are still substantially lower than those of fluorescent lights. The idea of SP enhanced light emission was proposed since 1990, and it has been applied to increase emission efficiencies of several materials which include InGaN QWs. Gontijo et al. reported the coupling of the emission from InGaN QW into the SP on silver thin firm, however they found that the PL intensities dramatically decreased by the SP coupling (Gontijo, et al., 1999). By using same sample structure, Neogi et al. confirmed that the recombination rate in an InGaN/GaN QW could be significantly enhanced by the time-resolved PL measurement (Neogi et al., 2002). However, in these early studies, light could not be extracted efficiently from the metal surface, and the SP coupling has been thought to be a negative factor for LEDs.

Recently, we have reported for the first time large photoluminescence (PL) increases from InGaN/GaN QW material coated with metal thin films (Okamoto et al., 2004). We obtained a 17-fold increase in the luminescence intensity along with a 7-fold increase in the internal quantum efficiency of light emission from InGaN/GaN QWs when nano-structured silver layers were deposited 10 nm above the QWs. We also observed a 32-fold increase in the spontaneous emission rate of InGaN/GaN at 440 nm probed by time-resolved PL measurements (Okamoto et al., 2005). Moreover, we obtained a huge enhancement of light emissions for silicon nanocrystals in silicon dioxide media (Okamoto et al, 2008). Usually the emission efficiencies of such indirect semiconductors are quite low, but by using the SP coupling, it is possible to increase these efficiencies up to values as large as those of direct compound semiconductors.

The SP-emitter coupling technique would lead to high-efficiency LEDs that offer realistic alternatives to conventional fluorescent light sources. However, detail mechanism and dynamics of the SP coupling have been still not so clear. We already achieved efficient blue emissions by using this technique. However, it has been still very difficult to obtain highly enhanced green emissions in spite of the importance of applications of the high-efficiency green LEDs. We try to control the SP coupling conditions by employing the metal nanostructures. Further optimizations of nanostructures should bring highly efficient LEDs and also light receiving devices, namely, solar cells.

3. Enhancement of photoluminescence of InGaN/GaN

Fig. 2 shows typical PL spectra from an InGaN/GaN QWs separated from Ag films by 10 nm GaN spacers. The PL peak intensities of uncoated samples were normalized to 1 and huge enhancements were observed by Ag coating especially at the shorter wavelength region. InGaN/GaN-based QW wafers were grown on a (0001) oriented sapphire substrate by a metal-organic chemical vapor deposition (MOCVD). The QW heterostructure consists of a GaN (4 μm) buffer layer, an InGaN QW (3 nm) and a GaN cap layer (10 nm). Silver films (50nm) were deposited on top of the surfaces of these wafers by a high vacuum thermal evaporation. The PL measurements were performed by exciting the QW with a 406nm diode laser and detecting the emission with a multi-channel spectrometer.

The wavelength dependences of the enhanced PL intensities were almost same for single QW and three QWs. These PL enhancements should be attributed to the SP coupling. A possible mechanism of the SP coupling was already proposed (Okamoto et al., 2005;

Okamoto & Kawakami, 2009). Electron-hole pairs in the QW couple to plasma oscillation of electrons at the metal/semiconductor interface when the energies of electron-hole pairs and of the SP frequency are similar. Then, electron-hole recombination may produce SPPs instead of photons or phonons, and this new recombination path increases the recombination rate and the internal quantum efficiency. If the metal surface is perfectly flat, the SPP energy would be thermally dissipated. By providing roughness or nanostructure of the metal layers, the SPP energy can be extracted as light. Such roughness allows SPPs of high momentum to scatter, lose momentum, and couple to radiative photon. In order to obtain the high photon extraction efficiencies, the few tens of nanometer sized structures at the metal surfaces were obtained by controlling the evaporation conditions.

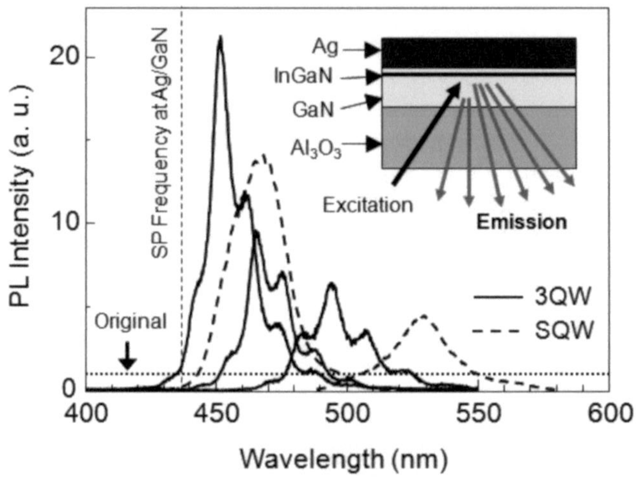

Fig. 2. SP enhanced PL spectra of InGaN/GaN SQW with peak wavelength at 470 and 530 nm (broken lines) and 3QW with peak wavelength at 450, 460, and 500 nm (solid lines) coated with Ag. The PL peak intensity of uncoated sample was normalized to 1. (Inset) Sample structure and excitation/emission configuration.

We photo-pump and detect emission from the backside of the samples through the transparent substrate by polishing the bottom surface. By employing such back-side access to the QWs, we can avoid an absorption loss at the metal layer and obtain an effective light extraction from SPP at the interface. Thus we can use very thick metal films. This thickness should be also very important factor to obtain a huge enhanced light emission. If the metal layer is thinner than the penetration depth of SPP, other SPP mode is generated at the air/metal interface of the opposite side of the metal layer. These SPP modes couple each other and form symmetric and anti-symmetric mode of the SPPs. This should modify the SP frequency and coupling condition and make the light extraction very difficult. The thick metal layer is also useful to avoid the oxidation of silver surface. Metal oxidation changes the surface roughness and SP mode. But the oxidation typically is generated only at air/metal interface and not at the metal/semiconductor interface. The thickness of metal films (50nm) is large enough to ensure that metal oxidation at air/metal interface does not influence the metal/semiconductor interface. It is very simple solution but the back-side access is the most important trick which enabled us to obtain light enhancements by the SP coupling for the first time.

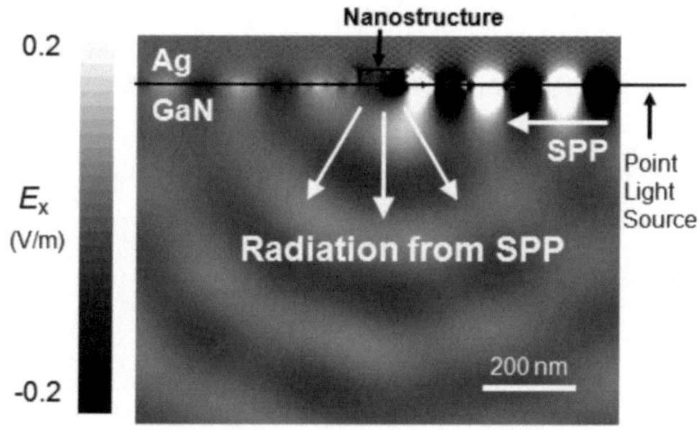

Fig. 3. 3D-FDTD simulations of generation and light extraction of SPPs. SSP was generated from a point light source located on the interface, and it was extracted as light at the gap in the metal layer.

In order to evaluate the SP coupling mechanism we proposed, we employed the 3-dimentional finite-difference time-domain (3D-FDTD) simulation. We used commercialized software "Poynting for optics" (Fujitsu Co.). Fig. 3 shows the calculated spatial distribution of the electromagnet field around the metal/semiconductor interface. The clear SPP mode appeared and propagated within the interface by the point light source located at the interface. A polarized plane wave with 525 nm wavelength and 1 V/m amplitude was used as a point light source which is as assumption of an electron-hole pair. This result suggests that the SPP mode can be generated easily by direct energy transfer from electron-hole pairs without any special structures. Usually, some special configurations are necessary to generate SPP mode such as a grating coupler or an Attenuated total reflection setting to satisfy a phase matting condition between SPPs and photons. If the light source is located near the metal/dielectric interface within wavelength scale, the SPP mode can be generated regardless of the phase matching condition. Also the light extraction processes can be reproduced by the simulation. The SPP mode can be coupled to photon if there is a nano-sized gap structure at the interface. Then, generated SPP can be extracted as light from the interface, and as a consequence, the emission efficiency is increased. These calculations support our proposed SP coupling model.

4. Enhancement of spontaneous emission rate

Fig. 2 shows that huge enhancements were observed by Ag coating especially at the shorter wavelength region. It was found that the enhancement effect became lower and lower with increasing of wavelength. This wavelength dependence of the SP enhancement effect is well correlated to the properties of the SPP. Fig. 4 shows the dispersion diagrams of SPP modes on metal/GaN surfaces calculated by Eq. (1). The SP frequency (ω_{SP}) at GaN/Ag is 2.84 eV (437nm). Thus, Ag is suitable for SP coupling to blue emission, and we attribute the large increases in the PL intensity from Ag-coated samples to such resonant SP excitation. By this reason, the SP coupling becomes remarkable when the energy is near to the SP frequency described as dotted line in Fig. 2. In contrast, ω_{SP} at GaN/Au is 2.462 eV (537 nm), and no measurable enhancement is observed in Au-coated InGaN emitters as the SP and QW energies do not match. In the case of Al, the ω_{SP} is 5.50 eV (225 nm), and the real part of the dielectric constant is negative over a wide wavelength region for visible light. Thus, a substantial and useful PL enhancement is observed in Al-coated samples, although the energy match is not ideal at 470 nm and a better overlap is expected at shorter wavelengths.

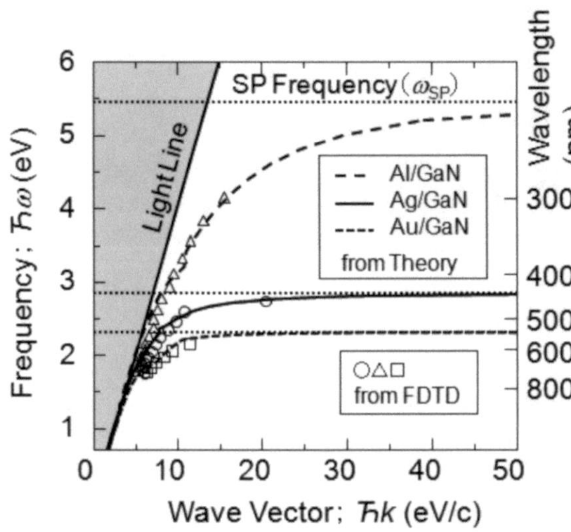

Fig. 4. Dispersion diagrams of the SPP at Al/GaN, Ag/GaN, and Au/GaN interfaces.

The external quantum efficiency (η_{ext}) of LED is given by a product of the light extraction efficiency (C_{ext}) and the internal quantum efficiency (IQE: η_{int}). The original η_{int} value is determined by the ratio of the radiative (k_{rad}) and nonradiative (k_{non}) recombination rates of excitons.

$$\eta_{ext}(\omega) = C_{ext}(\omega) \times \eta_{int}(\omega) = C_{ext}(\omega) \frac{k_{rad}(\omega)}{k_{rad}(\omega) + k_{non}(\omega)} \tag{2}$$

Under the existence of the SP coupling, the enhanced IQE value (η^*_{int}) can be described as follows,

$$\eta^*_{int}(\omega) = \frac{k_{rad}(\omega) + C'_{ext}(\omega) k_{SPC}(\omega)}{k_{rad}(\omega) + k_{non}(\omega) + k_{SPC}(\omega)} \approx C'_{ext}(\omega) \times \eta_{ex-sp}(\omega) \tag{3}$$

where k_{SPC} is the SP coupling rate and should be very fast because the density of states (DOS) of SPP is much larger than that of the excitons in the QW. C'_{ext} is the probability of photon extraction from the SPPs. C'_{ext} is decided by the ratio of light scattering and dumping of the SPPs through non-radiative loss. C'_{ext} should depend on the roughness and nanostructure of the metal surface. If the SP coupling is much faster ($k_{SPC} >> k_{rad}$), the η^*_{in} is given by the product of C'_{ext} and the exciton-SP coupling efficiency ($\eta_{ex\text{-}sp}$).

The increased emission rates were observed by the TRPL measurement. Fig. 3 (inset) shows the PL decay profiles of uncoated and Ag-coated InGaN/GaN QW samples emitters at 440nm. The PL decay rate (k_{PL}) is attributed to the radiative and nonradiative recombination rate of excitons as $k_{PL}=k_{rad}+k_{non}$. After Ag-coating, k_{PL} values were increased to k_{PL}^* by the SP coupling rate as $k_{PL}^{*}= k_{rad}+k_{non}+k_{SPC}$. Usually, k_{SPC} should be much faster than k_{rad} because of the higher DOS of SPP.

Theoretically, the SP coupling rate is given by the Fermi's Golden rule as follows (Gontijo, et al., 1999; Neogi et al., 2002),

$$k_{SPC}(\omega) = \frac{2\pi}{\hbar}\left|\vec{d}\cdot\vec{E}(\varpi)\right|^2 \rho(\varpi) \tag{4}$$

where $\hbar$ is the reduced Plank constant, $\vec{d}$ is the dipole moment of the electron-hole pair, $\vec{E}$ is the electric field of the SPP at the location of the active layer, and ρ is the DOS of the SSP. This equation suggests that the SP coupling rate should be proportional to the DOS.

We defined the enhancement factor (F) of the spontaneous emission rate as follows,

$$F(\omega) = \frac{k_{SPC}(\omega)}{k_{rad}(\omega)} \tag{5}$$

We plotted F value against wavelength in Fig. 5. The solid line in this figure shows slope of the dispersion diagram ($dk/d\omega$), which is proportional to DOS of SPP. We found that F and $dk/d\omega$ are almost same values. This suggests that the enhanced emission rates and IQEs are determined only by the DOS.

Fig. 6 shows the $\eta_{ex\text{-}sp}$ values estimated by

$$\eta_{ex-sp}(\omega) = \frac{k_{spc}(\omega)}{k_{rad}(\omega) + k_{non}(\omega) + k_{spc}(\omega)} \tag{6}$$

The η^{*}_{int} values can be estimated separately by the temperature dependence of the PL intensities by assuming IQE~100% at low temperature (Okamoto et al., 2004). Obtained η^{*}_{int} values were also shown in Fig. 6 (broken line). The ratios of η^{*}_{int} and $\eta_{ex\text{-}sp}$ give the SP-photon coupling efficiency (C'_{ext}) by Eq. (3) and were shown in Fig. 6 (solid line). Fig. 6 shows that all these efficiencies are reached to almost 100% at the shorter wavelength region. This suggests that if we can control the SP frequency and obtain the best SP coupling condition, we can develop super bright LEDs which have perfect efficiencies at any wavelength.

5. Tuning of the SP coupling

Our proposed mechanism suggests that very high efficiency may be achievable at any wavelength if we can control the SP frequency and obtain the best matching condition. Tuning of SP coupling should be attainable by choosing the appropriate metal or controlling nanostructures. For example, we fabricate several types of metal nanostructures shown in Fig. 7.

Fig. 7(a) shows the Scanning electron microscope (SEM) image of nano grating structure created by electron beam lithography and ion milling. By using this structure, we could enhanced the green emission from InGaN/GaN QW a few times larger. Fig. 7(b) shows the nano grain structure created by control of metal vapour deposition conditions. A few tens

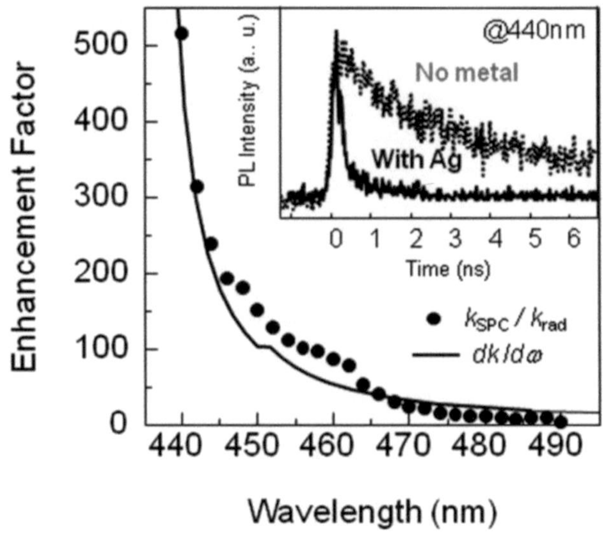

Fig. 5. Enhancement factor of the spontaneous emission obtained by the ratios of surface plasmon coupling rates and radiative recombination rates (k_{SPC}/k_{rad}). The solid line is $dk/d\omega$ of the SPP at Ag/GaN interface. (Inset) PL decay profiles of uncoated and Ag coated sample at 440nm.

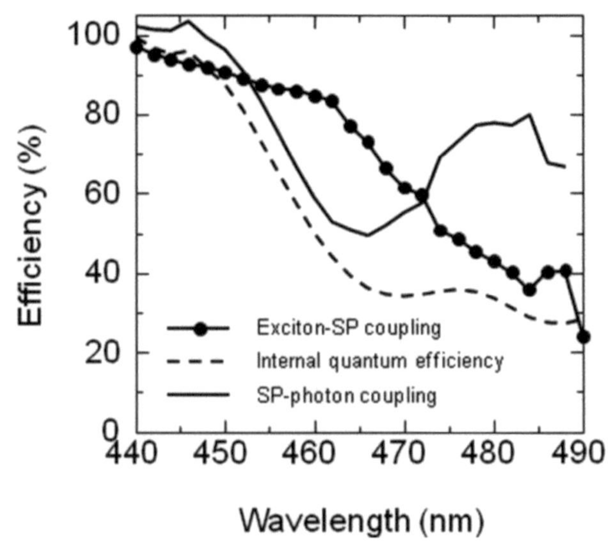

Fig. 6. Wavelength dependences of the exciton-SPP coupling efficiency (marks), the enhanced internal quantum efficiency (broken line), and the SSP-photon coupling efficiency (solid line).

nanometre sized metal grains ware generated under the slow deposition rate ~1 Å/s. This structure is very easy to fabricate and control, so we used this structure for the sample used in Fig. 2. Fig. 7(c) shows the nano particulate array structure created by thermal annealing under nitrogen atmosphere after metal thin film vapour deposition. We can control both the particle size and inter-particle distance independently by initial metal film thickness, annealing temperature, and annealing temperature. Fig. 7(d) shows the nano particle sheet structure created by the Langmuir-Blodgett (LB) technique at an air–water interface. Quite recently, we succeeded in fabrication of the beautiful periodic nanostructures by the bottom-up technique without E-beam lithography and found very interesting optical properties (Toma et al., 2011).

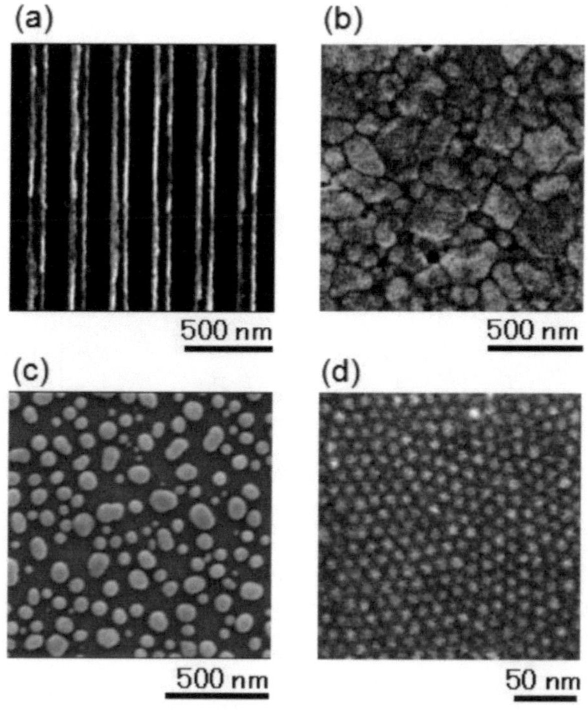

Fig. 7. Scanning electron microscope (SEM) image of various nanostructures of Ag. (a) Nano grating structure created by electron beam lithography and ion milling. (b) Nano grain structure created by control of metal vapour deposition conditions. (c) Nano particulate array structure created by thermal annealing under nitrogen atmosphere after metal thin film vapour deposition. (c) Nano particle sheet structure created by the Langmuir-Blodgett (LB) technique at an air–water interface.

By using the metal nanoparticle structures, we can use the localized surface plasmon (LSP) mode shown in Fig. 8(a). The LSP mode is tenable by the particle size and inter-particle distance. Fig. 8(b) shows the FDTD simulations of the LSP resonance spectra of Ag nano-particle with various diameters. Inset of Fig. 8(b) show the spatial distributions of the electric field of the LSP mode generated around the Ag nano-particle with 40 nm diameter. The electric fields are strongly localized near the metal particle surface, and the penetration depth is as long as the radii of the particles regardless of the wavelength. The LSP is the non-propagated mode of the electromagnetic field and the localized area is much smaller than the wavelength. Therefore the LSP mode should be applicable to the photonics with nanometer scale, such technology is called plasmonic nanophotonics.

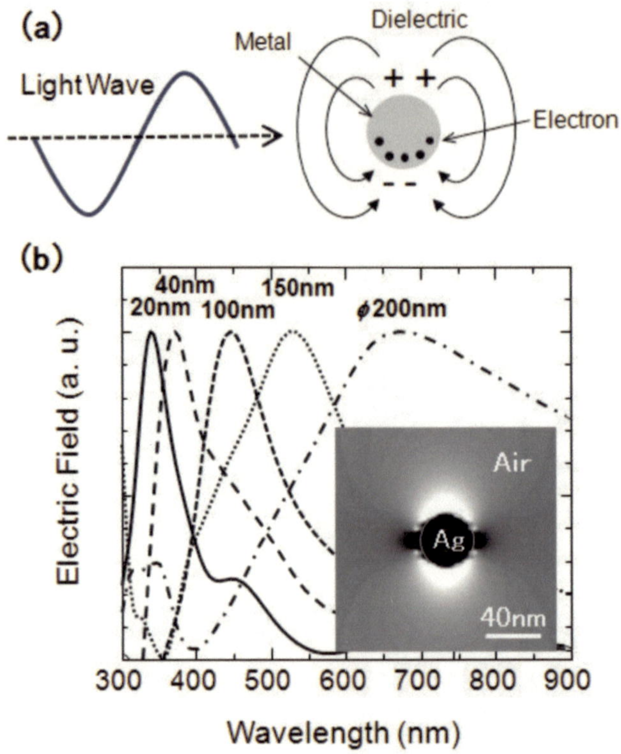

Fig. 8. (a) Schematic diagram of the localized surface plasmon (LSP) mode. (b) The FDTD simulations of resonance spectra of the LSP generated around the Ag nanoparticles with various diameters. (Inset) the spatial distributions of the electric field of the LSP mode.

Fig. 8(b) suggests that the LSP resonance is tunable within whole visible wavelength region by changing the Ag size, and 100~150 nm diameter should be suitable to couple to green light which has been very difficult to enhance. Fig. 9(a) shows the Scanning Probe Microscopic (SPM) image of Ag nanoparticle arrays fabricated on InGaN/GaN by a metal deposition and a thermal annealing. We could fabricate the Ag nano-particle array structure with about 100~150 nm diameter. The wavelength dependence of the PL enhancement ratios were shown in Fig. 9(b). Remarkable enhancement was observed at 500-520 nm with Ag particles, while the ratios were almost flat with Ag film. This difference should be due to the properties of the LSP mode and the propagating SPP mode. A huge enhancement of green emission, which has been very difficult to achieve, was observed at certain wavelength and special ranges by controlling the metal nanostructures. This result suggests that high efficiency light emitters can be achievable at various wavelength regions by further optimization of nanostructures.

6. Device applications using plasmonics

One of the most important targets of this study is device application of the SP coupling. The SP coupling technique increases IQEs by increasing spontaneous emission rates. This suggests that this should be applicable for electrical pumping because the IQEs do not

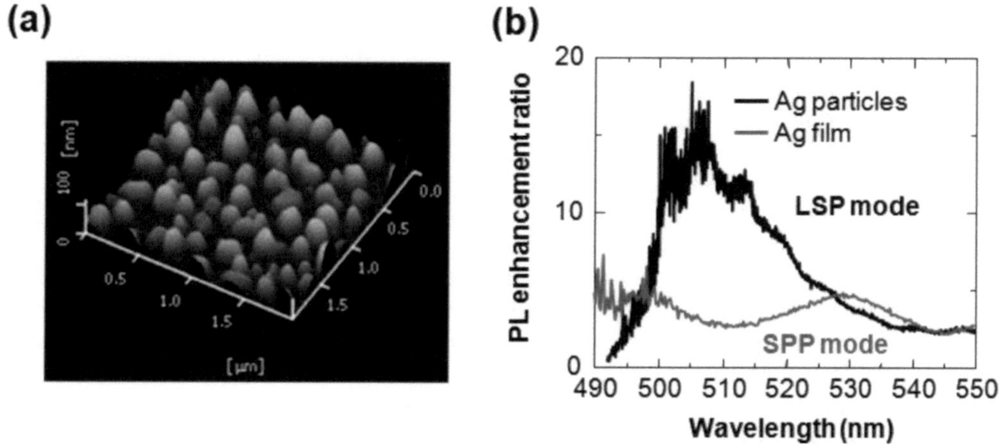

Fig. 9. (a) SPM image of the Ag nanoparticles array structure on InGaN/GaN. (b) PL enhancement ratio plotted against wavelength taken for InGaN/GaN QW with Ag particles and Ag thin film.

depend on the pumping method. So now we try to make super bright plasmonic LEDs by electrical pumping.

Fig. 10 shows the energy conversion scheme of the SP coupling and light emission. The SP-exciton and SP-photon coupling processes provide new emission pathways. The high efficiency LEDs should be achievable if the new emission path through the SP coupling is much faster than the original emission path [(2) + (11) > (3), (4) > (12) in Fig. 10]. By the similar way, plasmonics should be also able to improve high-efficiency solar cells, because the SP-exciton and SP-photon coupling processes are reversible. The sun-light can couple to the SP at the metal/dielectric interface and generate the excitons in the dielectric materials. This process should increase efficiencies of light absorption and photocurrent conversion if (8) > (7), (9) > (12) in Fig. 10.

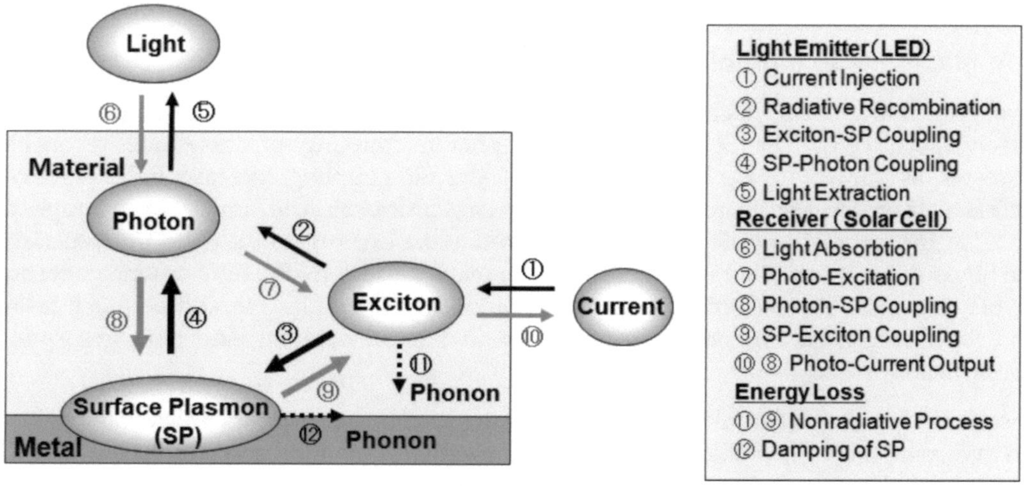

Fig. 10. Energy conversion schemes of the SP enhanced LEDs and solar cells.

Possible device structures of high efficient LEDs are shown in Fig. 11. Fig. 11(a) shows the simplest structure using a usual LED structure with a p-n junction. The metal layer can be used both as an electrical contact and for exciting plasmons. The important point of this structure is that the distance between the metal surface and the InGaN QW must be very close to get a good SP coupling. Therefore, the p-type GaN layer must be thinner than 10 nm. The PL enhancement ratios become exponentially decay with increasing of the thickness of the GaN spacer layer (Okamoto et al., 2004). This feature makes the device application of the SP coupling so difficult. We already fabricated the structure shown in Fig. 10(a) but we were not able to obtain a huge enhancement of emission. There are two reasons; first, p-doping was very difficult into 10 nm thick GaN layer. Second, we could not get a good ohmic contact because the p-GaN layer is too thin. Another possible structure of a plasmonic LED was shown in Fig. 11(b). In this structure, the metal layer for electrode and for SP coupling is different. The SP coupling should happen at the metal particles implanted just above a QW layer in a LED wafer. Fig. 11(c) shows another promising device structure which has a two-dimensional structure fabricated by the lithography and the dry etching processes. By using this structure, the electrons injection and the SP coupling can be well performed at the thick areas and the thin areas, respectively. This should enable both good ohmic contact and SP enhancement effects at the same time.

Recently a few groups reported about the SP enhanced LEDs based on our technique. Yeh et al. reported the SP coupling effect in an InGaN/GaN single-QW LED structure (Yeh et al., 2007). Their LED structure has a 10 nm p-type AlGaN current blocking layer and a 70 nm p-type GaN layer between the metal surface and the InGaN QW layer. The total distance is 80 nm, which is too far to obtain an effective SP coupling. By this reason, they obtained only 1.5 fold enhancement of the emission. Kwon, et al. also reported a plasmonic LED which has similar structure to Fig. 11(c) (Kwon et al., 2008). They put silver particles on the InGaN QW layer first, and over grew a GaN layer above the Ag particles. However, a large amount of Ag particles were gone by high temperature of the crystal growth and only 3% particles remained. Therefore, they obtained only 1.3-fold enhancement of the emission. These tiny enhancement ratios should be not good enough for device application. Therefore, a high efficient LED structure based on plasmonics is not yet achieved.

7. Applications to high-efficiency solar cells

Next very important application of plasmonics is high-efficiency light-resaving devices, namely, solar cells. The SP-exciton and SP-photon coupling processes are reversible processes as shown in Fig. 10. Therefore, if the SP coupling increase light emission processes, it should also increase the light receiving processes. The sun-light can couple to the SP at the metal/dielectric interface and generate the excitons in the dielectric materials. The SP coupling make giant electric field at the metal surface by the light-antenna effect of the SP. Therefore, the excitation processes through the SP coupling should be much faster than the direct excitation processes as shown in Fig. 10, and increase light absorption efficiencies.

Plasmonics has the potential to apply to high-efficiencies and ultra-thin solar cells, which can overcome the both problems of solar cells of efficiencies and costs. Until now, several types of the plasmonic solar cells have been reported by using random metal particle array structures (Stuart et al., 1996; Pillai et al., 2006) and top-down nanofabricated structures

(Atwater & Polman, 2010). However these palsmonic solar cells are still far from practical utilizations. Further optimization of the metal nanostructure and tuning of the SP coupling process are required in order to improve the plasmonic solar cell to the practical level

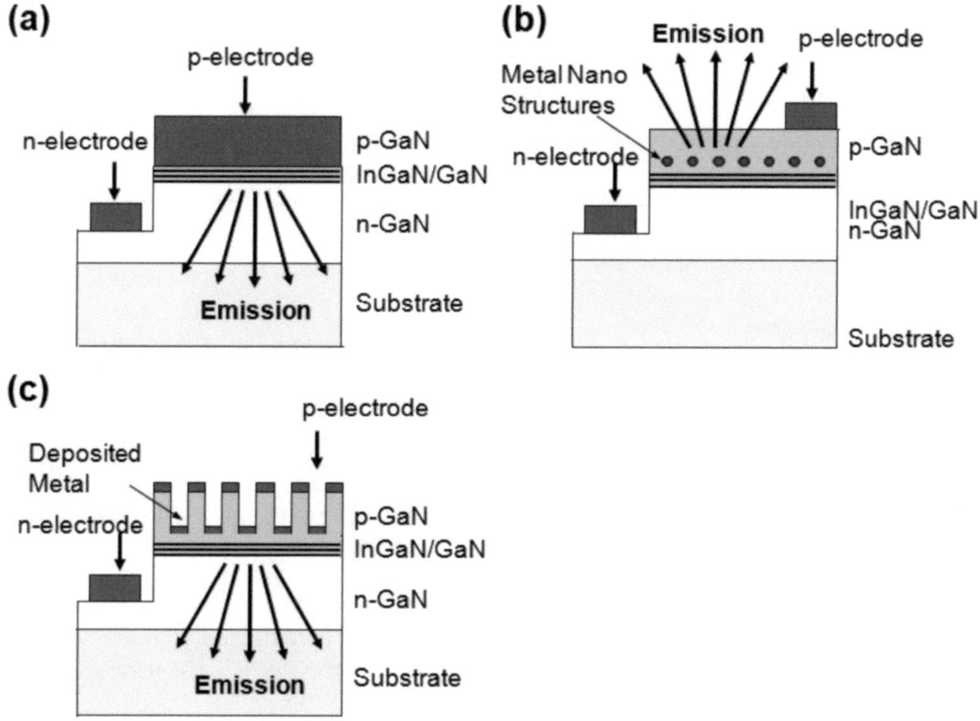

Fig. 11. Possible device structures of high efficient LEDs based on plasmonics with electrical pumping. (a) Metal electrode is located a few nm above the active layer. (b) Metal particles are embedded a few nm above the active layer, (c) p-GaN has 2-dimentional structures.

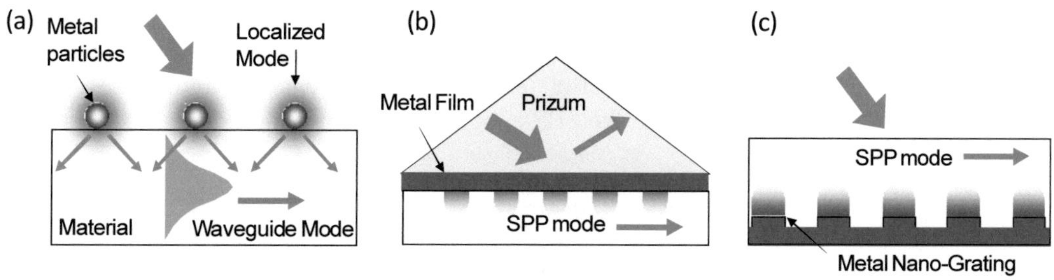

Fig. 12. The previously reported plasmonic solar cell structures based on (a) metal nano particles disposed on the materials, (b) attenuated- total-reflection (ATR) consignation with prism, and (c) nano periodic grating structure.

The previous reported plasmonic solar cells can be classified into three types shown in Fig. 12. Fig. 12(a) show the structure by using the metal nanoparticles, which were simply dispersed on the top of the solar cells. This structure is very easy to fabricate and useful to any type of the solar cells. But in this case, energy transfer from the LSP mode to material is difficult because the electrical field of the LSP mode is strongly confined around the metal nanoparticle. In many case the LSP mode have to couple to the waveguide mode in the materials. This fact limits the plasmonic enhancement effect, and the device structures must be thicker than the wavelength. On the other hand, the using of propagation SPP modes must be more useful to enhance the efficiency and making ultra-thin structures of solar cells. But the coupling between the light and the SPP is not easy. The wave vectors of the light and the SPP need to be matched in order to be coupled each other. Usually all-reflection setting with prism [Fig. 12(b)] or periodic nano-structures [Fig. 12(c)] are necessary to satisfy the matching condition of the wave vectors. These settings limit the plasmonic enhancement effect into some angle and wavelength of light.

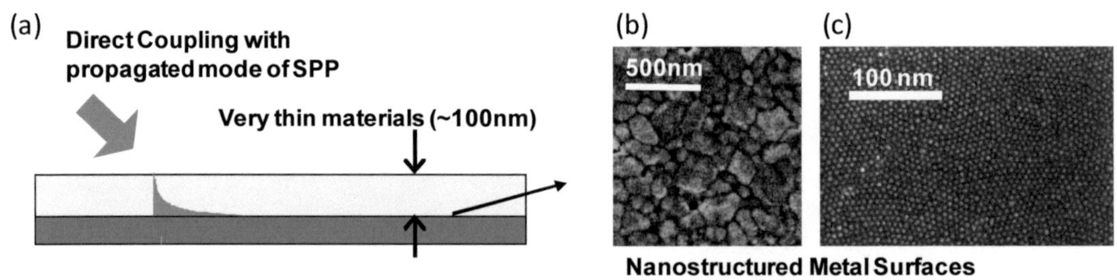

Fig. 13. (a) Our proposed advanced structure of plasmonic solar cells using propagating mode of the SPP. Sunlight can couple to the SPP mode directly by the metall nanostructures at the interface, such as (a) nano metal grain, and (b) metal nano particle sheet structures.

We believe that the structure shown in Fig.13 should be one of the promising device structures for palsmonic solar cells using propagating SPP mode. For the metal nanostructures, random Ag nano-grain structure shown in Fig. 13(a) [same as Fig. 7(b)] should be very useful. We can easily control the Ag nano-grain sizes of the random nanostructures by the metal-deposition conditions. Such structures were already used to extract light emission from the SPP for the light emission enhancements. In section 3, we described that the roughness allows SPPs of high momentum to scatter, lose momentum, and couple to radiative photon. By the opposite way, this structure should enable the direct coupling from sun-light with propagating SPP. Moreover, we can reduce the device thickness lower than 100nm because the SPP propagate within a few tens nm at the metal surface.

The other promising structure is 2D metal nanoparticle sheet structure shown in Fig. 13(c) [same as Fig. 7(d)]. This is high-dense packed structure of the Ag nanoparticles with 5 nm diameter fabricated by the LB technique at an air–water interface (Toma, et al. 2011). This structure enables more flexible tuning of the resonance spectra. Very strong and wide resonance spectra, which is almost overlapped to the solar spectrum, were found for the 2D sheet structure of Ag particles by the FDTD calculations. Moreover it was found that ~99% of the incident light can be confined into the metal nanosheet which is only ~10nm thickness by optical measurements and calculations. This structure should be promising to develop new type of palsmonic solar cells.

8. Conclusions

The SP coupling is very powerful method to enhance light emission efficiencies of various materials at wider wavelength regions. By using this technique, high-efficiency and high-speed light emission is predicted for optically as well as electrically pumped light-emitting devices, because the SP coupling increases the internal quantum efficiency, and this mechanism is not related to the pumping method. We found that both the exciton-SP and the SP-photon coupling efficiencies were reached almost 100% at the best matched wavelength. This suggests that if we can control the SP frequency and obtain the best SP coupling condition, we can develop super bright LEDs which have perfect efficiencies at any wavelength. It would bring full color devices and natural white LEDs. This technique can be applied not only to InGaN-based materials but also to various materials that suffer from low efficiencies. We believe that the high-efficient plasmonic LEDs should be obtainable by using silicon based materials or other earth-abundant materials. Such LEDs could be very cheap to make and easy to process, and would become widely used light source instead of fluorescent tubes in the near future. Similar plasmonic design should also be applicable to several other optical devices. Especially, the SP coupling technique has been expected to progress solar cell technology. The SP enhanced solar cell have been studied by several groups however such devices have not so far been used practically. We believe that further optimization of nanostructures and controlling the SP coupling would provide high efficient and ultra-thin plasmonics solar cells.

9. Acknowledgements

The author wish to thank Prof. Y. Kawakami (Kyoto University), Prof. K. Tamada (Kyushu University) and Prof. A. Scherer (California Institute of Technology) for valuable discussions and support. This work was supported by the Precursory Research for Embryonic Science and Technology (PRESTO) at Japan Science and Technology Agency (JST).

10. References

Atwater, H. A. (2007), The promise of plasmonics, *Scientific American*, pp. 56-63.

Atwater, H. A. & Polman, A. (2010), Plasmonics for improved photovoltaic devices, *Nat. Mater.*, 9, pp. 205-213.

Barnes, W. L.; Dereux, A. & Ebbesen, T. W. (2001), Surface plasmon subwavelength optics, *Nature*, vol. 424, pp. 824-830.

Gontijo, I.; Borodisky, M.; Yablonvitch, E.; Keller, S.; Mishra, U. K. & DenBaars, S. P. (1999), Enhancement of spontaneous recombination rate in a quantum well by resonant surface plasmon coupling, *Phys. Rev. B*, vol. 60, no. 16, pp. 11564 -11567.

Kwon, M.-K.; Kim, J.-Y.; Kim, B.-H.; Park, I.-K.; Cho, C.-Y.; Byeon, C. C. & Park, S.-J. (2008), Surface-Plasmon-Enhanced Light-Emitting Diodes" *Adv. Mate.* vol. 20, no. 7, pp. 1253-1257.

Neogi, A.; Lee, C.-W.; Everitt, H. O.; Kuroda, T.; Tackeuchi, A. & Yablonvitch, E. (2002) , Enhancement of spontaneous recombination rate in a quantum well by resonant surface plasmon coupling, *Phys. Rev. B*, vol. 66, 153305.

Okamoto, K.; Niki, I.; Shvartser, A.; Narukawa, Y.; Mukai, T. & Scherer, A. (2004), Surface-plasmon-enhanced light emitters based on InGaN quantum wells, *Nature Mater.*, vol. 3, pp. 601-605.

Okamoto, K.; Niki, I.; A. Scherer, Narukawa, Y.; Mukai, T. & Kawakami, Y. (2005), Surface plasmon enhanced spontaneous emission rate of InGaN/GaN quantum wells probed by time-resolved photoluminescence spectroscopy, *Appl. Phys. Lett.* vol. 87, 071102.

Okamoto, K.; Scherer, A. & Kawakami, Y. (2008), Surface plasmon enhanced light emission from semiconductor materials, *Phys. Stat. Sol. C*, vol. 5, no. 9. pp. 2822-2824.

Okamoto, K. & Kawakami, Y. (2009), High-Efficiency InGaN/GaN Light Emitters Based on Nanophotonics and Plasmonics, *IEEE J. Select. Top. Quantum Electron.* 15, pp. 1190-1209.

Pillai, S.; Catchpole, K. R.; Trupke, T. & Green, M. A. (2006), Surface plasmon enhanced silicon solar cells, *J. Appl. Phys.*, 101, 09310.

Raether, H. (1988), *Surface Plasmons on Smooth and Rough Surfaces and on Gratings*, Springer-Verlag.

Stuart, H. R. & Hall, D. G. (1996), Absorption Enhancement in Silicon-on-Insulator Waveguides using Metal Island Films, *Appl. Phys. Lett.*, 69, pp. 2327-2329.

Yeh, D.-M.; Huang, C.-F.; Chen, C.-Y.; Lu, Y.-C. & Yang, C. C. (2007), Surface plasmon coupling effect in an InGaN/GaN single-quantum-welllight-emitting diode, *Appl. Phys. Lett.* vol. 91, 171103.

Part 3

Photonics Applications

9

Lightwave Refraction and Its Consequences: A Viewpoint of Microscopic Quantum Scatterings by Electric and Magnetic Dipoles

Chungpin Liao, Hsien-Ming Chang, Chien-Jung Liao, Jun-Lang Chen and Po-Yu Tsai
National Formosa University (NFU)
Advanced Research and Business Laboratory (ARBL)
Chakra Energetics, Ltd.
Taiwan

1. Introduction

In optics, it is well-known that when a visible light beam, e.g., traveling from air (or more strictly, vacuum) into a piece of smooth flat glass at an angle relative to the normal of the air-glass interface, some proportion of the light will be bounced off at the reflection angle equal to the incident angle. However, when the light beam is with its oscillating electric field parallel to the plane-of-incidence (POI, i.e., the plane constituted by both propagation vectors of the incident and reflected light waves, as well as the interface normal vector) (called the p-wave), there is a particular incident angle at which no bounce-off would occur. This particular angle is known as the Brewster angle (θ_B) (Hecht, 2002). In contrast, when the light beam is with its electric field vector perpendicular to the plane-of-incidence (called the s-wave), no such angle exists (Hecht, 2002). In fact, this is only true for uniform, isotropic, and nonmagnetic (or equivalently, with its relative magnetic permeability (μ_r) equal to unity at the optic frequency of interest) materials such as the above glass piece. Indeed, it is known that for magnetic materials, there may instead exist Brewster angles for the s-waves, while none for the p-waves (as will be demonstrated in Section 2).

Traditionally, whichever the case, the Brewster angle is a solid property of the material in question with respect to a given light frequency of interest. Namely, there is a one-to-one correspondence between the Brewster angle and the incident light frequency. However, it is one of the purposes of this chapter to show that the Brewster angle of the material in hand can in principle be modified into a new controllable variable, even dynamically, if a post-process microscopic method called "dipole engineering" is applicable on that material. Among its predictions, the traditionally fixed Brewster angle of a specific material now not only becomes dependent on the density and orientation of incorporated permanent dipoles, but also on the incident light intensity (more precisely, the incident wave electric field strength). Further, two conjugated incident light paths would give rise to different refracted wave powers (Liao et al., 2006).

In order to reveal the intricacies of the mechanism of the proposed permanent dipole engineering subsequently, existing important result of Doyle (Doyle, 1985) is first thoroughly detailed. That is, the Fresnel equations and Brewster angle formula are to be arrived at intuitively and rigorously obtained, by viewing all light-wave-induced dipole moments (including both electric and magnetic dipoles) as the microscopic sources causing the observed macroscopic optical phenomena at an interface, as compared to the traditional academic "Maxwell" approach ignoring the dipole picture. Then, equipped with such-developed intuitive and quantitative physical picture, the readers are then ready to appreciate the way those optically-responsive, permanent dipoles are externally implemented into a selected host matter and their rendered effects. Namely, the Brewster angle of a selected host material becomes alterable, likely at will, and ultimately new optical materials, devices and applications may emerge.

2. Brewster angle and "scattering" form of Fresnel equations

Arising from Maxwell's equations (through assuming linear media and adopting monochromatic plane-waves expansion), Fresnel equations provide almost complete quantitative descriptions about the incident, reflected and transmitted waves at an interface, including information concerning energy distribution and phase variations among them (Hecht, 2002). Two of the Fresnel equations are relevant to reflections associated with both the p and s components (Hecht, 2002):

$$r^s = \frac{E_r^s}{E_i^s} = \frac{n_i \mu_{rt} \cos\theta_i - n_t \mu_{ri} \cos\theta_t}{n_i \mu_{rt} \cos\theta_i + n_t \mu_{ri} \cos\theta_t} \tag{1}$$

$$r^p = \frac{E_r^p}{E_i^p} = \frac{\mu_{ri} n_t \cos\theta_i - \mu_{rt} n_i \cos\theta_t}{\mu_{ri} n_t \cos\theta_i + \mu_{rt} n_i \cos\theta_t} \tag{2}$$

where E is the electric field, μ_r is the relative magnetic permeability, $n = (\varepsilon_r \mu_r)^{1/2}$ is the index of refraction (ε_r being the relative dielectric coefficient), superscripts "p" and "s" stand for the p-wave and s-wave components, while subscripts "i", "r" and "t" denote incident, reflected and transmitted components, respectively. When the incident angle (θ_i) is equal to a particular value (θ_B), one of the above reflection coefficients would vanish, then such value of the incident angle is known as the Brewster angle (θ_B). Note that for the most familiar case in which the light wave is incident from vacuum onto a linear nonmagnetic medium ($\mu_r = 1$), only the p-wave possesses a Brewster angle, not the s-wave.

In the following, to get ready for our proposed idea while without loss of generality, the medium on the incident side is designated to be vacuum (i.e., $n_i = 1$) for simplicity. In addition, to further facilitate our purpose, the Fresnel equations in the equivalent "scattering" form (due to Doyle) are retyped here (Doyle, 1985):

$$\frac{E_t^p}{E_i^p} = \left[\frac{-\sqrt{\mu_r}}{(\mu_r - 1)\sqrt{\varepsilon_r} + (\varepsilon_r - 1)\sqrt{\mu_r}\cos(\theta_t - \theta_i)}\right] \times \left[\frac{2\cos\theta_i \sin(\theta_t - \theta_i)}{\sin\theta_t}\right] \tag{3}$$

$$\frac{E_r^p}{E_i^p} = \left[\frac{(\mu_r - 1)\sqrt{\varepsilon_r} - (\varepsilon_r - 1)\sqrt{\mu_r}\cos(\theta_t + \theta_i)}{(\mu_r - 1)\sqrt{\varepsilon_r} + (\varepsilon_r - 1)\sqrt{\mu_r}\cos(\theta_t - \theta_i)}\right] \times \left[\frac{\sin(\theta_t - \theta_i)}{\sin(\theta_t + \theta_i)}\right] \tag{4}$$

$$\frac{E_t^s}{E_i^s} = \left[\frac{-\sqrt{\mu_r}}{(\varepsilon_r - 1)\sqrt{\mu_r} + (\mu_r - 1)\sqrt{\varepsilon_r}\cos(\theta_t - \theta_i)}\right] \times \left[\frac{2\cos\theta_i \sin(\theta_t - \theta_i)}{\sin\theta_t}\right] \tag{5}$$

$$\frac{E_r^s}{E_i^s} = \left[\frac{(\varepsilon_r - 1)\sqrt{\mu_r} - (\mu_r - 1)\sqrt{\varepsilon_r}\cos(\theta_t + \theta_i)}{(\varepsilon_r - 1)\sqrt{\mu_r} + (\mu_r - 1)\sqrt{\varepsilon_r}\cos(\theta_t - \theta_i)}\right] \times \left[\frac{\sin(\theta_t - \theta_i)}{\sin(\theta_t + \theta_i)}\right] \tag{6}$$

Namely, the right hand sides of Eq. (3)-(6) are in the form of D × S, with D being the first bracketed term, representing single dipole (electric and magnetic) oscillation; while S being the second term, depicting the collective scattering pattern generated by the whole array of dipoles. While S is nonzero, where D vanishes (Eq. (4) or (6)) is the condition for the Brewster angle to arise either for the p-wave or the s-wave. That is (Doyle, 1985),

$$\tan^2 \theta_B{}^p = \frac{\varepsilon_r(\varepsilon_r - \mu_r)}{\varepsilon_r \mu_r - 1} \tag{7}$$

$$\tan^2 \theta_B{}^s = \frac{\mu_r(\mu_r - \varepsilon_r)}{\varepsilon_r \mu_r - 1} \tag{8}$$

Note that if the medium is characterized by $\mu_r > \varepsilon_r$, only the s-wave may experience the Brewster angle; while in the $\varepsilon_r > \mu_r$ situation, only the p-wave can. Indeed, in the most familiar case of light going from vacuum into a piece of glass whose $\varepsilon_r > \mu_r = 1$, there is the Brewster angle only for the p-wave.

In the following, an alternative derivation of the "scattering" form Fresnel equations (Eq. (4)-(6)) will be reproduced from (Doyle, 1985) in somewhat details, which stems from the viewpoint of treating induced microscopic (electric and magnetic) dipoles as the effective sources of macroscopic EM waves at the interface. Then, effective ways to implement the proposed permanent dipoles on a host material will be proposed, which will in principle allow us to achieve variant Brewster angles and thus to create novel materials, devices, even new applications.

3. "Scattering" form Fresnel equations from the dipole source viewpoint

Microscopically, all matters including optical materials are made of atoms or molecules, each of which further consists of a positive-charged nucleus (or nuclei) and some orbiting negative-charged electron clouds. When subjected to the EM field of an impinging light wave, the positive and negative charges separate to form induced electric dipoles along the light electric field, while some electrons further move in ways to form induced magnetic dipole moments along the light magnetic field. Note that in this chapter, only far-fields generated by these dipoles are considered and Fig. 1 depicts the relevant orientations.

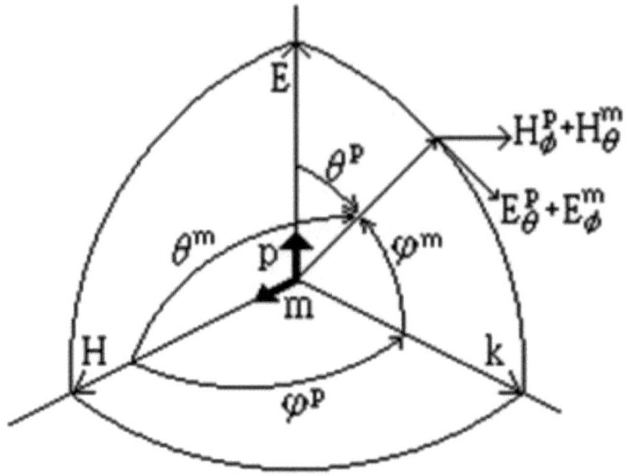

Fig. 1. Orientation of induced dipoles and their fields (Courtesy of W. T. Doyle (Doyle, 1985)).

3.1 The p-wave (E parallel to the plane-of-incidence) case

It is well-known from electromagnetism (Purcell, 1985) that if denoting the induced electric polarization by $\vec{P}$, the electric displacement is: $\vec{D} = \varepsilon_0 \vec{E} + \vec{P}$ for linear isotropic materials. Comparing it with $\vec{D} = \varepsilon_0 \varepsilon_r \vec{E}$, the induced polarization hence emerges as (all in space-time configuration) (Purcell, 1985):

$$\vec{P} = \varepsilon_0 (\varepsilon_r - 1) \vec{E}_t^p \tag{9}$$

When calculating the field contribution from the electric dipoles alone, it is assumed that $\mu_r = 1$ and thus $n = (\varepsilon_r)^{1/2}$. Putting this into an equivalent form of the Snell's law (Doyle, 1985):

$$n^2 - 1 = \frac{\sin(\theta_i + \theta_t)\sin(\theta_i - \theta_t)}{\sin^2 \theta_t} \tag{10}$$

and subsequently, together with Eq. (9), into the Fresnel equation relating E_t / E_i, we obtain an expression in which the incident electric field is expressed as a "consequence" of the microscopic sources -- the induced dipoles:

$$E_{ip}^p = \left[-\frac{P}{\varepsilon_0} \cos(\theta_t - \theta_i) \right] \times \left[\frac{\sin \theta_t}{2 \cos \theta_i \sin(\theta_t - \theta_i)} \right] \tag{11}$$

where the subscript "p" stands for contribution from the induced electric dipoles. Incorporating Eq. (11) into Eq. (2), the reflected electric field manifests itself as due to the electric dipoles sources:

$$E_{rp}^{p}=\left[\frac{P}{\varepsilon_0}\cos\left(\theta_t+\theta_i\right)\right]\times\left[\frac{\sin\theta_t}{2\cos\theta_i\sin\left(\theta_t+\theta_i\right)}\right] \tag{12}$$

Similarly, there are field contributions from the induced magnetic dipoles, which acted through the magnetization $\vec{M}$. Using the magnetic field strength $\vec{H}=\left(\vec{B}/\mu_0\right)-\vec{M}$ and the $E=\frac{\omega}{k}B=\nu B$ relation associated with general monochromatic plane waves, the magnetization becomes (Purcell, 1985):

$$M=\left(\mu_r-1\right)\sqrt{\frac{\varepsilon_0\varepsilon_r}{\mu_0\mu_r}}E_t^p \tag{13}$$

In considering the magnetic contribution, $\varepsilon_r=1$ and thus $n=\left(\mu_r\right)^{1/2}$ are adopted in Eq. (10). Then, using this together with Eq. (13), the Fresnel equation related to E_t/E_i (Eq. (3)), can be converted into an expression for the magnetic component of the incident electric field, in fact as a "consequence" of the microscopic induced magnetic dipoles:

$$E_{im}^{p}=\left[-\frac{M\sqrt{\mu_0}}{\sqrt{\varepsilon_0}}\right]\times\left[\frac{\sin\theta_t}{2\cos\theta_i\sin\left(\theta_t-\theta_i\right)}\right] \tag{14}$$

where the subscript "m" stands for contribution from the induced magnetic dipoles. Putting Eq. (14) into Eq. (2), the magnetic component of the reflected electric field appears as due to the induced magnetic dipoles too:

$$E_{rm}^{p}=\left[-\frac{M\sqrt{\mu_0}}{\sqrt{\varepsilon_0}}\right]\times\left[\frac{\sin\theta_t}{2\cos\theta_i\sin\left(\theta_t+\theta_i\right)}\right] \tag{15}$$

When a light wave impinges on an interface, it causes the excitation of electric and magnetic dipoles throughout the second medium, which in turn collectively give rise to reflected, transmitted, and formally, incident waves at the interface. As depicted in Fig. 1, as long as far fields are concerned, the induced electric and magnetic dipoles can be viewed as aligned along the incident electric and magnetic field vectors, respectively. All electric and magnetic fields of interface-relevant waves can be conceived as generated by the co-work of electric and magnetic dipoles. (Of course, for the incident wave, this is only formally true, that is, the incident fields are the "cause" not the "effect" of dipole oscillations.) We thus write:

$$E_i^p=E_{ip}^p+E_{im}^p \tag{16}$$

Namely, by adding Eq. (11) and (14), the incident electric field can be formally expressed as due to induced dipole sources P and M:

$$E_i^p = \left[-\frac{P}{\varepsilon_0}\cos(\theta_t - \theta_i) - \frac{M\sqrt{\mu_0}}{\sqrt{\varepsilon_0}} \right] \times \left(\frac{\sin\theta_t}{2\cos\theta_i \sin(\theta_t - \theta_i)} \right) \tag{17}$$

As a verification, if putting forms of these sources (i.e., Eq. (9) and (13) in which the induced P and M are expressed in term of $E_t^{\,p}$) back into Eq. (17), it is found that the obtained transmission coefficient of the p-wave is exactly that of Eq. (3).

Similarly, conceiving the reflected electric field as:

$$E_r^p = E_{rp}^p + E_{rm}^p \tag{18}$$

Adding Eq. (12) and (15), the reflected electric field appears as due to the induced dipole sources P and M:

$$E_r^p = \left[\frac{P}{\varepsilon_0}\cos(\theta_t + \theta_i) - \frac{M\sqrt{\mu_0}}{\sqrt{\varepsilon_0}} \right] \times \left(\frac{\sin\theta_t}{2\cos\theta_i \sin(\theta_t + \theta_i)} \right) \tag{19}$$

Similarly, putting forms of the dipoles sources (i.e., Eq. (9) and (13) in which P and M are expressed in terms of $E_t^{\,p}$) back into Eq. (19), and using the newly obtained $E_i^{\,p}$ vs. $E_t^{\,p}$ relation (i.e., Eq. (17)), it is found that the obtained reflection coefficient of the p-wave is exactly that of Eq. (4), in "scattering" form.

3.2 The s-wave (E perpendicular to the plane-of-incidence) case

Likewise, following Fig. 1 again, we can also express the incident and reflected electric fields of the s-wave as due to those induced electric and magnetic dipoles:

$$E_i^s = \left(-\frac{P}{\varepsilon_0} - \frac{M\sqrt{\mu_0}}{\sqrt{\varepsilon_0}}\cos(\theta_t - \theta_i) \right) \times \left(\frac{\sin\theta_t}{2\cos\theta_i \sin(\theta_t - \theta_i)} \right) \tag{20}$$

$$E_r^s = \left(-\frac{P}{\varepsilon_0} + \frac{M\sqrt{\mu_0}}{\sqrt{\varepsilon_0}}\cos(\theta_t + \theta_i) \right) \times \left(\frac{\sin\theta_t}{2\cos\theta_i \sin(\theta_t + \theta_i)} \right) \tag{21}$$

and that the transmission and reflection coefficients are indeed found to be those of Eq. (5) and (6), respectively, in "scattering" form.

4. The proposed permanent dipoles engineering

4.1 Observations and inspirations

From the above elaboration, it becomes obvious that the electric and magnetic dipoles can be much more than mere pedagogical tools for picturing dielectrics and magnetics, as indeed proven by Doyle (Doyle, 1985). In fact, treating them as microscopic EM wave sources from the outset, the "scattering" form of Fresnel equations (i.e., Eq. (3)-(6)), and consequently, the Brewster angle formulas (Eq. (7) and (8)) can all be reproduced. Then, emerging from such details come our inspired purposes. Namely, by acquainting ourselves with the role played by these *induced* dipoles (or, the microscopic scattering sources), it is then intuitively straightforward to learn how new macroscopic optical phenomena, such as the new Brewster angle may be generated if extra anisotropic optically-responsive *permanent* dipoles were implemented onto the originally isotropic host material in discussion (Liao et al., 2006). In other words, now the notions of P and M are further extended to include the total effects resulting from both the *induced* and *permanent* dipoles.

For instance, in Eq. (17), the $\cos(\theta_t - \theta_i)$ factor multiplying on P (but not on M) is due to the fact that the induced polarization P (along E_t^p) has only a fractional contribution to E_i^p determined by the vector projection as shown in Fig. 2, for the p-wave situation. Now, if an external polarization vector $P_0(\omega)$ (as the collective result of many imposed electric dipoles responding to the incident lightwave of radian frequency ω at the incident angle θ_i) is introduced within a host material, then, e.g., for the p-wave case, all electric dipoles' contribution to E_i^p (i.e., Eq. (17)) is now $P_{induced}\cos(\theta_i - \theta_t) + P_0\cos(\theta_0 + \theta_i)$, or $\varepsilon_0(\varepsilon_r - 1)E_t^p\cos(\theta_i - \theta_t) + P_0\cos(\theta_0 + \theta_i)$ (see Fig. 2). Namely, there is now an additional second anisotropic term resulting from *externally* imposed dipoles whose contribution may not necessarily be less than the induced dipoles of the original isotropic host. Note that the incident light-driven response of these externally imposed permanent dipoles is frequency dependent, and therefore, the above P_0 really stands for that amount of polarization at the relevant optical frequency of interest and the lightwave's incident angle θ_i. In other words, a "DC" polarization will never enter the above equation, and P_0/E_t^p would be constant for each specific θ_i if the added dipoles, and hence the resultant polarization, are linear.

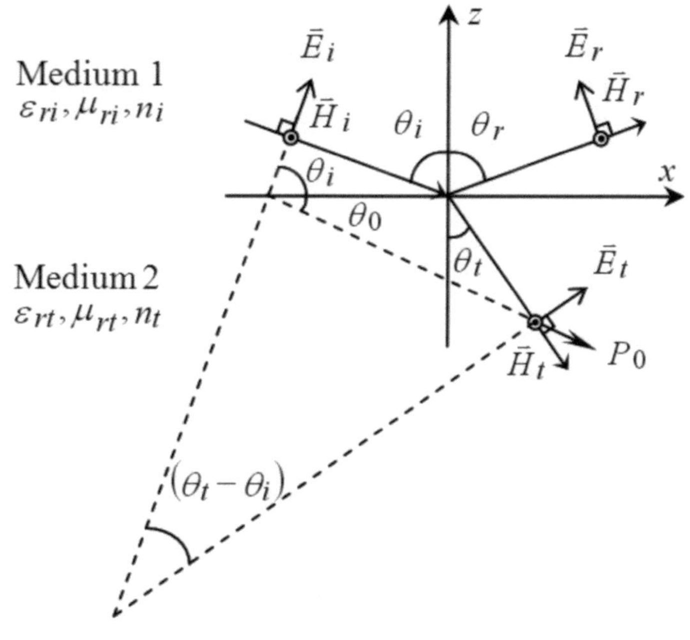

Fig. 2. P-wave configuration at the interface and the orientation of embedded permanent electric dipoles (Courtesy of W. T. Doyle (Doyle, 1985)).

Thus, if the dipole-engineered total contribution is recast in the traditional form, viz., $\varepsilon_0(\tilde{\varepsilon}_r - 1)E_t^p \cos(\theta_i - \theta_t)$, then it is clear that the modified relative dielectric coefficient is equivalently (Liao et al., 2006):

$$\tilde{\varepsilon}_r = \frac{P_0}{\varepsilon_0 E_t^p} \frac{\cos(\theta_i + \theta_0)}{\cos(\theta_i - \theta_t)} + \varepsilon_r \tag{22}$$

where θ_0 is the angle between the imposed extra polarization vector and the interface plane (see Fig. 2). Thus, by putting Eq. (22) into the p-wave Brewster angle formula (Eq. (7)), a new Brewster angle (θ_B) would then emerge:

$$\tan^2 \theta_B{}^p = \frac{\tilde{\varepsilon}_r(\tilde{\varepsilon}_r - \mu_r)}{\tilde{\varepsilon}_r \mu_r - 1} \tag{23}$$

4.2 Justification of the effectiveness and meaningfulness of implementing optically-responsive dipoles

A justification of the effectiveness of the proposed *permanent* dipole engineering is straightforward by noting the following fact. Namely, had the original host material been transformed into a new material by adding in a considerable amount of certain second substance, then P_0 in the above really would have stood for the extra induced dipole effect resulting from this second substance.

However, to this end, an inquiry may naturally arise as to whether the outcome of the proposed dipole-engineering approach being nothing more than having a material of multi-components from the outset. The answer is clearly no, and there are much more meaningful and practical intentions behind the proposed method. First of all, this is a controllable way to make new materials from known materials without having to largely mess around with typically complicated details of manufacturing processes pertaining to each involved material (if the introduced *permanent* dipoles are noble enough). Indeed, we have been routinely attempting to create various materials by combining multiple substances, and yet have also been very much limited by problems related to chemical compatibility, phase transition, in addition to many processing and economic considerations. Secondly, *permanent* dipole engineering would further allow delicate, precise means of manipulating the material properties, such as varying the dipole orientation to render desired optical performance on host materials of choice. Thirdly, all existing techniques known to influence dipoles can be readily applied on the now embedded dipoles to harvest new optical advantages, such as by electrically biasing the dipoles to adjust the magnitude of *permanent* dipole moment (in terms of P_0) in the frequency range of interest.

4.3 Different refracted wave powers on two conjugated incident light paths

If, instead of picking the incidence from the left hand side as depicted in Fig. 2, a conjugate path, i.e., from the right hand side, is taken (see, Fig. 3), then the formula for Eq. (22) becomes (Liao et al., 2006):

$$\tilde{\varepsilon}_r = \frac{P_0}{\varepsilon_0 E_t^p} \frac{\cos(\theta_i - \theta_0)}{\cos(\theta_t - \theta_i)} + \varepsilon_r \tag{24}$$

Unconventional Brewster angle can be found by Eq. (23) (p-wave case).

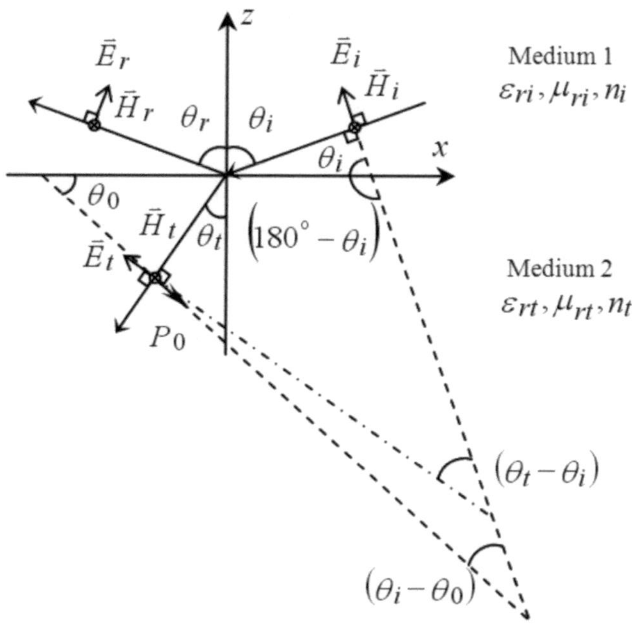

Fig. 3. P-wave configuration and the orientation of embedded permanent electric dipoles (Conjugated Incident Light Path, courtesy of W. T. Doyle (Doyle, 1985)).

In other words, the traditionally fixed Brewster angle of a specific material now not only becomes dependent on the density and orientation of incorporated permanent dipoles, but also on the incident light intensity (more precisely, the incident wave electric field strength). Further, two conjugated incident light paths would give rise to different refracted wave powers (Liao et al., 2006), (Haus & Melcher, 1989).

4.4 The surface embedded with a thin distributed double layer

The traditional Fresnel equations in the electromagnetic theory have been used in determining the light power distribution at an interface joining two different media in general. They are known to base upon the so-called "no-jump conditions" ((Haus & Melcher, 1989), (Hecht, 2002)) wherein the interface-parallel components of the electric and magnetic fields of a plane wave continue seamlessly across the interface, respectively:

$$\langle E_{||} \rangle = 0 \quad and \quad \langle H_{||} \rangle = 0 \tag{25}$$

where $\langle Q \rangle \equiv Q_a - Q_b$ stands for the discontinuity of the physical quantity Q by crossing from the a side to the b side of the interface, and | | (or $\perp$) is with respect to the interface plane. The general configuration of light incidence can be decomposed into the p-wave and s-wave situations. For the p-wave case, the lightwave's electric field is on the plane of incidence (POI) (see Fig. 4), and for the s-wave situation, it is pointing perpendicularly out of the POI.

When a light beam of frequency ω is incident from air (*a* side) onto a flat smooth dielectric (*b* side) embedded with a layer of distributed, incident-light-responsive electric dipoles near the surface (see Fig. 5), the above jump condition for the electric field becomes (Haus & Melcher, 1989):

$$E_{||}^{a} - E_{||}^{b} \equiv \langle E_{||} \rangle = -\frac{1}{\varepsilon_0}\frac{\partial \pi_s}{\partial x} \tag{26}$$

where ε_0 is the dielectric permittivity of free space. An interfacial double layer is composed of a top and bottom layers of equal but opposite surface charges ($\pm\sigma_s$) respectively and separated by a tiny distance (d). It is mathematically described by $\sigma_s \to \infty$ and $d \to 0$ such that $\pi_s \equiv \sigma_s \cdot d$ stands for the electric dipole moment per unit area on the interface. Eq. (26) is obtained through integrating Faraday's law $\nabla \times \vec{E} = -\partial \mu \vec{H} / \partial t$ (μ being the magnetic permeability) over a vanishingly thin strip area enclosing a section of the interface on the plane of incidence (see Fig. 5). Thus, with the integration on the right hand side being null, further applying Stokes' theorem ((Haus & Melcher, 1989), (Hecht, 2002)) on the left hand side reveals that the two normal sections (i.e., along $\vec{E}_{\perp}$) in the contour integration no longer cancel each other. This is because the vertical electric field distribution is now non-uniform in the presence of a distributed double layer (see Fig. 5). Thus, instead of continuing across smoothly, the jump in $E_{||}$ is now proportional to the spatial derivative of the electric dipole moment per unit area (π_s) on the interface, which in general is a space-time variable, i.e., $S(\vec{r},t)$ (Chen et al., 2008). In other words,

$$\begin{aligned} \langle E_{||} \rangle &\equiv E_{||}^{a} - E_{||}^{b} = \left(E_{i||} + E_{r||}\right) - E_{t||} = \\ -\frac{1}{\varepsilon_0}\frac{\partial \pi_s}{\partial x}(t) &\equiv S(\vec{r},t) = S_M(t)\cos\left(\vec{k}_s \cdot \vec{r} - \omega_s t\right) \end{aligned} \tag{27}$$

where the subscripts "i", "r", and "t" represent the incident, reflected, and transmitted components, respectively. In other words, the condition $\langle E_{||} \rangle \neq 0$ arises where there is non-uniform distribution of the dipole moment along the projection of the incident light's electric field on the interface. That is, the nonzero $E_{||}$ jump is effectively proportional to the displacement of a longitudinal (σ_s varying) or transverse (d varying) mechanical wave $S(t)$, which is excited by the incident wave and propagating along the double layer with wave vector $\vec{k}_s$ and frequency ω_s. Further, such a mechanical wave can additionally be modulated transversely in its dipole length (i.e., d) by a second wave of frequency ω_M such that $S_M(t) = S_{x0}\cos\omega_M t$ for the p-wave case, and $S_M(t) = S_{y0}\cos\omega_M t$ for the s-wave case. In the above, the modulating wave amplitudes are $S_{x0} = -\frac{1}{\varepsilon_0}\left(\frac{\partial \pi_s}{\partial x}\right)_0$ and $S_{y0} = -\frac{1}{\varepsilon_0}\left(\frac{\partial \pi_s}{\partial y}\right)_0$, respectively, with the subscript "0" standing for the root-mean-square amplitude.

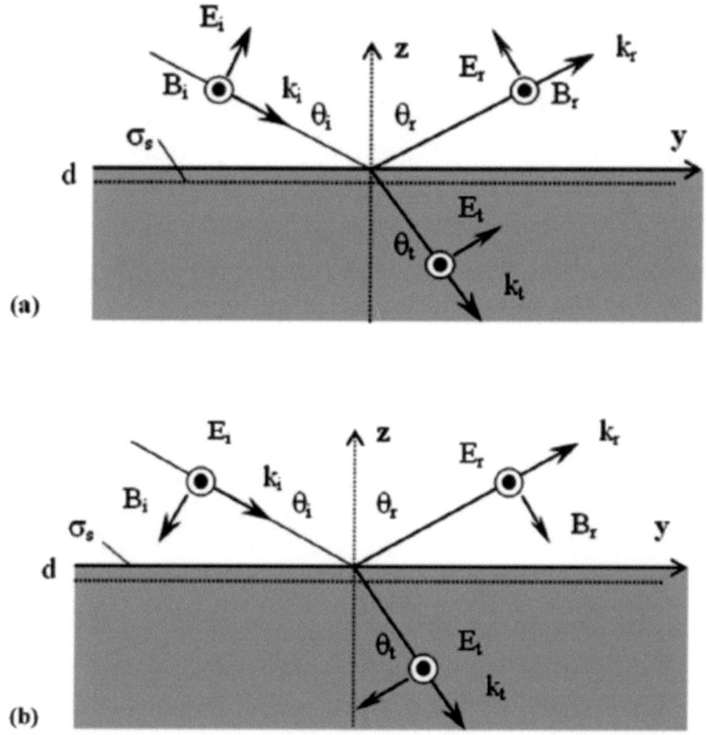

Fig. 4. (a) Refraction with p-wave incidence and (b) refraction with s-wave incidence.

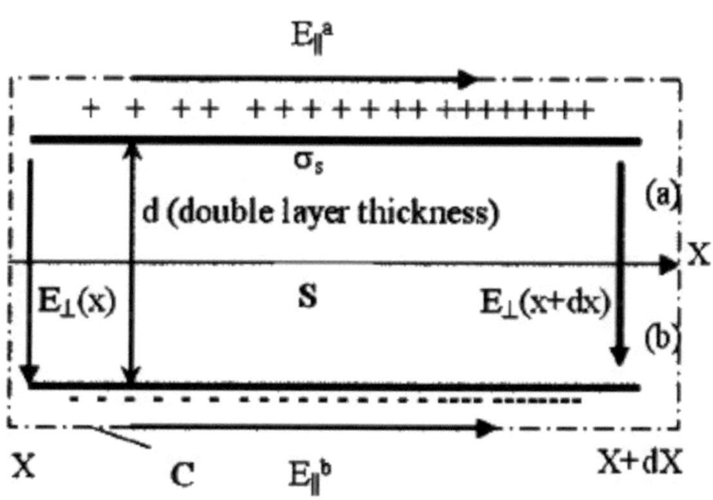

Fig. 5. Surface integration of Faraday's law enclosing a section of the double layer and regions above (a) and below (b) it.

By combining with the other jump condition, i.e., $\langle H_{||} \rangle = 0$, the modified Fresnel reflection coefficient for both the p- and s-wave configurations emerge readily after some algebraic manipulations (Chen et al., 2008):

$$r_p \equiv \left(\frac{E_{r0}}{E_{i0}} \right)_p = \overbrace{\frac{\left(\frac{n_t}{\mu_t}\cos\theta_i - \frac{n_i}{\mu_i}\cos\theta_t \right)}{\left(\frac{n_t}{\mu_t}\cos\theta_i + \frac{n_i}{\mu_i}\cos\theta_t \right)}}^{traditional} - \frac{\frac{n_t}{\mu_t} S_{y0}\cos\omega_M t}{\underbrace{\left(\frac{n_t}{\mu_t}\cos\theta_i + \frac{n_i}{\mu_i}\cos\theta_t \right) E_{i0}}_{due\ to\ distributed\ double\ layer}} \tag{28}$$

$$r_s \equiv \left(\frac{E_{r0}}{E_{i0}} \right)_s = \overbrace{\frac{\left(\frac{n_i}{\mu_i}\cos\theta_i - \frac{n_t}{\mu_t}\cos\theta_t \right)}{\left(\frac{n_i}{\mu_i}\cos\theta_i + \frac{n_t}{\mu_t}\cos\theta_t \right)}}^{traditional} + \frac{\frac{n_t}{\mu_t} S_{x0}\cos\omega_M t \cdot \cos\theta_t}{\underbrace{\left(\frac{n_i}{\mu_i}\cos\theta_i + \frac{n_t}{\mu_t}\cos\theta_t \right) E_{i0}}_{due\ to\ distributed\ double\ layer}} \tag{29}$$

The new power reflection coefficient is then $R = r^2$ and the new power transmission coefficient is: $T = 1 - R$, for both the p- and s-wave cases (Hecht, 2002). Therefore, in the presence of a light-responsive, distributed double layer at the interface (e.g., with $\partial\pi_s/\partial x > 0$ or $\partial\pi_s/\partial y > 0$ in Fig. 4), the reflected and transmitted lights are expected to be further modulated at the frequency ω_M. Additionally, this modification to the light reflectivity is incident-power-dependent. Namely, variation in reflectivity is more significant for dimmer incident lights, as implied by the presence of E_{i0} in the denominator of the 2nd term in both Eqs. (28) and (29). Note that these seemingly peculiar behaviors are by no means related to the well-known photoelectric effects, such as those manifested by light-guiding molecules used in liquid crystal displays.

In the absence of the second modulating light, the $\cos\omega_M t$ factor is reduced to unity in both Eqs. (28) and (29). As a consequence, the modified light reflection is no longer time-varying, but is either enhanced or decreased depending on the signs of S_{x0} or S_{y0} in the p- and s-polarized cases, respectively. More importantly, asymmetric reflection (or, refraction) would result should the path leading from the incidence to the reflection be reversed (see, e.g., Fig. 4), as the result of a sign change of the corresponding coordinate system. In the following, experimental investigations are conducted on the above-predicted power reflection asymmetry between conjugate light paths, as well as on the inverse dependence of reflectivity upon the incident power.

5. Possible implementations of altered Brewster angle demonstrated by quantum mechanical simulations

Numerical experiments for the p-wave case were conducted as an example to evidence the variation of Brewster angles rendered by the proposed dipole engineering. This task very much relied on the first-principle quantum mechanical software: CASTEP (Clark et al., 2005). CASTEP is an *ab initio* quantum mechanical program employing density functional theory (DFT) to simulate the properties of solids, interfaces, and surfaces for a wide range of materials classes including ceramics, semiconductors, and metals. Its first-principle calculations allow researchers to investigate the nature and origin of the electronic, optical, and structural properties of a system without the need for any experimental input other than the atomic number of mass of the constituent atoms.

The adopted simulation procedure was as follows (Liao et al., 2006). CASTEP first simulated the spectral dependence of the relative dielectric coefficient (ε_r , including the real (ε_{rR}) and imaginary (ε_{rI}) parts) of a chosen host material. Then, dipole engineering was exercised on this host lattice through artificially replacing some of its atoms with other elements, or with vacancy defects, hence resulting in the implementation of permanent dipoles of known orientation (θ_0) on the host. Due to this introduced anisotropy, we had to simulate the corresponding spectral dependence of the new relative dielectric coefficient ($\tilde{\varepsilon}_r$) for each incident angle (θ_i). Namely, $\tilde{\varepsilon}_r$ was then θ_i -dependent. Using the new medium refractive index ($n_t \approx \sqrt{\varepsilon_{rR}}$ for low ε_{rI} case; otherwise, $n_t = \sqrt{\varepsilon_{rR}^2 + \varepsilon_{rI}^2} \cdot \cos(\varphi/2)$ where $\tan(\varphi/2) = -\varepsilon_{rI}/\varepsilon_{rR}$ has to be used) and Snell's law, the corresponding refractive angle (θ_t) could thus be secured. (However, since CASTEP only simulates intrinsically, viz., it does not do Snell's law, we actually had to vary θ_t first instead and went backward to secure θ_i using Snell's law.) Then, using Eq. (22), the value of such-introduced permanent polarization at the incident angle θ_i, in terms of $P_0/\left(\varepsilon_0 E_t^p\right)$ at θ_i, was revealed. The resultant new Brewster angle was thus obtained through inspecting the modified reflection coefficient (i.e., Eq. (2), with ε_r replaced by ε_{rR}, and $n_i = 1$, and $\mu_{ri} = \mu_{rt} = 1$, becomes $r^p = \left(\sqrt{\tilde{\varepsilon}_{rR}} \cos\theta_i - \cos\theta_t\right) / \left(\sqrt{\tilde{\varepsilon}_{rR}} \cos\theta_i + \cos\theta_t\right))$ curve against θ_i varying from 0 to 90 degrees.

Two example situations are given here, where dipole engineering can noticeably alter the Brewster angles of a single-crystal silicon wafer under the incidence of a red and an infrared light, respectively (Liao et al., 2006). The red light is of energy 1.98 eV, or, vacuum wavelength λ = 0.63 μm. The infrared light is of energy 0.825 eV, or, λ = 1.5 μm. It is well-known that without the proposed dipole engineering treatment, the single-crystal silicon is opaque to the visible (red) light, while fairly transparent to the infrared light. In fact, for the latter reason, infrared light is routinely applied in the front-to-back side pattern alignment of wafers in microelectronic fabrications.

Here the Si single-crystal unit cell is modified by replacing 2 of its 8 atoms with vacancies (see Fig. 6, regions in dim color are the chopped-out sites). Note that the x-axis corresponds to $\theta_t = 0^\circ$ (and hence $\theta_i = 0^\circ$), while y or z-axis to $\theta_t = 90^\circ$. Fig. 6 shows that the defect-caused permanent polarization (maximum P_0) is most likely along the x-direction. The CASTEP-simulated curves of $\tilde{\varepsilon}_r$ vs. light energy (in eV) in the incident directions of x, y, and z, respectively, are given in Fig. 7. Hence, $\tilde{\varepsilon}_r$ is, as expected, θ_i -dependent. For both cases in which the incident light is red and infrared, the involvement of the introduced optically-responsive defect polarization is very much dependent on the its relative orientation with respect to the refractive p-wave's electric field (E_t^p) (see Figs. 8 and 9). Indeed, as evident from Fig. 2, it is most significant when E_t^p is in the direction of the permanent polarization (maximum P_0).

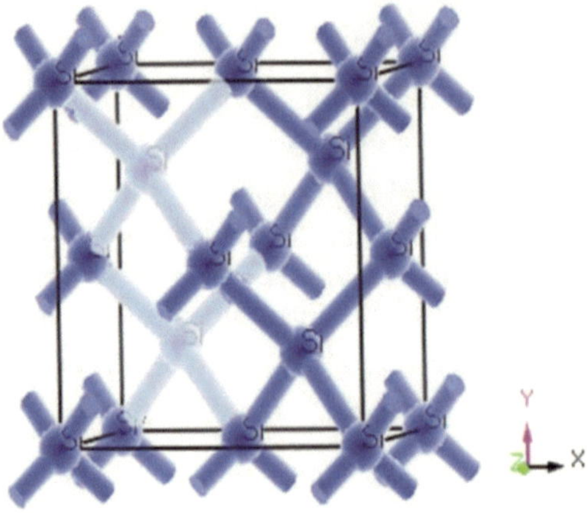

Fig. 6. Unit cell of the modified Si crystal with two vacancies (Space group: FD-3M (227); Lattice parameters: a: 5.43, b: 5.43, c: 5.43; α: 90º, β: 90º, γ: 90º).

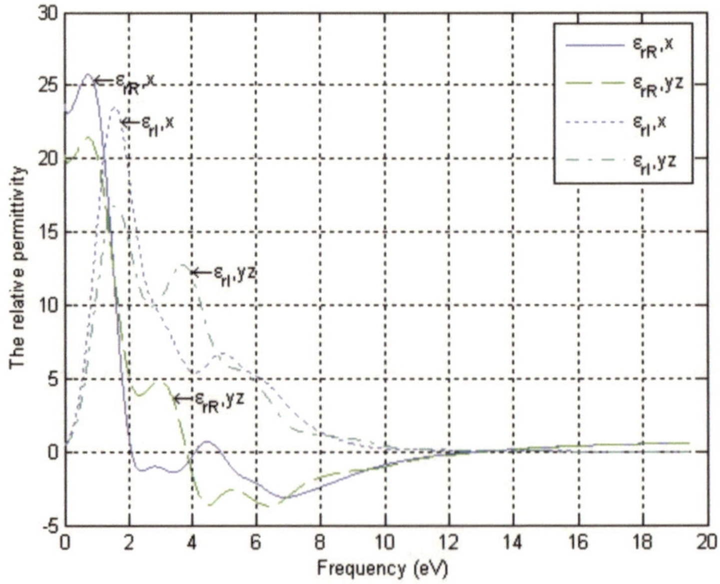

Fig. 7. The modified relative dielectric coefficient spectra ($\tilde{\varepsilon}_r = \tilde{\varepsilon}_{rR} + i\tilde{\varepsilon}_{rI}$), with the refractive light's propagation directions shown for x, y, and z axes, for $\theta_0 = 0^\circ$.

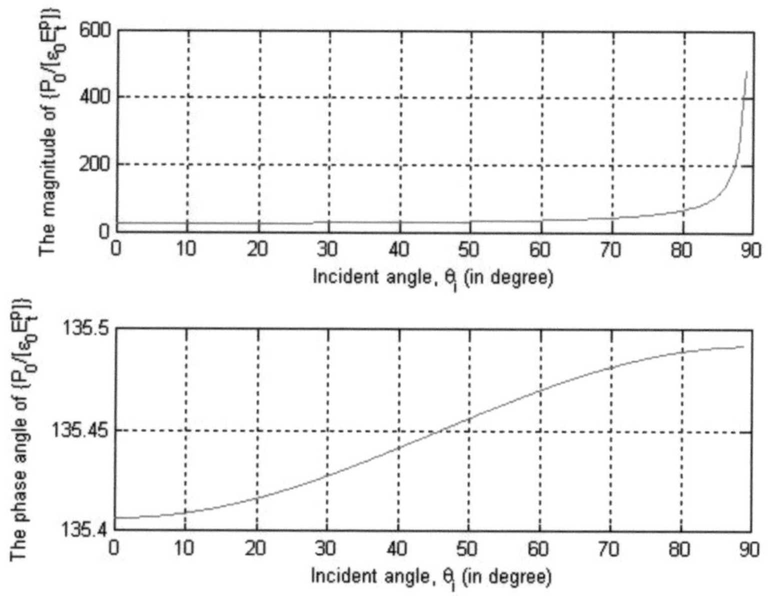

Fig. 8. The permanent defect polarization (at $\theta_0 = 0^\circ$) with respect to the incident angle of a red light of 1.98 eV (most significant when $\theta_i = 90^\circ$.)

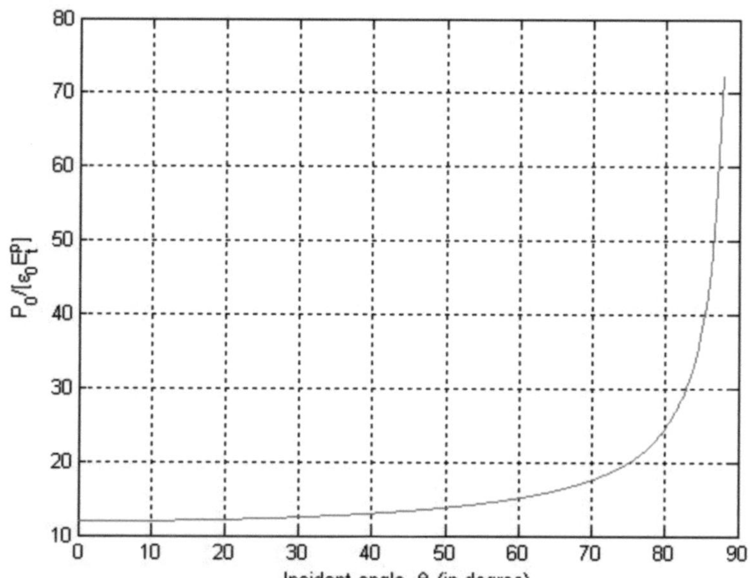

Fig. 9. The permanent defect polarization (at $\theta_0 = 0^\circ$) with respect to the incident angle of an infrared light of 0.83 eV energy (most significant when $\theta_i = 90^\circ$.)

Using the aforementioned calculation procedure, a considerable shift of the original Brewster angle of 77.2° to the new one of 72.5°, due to the vacancy defects replacement, is evident in Fig. 10 for the red light case. Note that the reflection for red light decreased somewhat post the dipole engineering treatment. For the second IR case, Fig. 9 shows the involved action of the permanent defect polarization. A noticeable shift of the Brewster angle from 74.9° to 78.7° results (see Fig. 11). Moreover, this new Si material manifests higher reflection (or, *opaqueness*) (see Fig. 11), in addition to relatively high dissipation (see Fig. 7) for the IR light, after the proposed treatment.

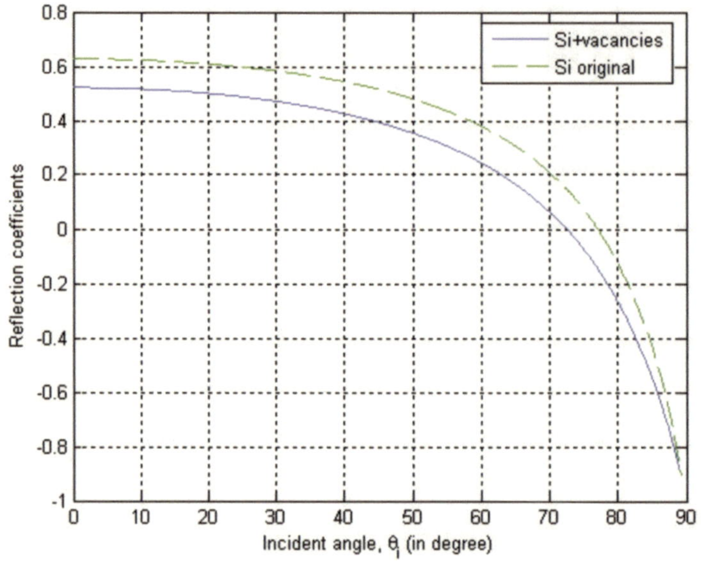

Fig. 10. Shift of the original Brewster angle (77.21°) to the new one (72.52°) for an incident light of energy 1.98 eV by the proposed vacancy defects introduction.

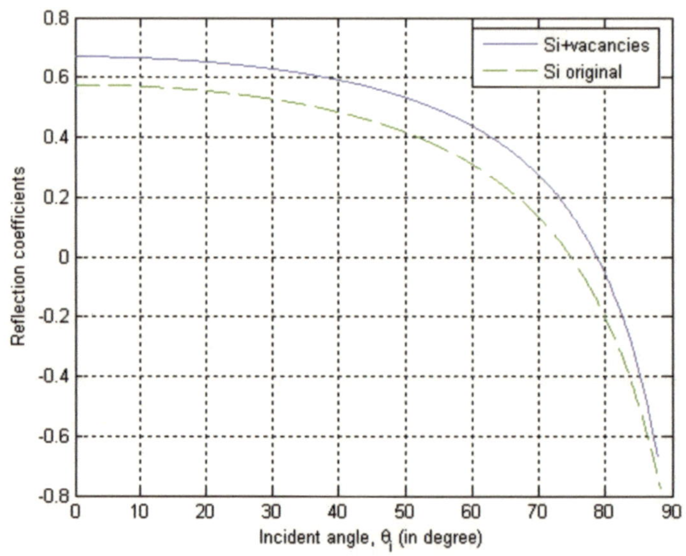

Fig. 11. Shift of the original Brewster angle (74.9°) to the new one (78.7°) for an incident light of energy 0.83 eV by the proposed vacancy defects introduction.

In the above, the shifted Brewster angles for both cases are associated with considerable dissipation caused by the introduced defects, as implied by Fig. 7. In particular, the defect-modified silicon wafer would absorb the total power of the p-wave incident at the Brewster angle. Nonetheless, this should not represent as sure characteristics for the general situations. For example, a θ_B -altering surface material (to be coated on the original host in a post-process manner) may have its permanent polarization implemented via replacing some of its atoms with other elements, instead of vacancies, and thus may not show such dissipating behavior.

6. Poled PVDF films and asymmetric reflection experiments

6.1 Background on the PVDF material ((Pallathadka, 2006), (Zhang et al., 2002)) and its preparations

Polyvinylidene difluoride, or PVDF (molecular formula: $-(CH_2CF_2)_n-$), is a highly non-reactive, pure thermoplastic and low-melting-point (170°C) fluoropolymer. As a specialty plastic material in the fluoropolymer family, it is used generally in applications requiring the highest purity, strength, and resistance to solvents, acids, and bases. With a glass transition temperature (T_g) of about -35°C, it is typically 50-60% crystalline at room temperature. However, when stretched into thin film, it is known to manifest a large piezoelectric coefficient of about 6-7 pCN^{-1}, about 10 times larger than those of most other polymers. To enable the material with piezoelectric properties, it is mechanically stretched and then poled with externally applied electric field so as to align the majority of molecular chains ((Pallathadka, 2006) , (Zhang et al., 2002)). These polarized PVDFs fall in 3 categories in general, i.e., alpha (TGTG'), beta (TTTT), and gamma (TTTGTTTG') phases, differentiated by how the chain conformations, of trans (T) and gauche (G), are grouped. FTIR (Fourier transform IR) measurements are normally employed for such differentiation purposes ((Pallathadka, 2006) , (Zhang et al., 2002)). With variable electric dipole contents (or, polarization densities) these PVDF films become ferroelectric polymers, exhibiting efficient piezoelectric and pyroelectric properties, making them versatile materials for sensor and battery applications.

In our experiments, PVDF films of Polysciences (of PA, USA) are subjected to non-uniform mechanical and electric polings to generate β-PVDF films of distributed dipolar regions. By applying infrared light beams on these poled β-PVDF films, the evidences of enhanced asymmetric refraction at varying incident angles as well as its inverse dependence on the incident power are sought for.

6.2 FTIR measurement setup

The adopted experimental setup takes full advantage of the original commercial FTIR measurement structure (Varian 2000 FT-IR) (Chen et al., 2008). The intended investigations are facilitated by putting an extra polarizer (Perkin-Elmer) in front of the detector and some predetermined number of optical attenuators (Varian) before the sample (see, Fig. 12). Figs. 13(a) and 13(b) show the detector calibration results under different numbers of attenuating sheets for both the p- and s-wave incidence, respectively. As is obvious, the detected intensities degraded linearly with the number of attenuators installed and the spectra remain morphologically similar.

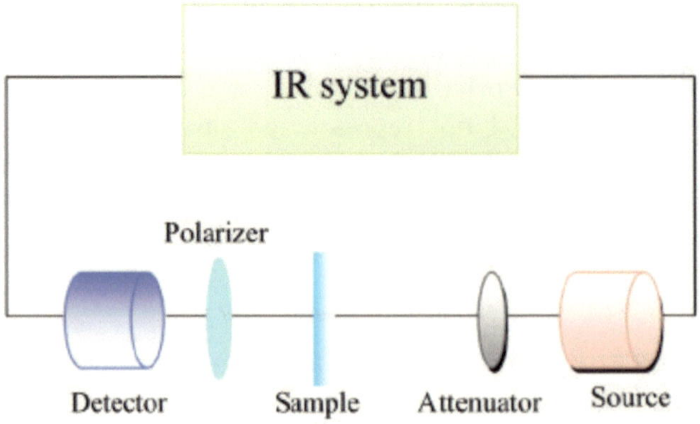

Fig. 12. FTIR measurement setup.

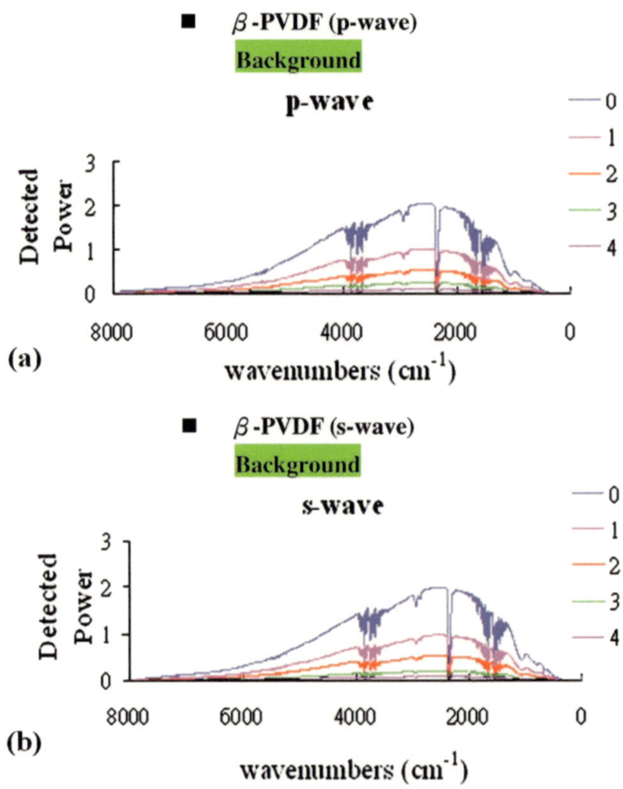

Fig. 13. Calibration on detected spectral effects when employing different numbers of attenuators under normal incidence of (a) p- and (b) s-polarized lights, respectively.

6.3 Evidence of asymmetric reflections and enhanced reflectivities for attenuated incident lights

In the reflectivity ($R \equiv r^2 = |E_{r0}/E_{i0}|^2$) experiments, each electrically-poled β-PVDF film (thickness 16-18 μm) is subjected to IR light irradiation at varying incident angles and beam intensities (via using the above attenuator films) (Chen et al., 2008). Even though a PVDF film encompasses many double layers along its thickness, the reason the above theoretical derivations based on a single double layer (see Fig. 5) should still apply is that there are non-uniform vertical polarizations (or, electric fields) at the surface. In other words, the resultant gradients in the incident-light-responsive planar dipole moment would then render nonzero interfacial jump of $E_{||}$ as addressed by Eq. (26).

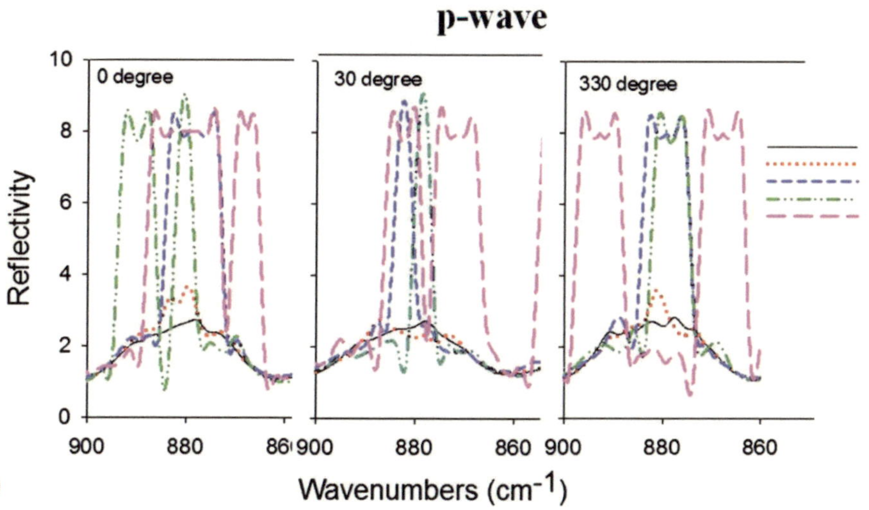

(a)

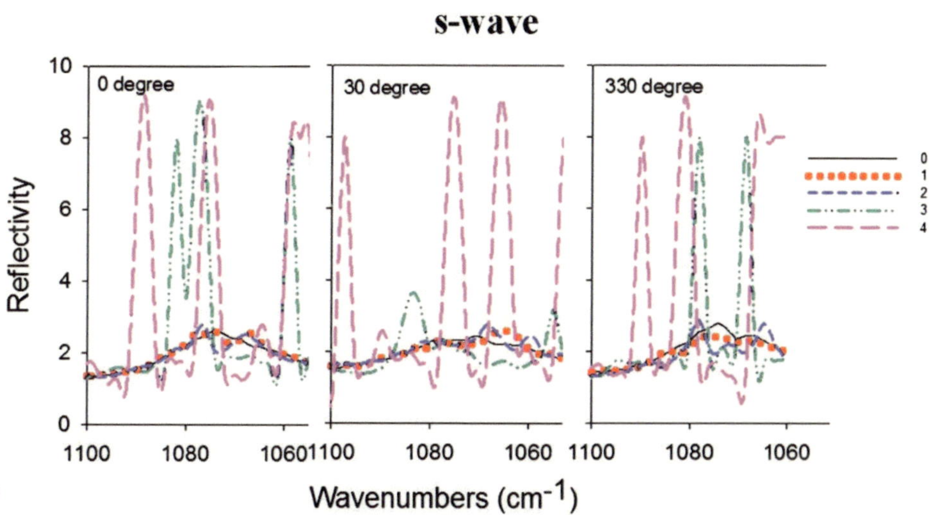

(b)

Fig. 14. Experimental results on asymmetric reflection among conjugate light paths and enhanced reflectivity at different numbers of attenuators.

Since it is the reflectivity that is of interest, at each incident angle and under a specified degree of attenuation, the reflection is normalized with respect to the corresponding reference value in the absence of a PVDF film as shown in Fig. 13 (a) and 13 (b) (Chen et al., 2008). Figs. 14 (a) and 14 (b) illustrate the arising of asymmetric reflections between conjugate incident paths (e.g., incidence at 30° versus 330°) for both the p- and s-polarized infrared lights under various degrees of intended attenuation (Chen et al., 2008). In the above, the incident angle is defined by rotating clockwise the poled-PVDF sample under top-view of the setup of Fig. 12. Hence, reversing the light reflection path indeed causes a different reflected power to arise, as predicted by the aforementioned theoretical exploration. Note that, however, in traditional FTIR measurements, decrease in the detected intensity has been routinely attributed to increased absorption by PVDF films. Nevertheless, for an obliquely incident light the beam path within PVDF is only slightly larger than that in a normal incident situation. Thus, the resultant infinitesimal increase in PVDF absorption should never be sufficient to account for the detected large difference in reflected power. Notably, the detected decrease in intensity should instead be attributed to enhanced reflection caused by distributed dipoles on the poled PVDF films.

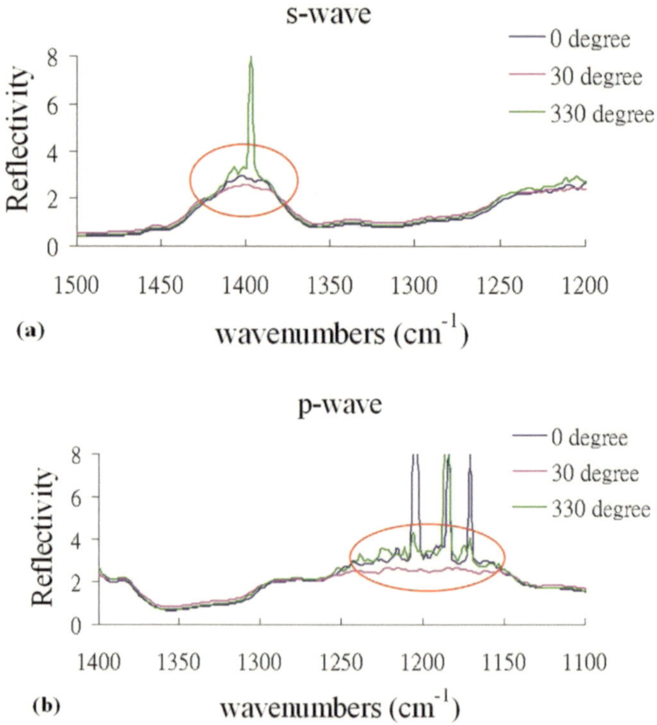

Fig. 15. Observed distinct asymmetric reflection among conjugate light paths under two attenuator films and at varying incident angles.

Furthermore, for both p- and s-polarized incident waves, the variations in reflectivity are more significant in the situations where more attenuation is imposed, complying with the above theoretical expectation. Namely, the second terms in the above-derived Eqs. (28) and (29) are more enhanced under the situation of smaller incident light electric field (or power). Lastly, even though the appearance of FTIR-detected spiky saturation peaks in Figs. 14 (a) and 14 (b) actually imply loss in the signal-to-noise ratios when under heavy attenuation of the incident power, several unsaturated features, e.g., in Figs. 15(a) and 15(b) still remain (wherein 2 sheets of attenuators are employed) (Chen et al., 2008). Such evidenced spectral features strongly endorse the above theoretical claim that the reflectivity variation of a dimmer light is more outstanding than that of a brighter one.

7. PVDF experiment on varying Brewster angle

7.1 PVDF new Brewster angle

The experimental set up is as arranged in Fig. 16, where a light beam of 0.686 mm radius from the He-Ne laser (of the wavelength of 632.8 nm) is converted into p-wave mode after getting through the polarizer (Tsai et al., 2011). Two double convex lenses, with focal lengths being 12.5 cm and 7.5 cm, respectively, are for shrinking down the beam radius to 0.19 mm to reduce the width of light reflection off the PVDF surface. The reflected light is then further focused by a lens (of 2.54 cm focal length) before reaching the diode power detector. An incident angle range is scanned from 50.5°to 59.5°(θ_i) with an accuracy of 0.015°, and then its conjugate range from -50.5°to -59.5°($-\theta_i$), while the incident light intensity is varied between 100% (8.54 mW) and 10% power by moving an attenuator into or withdrawing from the beam path.

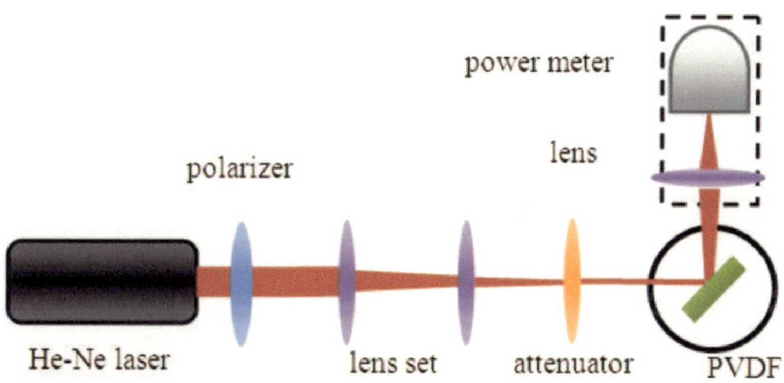

Fig. 16. Configuration of PVDF Brewster angle measurement.

The fitted curves for the measured Brewster angles for the two conjugate incident paths, under 100% and 10% laser beam intensities, on both the β- and poled-β films, are shown in Fig. 17 and Fig. 18, respectively (Tsai et al., 2011). It can be seen that Brewster angles measured via the two conjugate incident paths differ considerably. Such difference becomes more outstanding on the poled-β PVDF film, and in particular, when the laser beam is attenuated to 10%, as predicted by the theory of the authors (Liao et al., 2006). The typical data are given in Table 1.

Parameters	Beam intensity	$\theta_B(\theta_i)$	$\theta_B(-\theta_i)$	$\Delta\theta_B$
β-PVDF	100 %	54.7925°	55.1475°	0.355°
	10 %	54.695°	55.3875°	0.6925°
Poled-β PVDF	100 %	54.62°	55.77°	1.05°
	10 %	54.38°	56.245°	1.865°

Table 1. PVDF Brewster angles measurement.

It is noted that although even the intrinsic α phase PVDF possesses birefringence and this can lead to different Brewster angles as in the above too, the difference degree is at most around 0.129°, and hence may be ignored.

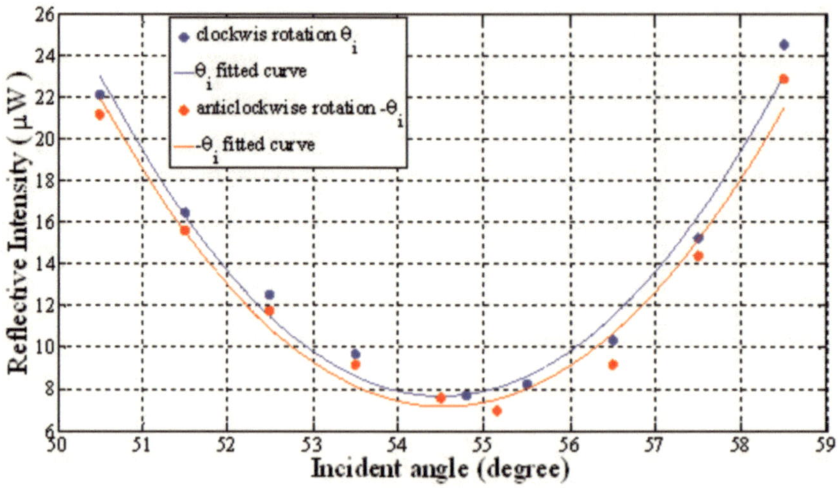

(a)

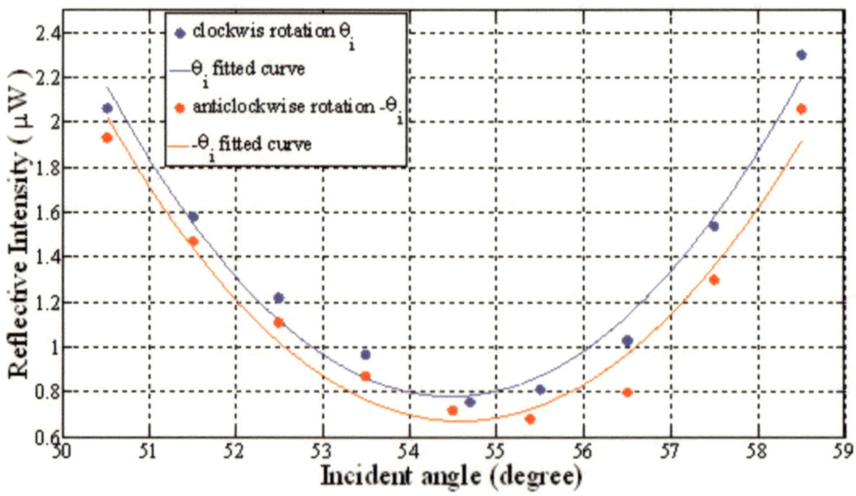

(b)

Fig. 17. Measured Brewster angles on · β-PVDF for two conjugate incident paths, under: (a) 100% laser intensity, (b) 10% laser intensity (only one-tenth of data points are shown)

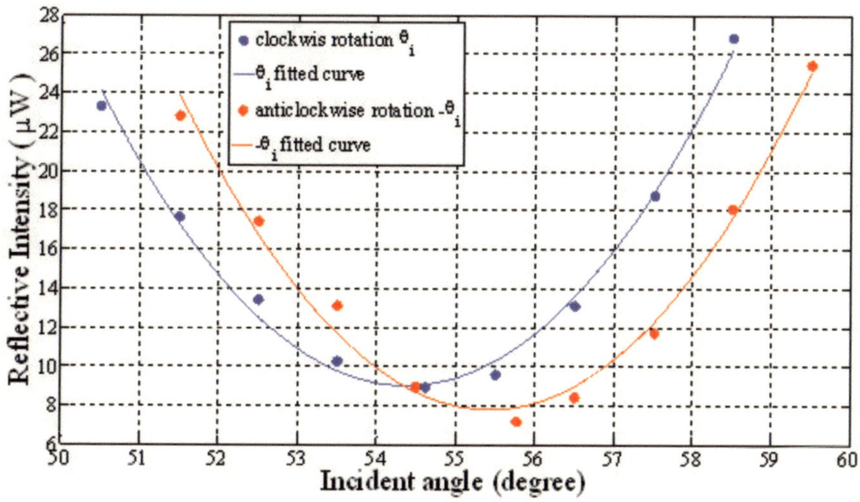

(a)

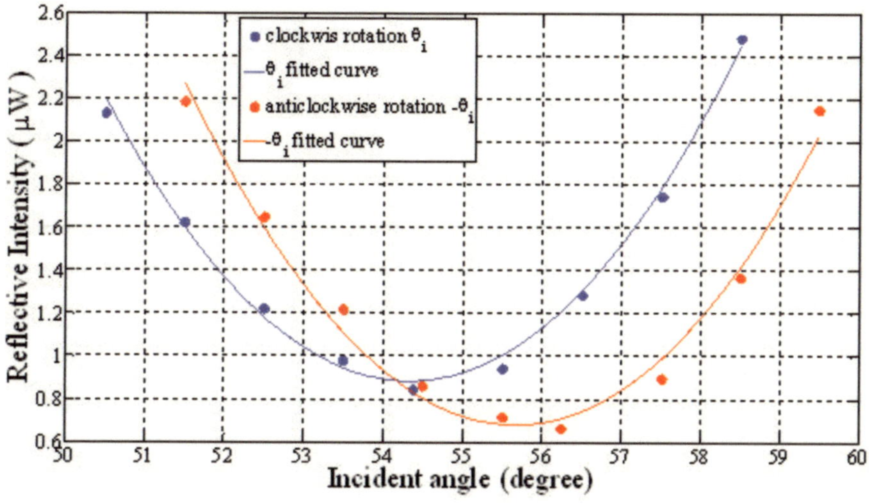

(b)

Fig. 18. Measured Brewster angles on poled • β-PVDF for two conjugate incident paths, under: (a) 100% laser intensity, (b) 10% laser intensity (only one-tenth of data points are shown)

This can be verified by putting into Eq. (23) the known ordinary and extraordinary refractive indices of PVDF and getting the Brewster angles of 54.814° and 54.933°, respectively. Further, the fact that the larger deviation is evidenced in the poled-β phase, as compared with that from the β phase, indicates that permanent dipoles are indeed the cause of such alteration in Brewster angles.

By putting the above experimental data (i.e., Table 1) into Eq. (23), the relative dielectric coefficients ($\tilde{\varepsilon}_r$) for both the β- and poled-β PVDF films are extracted and tabulated in Table 2 (Tsai et al., 2011).

Parameters	Beam intensity	$\tilde{\varepsilon}_r(\theta_i)$	$\tilde{\varepsilon}_r(-\theta_i)$
β-PVDF	100 %	2.008	1.994
	10 %	2.062	2.1
Poled-β PVDF	100 %	1.983	1.948
	10 %	2.16	2.239

Table 2. Relative dielectric coefficients through fitting experimental data.

Then, the averaged effective permanent polarization $P_0(\omega)$ and orientation θ_0 can be extracted through trial-and-error (see, Table 3) by putting these coefficients into Eqs. (22) and (24), and using the relations: $E_t^p = t_p \cdot E_i^p$ and $t^p = 2n_{air}\cos\theta_i / (n_{air}\cos\theta_t + n_t\cos\theta_i)$. In the above, $E_i^p = \sqrt{S/\varepsilon_0 C}$, $\theta_t = \sin^{-1}(n_{air}/n_t \sin\theta_i)$, with S, ε_0, and C being the irradiating light intensity per unit area, the vacuum permittivity (i.e., $8.85 \cdot 10^{-12}$ F/m), and the light speed in vacuum, respectively, and $\varepsilon_r = n_o^2$ (Tsai et al., 2011).

Parameters	θ_0	$P_0(\omega)$
β-PVDF	41.64°	$1.0226 \times 10^{-9} (C/m^2)$
Poled-β PVDF	61.62°	$1.7843 \times \times 10^{-9} (C/m^2)$

Table 3. Extracted effective permanent polarizations and orientations of dipole-engineered PVDF films.

It can be seen that the electro-poling has caused the permanent polarization to increase somewhat, and most of all, its orientation with respect to the interface to add around 20°.

7.2 Novel 2D refractive index ellipse

Owing to its intrinsic uniaxial birefringence property ((Matsukawa et al., 2006), (Yassien et al., 2010)), when a light is incident upon a PVDF film (as formed, without poling), the refracted light is decomposed into an ordinary wave and an extraordinary wave, which correspond to refractive indices of n_o and n_e, respectively. Namely, when the plane-of-incident is formed by the light's propagating direction vector $\vec{k}$ and the uniaxis $\hat{z}$ (see, Fig. 19), with the angle between them being θ, then a slice on the 3D refractive index ellipsoid cutting perpendicular to $\vec{k}$ will give rise to an elliptic contour which is of the minor axis n_o and major axis n_s in a relationship expressed as:

$$\frac{1}{n_s^2(\theta)} = \frac{cos^2\theta}{n_o^2} + \frac{sin^2\theta}{n_e^2} \tag{30}$$

However, the whole picture will change considerably in the presence of ordered permanent dipoles. Namely, unlike the traditional elliptic contour in red color in the polar diagram Fig. 20, unconventional contours in blue and green represent 2D (two dimension) refractive index surfaces of the situation on β-PVDF (i.e., $\theta_0 = 41.64\,°$) under 100 % and 10 % laser power, respectively; and those in purple and orange colors are on poled-β PVDF (i.e., $\theta_0 = 61.62\,°$) under 100 % and 10 % laser power, respectively (Tsai et al., 2011). Note that, in Fig. 20, as the ordinate dimension is along the direction of interface (in green) and the abscissa along the norm in real setup, both the I and III quadrants describe the refraction in the incident angle range of 0°~ 90° (i.e., θ_i), and quadrants II and IV depicts that of 0°~ -90° (i.e., $-\theta_i$). Hence, the dipole-engineered ones would demonstrate open splittings near the traditional incident angles. Among them, the deviation should be more outstanding for the case with the test film being poled-β than β-PVDF, and especially when at lower incident laser power (Tsai et al., 2011).

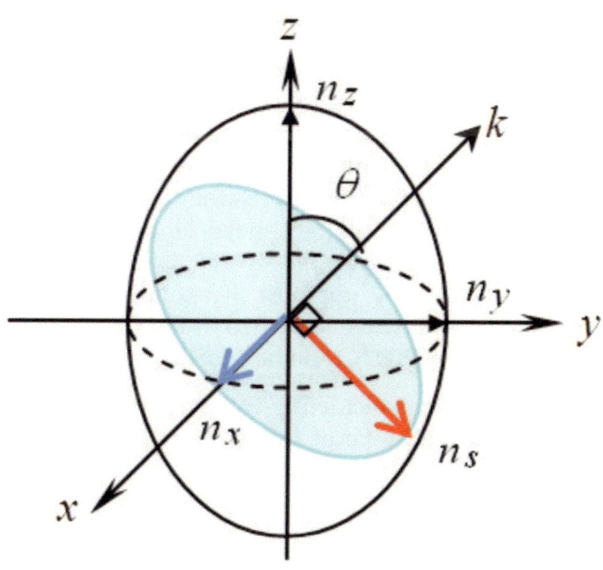

Fig. 19. Construction of the refractive index surface.

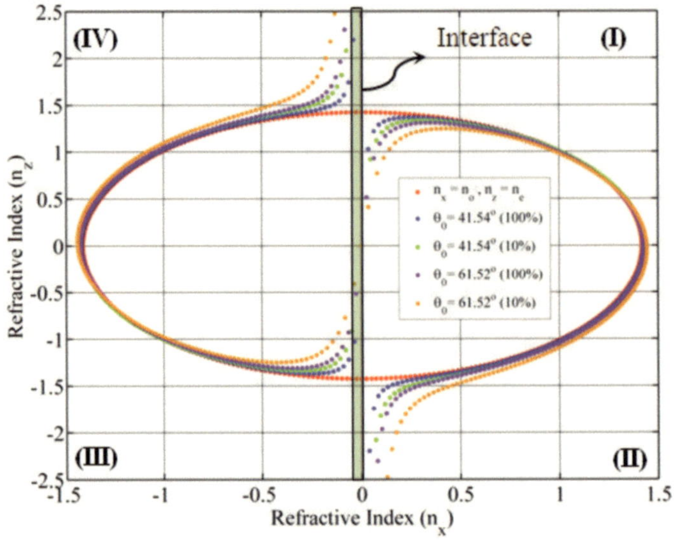

Fig. 20. The polar diagram for refractive index distribution.

Experimentally, the following tendency was observed when the test samples were altered from normal to β- and then to poled-β PVDF films. When the incident angles was within the I- and III- quadrants (i.e., θ_i), the evidenced Brewster angles shrunk and the refractive index became smaller (i.e., $\tilde{\varepsilon}_r < \varepsilon_r$). On the other hand, when the laser was incident in the II- and IV- quadrants (i.e., $-\theta_i$), both the Brewster angles and the refractive index ($\tilde{\varepsilon}_r > \varepsilon_r$) switched to larger values. This tendency apparently goes with the above theoretical prediction.

7.3 Notable indication on traditional SPR measurements

Facilitated by its very high-Q resonance angle, the surface plasmon resonance (SPR) type of techniques, and their variations, are known to be very sensitive tools for measurements of refractive indices ($n = \sqrt{\varepsilon_r \mu_r}$). Among them, the one using prism coupling and in the so-called "Kretschmann-Raether (KR) configuration" ((Maier, 2007), (Raether, 1988)), is probably the most widely adopted practice. That is, the KR- configured materials from top down is arranged to be: prism (the "0" matter), metal film (the "1" matter), air gap, target under test (the "2" matter), and buffer layer (the "3" matter). When a p-wave is incident at a so-called "resonance angle" onto the topmost KR-configured plane (i.e., at "0"-"1" interface), a surface resonant plasma wave is excited at the metal-air interface, leading to a minimum in light reflection ((Maier, 2007), (Raether, 1988)).

The overall reflection coefficient (of an incident p-wave) off this 4-material KR configuration can be derived from using the Fabry-Perot interference principle and is ((Maier, 2007), (Raether, 1988)):

$$r_{0123}^{p} = \frac{r_{01}^{p} + \dfrac{r_{12}^{p} + r_{23}^{p} e^{i2\phi_2}}{1 + r_{12}^{p} r_{23}^{p} e^{i2\phi_2}} e^{i2\phi_1}}{1 + r_{01}^{p} \dfrac{r_{12}^{p} + r_{23}^{p} e^{i2\phi_2}}{1 + r_{12}^{p} r_{23}^{p} e^{i2\phi_2}} e^{i2\phi_1}} \qquad (31)$$

And, the overall reflectivity in power is: $R = \left| r_{0123}^{p} \right|^2$, where r_{01}^{p}, r_{12}^{p}, r_{23}^{p} are reflection coefficients at "0"-"1", "1"-"2", and "2"-"3" interfaces according to the traditional Fresnel equations; and $\phi_i = k_0 \sqrt{\varepsilon_i} cos\, \theta_i d_i$ are phase angles associated with matters "1" and "2", respectively; and k_0 is incident wave vector; and ε_1 / d_1 and ε_2 / d_2 are relative dielectric coefficient / layer thickness of the metal film and material under test, respectively.

However, it is found in the above experiments that in the presence of permanent dipoles, not only is the Brewster angle dependent on the incident light power as well as the dipole orientation, but also that two conjugate incident light paths result in distinctively different refractions. Therefore, although the form of Eq. (31) remains the same in the presence of permanent dipoles, values of local reflection coefficients involved can vary considerably from those of their classical counterparts. In other words, the traditional confidence in SPR type of measurements may be in jeopardy when the material under test is embedded with permanent dipoles, as will be shown in what follows.

Consider a KR-configured SPR measurement setup as an example. It includes: a lens (SF 11) of relative dielectric coefficient of $(1.7786)^2$, a silver metal film of 52 nm (d_1) thickness of a relative dielectric coefficient of $-17.6 + 0.67i$ (Raether, 1988), a PVDF film (as grown, or β, or poled-β) as the material under test of thickness of about 15 μm (d_2), with its original relative dielectric coefficients being n_o^2 and n_e^2, and a buffer layer of air of a relative dielectric coefficient of about 1. This configuration is then subjected to the irradiation of a light beam , from a 632.8 nm wavelength He-Ne laser, of the incident angles ranging within 50°~70°(θ_i) and its conjugate counterpart paths within the angle range -50°~ -70°($-\theta_i$).

The numerical calculation result based on the above setup is shown in Fig. 21 and indicates the following (Tsai et al., 2011). The birefringence (i.e., n_e besides n_o) of as-grown PVDF suffices to give a maximal SPR resonance angle deviation of about 0.5° (away from 58°). This deviation of resonance angles is considerably amplified in the β-(2.5°) and poled-β (4°) cases, owing to the increase of permanent polarization density and alignment. In such sensitive SPR type of measurements, these large deviations of resonance angles represent large distortions in the light reflectivity, as illustrated in Fig. 21. Notably, it also affirms the theoretical prediction (see, Eq. (24)) that the reflectivity coefficient is inversely proportional to the strength of the incident (or, transmitted) light electric field. All these findings indicate that traditional SPR type of measurements needs to exercise precaution when the material under test is embedded with permanent dipoles. For example, most living cells are with cell walls made of two opposite double layers of dipolar molecules (Alberts et al., 2007).

8. The quasistatic macroscopic mixing theory for magnetic permeability

Although mixing formulas for the effective-medium type of approximations for the dielectric permittivities in the infinite-wavelength (i.e., quasistatic) limit (Lamb et al., 1980), such as the Maxwell Garnett formula (Garnett, 1904), have been popularly applied in the whole spectral range of electromagnetic fields, their magnetic counterpart has seldom been addressed up to this day. The current effort is thus to derive such an equation to approximately predict the final permeability as the result of mixing together several magnetic materials (Chang & Liao, 2011).

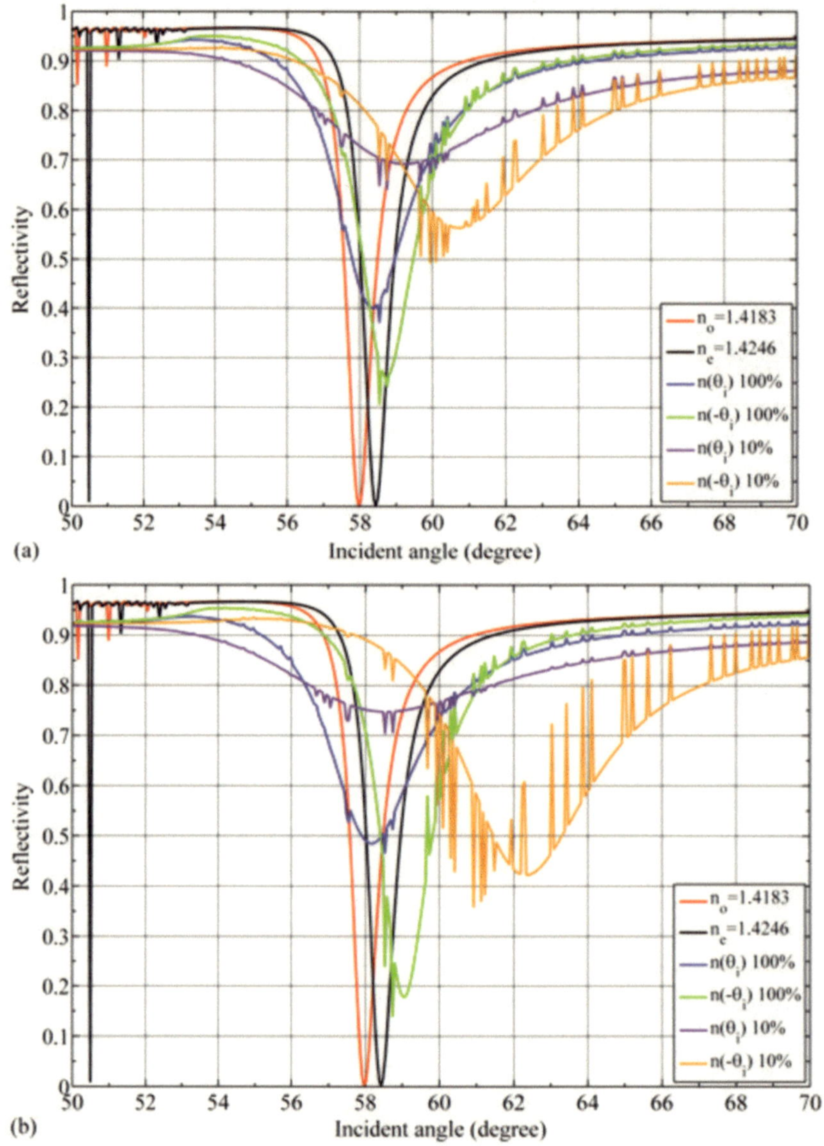

Fig. 21. Calculated SPR reflectivities for: (a) β-PVDF, (b) poled-β-PVDF.

8.1 A brief review of the derivation leading to the Maxwell Garnett and Bruggeman formulas

Historically, an isotropic host material was hypothesized to embed with a collection of spherical homogeneous inclusions. With the molecular polarization of a single molecule of such inclusions being denoted a, the following relation was established within the linear range (Cheng, 1989):

$$\vec{p}_m = \alpha \varepsilon_0 \vec{E}_m \tag{32}$$

where $\vec{p}_m$ was the induced dipole moment and $\vec{E}_m$ was the polarizing electric field intensity at the location of the molecule. Since the treatment was aiming for uniform spherical inclusions, the polarizability became a scalar, such that $\vec{E}_m$ was expressed as (Purcell, 1985):

$$\vec{E}_m = \vec{E} + \vec{E}_p + \vec{E}_{near} \tag{33}$$

Here $\vec{E}$ was the average field within the bulk host, $\vec{E}_p$ was the electric field at this molecular location caused by all surrounding concentric spherical shells of the bulk, and $\vec{E}_{near}$ was due to asymmetry within the inclusion. In those cases of interest where either the structure of the inclusion was regular enough, such as a cubical or spherical particulate, or all incorporated molecules were randomly distributed, $\vec{E}_{near}$ became essentially zero. It was further approximated that $\vec{E}_m = \vec{E} + \vec{P}/(3\varepsilon_0)$ (Purcell, 1985), to be elaborated later, with $\vec{P}$ being the polarization density associated with a uniformly polarized sphere, and ε_0 being the permittivity in free space. Hence, given Eq. (32), with the number density of such included molecules denoted as n, and $\vec{P} = n\vec{p}_m$ (Cheng, 1989), the polarization density was further expressed as:

$$\vec{P} = n\alpha\varepsilon_0\left[\vec{E} + \vec{P}/(3\varepsilon_0)\right] \tag{34}$$

However, it was well-known that for isotropic media $\vec{P} = (\varepsilon_r - 1)\varepsilon_0\vec{E}$ where ε_r stood for the relative permittivity (i.e., the electric field at the center of a uniformly polarized sphere (with $\vec{P}$ being its polarization density) was $-\vec{P}/(3\varepsilon_0)$). Then, a relation known as the Lorentz-Lorenz formula readily followed ((Lorenz, 1880), (Lorentz, 1880)):

$$\alpha = \frac{3(\varepsilon_r - 1)}{n(\varepsilon_r + 2)} \tag{35}$$

In those special cases where the permittivity of each tiny included particle was ε_s and the host material was vacuum (ε_r =1), such that $n = V^{-1}$ (V being the volume of the spherical inclusions), and Eq. (35) would have to satisfy (Garnett, 1904):

$$\alpha = 3V\frac{\varepsilon_s - \varepsilon_0}{\varepsilon_s + 2\varepsilon_0} \tag{36}$$

Combining Eqs. (35) and (36) gave the effective permittivity (ε_{eff}) of the final mixture (Garnett, 1904):

$$\varepsilon_{eff} = \varepsilon_r\varepsilon_0 = \varepsilon_0 + 3f\varepsilon_0\frac{\varepsilon_s - \varepsilon_0}{\varepsilon_s + 2\varepsilon_0 - f(\varepsilon_s - \varepsilon_0)} \tag{37}$$

with $f = nV$ being the volume ratio of the embedded tiny particles ($0 \le f \le 1$) within the final mixture. If, instead of vacuum, the host material was with a permittivity of ε_h, Eq. (37) was then generalized to the famous Maxwell Garnett mixing formula:

$$\varepsilon_{eff} = \varepsilon_h + 3f\varepsilon_h\frac{\varepsilon_s - \varepsilon_h}{\varepsilon_s + 2\varepsilon_h - f(\varepsilon_s - \varepsilon_h)} \tag{38}$$

For the view in which the inclusion was no longer treated as a perturbation to the original host material, Bruggeman managed to come up with a more elegant form wherein different ingredients were assumed to be embedded within a host (Bruggeman, 1935). By utilizing Eqs. (35) and (36), he had:

$$\frac{\varepsilon_{eff}-\varepsilon_0}{\varepsilon_{eff}+2\varepsilon_0}=\sum_i f_i\frac{\varepsilon_i-\varepsilon_0}{\varepsilon_i+2\varepsilon_0} \tag{39}$$

where f_i and ε_i are the volume ratio and permittivity of the i-th ingredient.

8.2 The magnetic flux density at the center of a uniformly magnetized sphere

surface current can be expected to appear on the surface of a uniformly magnetized sphere (wherein $\vec{M}$ is the finalized net anti-responsive magnetization vector, see Fig. 22).

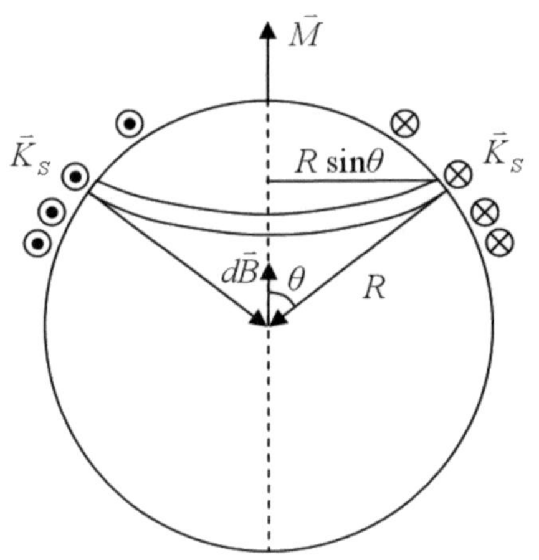

Fig. 22. Situation for calculation of the central magnetic flux density on a uniformly magnetized sphere.

In Fig. 22, $\vec{K}_S$ is the induced anti-reactive surface current density (in A/m) on the sphere's surface. By integrating all surface current density on strips of the sphere's surface the magnetic flux density ($\vec{B}_c$) at the center of a uniformly magnetized sphere is obtained to be (Lorrain & Corson, 1970):

$$\vec{B}_c=\frac{2\vec{M}\mu_0}{3} \tag{40}$$

8.3 Magnetic permeabilities formula mixing

Now, this time consider an isotropic host material embedded with a collection of spherical homogeneous magnetic particles. Given the magnetic flux density at the location of a single molecule of the inclusions being $\vec{B}_m$, the following relation holds in general:

$$\vec{B}_m = \vec{B} - \vec{B}_c + \vec{B}_{near} \tag{41}$$

where $\vec{B}$ is the average magnetic flux density within the bulk host and $\vec{B}_{near}$ is due to the asymmetry in the inclusion. In those cases of interest where either the structure of the included particles is regular enough, such as a cubical or spherical particulate, or all incorporated molecules are randomly distributed, $\vec{B}_{near}$ can be taken as zero.

If the magnetic field intensity at the location of the molecule is denoted $\vec{H}_m$, the induced magnetic dipole moment (m_m) is:

$$\vec{m}_m = \kappa_m \vec{H}_m \tag{42}$$

where κ_m is the molecular magnetization of the molecule. Because $\vec{M}$ equals $n\vec{m}_m$, we have (Cheng, 1989)

$$\vec{M} = n\kappa_m \vec{H}_m = \chi_m \vec{H}_m \tag{43}$$

where χ_m is known as the magnetic susceptibility. Hence, $\vec{B}_m$ can be further expressed as (Cheng, 1989)

$$\vec{B}_m = \mu_0 \mu_r \vec{H}_m = \mu_0 (1 + n\kappa_m) \vec{H}_m = \mu_0 (1 + \chi_m) \vec{H}_m \tag{44}$$

with μ_r being the relative permeability. By incorporating Eq. (41) and Eq. (44) into Eq. (43) we obtain:

$$M = \frac{n\kappa_m}{\mu_0 (1 + n\kappa_m)} \left(B - \frac{2M\mu_0}{3} \right) \tag{45}$$

Further, for isotropic magnetized materials (Cheng, 1989):

$$M = (\mu_r - 1) H = \frac{(\mu_r - 1) B}{\mu_0 \mu_r}, \text{ or}$$

$$B = \frac{\mu_0 \mu_r}{(\mu_r - 1)} M \tag{46}$$

Substituting Eq. (46) into Eq. (45) gives (Chang & Liao, 2011)

$$\kappa_m = \frac{3}{n} \left(\frac{\mu_r - 1}{-2\mu_r + 5} \right) \tag{47}$$

In the special case where the host material is vacuum ($\mu_r = 1$) and the permeability of the spherical particles is μ_s, $n = V^{-1}$ (V being the volume of a spherical particle), and Eq. (47) is satisfied by:

$$\kappa_m = 3V \frac{\mu_s - \mu_0}{-2\mu_s + 5\mu_0} \tag{48}$$

Combining Eqs. (47) and (48) gives the effective permeability (μ_{eff}) of the final mixture, i.e. (Chang & Liao, 2011),

$$\mu_{eff} = \mu_r \mu_0 = \mu_0 + 3f\mu_0 \frac{\mu - \mu_0}{-2\mu + 5\mu_0 + 2f(\mu - \mu_0)} \tag{49}$$

Where $f = nV$ is the volume ratio of the embedded particles within the mixture ($0 \le f \le 1$). In the more general situations where the host is no longer vacuum but of the permeability μ_h, then the more general mixing formula of permeabilities becomes (Chang & Liao, 2011):

$$\mu_{eff} = \mu_h + 3f\mu_h \frac{\mu_s - \mu_h}{-2\mu_s + 5\mu_h + 2f(\mu_s - \mu_h)} \tag{50}$$

As with Bruggeman's approach for dielectrics (Bruggeman, 1935), the derived magnetic permeability formula can be generalized to the multi-component form (Chang & Liao, 2011):

$$\frac{\mu_{eff} - \mu_0}{-2\mu_{eff} + 5\mu_0} = \sum_i f_i \frac{\mu_i - \mu_0}{-2\mu_i + 5\mu_0} \tag{51}$$

where f_i and μ_i denote the volume ratios and permeabilities of the involved different inclusions, respectively. Or (Chang & Liao, 2011),

$$\frac{\mu_{reff} - 1}{-2\mu_{reff} + 5} = \sum_i f_i \frac{\mu_{ri} - 1}{-2\mu_{ri} + 5} \tag{52}$$

Although the actual mixing procedures can vary widely such that substantial deviations may result between the theoretical and measured values, Eq. (52) should still serve as a valuable guide when designing magnetic materials or composites.

9. A practical method to secure magnetic permeability in optical regimes

As widely known, electronic polarization is involved in the absorption of electromagnetic wave within materials and this mechanism is represented by permittivity even in light-wave frequencies. However, refractive index which describes light absorption and reflection is calculated in terms of permittivity and permeability (accounting for magnetization). But permeability spectra in light frequencies are hardly available for most materials. In contrast with it, permeabilities for various materials are fairly well documented at microwave frequencies (see, for example, (Goldman, 1999) and (Jorgensen, 1995)). It is noted that more often than not an electronic permittivity spectrum ($\varepsilon_r(\omega)$) is secured by measuring its corresponding refractive index ($N(\omega)$) while bluntly assuming its relative permeability (μ_r) to be unity. Obviously, this approach is one-sided and inappropriate. In particular, as we are entering the nanotech era, many new possibilities should emerge and surprise us with their novel optical permeabilities, for example, the originally nonmagnetic manganese crystal can be made ferromagnetic once its lattice constant is varied (Hummel, 1985). In this section, a method is proposed such that reliable optical permeability values can be obtained numerically.

The magnetic permeability of a specific material emerges fundamentally from wavefunctions of its electrons. In the popular density functional theory (DFT) approach, the role of these electron wavefunctions is taken by the one-electron spinorbitals, called the Kohn-Sham orbitals ($|\psi_{os}\rangle$) (see, e.g., (Shankar, 1994) and (Atkins & Friedman, 1997)). A spinorbital is a product of an orbital wavefunction and a spin wavefunction:

$$|\psi_{os}\rangle = |\psi_o\rangle \otimes |\psi_s\rangle \tag{53}$$

where $|\psi_o\rangle$ is the orbital part, and $|\psi_s\rangle$ is the spin part of $|\psi_{os}\rangle$.

In the density functional theory, these spinorbitals are solutions to the equation:

$$\rho(\vec{r}) = \sum_{i=1}^{n} \left|\psi_{os_i}(\vec{r})\right|^2, \tag{54}$$

where $\rho(\vec{r})$ is the local charge probability density function. Then, the exact ground-state electronic energy as well as other electronic properties of this n-electron system are known to be unique functions of ρ ((Shankar, 1994), (Atkins & Friedman, 1997)). Further, the overall wavefunction satisfying the Pauli exclusion principle is often expressed in terms of the Slater determinant ((Shankar, 1994), (Atkins & Friedman, 1997)) as:

$$\psi_{total}(\vec{r}) = (n!)^{-1/2} \det\left|\psi_{os_1}(\vec{r}) \cdot \psi_{os_2}(\vec{r}) \cdot \psi_{os_3}(\vec{r}) \cdots \psi_{os_n}(\vec{r})\right|, \tag{55}$$

Hence, the primitive transition rate of a typical electronic excitation within a material of interest can now be expressed in terms of the Fermi's golden rule ((Shankar, 1994), (Atkins & Friedman, 1997)) as:

$$R_{total} = C \cdot \sum_{f,i} \left|\langle \psi_{os}^{f0} | H^1 | \psi_{os}^{i0} \rangle\right|^2 \delta\left(E_{os}^{f0} - E_{os}^{i0} - \hbar\omega\right), \tag{56}$$

where H^1 is the first order perturbation to the Hamiltonian of electrons caused by a light wave propagating within; ψ_{os}^{i0} (of E_{os}^{i0}) and ψ_{os}^{f0} (of E_{os}^{f0}) are the electron states (energies) before and after the perturbation (H^1) takes place, respectively; ω is the radian frequency of the propagating light wave; $\hbar$ is Planck constant divided by 2π; and C is a proportional constant. H^1 in Eq. (56) can be expressed as (Shankar, 1994):

$$H^1 = \frac{e}{2mc} \vec{A}_0 \cdot \vec{P}_M \tag{57}$$

where $\vec{A}_0$ is the vector potential of the injected light wave; $\vec{P}_M$ is the momentum operator of electrons in the material of interest. $\vec{P}_M$ represents electrons' straight-line motion which constitute electronic polarization. Thus substituting Eq. (57) into Eq. (56) will result in R_{total} proportional to the imaginary part of ε_r (i.e. ε_{rI}). At the same time (Shankar, 1994),

$$H^1 = \frac{e}{2mc} \vec{B}_0 \cdot \vec{\mu}_M \tag{58}$$

where $\vec{B}_0$ is the magnetic flux density of the injected light wave; $\vec{\mu}_M$ is the magnetic moment operator of electrons in the material of interest. $\vec{\mu}_M$ represents electrons' angular momentum which constitutes magnetization. Thus substituting Eq. (58) into Eq. (56) will result in R_{total} proportional to the imaginary part of μ_r (i.e. μ_{rI}).

Because

$$\vec{P}_M \left| \psi_s \right\rangle = 0 \tag{59}$$

Thus according to Eq. (57), the spin part of $\left| \psi_{os} \right\rangle$ and ε_{rI} are irrelative. But $\left| \psi_s \right\rangle$ contains angular momentum and therefore must be involved in μ_{rI}. From Eq. (56) and Eq. (58), we can obtain μ_{rI} as follows:

Applying a spin-orbit decomposition (see Eqs. (53) and (55)) on the kernel of the transition rate expression in Eq. (56) suggests the convenience of defining parameters as follows:

$$\left\langle \psi_{os}^{f0} \right| H^1 \left| \psi_{os}^{i0} \right\rangle = \left(\left\langle \psi_o^{f0} \right| \otimes \left\langle \psi_s^{f0} \right| \right) \cdot H^1 \cdot \left(\left| \psi_o^{i0} \right\rangle \otimes \left| \psi_s^{i0} \right\rangle \right) \equiv t_{total} , \tag{60a}$$

$$\left\langle \psi_o^{f0} \right| H^1 \left| \psi_o^{i0} \right\rangle \equiv t_o , \tag{60b}$$

$$\left\langle \psi_s^{f0} \right| H^1 \left| \psi_s^{i0} \right\rangle \equiv t_s , \tag{60c}$$

where through explicit matrix operations, it can be shown that

$$t_{total} = t_o t_s , \tag{61}$$

As a result, Eq. (56) can be rewritten as:

$$R_{total} = C \cdot \sum_{f,i} t_{os} \cdot \delta\left(E_{os}^{f0} - E_{os}^{i0} - \hbar\omega\right) = C \cdot \sum_{f,i} t_o t_s \cdot \delta\left(E_{os}^{f0} - E_{os}^{i0} - \hbar\omega\right) = R_o \cdot R_s , \tag{62}$$

where

$$R_o = C_o \sum_{f,i} \left| \left\langle \psi_o^{f0} \right| H^1 \left| \psi_o^{i0} \right\rangle \right|^2 \delta\left(E_o^{f0} - E_o^{i0} - \hbar\omega\right) , \tag{63}$$

$$R_s = C_s \sum_{f,i} \left| \left\langle \psi_s^{f0} \right| H^1 \left| \psi_s^{i0} \right\rangle \right|^2 \delta\left(E_s^{f0} - E_s^{i0} - \hbar\omega\right) , \tag{64}$$

with C_o and C_s being constant coefficients.

As aforementioned, despite being termed "electronic transition rate" to account for the lightwave absorption within a material, the primitive transition rate R_{total} is in essence a series of delta functions situated at varying frequencies (or energies, see Eq. (56)), and thus is too spikey to be real. In fact, this spikey nature results from our attempt to describe the dynamics of multitudes of electrons by a limited number of Kohn-Sham orbitals. Therefore, R_{total} ought to be "smoothed" prior to being converted to a realistic absorption spectrum. Here, Gaussian functions are adopted to replace all delta functions. The smoothed R_{total}, R_o and R_s are now denoted as R'_{total}, R'_o and R'_s, respectively, and thus from Eq. (62) we have:

$$R'_{total} = R'_o R'_s \,, \tag{65}$$

It is a common practice in the first-principle quantum mechanical calculations (such as by the computer codes CASTEP and DMol[3] ((Clark et al., 2005), (Delley, 1990)) that R'_o is normally set equal to ε_{rI} simulated with the "spin polarized" option in these kind of codes turned off (as in the nonmagnetic case).

Substituting Eq. (58) into Eq. (64) gives R'_s. And then the product of R'_o and R'_s, R'_{total}, is proportional to μ_{rI} (the scaling factor is defined as C_M). With μ_r being linear and causal, there is an exact one-to-one correspondence between its real and imaginary parts as prescribed by the Kramers-Kronig relation (see, e.g., (Landau & Lifshitz, 1960)). Thus, the real part of μ_r (i.e., μ_{rR}) is readily available once μ_{rI} is numerically obtained. To make all this happen, the calculation of R'_s is now in order.

Within the formulation of R'_s spectrum (see its unsmoothed form in Eq. (64)), physical quantities $\left|\psi_s^{f0}\right\rangle$, $\left|\psi_s^{i0}\right\rangle$, E_s^{f0} and E_s^{i0} would emerge from CASTEP or DMol[3] ((Clark et al., 2005), (Delley, 1990)) simulations by selecting the "spin polarized" option. In evaluating the bracketed term in Eq. (64), we first have:

$$t_s \equiv \left\langle \psi_s^{f0} \middle| H^1 \middle| \psi_s^{i0} \right\rangle = \left\langle \psi_s^{f0} \middle| \frac{1}{2} \vec{\mu}_M \cdot \vec{B}_0 \middle| \psi_s^{i0} \right\rangle = \left\langle \psi_s^{f0} \middle| \frac{e}{2mc} \vec{S} \cdot \vec{B}_0 \middle| \psi_s^{i0} \right\rangle , \tag{66}$$

where $\vec{\mu}_M = \gamma \vec{S}$ ($\vec{S}$ being the spin angular momentum), with the gyromagnetic ratio $\gamma = -\dfrac{ge}{2mc}$ and g = 2 (Shankar, 1994), such that $\dfrac{e}{2mc}\vec{S}\cdot\vec{B}_0 = \dfrac{e}{2mc}\left(S_x B_{0x} + S_y B_{0y} + S_z B_{0z}\right)$ in Cartesian coordinate. Further, it is known that the spin angular momentum around the z axis (S_z) (i.e., eigenvectors of $\left|\psi_s^{f0}\right\rangle$ and $\left|\psi_s^{i0}\right\rangle$ possess only two possible eigenvalues: either $+\hbar/2$ (spin up) or $-\hbar/2$ (spin down), with $\hbar$ being the Planck's constant divided by 2π. Hence, the two possible states of $\left|\psi_s^{f0}\right\rangle$ and $\left|\psi_s^{i0}\right\rangle$ expressed on the eigenbasis of S_z are (Shankar, 1994):

Spin up: $\left|\psi_s\right\rangle \leftrightarrow \begin{bmatrix} 1 \\ 0 \end{bmatrix}$, spin down: $\left|\psi_s\right\rangle \leftrightarrow \begin{bmatrix} 0 \\ 1 \end{bmatrix}$, while the spin angular momenta S_x, S_y and S_z are: $S_x \leftrightarrow \dfrac{\hbar}{2}\begin{bmatrix} 0 & 1 \\ 1 & 0 \end{bmatrix}$, $S_y \leftrightarrow \dfrac{\hbar}{2}\begin{bmatrix} 0 & -i \\ i & 0 \end{bmatrix}$, $S_z \leftrightarrow \dfrac{\hbar}{2}\begin{bmatrix} 1 & 0 \\ 0 & -1 \end{bmatrix}$ (Shankar, 1994).

Hence, the value of the bracketed term t_s of Eq. (66) varies according to the four possible combinations of $\left|\psi_s^{f0}\right\rangle$ and $\left|\psi_s^{i0}\right\rangle$, namely:

1. $\left|\psi_s^{f0}\right\rangle$ and $\left|\psi_s^{i0}\right\rangle$ both spin up such that

$$t_s \equiv \left\langle \psi_s^{f0} \middle| \frac{e}{2mc} \vec{S} \cdot \vec{B}_0 \middle| \psi_s^{i0} \right\rangle = \frac{e}{2mc} \left\langle j'm' \middle| S_x B_{0x} + S_y B_{0y} + S_z B_{0z} \middle| jm \right\rangle$$

$$\leftrightarrow \frac{e}{2mc}\begin{bmatrix} 1 & 0 \end{bmatrix}\left(\frac{\hbar B_{0x}}{2}\begin{bmatrix} 0 & 1 \\ 1 & 0 \end{bmatrix} + \frac{\hbar B_{0y}}{2}\begin{bmatrix} 0 & -i \\ i & 0 \end{bmatrix} + \frac{\hbar B_{0z}}{2}\begin{bmatrix} 1 & 0 \\ 0 & -1 \end{bmatrix} \right)\begin{bmatrix} 1 \\ 0 \end{bmatrix} = \frac{e\hbar}{4mc} B_{0z} \equiv t_z$$

2. $\left|\psi_s^{f0}\right\rangle$ spin up, and $\left|\psi_s^{i0}\right\rangle$ spin down such that $t_s = \frac{e\hbar}{4mc}\left(B_{0x} - iB_{0y}\right) \equiv t_{xy}$.

3. $\left|\psi_s^{f0}\right\rangle$ spin down, and $\left|\psi_s^{i0}\right\rangle$ spin up such that $t_s = \frac{e\hbar}{4mc}\left(B_{0x} + iB_{0y}\right) \equiv t_{xy}^*$.

4. $\left|\psi_s^{f0}\right\rangle$ and $\left|\psi_s^{i0}\right\rangle$ both spin down such that $t_s = -\frac{e\hbar}{4mc}B_{0z} \equiv -t_z$.

It is obvious that the values of $|t_s|^2$ for cases 1 and 4 are identical, i.e., $|t_z|^2 \equiv T_z$; and those for cases 2 and 3 are the same and are $|t_{xy}|^2 \equiv T_{xy}$.

To simplify the R'_s calculation without loss of generality, the magnetic flux density vector of a linearly polarized electromagnetic wave is oriented to parallel to the z axis, i.e., $\vec{B}_0 = \vec{B}_{0z}$ ($\vec{B}_{0x} = \vec{B}_{0y} = 0$). As a result, only T_z is relevant in the R'_s calculation. Further, as implied by its representation, T_z is invariant in value regardless of what the initial and final energy levels (E_s^{f0} and E_s^{i0}) are in a transition, so long as Fermi's golden rule is satisfied. This property greatly facilitates the calculation of R'_s, in that R'_s now simply becomes proportional to the number of identical-spin transition electron pairs. Within each transition pair there is an energy difference of $\hbar\omega$ while being irradiated by a linearly polarized light of frequency ω.

In first-principle quantum mechanical simulation codes, such as the CASTEP and DMol^3, a finite number of E_s^{f0} and E_s^{i0} are generated to approximate the transitions of a multitude of electrons within a material of interest. In other words, each output energy value actually stands for a narrow continuous band of states centered at this specific value. Therefore, to simulate more closely to the reality, all delta functions of E_s^{f0} and E_s^{i0} are replaced with Gaussian functions prior to being added up into continuous density spectra, for both the "all spin-up" (case 1) and "all spin-down" (case 4) states, respectively. Since the magnetic properties are manifested by unpaired spins, and on the same energy level a spin-up is neutralized by a spin-down (Pauli's exclusion principle), the net spin density spectrum is thus settled by subtracting that of the spin-downs from that of the spin-ups.

With all inner work laid out, the detailed procedure for evaluating R'_s is outlined as follows:

1. Subtract the density spectrum of the spin-down states from that of the spin-ups to result in the net spin density spectrum. The positive part of it is the net spin-up density spectrum, and the absolute value of the other part is the net spin-down density spectrum.
2. Randomly sample the net spin-up density spectrum at each energy of interest and denote the sampled energy value E_i if it is lower than the Fermi level, otherwise, name it E_j.
3. Define $P_{i,j}$, as the product of n_i and n_j, where n_i and n_j are the density-of-states at E_i and E_j, respectively. Namely, $P_{i,j}$ is proportional to the number densities associated with a transition pairs of net spin-up electrons linked by an energy difference of $E_{i,j} \equiv (E_j - E_i)$.
4. Calculate $P_{i,j}$'s and $E_{i,j}$'s for each (E_i, E_j) pair to get the net spin-up $P_{i,j}$ vs. $E_{i,j}$ collection.
5. Obtain the $P_{i,j}$ vs. $E_{i,j}$ collection for the net spin-down states in a similar fashion.
6. Then, obtain the union of the $P_{i,j}$ vs. $E_{i,j}$ collection of the net spin-downs and that of the net spin-ups to result in the total $P_{i,j}$-$E_{i,j}$ collection.
7. Replace all $P_{i,j}$ delta peaks by Gaussian functions to arrive at the desired continuous spectrum of R'_s.

As mentioned, in those cases where the "spin polarized" option in, e.g., CASTEP is turned off, the resultant ε_{rI} spectra actually gives R'_o in Eq. (65). With both R'_s and R'_o being revealed, the optical μ_r spectrum of interest can be secured within a proportional constant C_M, and then the Kramers-Kronig relation. Finally, this universal constant C_M is uncovered by comparing the erected μ_{rI} spectrum with existing data covering from low to lightwave frequencies (Lide & Frederikse, 1994).

The calculated refractive index spectrum of iron crystal from using the proposed approach is shown in Fig. 23. A comparison with the known refractive index spectrum of iron crystal exposed by a linearly polarized light (Lide & Frederikse, 1994) (see Fig. 24) reveals that both figures are relatively close in features. Accordingly, the universal proportional constant C_M obtained from comparing the two N_I curves is about 13.7. The completed iron μ_r spectrum is provided in Fig. 25 after applying $R'_o = \varepsilon_{rI}$, Eq. (66) and then the Kramers-Kronig relation. Therefore, the optical μ_r and refractive index spectra of all materials, including those to be developed, can now be fully explored for the first time.

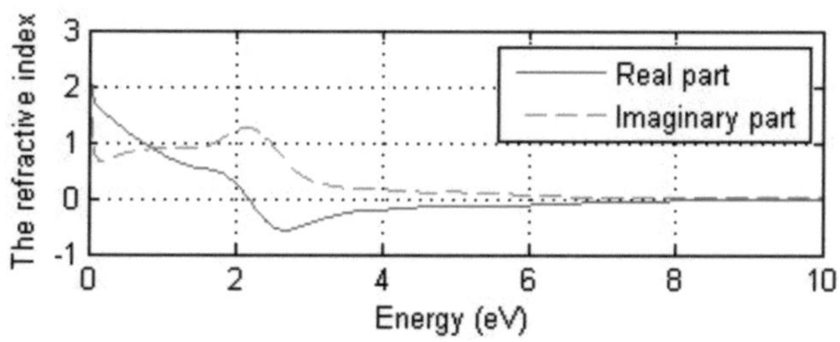

Fig. 23. The calculated relative refractive index spectrum of the Fe crystal ($C_M = 1$).

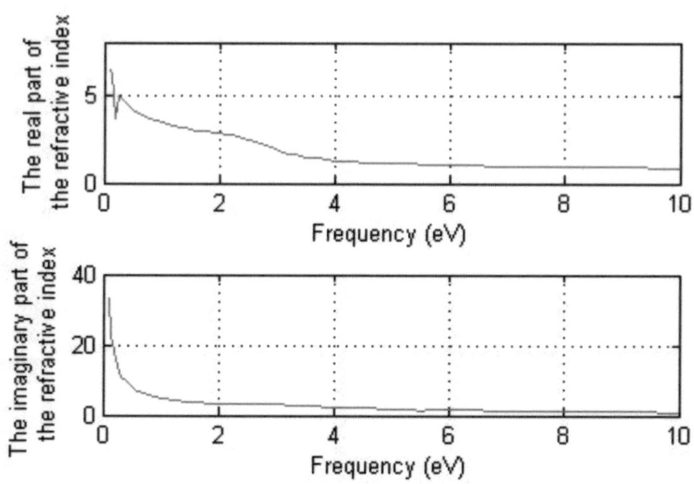

Fig. 24. The refractive index spectrum of Fe crystal in CRC Handbook.

Incidentally, to further challenge the proposed procedure, we erroneously leave the "spin polarized" action of CASTEP on when simulating the copper crystal, a knowingly nonmagnetic material. It turned out that the numerically converged spin density was a minimal 1.35 x 10^{-4} μ_B/atom, showing that the proposed procedure is robust against erroneous initial conditions.

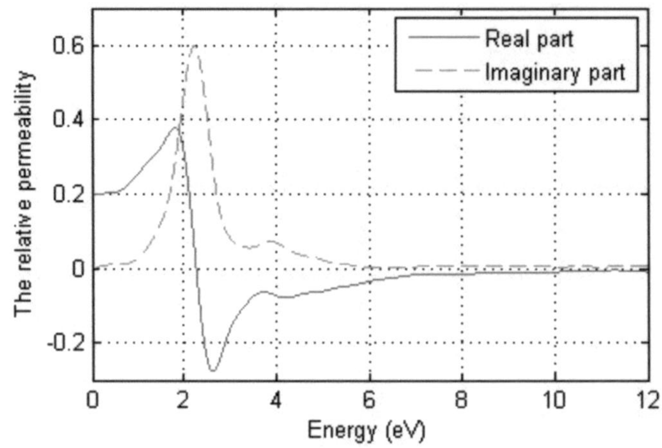

Fig. 25. The optical μ_r spectrum of Fe crystal obtained from the proposed procedure.

10. Conclusion

This chapter contains a clear, yet rigorous, picture as to how the real physical causes behind all these macroscopic optical phenomena – i.e., the microscopic electric and magnetic dipoles-work to come up with such macroscopic results. Both the electric and magnetic dipoles react to the imposed EM wave in such a way that actually results in the Fresnel equations.

In arriving at the more intuitive "scattering form" of the Fresnel equations, microscopic EM-induced physical electric and magnetic dipoles were rigorously employed as the source of electromagnetic waves by Doyle et al. Motivated by such an approach, the authors started to speculate how the incorporation of permanent dipoles might affect many macroscopic optical phenomena, e.g., the Brewster angle of a specific optical material. Among its predictions, the traditionally fixed Brewster angle of a specific material now not only becomes dependent on the density and orientation of incorporated permanent dipoles, but also on the incident light intensity (more precisely, the incident wave electric field strength). Further, two conjugated incident light paths would give rise to different refracted wave powers. This kind of microscopic approaches are called "dipole engineering".

Theoretical elaboration and then IR experiments on poled polyvinylidene fluoride (PVDF) films were conducted to verify the emergence of asymmetric reflections at varying incident angles, as well as the inverse dependence of reflectivity upon the impinging light intensity. In addition, experiments on dipole-engineered PVDF films show that by way of adding/reducing permanent dipole density and varying orientations, the aforementioned theoretical predictions can be evidenced unambiguously in the visible light range. Further, effective polarization density can be quantified from the above experiments subjected to different dipole engineering processes. As a result, the traditionally elliptic contour of a slanted two dimensional section of the refractive index ellipsoid now manifests symmetric open splittings at near the traditional incident angle. It implies that severe challenge to the accuracy of traditional surface plasmon resonance measurements may arise in the presence of permanent dipoles of various morphologies, such as in the forms of nano-particles or membrane double layers.

Traditional approach of obtaining electric permittivity through directly measuring a material's refractive index begins to appear flawed at dawn of the nanotech era. On the other hand, achieving novel magnetics and thus new materials of variable refractive indices starts to become meaningful. However, unlike its electric counterpart, data for optical magnetic permeability is hardly available for most materials. An effective method is proposed to secure practically accurate optical permeabilities through manipulating electronic transition rates generated by first-principle quantum mechanical simulations.

11. References

Alex J. Goldman. (1999). *Handbook of Modern Ferromagnetic Materials*, Springer, Kluwer Academic Publishers, Norwell, MA. (Goldman, 1999)

B. Alberts, A. Johnson, J. Lewis, M. Raff, K. Roberts. (2007). *Molecular Biology of the Cell*, Garland Science, New York. (Alberts et al., 2007)

B. Delley. (1990). "An all-electron numerical method for solving the local density functional for polyatomic molecules", *The Journal of Chemical Physics*, vol. 92, (1990), 508. (Delley, 1990)

Chungpin Liao, Hsien-Ming Chang, and Chien-Jung Liao. (2006). "Manipulating the Brewster angles by using microscopically coherent dipole quanta and its possible implications," *IEEE Journal Lightwave Technology*, vol. 24, no. 8, (Aug 2006), pp. 3248-3254. (Liao et al., 2006)

D. A. G. Bruggeman. (1935). „Berechnung verschiedener physi-kalischer Konstanten von heterogenen Substanzen. I. Di-elektrizitätskonstanten und Leitfähigkeiten der Mischkörper aus isotropen Substanzen", *Annalen der Physik*, vol. 416, (1935), pp. 665-679. (Bruggeman, 1935)

D. R. Lide, H. P. R. Frederikse. (1994). *CRC Handbook of Chemistry and Physics*, pp.123-124, CRC-Press, Boca Raton, FL. (Lide & Frederikse, 1994)

David K. Cheng. (1989). *Field and Wave Electromagnetics*, second ed., Addison-Wesley Pub. Co., Reading, MA. (Cheng, 1989)

E. Hecht. (2002). *Optics*, 4th Ed., Addison Wesley, New York. (Hecht, 2002)

Edward M. Purcell. (1985). *Electricity and Magnetism*, 2nd Ed., Education Development Center, Inc., MA. (Purcell, 1985)

Finn Jorgensen. (1995). *The Complete Handbook of Magnetic Recording*, McGraw-Hill, New York. (Jorgensen, 1995)

H. A. Haus and J. R. Melcher. (1989). *Electromagnetic Fields and Energy*, Prentice-Hall, New York. (Haus & Melcher, 1989)

H. A. Lorentz. (1880). „Ueber die Beziehung zwischen der Fortpflanzungsgeschwindigkeit des Lichtes und der Körperdichte", *Annalen der Physik*, vol. 9, (1880), pp. 641-665. (Lorentz, 1880)

H. Raether. (1988). "Surface Plasmons on Smooth and Rough Surfaces and on Gratings," In: *Springer Tracts in Modern Physics*, Springer, Berlin. (Raether, 1988)

Hsien-Ming Chang and Chungpin Liao. (2011). "A Parallel Derivation to the Maxwell-Garnett Formula for the Magnetic Permeability of Mixed Materials", *World Journal of Condensed Matter Physics*, in press. (Chang & Liao, 2011)

J. C. Maxwell Garnett. (1904). "Colours in Metal Glasses and in Metallic Films", *Philosophical Transactions of the Royal Society of London*. Series A, Containing Papers of a Mathematical or Physical Character, vol. 203, (1904), pp. 385-420. (Garnett, 1904)

Jun-Lang Chen, Min-Yen Shieh, Hsin Her Yu, Chungpin Liao, Hsien-Ming Chang, Bin-Huang Yang, and Zi-Peng Zhao. (2008). "Asymmetric reflectivity from surfaces with distributed dipoles and IR experiments on poled polyvinylidene difluoride films", *APPLIED PHYSICS LETTERS*, vol. 93, 011902, (July 2008). (Chen et al., 2008)

K.M. Yassien, M. Agour, C. von Kopylow and H.M. EI-Dessouky. (2010). "On the digital holographic interferometry of fibrous material, I: Optical properties of polymer and optical fibers," *Optics and Lasers in Engineering* , Vol. 48, Issue 5, (2010), pp. 555-560. (Yassien et al., 2010)

L. Lorenz. (1880). „Ueber die Refractionsconstante", *Annalen der Physik*, Vol. 11, (1880), pp. 70-103. (Lorenz, 1880)

L.D. Landau, E.M. Lifshitz. (1960). *Electrodynamics of Continuous Media*, pp. 260-263, Pergamon Press, Oxford. (Landau & Lifshitz, 1960)

M. Matsukawa, K. Shintani, S. Tomohiro and N. Ohtori. (2006). "Application of Brillouin scattering to the local anisotropy and birefringence measurements of thin layers," *Ultrasonics*, Vol.44, Suppl. 1, (2006), pp. e1555-e1559. (Matsukawa et al., 2006)

P. W. Atkins and R. S. Friedman. (1997). *Molecular quantum mechanics*, Oxford University Press, New York. (Atkins & Friedman, 1997)

Paul Lorrain, Dale R. Corson. (1970). *Electromagnetic Fields and Waves*, second ed., W.H.Freeman & Co Ltd, San Francisco. (Lorrain & Corson, 1970)

Po-Yu Tsai, Chien-Jung Liao, Wen-Kai Kuo, Chungpin Liao. (2011). "Birefringence caused by the presence of permanent dipoles and its possible threat on the accuracy of traditional surface plasmon", *World Journal of Condensed Matter Physics*, in press. (Tsai et al., 2011)

Pramoda K. Pallathadka. (2006). "Solid state ^{19}F NMR study of crystal transformation in PVDF and its nanocomposites (polyvinylidene fluoride)", *Polymer Engineering and Science*, 46(12), 1684-7, (2006). (Pallathadka, 2006)

Q. M. Zhang, V. Bharti, G. Kavarnos. (2002). "Poly (Vinylidene Fluoride) (PVDF) and its Copolymers", In: Encyclopedia of Smart Materials, M. Schwartz (Ed.), Volumes 1-2, pp. 807-825, John Wiley & Sons, and references therein. (Zhang et al., 2002)

R. Shankar. (1994). *Principles of Quantum Mechanics*, Springer Science & Business Media, New York. (Shankar, 1994)

Rolf E. Hummel. (1985). *Electronic Properties of Materials*, pp. 243-247, Springer-Verlag, New York. (Hummel, 1985)

S. A. Maier. (2007). *Plasmonics: Fundamental and Applications*, Springer, New York. (Maier, 2007)

Stewart J. Clark, M.D. Segall, Christopher J Pickard, P.J. Hasnip, M. I. J. Probert, Keith Refson, and Mike C. Payne. (2005). "First principles methods using CASTEP", *Zeitschrift Fuer Kristallographie*, vol. 220, (2005), pp. 567-570. (Clark et al., 2005)

W. Lamb, D. M. Wood, and N. W. Ashcroft. (1980). "Long-wavelength electromagnetic propagation in heterogeneous media", *Physical Review B*, vol. 21, no. 6, (March 1980), pp. 2248-2266. (Lamb et al., 1980)

W. T. Doyle. (1985). "Scattering approach to Fresnel's equations and Brewster's law," *Am. J. Phys.*, 53 (5), pp. 463-468. (Doyle, 1985)

10

High Resolution Laser Spectroscopy of Cesium Vapor Layers with Nanometric Thickness

Stefka Cartaleva[1], Anna Krasteva[1], Armen Sargsyan[2], David Sarkisyan[2], Dimitar Slavov[1], Petko Todorov[1] and Kapka Vaseva[1]
[1]*Institute of Electronics, Bulgarian Academy of Sciences, Sofia*
[2]*Institute for Physical Research, National Academy of Sciences of Armenia, Ashtarak*
[1]*Bulgaria*
[2]*Armenia*

1. Introduction

High resolution laser spectroscopy of alkali vapor contained in conventional thermal optical cells with centimeter dimensions is widely used for various applications: among them wavelength references, atomic clocks, precise optical magnetometers, slow and stored light etc. For all these photonic sensors, the reduction of their dimensions is of significant importance. One of the main concerns is to keep the parameters of the photonic sensor when reducing its size. In this chapter are presented the obtained by authors experimental and theoretical results concerning high-resolution spectroscopy of Cs vapor layer with nanometric thickness. The thickness of the vapor layer varies from 100 nm to about 5000 nm. The practical importance of this study is accompanied by numerous new peculiarities of atomic spectra of 1 D confined atoms, when the nanometric dimension approaches the wavelength of the irradiating light. These peculiarities in the absorption and fluorescence spectra represent a basic importance as well.

2. Unique optical cells for confinement of Cs atomic layers with nanometric thickness

2.1 Main characteristics of atomic confinement

In this chapter, the high resolution laser spectroscopy is concerned of alkali vapor confined in unique optical cell with nanometric thickness [Sarkisyan, 2001], further on called Extremely Thin Cell (ETC). The transversal and longitudinal dimensions of such cell (Fig.1) differ significantly. The distance between the high-quality ETC windows L varies from 100 nm to (1-3) μm. At the same time, the cell window diameter is about 2 cm (Fig.1b). Therefore, a strong spatial anisotropy is present for the time of interaction between atoms confined in the ETC and the laser radiation used for spectroscopy performed with such optical cell.

Let us consider Cs atoms flying orthogonally to the cell windows (Fig.1a, atoms denoted by $v_\perp$), which average thermal velocity at room temperature is about 200 m/s. Those atoms will pass the L = 1 μm distance for 5 ns. Hence the time of flight of atoms is much shorter than the lifetime of the excited atomic state. Such a limit is not imposed on the atoms (Fig.1a, denoted by v_{II}), moving parallel to the windows of the ETC. The second group of atoms will interact with the laser radiation for a time determined by the diameter of the laser beam D (D >> L). When the ETC is irradiated by a laser beam propagating in direction orthogonal to the ETC window surfaces, the atoms with velocity direction close to parallel to the window surface can be considered as "slow" atoms, i.e. atoms with very small velocity projection on the laser beam direction. Hence two groups of atoms can be mainly distinguished – "slow" (moving parallel to the windows) and "fast" (moving parallel to the laser beam propagation direction, in the extremely narrow space between the two windows of the ETC). As a first result of the light interaction with those atomic groups, a strong reduction of the Doppler effect influence occurs and of the related Doppler broadening of spectral lines as well.

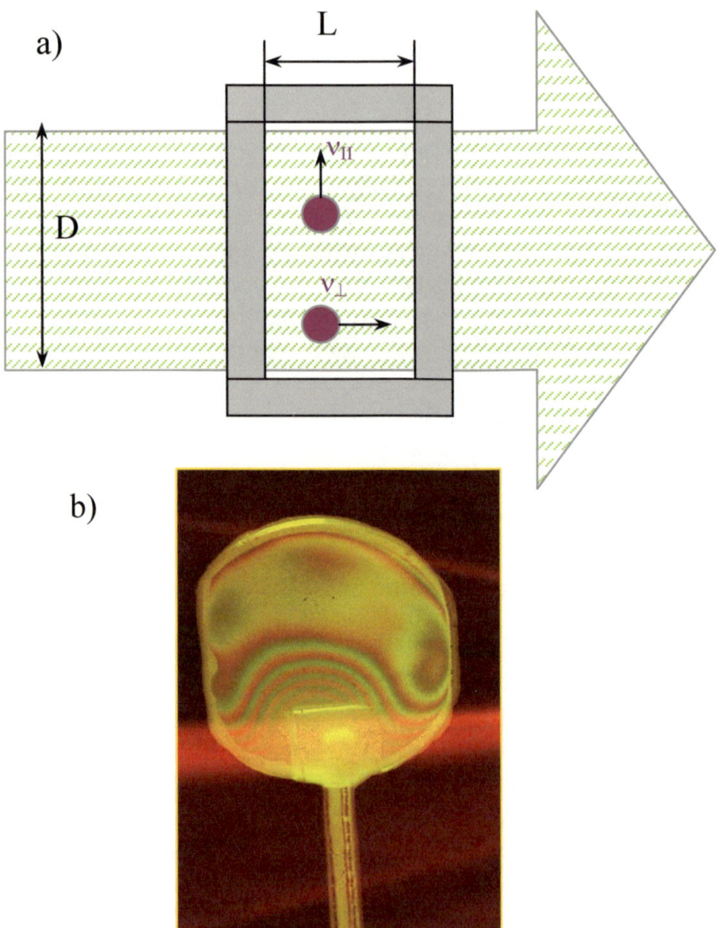

Fig. 1. (a) Atomic movement in cell of nanometric thickness L: v_{II} – velocity component parallel to the cell windows; $v_{\perp}$ - atomic velocity component orthogonal to the windows and along the laser beam, (b) Practical realization of nanometric cell.

2.2 Hyperfine transitions on the D_2 line of Cs atoms

In this chapter, the spectral properties of Cs atoms confined in ETC are particularly concerned. The diagram of ^{133}Cs energy levels with the hyperfine transitions on the D_2 line is shown in Fig.2.

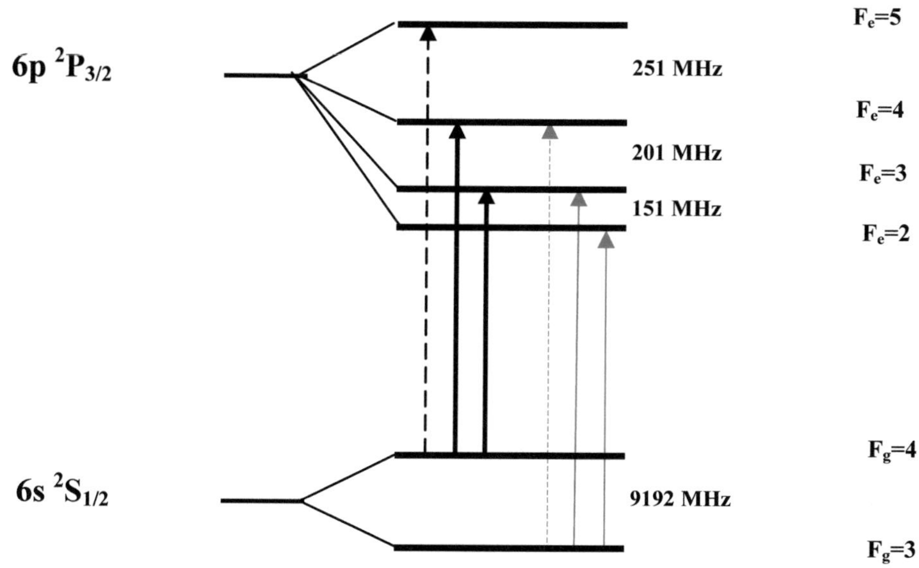

Fig. 2. Energy-level diagram for D_2 line of ^{133}Cs. $F_g \to F_e \le F_g$ transitions (solid line) are distinguished from $F_g \to F_e > F_g$ transitions (dashed line). The wavelength of the D_2 line is $\lambda = 852nm$.

Cesium D_2 line consists of two sets of hyperfine transitions, forming two absorption (fluorescence) lines:

- $F_g = 3$ set, involving three hyperfine transitions starting from the $F_g = 3$ ground level to the respective $F_e = 2, 3, 4$ excited levels and
- $F_g = 4$ set of transitions - the $F_g = 4 \to F_e = 3, 4, 5$ hyperfine transitions.

In the widely used optical cells with centimeter dimensions (further on called conventional cells), the hyperfine transitions starting from single ground level completely overlap due to the Doppler broadening (the spectral width ~ 400 MHz, for individual hyperfine transition), which is larger than the separation between the excited state hyperfine levels. Thus, the two types of transitions ($F_g \to F_e \le F_g$ or $F_g \to F_e > F_g$) involved in the absorption line can not be resolved. However, they have different properties. The first type ($F_g \to F_e \le F_g$ transition) can suffer population loss from the excited by the light ground level due to hyperfine optical pumping to the other ground hyperfine level, which does not interact with the light. When this type of transition is closed for hyperfine optical pumping, it intrinsically will exhibit Zeeman optical pumping to ground-state Zeeman sublevels non-interacting with the light. In opposite, if the transition of the second type ($F_g \to F_e > F_g$ transition) is closed for hyperfine optical pumping, it can be considered as a completely closed one [Andreeva, 2007, a; Andreeva, 2007, b; Andreeva, 2002], i.e. the atomic population is cycling only between the energy levels involved in the atomic transition, with all Zeeman sublevels of the ground level excited by the light.

The situation is different in the case of ETC. Here the hyperfine transitions within a single absorption (fluorescence) line can be resolved [Andreeva, 2007, a; Andreeva, 2007, b] (see Fig.3). Hence we can investigate separately the two types of hyperfine transitions, which are involved in the absorption (fluorescence) line by means of very simple single beam spectroscopy. The $F_g = 4 \rightarrow F_e = 3, 4, 5$ set of transitions involves the $F_g = 4 \rightarrow F_e = 5$ transition, which is the only completely closed one on the D_2 line.

The presence of the two main groups of atoms, determined by the ETC anisotropy, leads also to the observation of a significant difference between the fluorescence and transmission spectra (Fig. 3). In the ETC, the basic contribution to the creation of fluorescence signal belongs to the atoms moving parallel to the windows, i.e. "slow" atoms. In fact, the "fast" atoms moving along the laser beam direction do not have enough time to perform a complete absorption-emission cycle, i.e. to fluoresce. Hence, strong narrowing of the fluorescence profiles occur and the hyperfine optical transitions are completely resolved in the fluorescence case.

The situation with the absorption is different. Here, together with the "slow" atoms, the "fast" atoms are also able to contribute to the absorption spectrum of the nanometric atomic vapor layer, because the absorption of a quantum of light is a process a lot faster than the fluorescence. In this way, the absorption profile "suffers" much more from the Doppler effect. Hence, because different velocity groups of atoms contribute to the fluorescence and absorption profile formation, a big difference in the width of the corresponding optical transitions occurs (Fig.3).

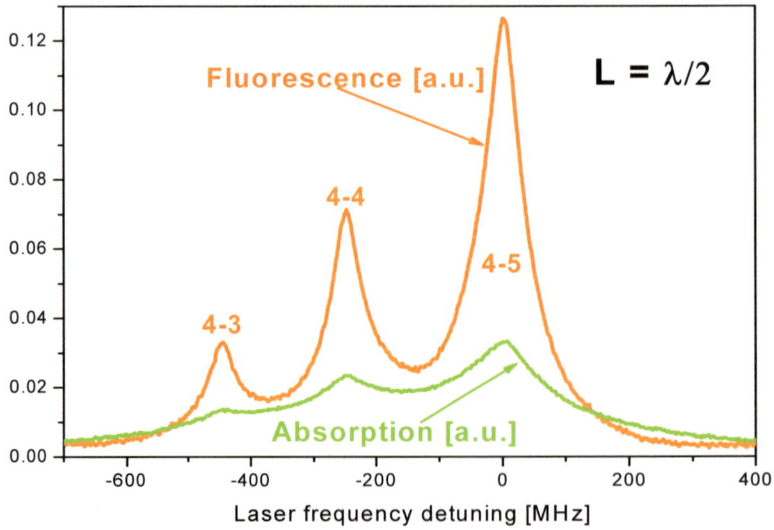

Fig. 3. Difference between the spectral widths of fluorescence and absorption profiles; the illustration is on the D_2 line of Cs, for optical transitions starting from the ground level $F_g = 4$; the 4-3, 4-4 and 4-5 are the respective hyperfine transitions. $\lambda = 852$nm.

3. Main properties of the experimental fluorescence and absorption spectra of Cs atoms confined in nanometric cells

As above mentioned, in the conventional thermal cells the different hyperfine Cs transitions starting from single ground level F_g are completely overlapped and cannot be resolved at all. Consequently, the application of the nanometric cell provides a new opportunity for significant enhancement of the resolution in laser spectroscopy of thermal cell, without application of complex atomic beam or laser cooling systems. Moreover, this simple tool makes it possible to study also the dynamic processes in the absorption and fluorescence, as well as in the electromagnetically induced transparency and absorption.

3.1 Sub-Doppler resonances in the absorption (transmission) of confined in nanometric cell atoms – Experimental observations

3.1.1 Experimental set up

A schematic drawing of the experimental set up is presented in Fig.4. Three different radiation sources emitting at $\lambda = 852$ nm were used in the experiments presented here:

- a free running diode laser with linewidth of about 15 MHz,
- a cw Extended Cavity Diode Laser (ECDL) operating at single-frequency mode with linewidth of about 3 MHz, and
- a Distributed Feedback Laser (DFB) with very low-noise current controller and linewidth of about 2 MHz.

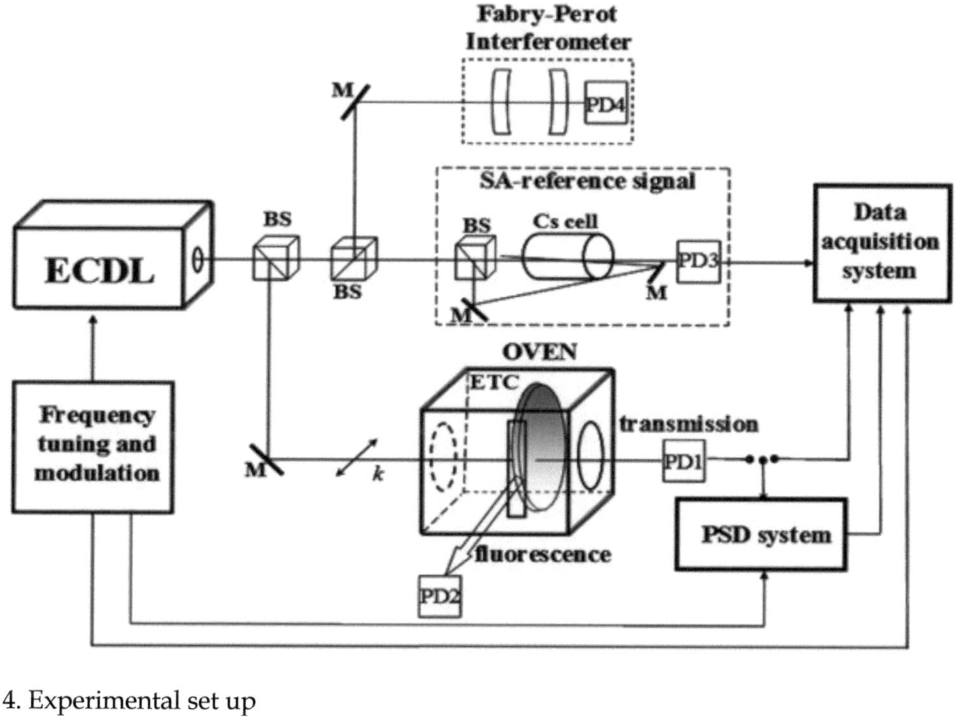

Fig. 4. Experimental set up

For all laser systems, the laser light is linearly polarized and its frequency is scanned in the region of the hyperfine transitions starting from ground-state levels with quantum numbers $F_g = 3$ and $F_g = 4$, at the D_2 line of Cs (see Fig.2).

The main part of the laser beam is directed at normal incidence onto the ETC filled with Cs vapor from a side-arm source. The construction of the ETC with a wedge-shaped (tunable) gap between high optical quality windows is similar to that reported in Ref. [Sarkisyan, 2001], but in some of the experiments the thickness of the vapor layer may vary in the range of (350–5000) nm. This enhancement of the thickness variation range is achieved by a preceding deposition of about 5000 nm thick Al_2O_3 layer onto the surface of one of the ETC windows in its lower part. The wedge-shaped (along the vertical direction) thickness of the ETC was measured by the interference technique described in Ref. [Dutier, 2003, b]. The dimensions of the ETC windows are 20 mm x 30 mm x 2.5 mm. A hole of 2 mm in diameter was drilled in the bottom of the windows, into which a tube of the same diameter and 50 mm long, made of commercial sapphire, was inserted. Then, the entire construction was assembled and glued in a vacuum furnace. After the gluing, a glass extension was sealed in the sapphire tube, and the ETC was filled with Cs metal, as it is usually done for conventional glass cells. The transmitted through the ETC beam is measured by a photodiode PD1 in dependence on the laser frequency, and the fluorescence is registered by a photodiode PD2.

The remaining part of the laser beam is used for laser frequency control:

- one beam is sent to a scanning Fabry-Perot interferometer for monitoring the single-mode operation of the ECDL;
- the second one – to an additional branch of the set-up including conventional, 3-cm long Cs cell with 2 cm diameter, for registration of the Saturated Absorption (SA) spectrum (PD-3).

The SA spectrum is recorded simultaneously with the signal from the ETC branch, thus ensuring precise frequency reference and scaling for the laser system frequency tuning. The sub-Doppler spectra in the transmitted beam and in fluorescence are recorded for different laser light intensities and ETC thicknesses.

3.1.2 Absorption spectra of atoms confined in ETC with $L = \lambda/2$ and $L = \lambda$

For illustration of the narrowing of absorption spectra of atomic transitions in ETC and their strong dependence on the cell thickness, first the experimentally observed spectrum only at two cell thicknesses is discussed.

In Fig.5a, the absorption spectrum of $6S_{1/2}(F_g = 4) \rightarrow 6P_{3/2}(F_e = 3, 4, 5)$ hyperfine transitions is presented, for three different irradiating light intensities and $L = \lambda/2$. The strong narrowing of each hyperfine transition is clearly seen, which is due to the absorption enhancement in the transition center relative to that in the wings. The origin of the narrowing of the hyperfine transition profiles is attributed to the anisotropy of the atom-light interaction time and to the Dicke effect [Romer]. Processes responsible for the coherent Dicke narrowing of the hyperfine transition profile at $L = \lambda/2$ can be briefly summarized as follows [Dutier, 2003, a; Sarkisyan, 2004; Maurin]. If an atom at the moment of leaving the cell wall is excited by resonant light, the excitation will start to precess in phase with the exciting

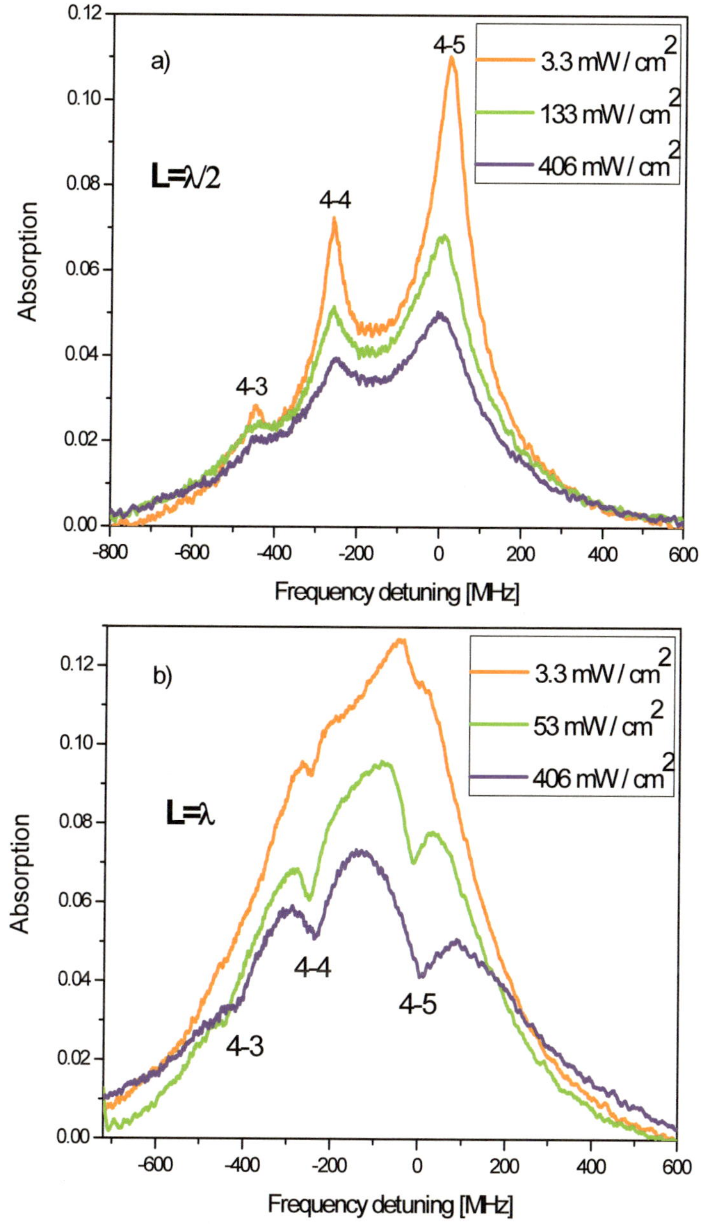

Fig. 5. Absorption spectra for the $F_g = 4$ set of transitions, at three light intensities: (a) $L = \lambda/2$ and (b) $L = \lambda$.

electromagnetic field at the wall position. However, with atomic motion the excitation will go gradually out of phase with the local exciting field. The phase mismatch appearing on the line center under a weak exciting field is independent of the atomic velocity and for a cell thickness up to λ/2 all regions of the cell interfere constructively, leading to a strong absorption enhancement at the hyperfine transition center. However, if the exciting light is detuned from the hyperfine transition center, the angular precession of the atomic excitation becomes velocity-dependent resulting in smooth reduction of the absorption in the wings of the hyperfine profile. From Fig.5a, we can also see the power broadening of the observed sub-Doppler-width profiles, together with a reduction in the overall absorption with irradiation light power density.

In Fig.5b, the absorption spectra at $6S_{1/2}(F_g = 4) \rightarrow 6P_{3/2}(F_e = 3, 4, 5)$ set of transitions are shown, for $L = \lambda$. One could immediately notice the significant difference observed between the absorption spectrum at $L = \lambda/2$ and that at $L = \lambda$. For $L = \lambda$, the coherent Dicke narrowing vanishes and as a result of the Doppler broadening the hyperfine structure is not resolved at very low intensity. However, rising the light intensity, it is possible to observe at cell thickness equal to λ well pronounced narrow dips of velocity selective reduced absorption, centered at the hyperfine transitions. The origin of these dips is related to the fact that processes like the two-level atomic system saturation and optical pumping can be completed only for atoms with large enough time of interaction with the laser light, i.e. "slow" atoms. Thus, the atoms flying nearly parallel to the ETC windows undergo optical pumping process (for the open transitions) and a saturation process (for all transitions) and this gives rise to highly velocity-selective dips in the absorption spectra. The amplitude of the reduced-absorption dips increases with increasing the power density, while the amplitude of the absorption profile as a whole decreases due to the saturation of the transition.
It is worthy to mention another interesting property of atomic absorption resulting from Dicke effect observed in the optical domain of the spectrum. The first spectrum shown in Fig.6 represents the absorption of Cs layer with thickness $L = \lambda/2$, and the second for $L = \lambda$. Interesting peculiarity is that the twofold enhanced thickness of Cs layer does not result in the absorption doubling at the centers of the optical transitions. Even in opposite - at the center of the strongest 4-5 transition, the absorption at $L = \lambda/2$ is larger than that at $L = \lambda$, under the condition of atomic transition saturation. The reason for that is the coherent response of the medium as a result of the Dicke effect, which is the strongest at $L = \lambda/2$ and the hyperfine transition saturation at $L = \lambda$.

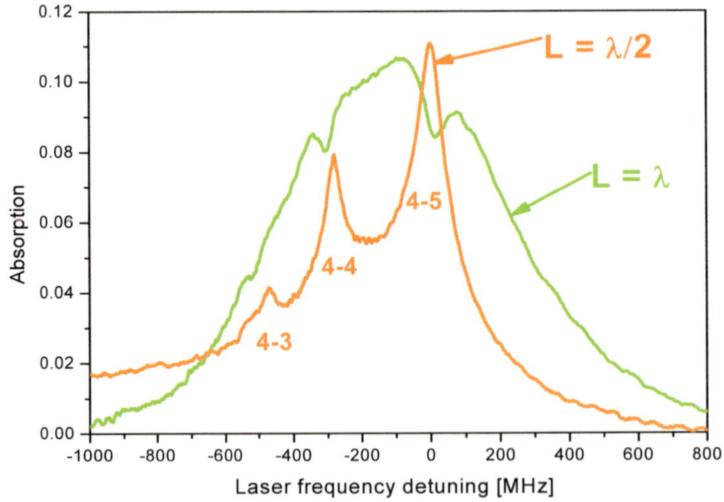

Fig. 6. Absorption spectra of two nanometric vapor layers with thicknesses $L = \lambda/2$ and $L = \lambda$. Nanometric change in the ETC thickness results in a principle modification of the absorption spectrum.

For both ETC thicknesses ($L = \lambda/2$ and $L = \lambda$), a detailed experimental study is performed [Varzhapetyan], related to the saturation with light intensity of all hyperfine transitions of the D_2 line of Cs. Sub-Doppler features centered at the resonance frequency of the hyperfine transitions have been observed for all used light intensities. Substantial changes in the amplitude and width of the sub-Doppler resonance for individual hyperfine transitions occur as a function of the intensity of the incident laser radiation. The absorption spectra of the transitions $6S_{1/2}(F_g = 3) \rightarrow 6P_{3/2}(F_e = 2, 3, 4)$ are presented in Fig.7a,b, and of the transitions $6S_{1/2}(F_g = 4) \rightarrow 6P_{3/2}(F_e = 3, 4, 5)$ – in Fig.8a,b.

For each value of light intensity, one could immediately note the significant difference observed between the absorption spectra at $L = \lambda/2$ (a) and these at $L = \lambda$ (b). Confirming the results of Ref. [Dutier, 2003, a; Sarkisyan, 2004], at $L = \lambda/2$ strong narrowing of the hyperfine transition absorption is obtained. All hyperfine transitions are well resolved. Fig.7a and Fig.8a show significant power broadening of the observed sub-Doppler structures.

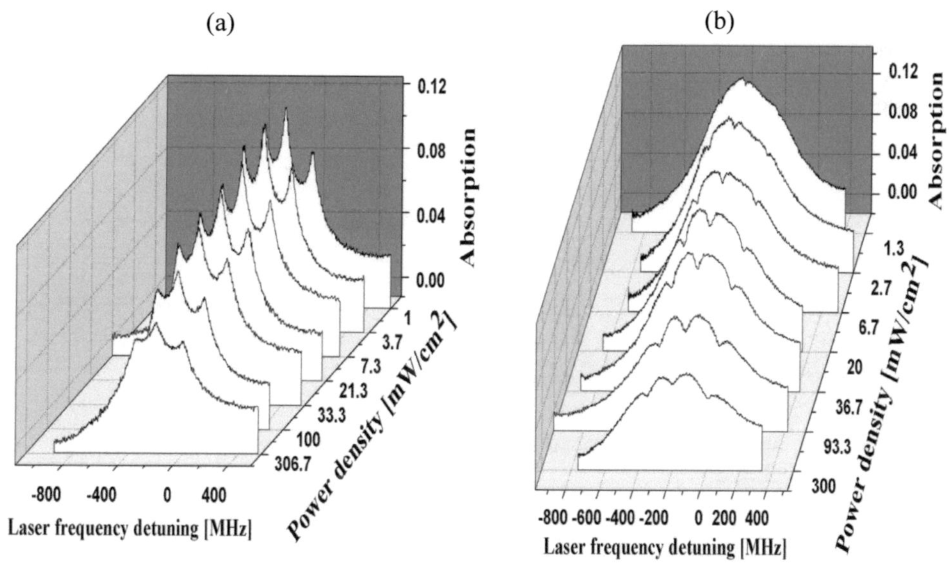

Fig. 7. Absorption spectra of the $F_g = 3$ set of hyperfine transitions obtained for different laser intensities at $L = \lambda/2$ (a) and $L = \lambda$ (b). The atomic source temperature is T=100°C. ECDL is used with spectral width of 3 MHz.

For $L = \lambda$, the coherent Dicke narrowing vanishes and as a result of the Doppler broadening the hyperfine structure is not resolved, which is evident at the lowest light intensities. At ETC thickness equal to λ, dips of reduction of the absorption appear. The origin of these dips is related to the fact that processes like saturation of the atomic transition and optical pumping to the ground level non-interacting with the laser light can be completed only for atoms with large enough time of interaction with the laser light, giving rise to highly velocity-selective dips in the absorption spectra (Fig.7b and Fig.8b). These sub-Doppler resonances of reduced absorption suffer strong power broadening. For the hyperfine transitions starting from $F_g = 3$ level and for laser power density of 300 mW/cm^2 the width of the reduced absorption resonances is about 60 MHz.

Consequently, by utilizing ETC and low-intensity irradiation, it is possible to distinguish two very important cases: (i) cell thickness L = λ/2 and (ii) L = λ. In the first case, the coherent response of the atomic dipoles to the light results in maximal narrowing of the hyperfine transitions profile, while in the second, the contribution of the slow atoms is concealed by destructive interference, leaving a Doppler-broadened absorption spectrum only.

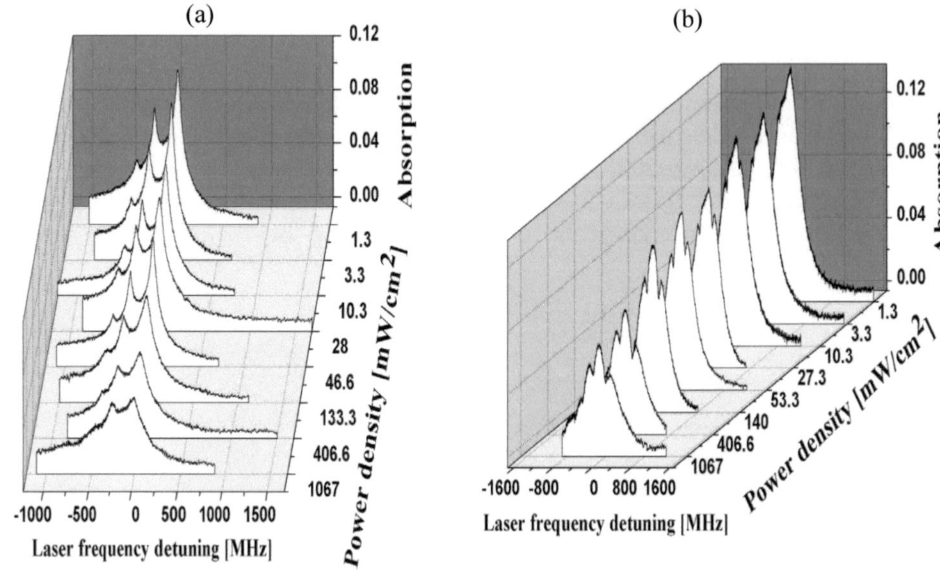

Fig. 8. Absorption spectra of the $F_g = 4$ set of hyperfine transitions obtained for different laser intensities at L = λ/2 (a) and L = λ (b). The atomic source temperature is T=119°C. ECDL is used with spectral width of 3 MHz.

3.1.3 Absorption spectra of atoms confined in ETC with L = mλ (m = 0.5, 1, 1.5, 2, 2.5, 3)

In this section, the discussion is expanded to ETCs with larger thickness - up to several microns. As discussed above the Dicke narrowing vanishes for L = λ. However further enlargement of L can result in the Dicke narrowing revival. Collapse and revival of the coherent Dicke narrowing of atomic transition absorption profiles were demonstrated in Ref. [Dutier, 2003, a; Sarkisyan, 2004], revealing the transition width quasi periodicity as a function of optical cell thickness L, with minima at L = (2n+1) λ/2 (n – integer; λ - light wavelength).

The Cesium-vapor-layer transmission spectrum is measured for different cell thicknesses and low and high intensities of the light [Cartaleva, 2009]. In Fig.9, the transmission spectrum of the $6S_{1/2}(F_g = 4) \rightarrow 6P_{3/2}(F_e = 3, 4, 5)$ set of hyperfine transitions is presented, for the ETC thickness L = mλ (m = 0.5, 1, 1.5, 2, 2.5, 3).

For low light intensity (0.2 mW/cm^2, where the saturation of optical transitions can be neglected) and m = 0.5, one can see (Fig.9a) that all hyperfine transitions are well resolved and the Doppler broadening of the transmission (absorption) profiles is very small. Keep in mind that the Doppler width of the hyperfine transitions of Cs in conventional cells at room temperature is about 400 MHz. In the L = 0.5 λ case, the absorption at the resonance frequency is strongly enhanced due to the existence of two similar in their result but different in their nature processes. As above discussed, the first process is connected to the anisotropy of atom-light interaction. As in our experiment the laser beam diameter (about 1 mm) largely exceeds the cell thickness, the time interaction between an atom and the laser radiation is different depending on the atomic velocity direction. For the ETC, it is assumed that atoms lose their excitation when hitting the cell wall. The second process arises as a consequence of the coherent nature of the atomic ensemble emission, which leads to relative enhancement of absorption at the optical transition center compared to its wings.

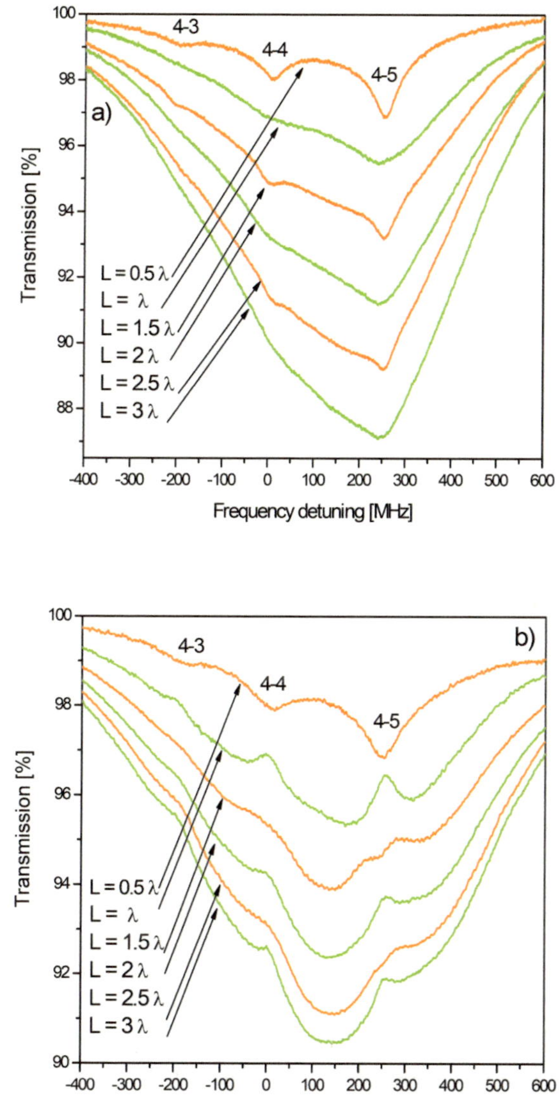

Fig. 9. Transmission spectra of the $F_g = 4$ set of transitions, for low (a, 0.2 mW/cm^2) and high (b, 20 mW/cm^2) light intensities. Monomode laser used is with spectral width of 15 MHz.

The quasi periodic nature of the second process - periodical revival of Dicke narrowing (its relative contrast decreases with respect to the broad pedestal, which increases proportionally to L) with the ETC thickness enlargement - is well recognizable in Fig.9a (for m = 0.5, 1.5, 2.5; orange traces). While the Dicke coherent narrowing has it maximum at m = 0.5, for the first Dicke revival case (m = 1.5), the well pronounced narrow, enhanced absorption peaks are superimposed on a large pedestal. Simple calculation shows that the revival of narrowing should be observable up to thickness $L_{max} = ut$, where u is the mean thermal velocity of Cs atoms (~250 m/s), and t is the excited state lifetime (~30 ns), that is L_{max} ~ 7.5 μm [Briaudeau, 1998].

At m = 1, 2, 3 (green traces) however, the coherent Dicke narrowing vanishes, and because of that these ETC thicknesses are determined as positions of Dicke effect collapsing. Hence, due to the Doppler broadening the optical transitions are not resolved even at low light intensity [Briaudeau, 1996].

Increasing the irradiation power (Fig. 9b, m = 0.5), some broadening of the sub-Doppler absorption profiles is observed due to saturation. At m = 1, 2, 3 (green traces), well pronounced narrow peaks in transmission (reduced absorption dips) are observed at the hyperfine transition centers. The amplitude and contrast of the narrow transmission peaks reduce with the ETC thickness enlargement. It has been shown that the transmission peak amplitude increases with increasing laser power density [Andreeva, 2007, b; Sarkisyan, 2004].

For m = 1.5, 2.5 and for the completely closed $F_g = 4 \rightarrow F_e = 5$ transition, a small absorption peak (i.e. dip in the transmission) appears superimposed on the velocity selective absorption dip. Such a narrow absorption peak (Dicke signal) does not appear for the open transitions under the same experimental conditions. This difference can be related to the fact that for the open transitions part of the atoms is transferred to the $F_g = 3$ level and this part consists mainly of slow atoms. Thus, slow atoms are lost for the absorption and they can not participate in the formation of the coherent signal. One can conclude that for open transitions, mainly fast atoms contribute to the Dicke effect. In opposite, for the completely closed transition the slow atoms give enhanced contribution to Dicke signal, resulting in increased absorption in a narrow spectral region, centered at the optical transition. In support of this statement one can notice that due to the contribution of Dicke effect to the $F_g = 4 \rightarrow F_e = 4$ transition absorption (Fig.9b, m = 1.5), the amplitude of the velocity selective absorption dip is sufficiently less and its width is larger than that for the same transition but at m = 1, 2. Nevertheless, this contribution is not enough for narrow absorption peak formation, as it happens for the $F_g = 4 \rightarrow F_e = 5$ transition (Fig.9b, m = 1.5) under the same experimental conditions.

To analyze in more precise way the Dicke effect for open transitions, the transmission spectra are presented of the $6S_{1/2}(F_g = 3) \rightarrow 6P_{3/2}(F_e = 2, 3, 4)$ set of transitions (Fig.10), which contains only open transitions, i.e. atomic transitions suffering population loss to level non-interacting with the laser field. The population loss is caused by hyperfine and/or Zeeman optical pumping processes, based on the spontaneous decay from the excited level. Moreover, a very recently developed laser system is used with DFB laser source and extremely low noise current controller. This system allowed us to achieve strong improvement of the spectral resolution and signal-to-noise ratio.

From Fig.10 it can be seen that narrow and well pronounced enhanced transmission peaks can be observed at much lower than in Fig.9b light intensity. This makes possible simultaneous observation of Dicke enhanced absorption narrow features at m = 1.5 (first Dicke revival) and enhanced transmission peaks for m = 1, 2, 2.5, 3. Note also the Dicke effect contribution to some broadening and amplitude reduction of resonances at m = 2.5.

Consequently, based on the transmission spectrum of ETC with thickness of few microns, a frequency reference can be developed with sub-Doppler-width precision. It should be stressed that the transmission spectrum is measured by extremely simple and robust, single beam optical system. As the narrow resonances are centered at the hyperfine transitions, the potential accuracy of the frequency reference is high.

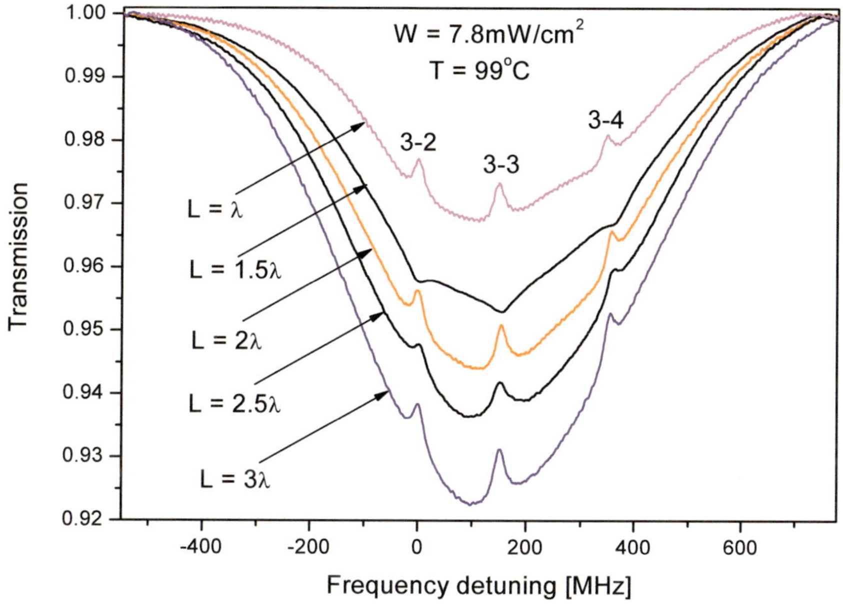

Fig. 10. Transmission spectra of the $F_g = 3$ set of transitions, for five different ETC thicknesses and W = 7.8 mW/cm². The low intensity noise DFB laser system is used with spectral width of 2 MHz.

As the contrast of the enhanced transmission peaks (Fig.10) is largest at L = λ, a measurement is performed of the broadening by the light intensity of the enhanced transmission peaks observed in the transmitted spectrum, for three light intensities (Fig.11). It can be seen that the resonance broadening with light intensity is not fast, and very good signal to noise ratio can be provided at resonance width of about 10 MHz. Note that the measured width is about 40 times less than the Doppler width of the hyperfine transition and close to the natural width of the transition (6 MHz). It is worthy to point out that the lifetime of the excited state for atoms flying close to parallel to laser beam is significantly reduced due to the small ETC thickness, determining about 50 MHz homogeneous width of atomic transition. However, based on "slow" atoms contribution much more narrow experimental structure is observed.

Recently developed laser system with low intensity noise in the light emission and narrow spectral width allowed us to study in more detail the difference between the saturation of open and closed transitions under conditions of very well expressed Dicke revival, namely at L = 1.5 λ. The experimentally obtained spectra are shown in Fig.12, for larger interval of used light intensities than that presented in Fig.9. For an intensity of 1.8 mW/cm², one can distinguish all three hyperfine transitions due to the small Dicke dip in transmission, centered at each atomic transition (Fig.12). With the light intensity enhancement, a reduction of the amplitude of the narrow Dicke features occurs for both open $F_g = 4 \rightarrow F_e = 3, 4$ transitions. Further on saturation dips around the atomic transition centers start to appear. These dips grow in amplitude and their width with the light intensity.

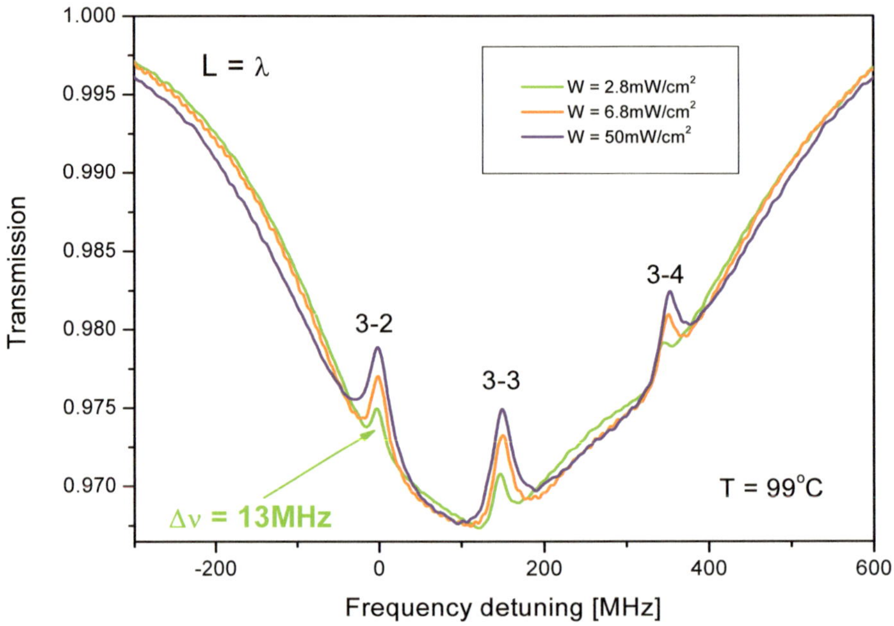

Fig. 11. Transmission spectra at three light intensities, for the $F_g = 3$ set of transitions and L = λ. DFB laser system is used with spectral width of 2 MHz.

However, for the closed $F_g = 4 \rightarrow F_e = 5$ transition, the saturation dip is not the only sub-Doppler feature that is recognizable in the absorption spectrum. In the position of the transition center, one can see remaining even at highest used intensity the Dicke absorption peak superimposed on the saturation dip. This peak is observable at all intensities, within the intensity range explored in the experiment. As above discussed, the narrow absorption peak originates from the coherent atom light interaction, and it is associated with the Dicke narrowing. Increasing the light intensity, it can be seen that the Dicke signal does not change notably its amplitude even it falls in the saturation dip of the transition, which grows in amplitude. Note also that the width of Dicke enhanced absorption resonance does not change significantly with light intensity.

When comparing the open and closed (in terms of optical pumping) transitions, one can see that the saturation peak of enhanced transmission appears simultaneously for both types of

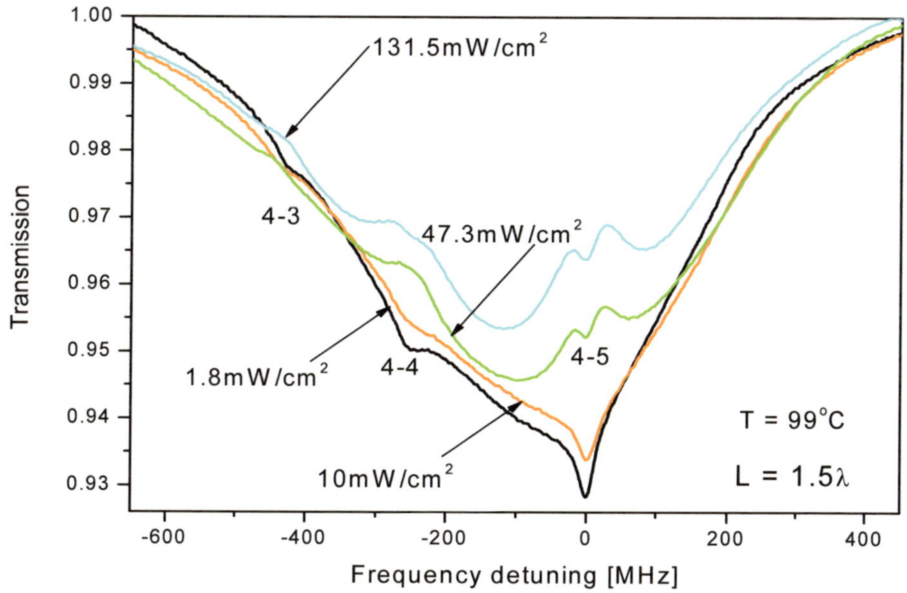

Fig. 12. Transmission spectra of the $F_g = 4$ set of transitions, different saturation behavior of open and closed atomic transitions under condition of Dicke effect revival. DFB laser system is used with spectral width of 2 MHz.

transitions. The behavior of the peak amplitude and width dependences on the light intensity is also similar. However, for higher light intensity, the narrow Dicke signal is not pronounced for the open transitions, while the narrow Dicke dip in transmission is always present, in the closed transition case. The reason for such difference could be found in the transit time effects of atom light interaction, intrinsic to the nature of the ETC, combined with the optical pumping (in three-level system) or with the saturation effects in two-level system. It is well known [Dutier, 2003, a; Sarkisyan, 2004; Briaudeau, 1998] that the Dicke narrowing results in the appearance of a narrow Dicke signal over Doppler broaden pedestal, where Dicke signal originates from the transit times effects. The significant contribution to the narrow Dicke signal comes from the slow atoms. However, namely the slow atoms suffer the highest loss due to the optical pumping to the ground-state level non-interacting with the light, which occurs in case of the open transitions. Hence at the open transitions, the Dicke signal is missing, for higher light intensities.

3.2 Sub-Doppler resonances in the fluorescence – Experimental results

3.2.1 Monotonic broadening of sub-Doppler fluorescence profile with ETC thickness enlargement

The investigation of the absorption and fluorescence of such confined media has started in the weak-intensity regime [Dutier, 2003,b], and the work was concentrated mainly on the investigation of the absorption spectra, considering that the fluorescence profiles exhibit only monotonic broadening with the cell thickness [Sarkisyan, 2004].

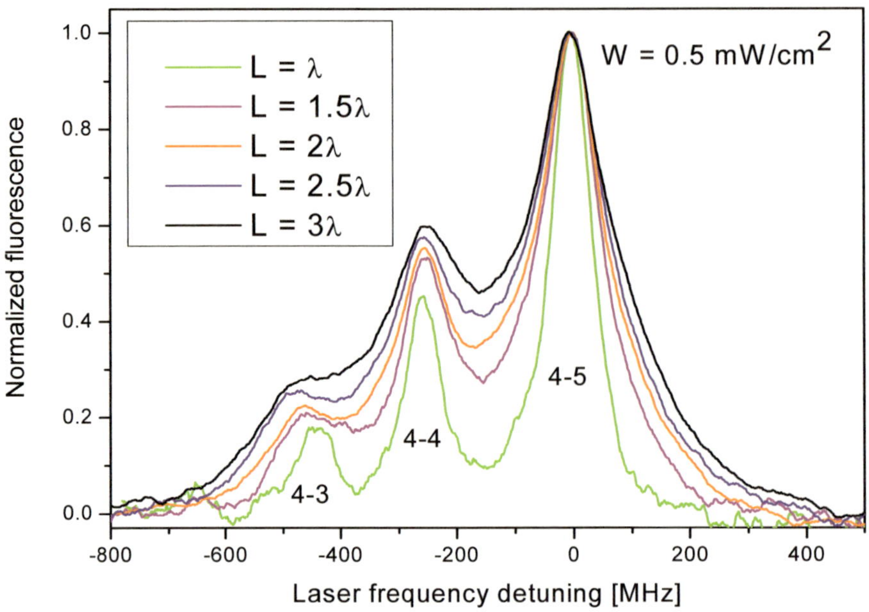

Fig. 13. Normalized fluorescence spectra of the $F_g = 4$ set of transitions, at five different ETC thicknesses. The fluorescence profile broadening with L is clearly seen.

For low light intensity, our investigations confirm the monotonic broadening of the fluorescence profiles when increasing the ETC thickness. In Fig.13, the normalized fluorescence spectra of the $6S_{1/2}(F_g = 4) \rightarrow 6P_{3/2}(F_e = 3,\ 4, 5)$ transitions are presented, for low irradiating power and $L = m\lambda$ with m = 1, 1.5, 2, 2.5, 3. Due to the small ETC thickness (for m = 1), atoms with velocity normal to the cell windows give very small contribution to the fluorescence signal compared to atoms flying parallel to the ETC windows, which have much larger time of interaction with the laser beam. Confirming previous investigations [Sarkisyan, 2001; Andreeva, 2007, b; Sarkisyan, 2004] we present here very well resolved fluorescence profiles of the three hyperfine transitions, for $L = \lambda$. The reason is that the fluorescence signal comes mainly from "slow" atoms, whose velocity projection on the light beam is small enough to allow time for absorption of a photon and subsequent spontaneous emission before collisions with the cell wall. With the ETC thickness enhancement broader velocity class atoms have enough time to complete the fluorescence emission, which leads to monotonic rising of the fluorescence profile width. In this way, the resolution of the hyperfine transitions at $L = 3\lambda$ is much worse than that for $L = \lambda$.

3.2.2 Appearance of a narrow structure in the sub-Doppler fluorescence profiles: Different behavior of open and closed transitions

Further study of the ETC fluorescence spectra at $L = \lambda$ has demonstrated [Andreeva, 2007, b] that tiny saturation dip of reduced fluorescence appears in the narrow fluorescence profiles when irradiating the ETC with higher-intensity laser light. These saturation dips appear for all open, in terms of hyperfine and Zeeman optical pumping, hyperfine transitions of Cs D_2 line. No saturation dip was reported there in the fluorescence of the $F_g = 4 \rightarrow F_e = 5$

transition, which is the only completely closed transition on the D_2 line of Cs. As for ETC with L = λ the fluorescence dips are of extremely small amplitude (slightly higher than the experimental noise), in Ref. [Andreeva, 2007, b; Varzhapetyan] their examination has been performed by phase sensitive registration. Later on, it has been shown that with the enhancement of ETC thickness the reduced fluorescence dips increase their amplitudes [Cartaleva, 2009]. However, there the spectral width of the emission of the used diode laser system was about 15 MHz, which was a reason for significant broadening of the dips observed in the fluorescence and strong reduction of their amplitude. Due to this, we present here our recent experimental results, obtained by means of the narrow band and low noise laser system. The significant difference between the behavior of the open and closed transitions is illustrated in Fig.14, for ETC with thickness L = 1.5λ. It can be seen that at both $F_g = 4 \rightarrow F_e = 3, 4$ open hyperfine transitions good-amplitude and narrow reduced fluorescence dips occur.

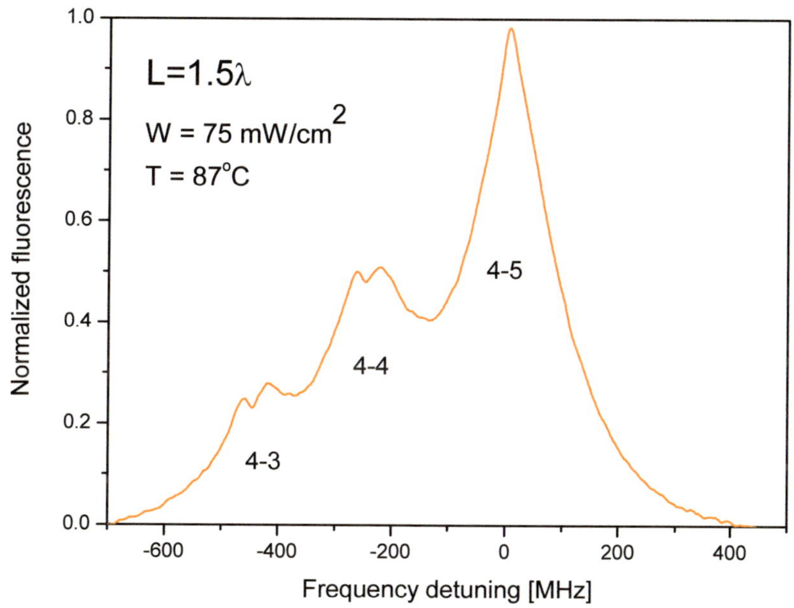

Fig. 14. Fluorescence spectrum, observed in ETC with thicknesses L = 1.5λ, at the $F_g = 4$ set of transitions. The 4-5 transition is the closed one, while the 4-4 and 4-3 transitions are open transitions.

Similar to the absorption at lower light intensity, the fluorescence on the open transitions suffers loss from the velocity selective optical pumping, which results in narrow reduced-fluorescence dips appearing at the centre of the transitions. Thus, the ETC fluorescence profiles, which are intrinsically narrower than the absorption profiles, exhibit velocity selective dips for higher light intensity values.

The situation is different for the completely closed $F_g = 4 \rightarrow F_e = 5$ transition, namely no dip occurs at the atomic transition centre (Fig.14). In opposite, the top of the highest probability optical transition is very sharp, and one can assume that the fluorescence profile of the closed transition involves a small narrow peak at its centre.

The processes leading to the dip and the peak formations can be illustrated as follows.

For the open transitions (see as an example the $F_g = 3 \rightarrow F_e = 2$ transition, Fig.15a), both two-level-system saturation and three-level-system optical pumping deplete the atomic population of the coupled by the light ground Zeeman sublevels. The result of this velocity selective depleting is the narrow, reduced fluorescence dip centered at the open transition.

The situation is different for the completely closed $F_g = 4 \rightarrow F_e = 5$ transition (Fig.15b). Here, the two-level-system saturation process decreases the ground Zeeman sublevel populations, while the fluorescence increases the sublevel population in a narrow frequency interval. Moreover, the fluorescence is enhanced by the fact that the light accumulates atomic population on Zeeman sublevels with the largest absorption probability [Andreeva, 2002] i.e. a large portion of the redistributed by the light atoms stays in the most absorbing magnetic sub-level of $F_g = 4$ level. Thus, the summary contributions of both processes result in a small narrow peak formation.

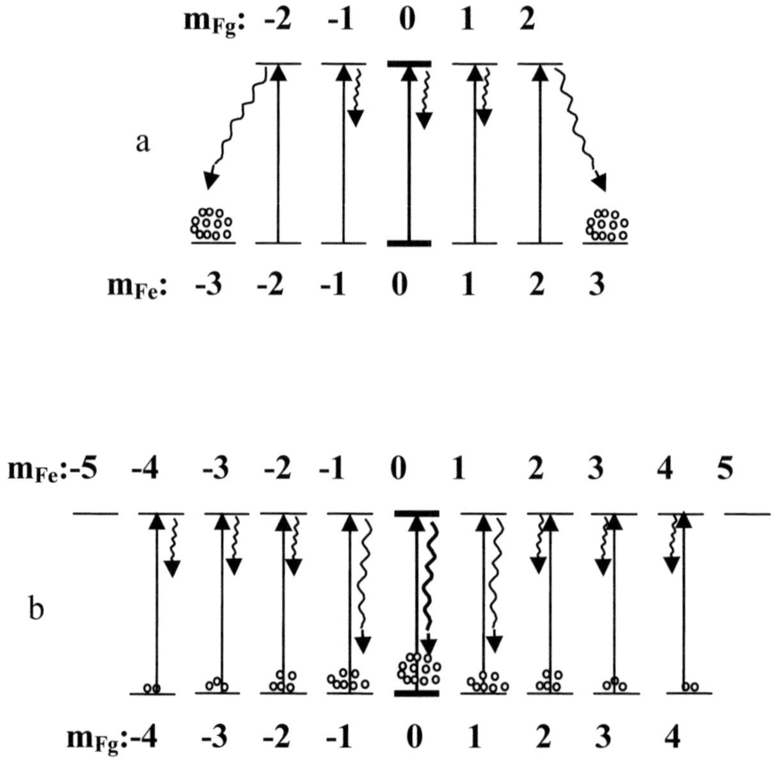

Fig. 15. Illustration of saturation and optical pumping for the open in terms of Zeeman optical pumping $F_g = 3 \rightarrow F_e = 2$ transition (a) and the completely closed $F_g = 4 \rightarrow F_e = 5$ transition (b).

3.2.3 The behavior of the narrow structure in the fluorescence profiles with the ETC thickness enlargement

In Ref. [Cartaleva, 2009], the examination of the saturated regime in the fluorescence was expanded to ETC thickness up to L = 3λ. It has been shown that the amplitude of the narrow dip in the fluorescence profile of the open transitions increases with the ETC thickness. Starting from L = 2λ, the fluorescence profile of the completely closed transition exhibits small plateau, centered at the optical transition.

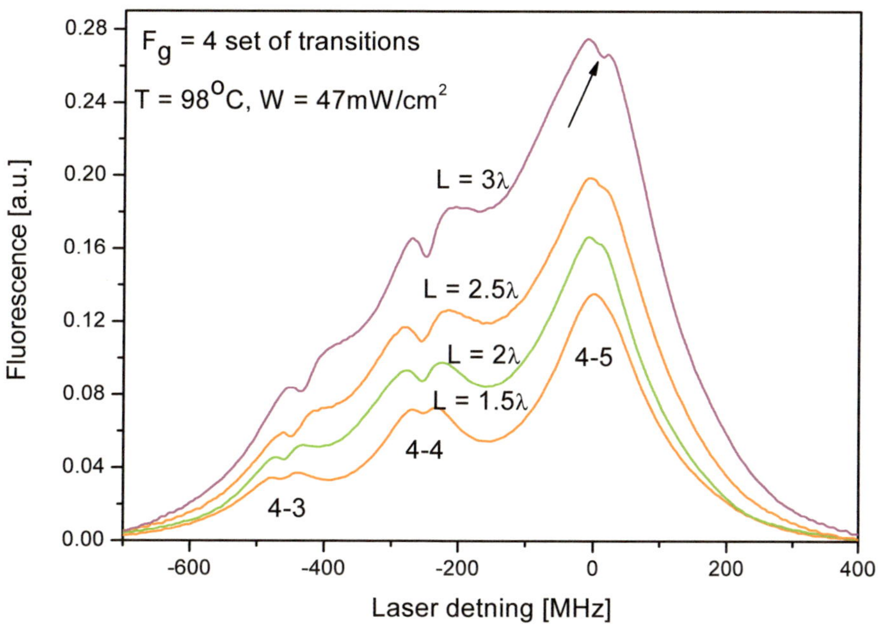

Fig. 16. Fluorescence spectra at the $F_g = 4$ set of transitions, observed in ETC with thickness L = 1.5λ, 2λ, 2.5λ, 3λ. The ETC temperature T = 98°C and the light intensity W = 47 mW/cm². The new dip in the profile of the $F_g = 4 \rightarrow F_e = 5$ transition is marked by an arrow.

Very recently, the first observation of a narrow, reduced fluorescence feature also in the profile of the completely closed $F_g = 4 \rightarrow F_e = 5$ transition has been reported [Cartaleva, 2011], obtained by means of the developed narrow-band laser system. To illustrate the new structure, the ETC fluorescence spectra for the $F_g = 4$ set of transitions are presented, at cell thickness L = 1.5λ, 2λ, 2.5λ, 3λ (Fig.16). The two open $F_g = 4 \rightarrow F_e = 3$, 4 transitions show reduced fluorescence dips at the central frequency of each fluorescence profile, for all examined ETC thicknesses. The dip amplitudes increase noticeably with the cell thickness. This result differs from that reported in Ref.[Cartaleva, 2009], where the dip formation starts from L = 2λ, for the open transitions. We attribute this difference to the fact that in the present work the spectral width of the laser line is significantly narrower (2 MHz) than that used in [Cartaleva, 2009] (15 MHz). Thus, the observation of the narrow dip in the fluorescence depends very critically on the laser line width, which was not as critical for the transmission spectra. The amplitude of the dips increases with the cell thickness, due to the contribution of larger number of atoms to its formation. At larger cell thickness, the time of flight of the atoms between the cell walls increases, which broadens the atomic velocity interval of light-emitting atoms. It can be seen that the emitting atom number enlargement with L makes higher contribution to the dip amplitude than to its width.

The used in the present experiment DFB laser system allows observation of a new dip in the profile of the fluorescence of the completely closed $F_g = 4 \rightarrow F_e = 5$ transition (Fig.16, L = 3λ). With a similar set of the ETC thicknesses [Cartaleva, 2009] only a small plateau has been observed at the $F_g = 4 \rightarrow F_e = 5$ transition up to L = 3λ. In Fig.16, similar plateau can be seen for L = 2λ, 2.5λ. In Ref. [Cartaleva, 2009], the formation of the plateau for the completely closed transition instead the dip for open transitions has been explained by the saturation and optical pumping processes. For the open transitions, both processes deplete the atomic population of ground Zeeman sub-levels coupled by the light. However, in case of the $F_g = 4 \rightarrow F_e = 5$ transition, while the two-level-system saturation process decreases the ground Zeeman sublevel populations, the fluorescence increases the sub-level population in a narrow frequency interval. The summary contributions of both processes can result in the small plateau formation (see Fig.16, L = 2λ, 2.5λ).

Concerning the new dip formation [Cartaleva, 2011], the assumed physical processes behind this are based on the degeneracy of the two-level system. As has been shown in Ref.[Andreeva, 2002], if the $F_g = 4 \rightarrow F_e = 5$ transition is excited by linearly polarized light (as in our experiment), Cs atoms accumulate on the $F_g = 4$ Zeeman sublevels with the highest probability of excitation, i.e. the highest fluorescence is expected for slow atoms [Cartaleva, 2009]. However, in the case of depolarization of the excited level, a significant portion of slow atoms will be accumulated on the $F_g = 4$ Zeeman sublevels with the lowest probability of excitation [Andreeva, 2002]. Hence if the two-level system is a degenerate one, the excited state depolarization will transform the completely closed system to one with effective loss in the excitation process. The reducing of the optical transition excitation rate will be the most significant for slow atoms, resulting in narrow dip formation in the fluorescence profile. The assumed depolarization of the excited level can be caused by the collisions between Cs atoms. Another reason for the excited state depolarization could be the influence of the cell window, as has been suggested in [Andreeva, 2007, a].

3.2.4 Frequency reference based on the ETC fluorescence spectrum

While the study of the difference between the open and closed transitions is important from basic point of view, the set of the $6S_{1/2}(F_g = 3) \rightarrow 6P_{3/2}(F_e = 2, 3, 4)$ transitions is very interesting for the development of practical frequency reference. This group consists of open transitions only. Hence, for each hyperfine transition, well pronounced reduced fluorescence dip can be formed, which is seen from Fig.17. While at L = λ the amplitude of dip is very small, it grows up significantly at L = 3λ. As the dips are observed at the centres of the optical hyperfine transitions, they can be used as frequency reference in the field of laser spectroscopy and for development of highly stabilized laser systems.

Our theoretical and experimental investigations [Vaseva] have shown that further enlargement of the ETC thickness up to L = 6λ results in even better parameters (contrast and spectral width) of the narrow dip centered in the fluorescence profile. The enhancement of the ETC thickness up to L = 6λ not only provides the possibility for strong reduction of the used light intensity but also allows working at much lower atomic source temperature, which is of significant importance for operating of practical frequency reference devices. In

Fig.18, the ETC transmission and fluorescence spectra are presented for reduced atomic source temperature down to 49°C. An additional advantage of the ETC with L = 6λ is that it exhibits even lower width of the dips observed in the fluorescence spectrum than that of the peaks in the transmission spectrum (Fig.18).

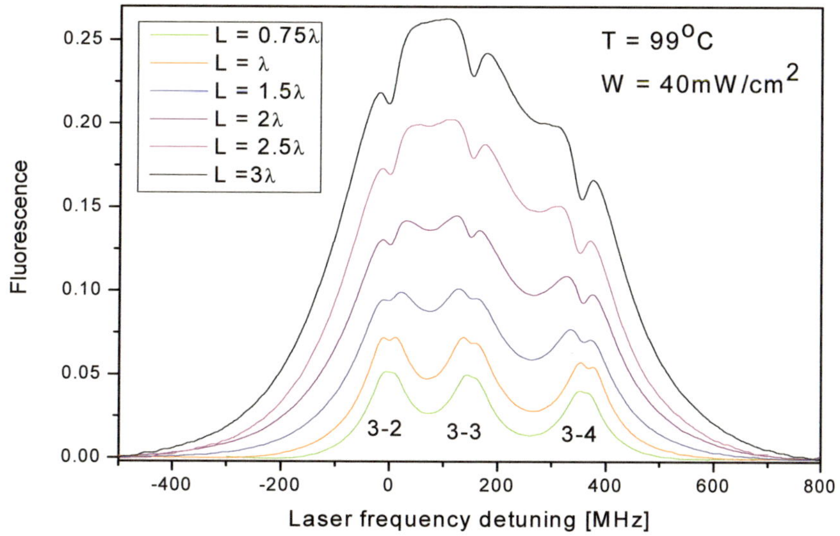

Fig. 17. Fluorescence spectra at the F_g = 3 set of transitions, observed in ETC with thickness L = 0.75λ, λ, 1.5λ, 2λ, 2.5λ, 3λ. The ETC temperature T = 99°C and the light intensity W = 40 mW/cm^2.

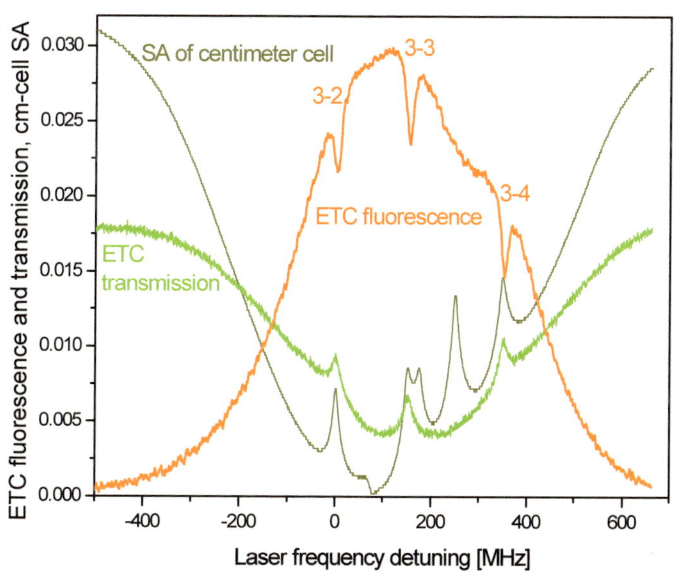

Fig. 18. Experimental fluorescence and transmission spectra (W = 20 mW/cm^2), for ETC with L = 6λ. Cs source temperature T = 49°C. SA: Saturated absorption spectrum for conventional cell with L = 2.5 cm, at room temperature.

For comparison with the ETC spectra, in Fig.18 a saturated absorption spectrum is shown, measured by the conventional two-counter-propagating-beam technique (see Fig.4). It is clearly seen that the saturated absorption spectrum observed in the conventional cell is more complicated than the three-narrow-resonance spectra in the ETC transmission and fluorescence. Note that even at enhanced thickness of $L = 6\lambda$, the anisotropy in the time of atom-light interaction still suppress the formation of crossover resonances, in the case of ETC. It has been shown [Sargsyan, 2008, b] that weak crossover resonance formation by single beam irradiation of micrometric cell is possible due to the reflected by the second window laser beam.

4. Theoretical simulation of absorption (transmission) and fluorescence spectra in ETC

The theoretical model in use is based on the Optical Bloch Equations for two-level system (closed and open), presented in Ref. [Andreeva, 2007, b]. For the open system, atomic population losses are introduced by means of a spontaneous emission to a "third level" non-excited by the light. No atomic population losses are assumed for the closed two-level system. The atomic system consists of two levels denoted as 1 and 2, which are coupled by a laser light with frequency ω detuned by Δ from the resonant transition frequency ω_{21} ($\Delta = \omega - \omega_{21}$).

The control parameters are chosen to be applicable to a realistic alkali system, namely: γ_{21} = 5MHz, ku = 250MHz, and $\alpha = 1$ (closed system) or $\alpha = 0.6$ (open system). Here γ_{21} is half the transition rate from the excited state, ku is the most probable Doppler shift and α is the probability of decay from level 2 to level 1. The Rabi frequency of the optical transition is denoted by Ω_R.

The used system of Bloch equations is the following:

$$v\frac{d\sigma_{21}}{dz} + D_{21}\sigma_{21} - i\frac{\Omega_R}{2}(\sigma_{11} - \sigma_{22}) = 0 \quad (1)$$

$$v\frac{d\sigma_{22}}{dz} + \gamma_2\sigma_{22} - \Omega_R \operatorname{Im}\sigma_{21} = 0 \quad (2)$$

$$v\frac{d\sigma_{11}}{dz} - \alpha\gamma_2\sigma_{22} + \Omega_R \operatorname{Im}\sigma_{21} = 0, \quad (3)$$

where $D_{21} = \gamma_{21} + ikv - i\Delta$, v is the atomic velocity and σ_{ij} are the reduced density matrix elements in the rotating frame.

Atomic absorption is given by:

$$A = \int_0^\infty G(v)\exp\left[-\left(\frac{kv}{ku}\right)^2\right]dv, \text{ where } G(v) = \int_0^L \operatorname{Im}\left[\sigma_{21}(z,v)\right]dz \quad (4)$$

The fluorescence is proportional to the quantity U expressed as:

$$U = \int_0^\infty Q(v)\exp\left[-\left(\frac{kv}{ku}\right)^2\right]dv, \text{ where } Q(v) = \int_0^L \left[\sigma_{22}(z,v)\right]dz\,. \quad (5)$$

4.1 Simulated transmission (absorption) spectra for Cs atoms confined in ETC with $L = m\lambda$ ($m = 0.5, 1, 1.5, 2, 2.5, 3$)

Based on the briefly presented model, theoretical simulation of transmission spectra concerning open and closed atomic transitions is performed for two important cases - low and high light intensity. At low light intensity (0.2 mW/cm^2), the calculated transmission (absorption) spectra of both open (Fig.19a) and closed (Fig.19b) transitions show collapse (for m = 1, 2, 3) and revival (for m = 1.5, 2.5) of the coherent Dicke narrowing: the narrow Dicke structure is superimposed on a broad pedestal. Note that at low light intensity, only minor difference can be found between the profiles of open and closed transitions. This can be attributed to the negligible contribution of the optical pumping effect. Comparing the experimental results presented in Fig.9a with the theoretical spectra shown in Fig.19a,b, the very good agreement can be pointed out, both for open and closed atomic transitions.

In the case of saturation regime, the behavior of the simulated closed transition differs significantly from that of the open transitions. At higher light intensity (20 mW/cm^2), in case of open transitions and $L > \lambda/2$, our calculation shows peak in the transmission (reduced absorption) formed at the transition profile center (Fig.19c). The contrast of the theoretical absorption dips in the open transitions increases with ETC thickness, which differs from the experimental spectra (Fig.9b), where the dip contrast is lower and reduces with cell thickness. It should be pointed out that while in the theoretical simulation the laser light is considered monochromatic, in the experimental realization presented in Fig.9b the laser emission is with spectral width of 15 MHz, which is a reason for the experimental dips broadening. However, even significantly narrower, the experimental spectra at the F_g=3 set of transitions (Fig.10) also do not demonstrate strong enhancement of the absorption dip amplitude with the cell thickness.

In agreement with the experimental results (Fig.9b; m = 1.5, 2.5), the modeling (Fig.19c; m = 1.5, 2.5) shows that in case of open transitions, the Dicke effect results only in some reduction of the absorption dip amplitude.

The closed transition behavior differs from that of the open transitions: together with the absorption dips (Fig.19d; m = 1, 2, 3) also narrow absorption peaks superimposed on broader dips are obtained (Fig.19d; m = 1.5, 2.5). These absorption peaks present the so called Dicke revival, which is well pronounced both in the experiment (see Fig.9b and Fig.12) and theory, in the saturated regime.

4.2 Simulated fluorescence spectra for Cs atoms confined in ETC with $L = m\lambda$ ($m = 0.5, 1, 1.5, 2, 2.5, 3$)

Simulated fluorescence spectra for open and closed atomic transitions are shown in Fig.20, for six values of the ETC thickness. Comparing Fig.19 and Fig.20 one can notice that the calculated fluorescence profiles are narrower than those in the transmitted light. Fluorescence spectra exhibit narrow features, centered at the transition profiles. For the open transitions and $L > \lambda/2$, a narrow dip in the fluorescence (Fig. 20a) appears, superimposed on the top of the sub-Doppler fluorescence profile. From the experimental fluorescence spectra presented in Fig.16 and Fig.17 one can conclude that the contrast of the experimentally observed dip in the fluorescence profile is lower than the theoretically obtained (Fig.20a). However, in the case of fluorescence, the contrast of both experimental and theoretical dips increases with ETC thickness.

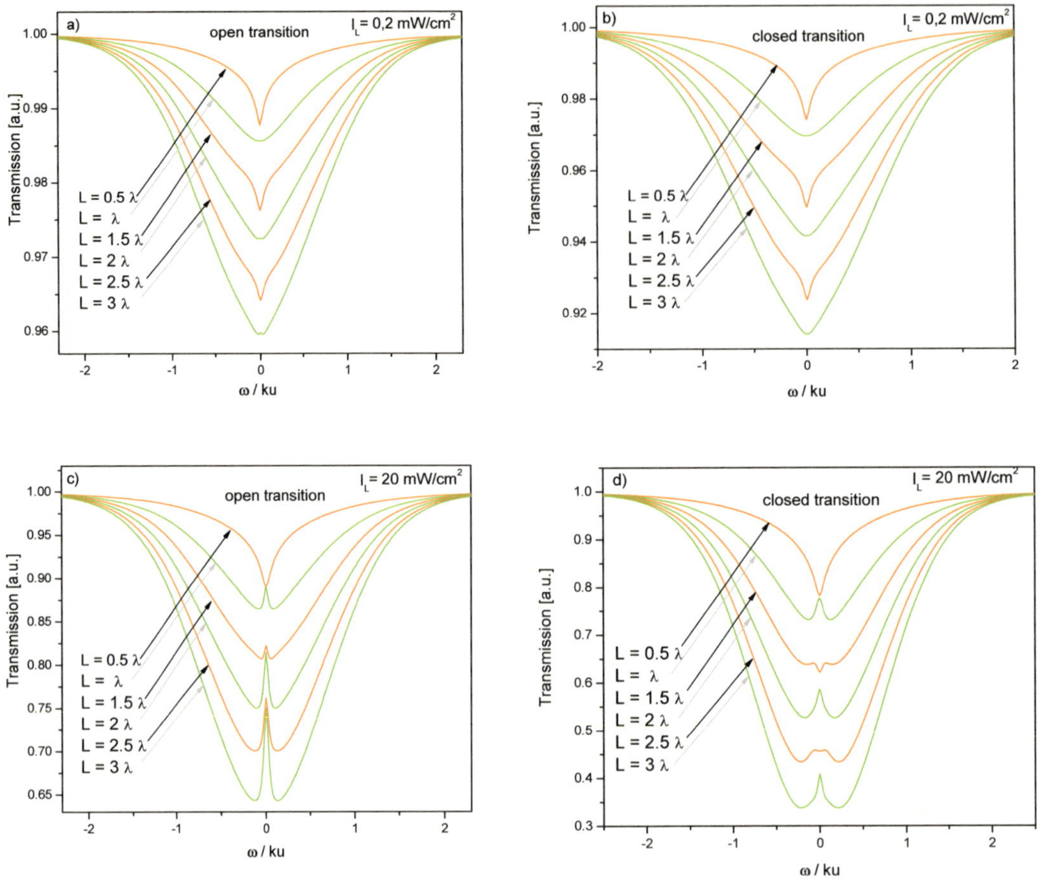

Fig. 19. Theoretical simulation of the transmission (absorption) spectra for the open (a,c) and closed (b,d) transitions performed at low (a,b) and high (c,d) laser light intensities.

For the closed transition (Fig.20b), no dip formation can be seen in the theoretical fluorescence profile, but we observe an interesting small feature at the transition center. The fluorescence profile modification with ETC thickness starts with very small plateau on the top of the profile for $m = 1$. For $m \geq 1.5$ however, a tiny narrow peak superimposed on the top of the broader fluorescence profile is clearly seen. The basic (for $L > \lambda/2$) and increasing with ETC thickness difference of saturation behavior between the open and closed transitions is evident (Fig.20).

This difference can be related to the ground level population redistribution caused by the fluorescence, which is emitted by the "slow" atoms in a very narrow spectral interval. For the open transitions, this fluorescence leads to depletion of the slowest atoms from the ground Zeeman sublevels coupled to the light (see Fig.15a). The situation is opposite in case of the closed transition (see Fig.15b). Here, due to the fluorescence the slowest atoms are coming back to the coupled to the light ground Zeeman sublevels. Thus, for the open transition dip in the fluorescence is observed, while for the closed one - a tiny peak.

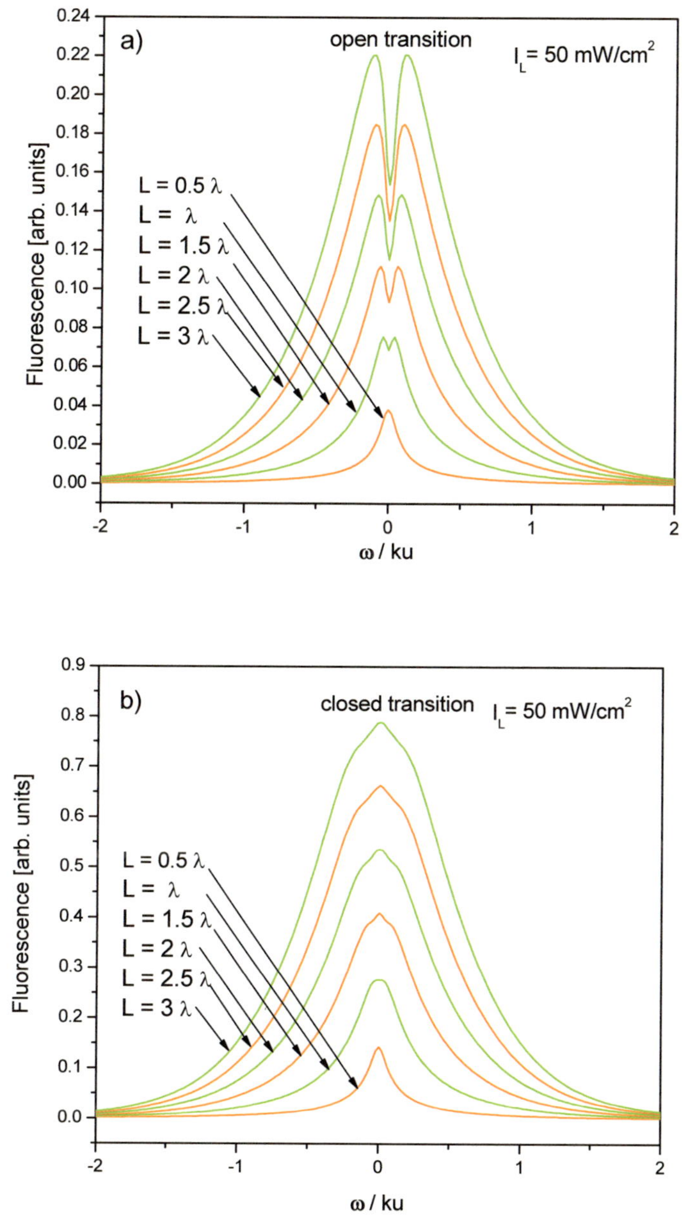

Fig. 20. Theoretical simulation of the fluorescence profile for open (a) and closed (b) transitions, at laser light intensity of 50 mW/cm^2.

The experimentally observed difference between open and closed hyperfine transitions is well illustrated by Fig.14. Here, the fluorescence spectrum is measured for L = 1.5λ and for relatively low temperature of the atomic source (87°C). Further on with the ETC thickness enlargement and Cs atom concentration enhancement (Fig.16, atomic source temperature 98°C and $L = 3\lambda$), new tiny dip appears also in the fluorescence profile of the closed $F_g = 4 \rightarrow F_e = 5$ transition. Experimental and theoretical study is in progress, in order to clarify the physical process behind this new feature.

4.3 Comparison of simulated absorption and fluorescence spectra for Cs atoms confined in ETC with L = mλ (m = 1, 2, 3, 4, 5, 6)

Theoretical analysis is performed to clarify the potential of narrow dips observed in the fluorescence and absorption of open transitions for development of frequency reference. The result of the theoretical simulation is presented in Fig.21. In Fig.21a, the absorption spectra are shown, for the open transition at ETC thicknesses $L = m\lambda$ ($m = 1, 2, 3, 4, 5, 6$) and light intensity of 20 mW/cm^2. Due to the population loss introduced by the hyperfine and Zeeman optical pumping, well pronounced narrow dip is observed, at the centre of the optical transition for each ETC thickness under consideration. The amplitude and the contrast of the dips are growing with the cell thickness. In case of the fluorescence (Fig.21b), one can see that the transition profiles are narrower compare to those in absorption. As discussed above, the reason is that as the laser beam diameter largely exceeds the cell thickness, the time interaction between an atom and the laser radiation is different depending on atomic velocity direction. In the narrower fluorescence profile also narrow dip occurs, exhibiting fast amplitude enhancement with the ETC thickness.

From the theoretical spectra presented in Fig.21, two important parameters were determined: the contrast and the Full Width at Half Maximum (FWHM) of the dip formed at the optical transition centre. The dependence of the dip contrast on the ETC thickness is presented in Fig.22a. It can be seen that the dip contrast observed in the fluorescence is generally lower but increases with L faster than that in the absorption profile. More specifically, the ratio R_C of the dip contrast in the fluorescence C_{Fl} to the dip contrast in the absorption C_{Abs}, $R_C = C_{Fl}/C_{Abs}$, increases 2.75 times with ETC thickness enhancement from L = 1000nm to L = 5000nm.

The FWHM of the dip also increases with the cell thickness (Fig.22b), but all the time the feature observed in the fluorescence is narrower than that in the absorption profile.

From the theoretical results the following conclusions can be made. A small enhancement of the ETC thickness up to L = 5000nm will result in about 10% rising of the contrast to FWHM ratio, in case of the fluorescence spectrum, while for the dip observed in the absorption spectrum, this ratio reduces by 30%. As in general this ratio mainly determines the merit of the frequency reference, there is a strong motivation to study the fluorescence spectra for ETCs with thickness of 5-7 light wavelengths.

Consequently, both narrow dips in the absorption and fluorescence can be used for development of frequency reference. The approach based on the absorption spectrum will have better parameters at ETC with $L = \lambda$, will use very low light intensity and will require atomic source heating to about 100°C. Using the fluorescence spectrum will allow two-time reduction of the source temperature and will relax to some extent the complexity of ETC building, requiring slightly higher light intensity.

Some possible applications such as magnetometer with submicron local spatial resolution and tunable atomic frequency references based on the ETC are described in Ref. [Sargsyan, 2008, a].

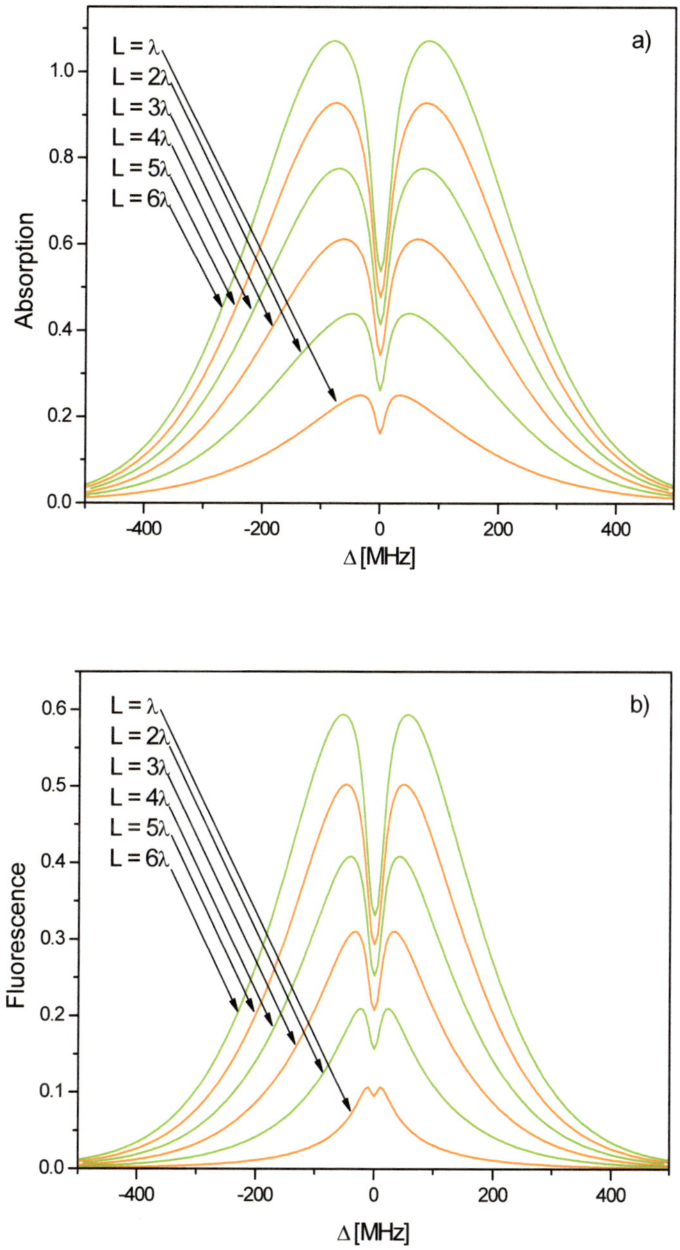

Fig. 21. Theoretical calculations of absorption (a) and fluorescence (b) spectra for open transition, and for L = mλ, with m = 1, 2, 3, 4, 5, 6. The light intensity is of 20 mW/cm^2.

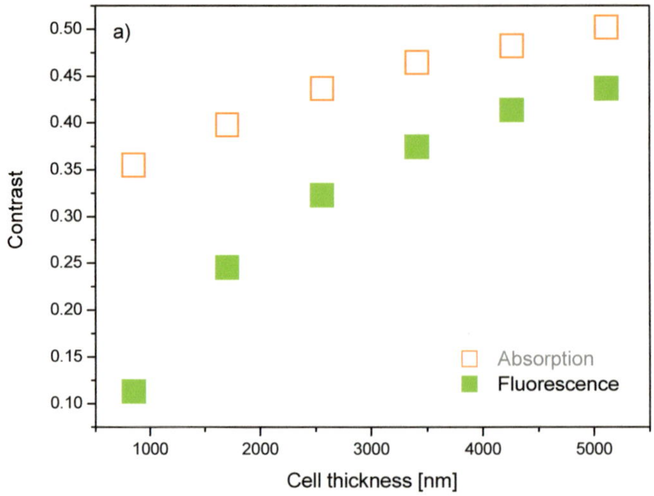

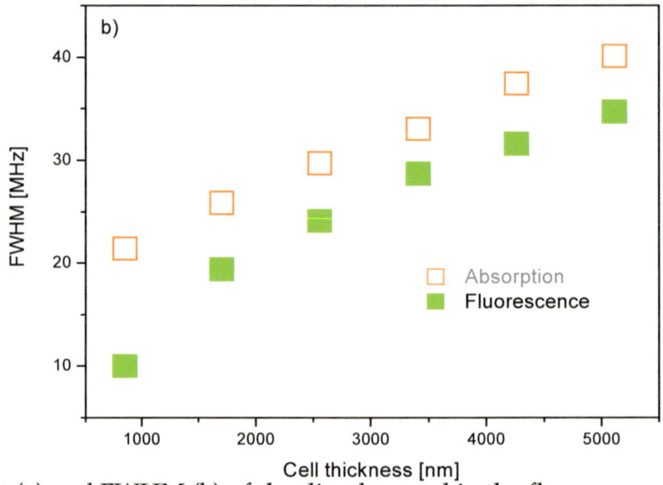

Fig. 22. Contrast (a) and FWHM (b) of the dip observed in the fluorescence and absorption.

5. Conclusion

The application of the nanometric cell provides a new opportunity for significant enhancement of the resolution in laser spectroscopy of thermal cell, without application of complex atomic beam or laser cooling systems. This simple approach makes it possible to study also the dynamic processes in the absorption and fluorescence. It is shown that when the Cs vapor layer thickness is about the wavelength of the irradiating light, the fluorescence spectrum is significantly narrower than the absorption one and the absorption coefficient does not grow linearly with the thickness of the atomic layer. The observed in the optical domain coherent Dicke narrowing of atomic transitions is strongest at $L = \lambda/2$, while further expanding of vapor layer thickness results in periodical collapse and revival of this coherent effect. Demonstrated and analyzed is the significant difference between the saturation of open and closed transitions under conditions of very well expressed Dicke revival (L = 1.5 λ). The appearance of a narrow structure in the sub-Doppler fluorescence profile is studied both experimentally and theoretically, distinguishing the difference between open and closed transitions. Theoretical simulations have predicted a significant enhancement of the amplitude of the narrow dip observed in the fluorescence profile with the ETC thickness enlargement. The prediction is confirmed experimentally, demonstrating that the narrow structure in the fluorescence at L = 6λ has a very good potential for development of frequency references.

6. Acknowledgement

This chapter is dedicated to the memory of our colleague and friend Prof. Solomon Saltiel, great teacher in linear and non-linear spectroscopy.

The work is partially supported by the Bulgarian Fund for Scientific Research (grant No: DO 02-108/22.05.2009 and grant DMU 02/17 - 18.12.2009) and Indian -Bulgarian (BIn-2/07) bilateral contract.

7. References

Andreeva, C., Atvars, A., Auzinsh, M., Bluss, K., Cartaleva, S., Petrov, L. & Slavov, D. (2007, a) Ground-state magneto-optical resonances in Cesium vapour confined in an extremely thin cell, *Physical Review A,* Vol.76, 063804.

Andreeva, C., Cartaleva, S., Dancheva, Y., Biancalana, V., Burchianti, A., Marinelli, C., Mariotti, E., Moi, L. & Nasyrov, K. (2002) Coherent spectroscopy of degenerate two-level systems in Cs, *Physical Review A,* Vol.66, 012502.

Andreeva, C., Cartaleva, S., Petrov, L., Saltiel, S.M., Sarkisyan, D., Varzhapetyan, T., Bloch, D. & Ducloy, M. (2007, b) Saturation effects in the sub-Doppler spectroscopy of Cesium vapor confined in an extremely thin cell, *Physical Review A,* Vol.76, 013837.

Briaudeau, S., Bloch, D. & Ducloy, M. (1996) Detection of slow atoms in laser spectroscopy of a thin vapor film, *Europhysics Letters,* Vol.35, pp. 337-342

Briaudeau, S., Saltiel, S., Nihenius, G., Bloch, D. & Ducloy, M. (1998) Coherent Doppler narrowing in a thin vapor cell: Observation of the Dicke regime in optical domain, *Physical Review A,* Vol.57, pp. R3169-R3172

Cartaleva, S., Saltiel, S., Sargsyan, A., Sarkisyan, D., Slavov, D., Todorov, P., & Vaseva, K. (2009) Sub-Doppler spectroscopy of cesium vapor layers with nanometric and micrometric thickness, *Journal of the Optical Society of America B: Optical Physics,* Vol.26, pp.1999-2006.

Cartaleva, S., Sargsyan, A., Sarkisyan, D., Slavov, D. & Vaseva, K. (2011) New narrow resonance in the fluorescence of closed optical transition observed in nanometric Cs - vapor layers, *Proceedings of SPIE - The International Society for Optical Engineering,* Vol.7747, 77470H

Dutier, G., Saltiel, S., Bloch D. & Ducloy M. (2003, a) Revisiting optical spectroscopy in a thin vapor cell: mixing of reflection and transmission as a Fabri-Perot microcavity effect, *Journal of the Optical Society of America B: Optical Physics,* Vol.20, 793-800.

Dutier, G., Yarovitski, A., Saltiel, S., Papoyan, A., Sarkisyan, D., Bloch, D. & Ducloy, M. (2003, b) Collapse and revival of a Dicke-type coherent narrowing in a sub-micron thick vapor cell transmission spectroscopy, *Europhysics Letters*, Vol.63, pp.35-41.

Maurin, I., Todorov, P., Hamdi, I., Yarovitski, A., Dutier, G., Sarkisyan, D., Saltiel, S., Gorza, M.-P., Fichet, M., Bloch, D. & Ducloy, M. (2005) Probing an atomic gas confined in a nanocell, *Journal of Physics: Conference Series*, Vol.19, pp.20-29.

Romer, R. & Dicke, R. (1955)New technique for high-resolution microwave spectroscopy, *Physical Review,* Vol.99, pp.532-536.

Sargsyan, A., Hakhumyan, G., Papoyan, A., Sarkisyan, D., Atvars, A., & Auzinsh, M. (2008, a) A novel approach to quantitative spectroscopy of atoms in a magnetic field and applications based on an atomic vapor cell with L=λ, Applied Physics Letters, Vol.93, 021119

Sargsyan, A., Sarkisyan, D., Papoyan, A., Pashayan-Leroy, Y., Moroshkin, P., Weis, A., Khanbekyan, A., Mariotti, E. & Moi, L. (2008, b) Saturated absorption spectroscopy: elimination of crossover resonances by use of a nanocell, *Laser Physics*, Vol.18, 749-755.

Sarkisyan, D., Bloch, D., Papoyan, A. & Ducloy, M. (2001) Sub-Doppler spectroscopy by sub micron thin Cs vapour layer, *Optics Communications,* Vol. 200, pp.201-208.

Sarkisyan, D., Varzhapetyan, T., Sarkisyan, A., Malakyan, Yu., Papoyan, A., Lezama, A., Bloch, D. & Ducloy, M. (2004) Spectroscopy in an extremely thin cell: Comparing the cell-length dependence in fluorescence and in absorption techniques, *Physical Review A,* Vol.69, 065802.

Todorov, P., Vaseva, K., Cartaleva, S., Slavov, D., Maurin, I. & Saltiel, S. (2008) Absorption and fluorescence in saturation regime of Cs-vapour layer with thickness close to the light wavelength, *Proceedings of SPIE - The International Society for Optical Engineering,* Vol.7027, 7027R

Varzhapetyan, T., Sarkisyan, D., Petrov, L., Andreeva, C., Slavov, D., Saltiel, S., Markovski, A., Todorov, G. & Cartaleva, S. (2005) Sub-Doppler spectroscopy and coherence resonances in submicron Cs vapour layer, *Proceedings of SPIE - The International Society for Optical Engineering,* Vol.5830, pp.196-200.

Vaseva, K., Slavov, D., Todorov, P., Taslakov, M., Saltiel, S. & Cartaleva, S. (2011) High resolution spectroscopy of Cesium-vapor layer with micrometric thickness for development of frequency reference, *submitted for publication*

Confocal White Light Reflection Imaging for Characterization of Nanostructures

C. L. Du[1], Y. M. You[2, 3], Z. H. Ni[4], J. Kasim[2] and Z. X. Shen[2]
[1]*College of Science, Nanjing University of Aeronautics and Astronautics, Nanjing*
[2]*Division of Physics and Applied Physics, School of Physical and Mathematical Sciences Nanyang Technological University*
[3]*Department of Chemistry, Yale University, CT*
[4]*Department of Physics, Southeast University, Nanjing*
[1,4]*PR China*
[2]*Singapore*
[3]*USA*

1. Introduction

The ability to image nanostructures with a high spatial resolution as well as spectral resolution is very important for a host of both fundamental and practical studies (Grigorenko et al., 2005; Dixon et al., 1991; Verveer et al., 2007; Yoshifumi et al., 2006; Singh et al., 2007; Patel & McGhee, 2007; Laurent et al., 2006). Recently, optical imaging and spectroscopic studies of metal nanoarrays, individual metal nanostructures, and graphene (one monolayer thick carbon atoms packed into a two-dimensional honeycomb lattice, which is the basic building block for other sp^2 carbon nanomaterials.) sheet have attracted much attention (Du et al., 2008, 2010; Laurent et al., 2005, 2006; Ni et al., 2007; Wang et al., 2010). As an example, scanning near-field optical microscope (SNOM) has provided high resolution and been widely used in nanostructure study. However, collecting an image by SNOM is very time-consuming and relies heavily on the equipment as well as the skill of the operator. SNOM is also ill-suited for spectroscopic measurements due to the weak signals. Comparatively, far-field techniques are simpler and generate much stronger signals, which have been successfully used to study localized surface plasmons (LSPs) of gold nanoparticle arrays (Laurent et al., 2005, 2006). Moreover, far-field white light scanning is a simple and low cost method, which also offers multiple-wavelength advantage and is suitable to study spectral properties. White light confocal scanning microscopy has also been used to characterize material morphology, refractive index profile of fibers, etc (Ribes et al., 1995; Youk & Kim, 2006), where aperture or fiber were used as confocal pinhole. The best spatial resolution for normal confocal white light scanning optical microscope (not including that from a super continuum light source (Lindfors et al., 2004)) has been improved from 1.500 μm to about 0.800 μm (Youk & Kim, 2006). However, improvement of the spatial resolution is still much desired for the study of small-scale materials.

In this chapter, a new confocal white light reflection imaging technique is proposed by combining a confocal white light scanning microscope with a spectrometer. By decreasing the diameters of the incident light beam and the collection fiber, a spatial resolution of about 0.410 μm was achieved, which doubly enhances the previously reported best spatial resolution (~0.800 μm) of white light scanning and is even higher than those of laser scanning techniques (Rembe & Dräbenstedt, 2006; Gütay and Bauer, 2007). This system can provide both sample images extracted from reflection within a selective range of wavelength and their white light reflection spectra at each point. The simplicity in carrying out experiments makes this technique attractive, easy and fast. Metal nanoarrays, individual metal nanostructures (including single, dimer gold nanospheres and silver nanowires) and graphene sheet were characterized by the proposed system, demonstrating the strong capabilities in resolving nanometre structures, distinguishing different LSP resonant energies between different individual metal nanostructures and determining the graphene number layers, even the refractive index information of graphene.

2. Instrumentation and experiment

The schematic diagram of the experimental setup was shown in Figure 1. Light from a normal white light source (Xenon lamp) was polarized after passing through a polarizer, which serves as the incident light and was focused onto the sample through a holographic beam splitter and an OLYMPUS microscope objective lens (100X, NA=0.95). A tuneable aperture with a minimum diameter of 200 μm was introduced in the incident light path to tune the spatial resolution of the optical system. Different nanostructure samples were placed on a translation stage which provides coarse movement along the *x* and *y* axes, while the fine movement is offered by a piezostage with 100 μm travel distance along the *x* and *y* directions and 20 μm along the *z* direction. The piezostage also works as a mapping stage.

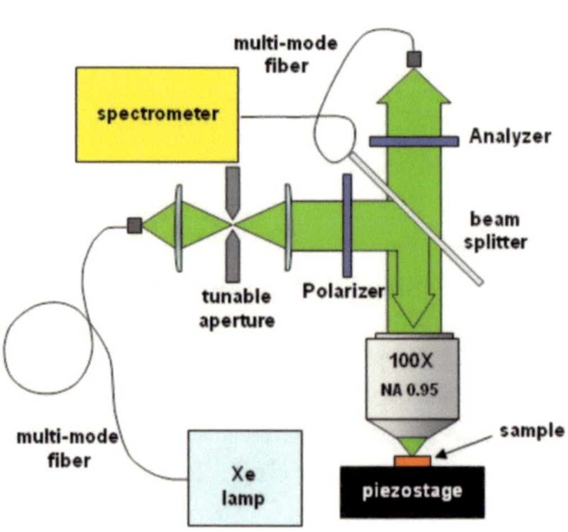

Fig. 1. Schematic diagram of the proposed confocal white light reflection imaging system. (Du et al., 2008).

The reflected light from the sample was collected by the same lens and directed to a spectrometer through a fiber. Fibers with various core diameters of 100, 50 and 25 μm were adopted in this work. The collection fiber also works as a pinhole, which is confocal with the illuminated spot on the sample. The reflected light was directed to a 150 grooves/mm grating and detected by a TE-cooled charge-coupled device (CCD). Typical integration time for imaging was 100 ms/pixel. The stage movement and data acquisition were controlled using ScanCtrl Spectroscopy Plus software from WITec GmbH, Germany.

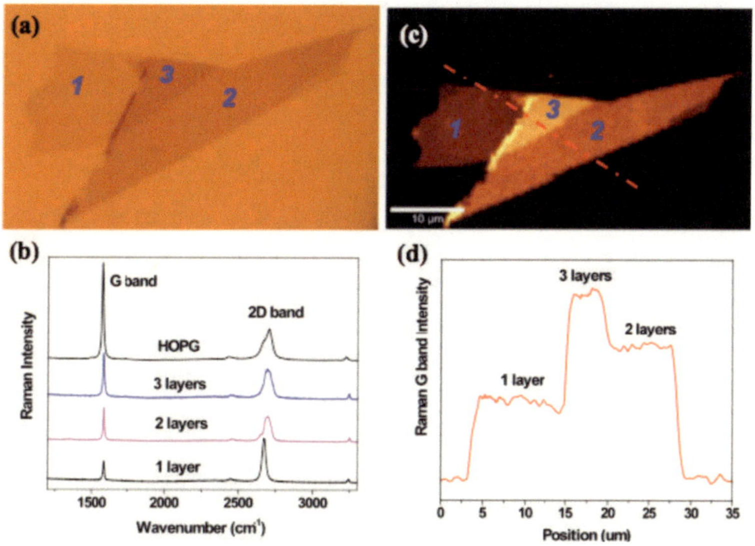

Fig. 2. (a) optical image of 1-, 2-, and 3-layer graphene. (b) Raman spectra of different layer graphene sheets and HOPG. (c) Raman image plotted by the intensity of G-band. (d) Cross section of Raman image, which corresponds to the dashed line in (c).

To determine the spatial resolution of the confocal reflection imaging system, we referred to the scanning knife edge method (Veshapidze et al., 2006) by using a two-layer graphene sheet on SiO_2/Si substrate as the edge (Ni et al., 2007). Thickness of a single layer graphene sheet is ~0.34 nm. The graphene sample was prepared by micromechanical cleavage on a silicon wafer with a 300 nm of SiO_2 capping layer (Novoselov et al., 2004). The optical microscope (Figure 2a) was used to locate the graphene sheet, which thickness was further confirmed by Raman spectroscopy. Figure 2b gives the Raman spectra of the graphene sample while its Raman image plotted by the intensity of G band was shown as Figure 2c. Figure 2d plots the cross section of the Raman image from the dashed line in Figure 2c, showing distinct difference in the Raman G-band intensity from the different thicknesses of graphene sample. Hence, the 2-layer graphene sheet can be identified easily, which provides an ideal edge sample because it has strong enough Raman signal with sharp edge, and most importantly it is thin such that there is no ambiguity in determining the spatial resolution caused by the edge effect.

The scanning confocal white light reflection spectrum by using 25 µm core diameter collection fiber and 200 µm diameter aperture was shown in Figure 3, which was fitted quite well with the following equation:

$$I(x) = \frac{P}{2}\left\{1 + erf\left(\frac{\sqrt{2}(x - x_0)}{w}\right)\right\} \tag{1}$$

where P is the total power for the incident white light beam, x_0 is the centre of the incident light beam and w is the desired $1/e^2$ half width. Thus the full width at half maximum (FWHM) of the incident beam (spot size) can be obtained from $\sqrt{2\ln 2}w$, which was defined as the spatial resolution.

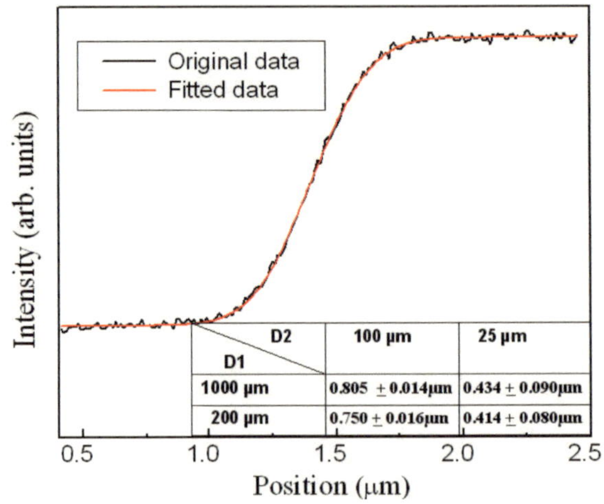

Fig. 3. Typical intensity versus graphene edge position data with the best fitting to Eq. (1) superimposed to determine the white light spot size. The inset gives the spatial resolution fitting results along with the fitting errors by using different aperture and collection fiber sizes.

Through tuning the collection fiber core diameter (D1) and the aperture diameter (D2), different spatial resolutions were obtained from Eq. (1) as shown in the inset table of Figure 3. It reveals the best spatial resolution about 0.410 µm, obtained using the 25 µm core diameter collection fiber and setting the aperture diameter to 200 µm. This doubles the previously reported best spatial resolution (~0.800 µm) for white light scanning (Youk and Kim, 2006) and is even better than those of laser scanning techniques (Rembe & Dräbenstedt, 2006; Gütay and Bauer, 2007). It also indicates that the aperture size, and especially the diameter of collection pinhole (the diameter of the collection fiber in our case), plays a significant role in improving the system resolution. All the white light reflection images shown below were obtained by setting D1 and D2 to 200 µm and 25 µm, respectively, unless stated otherwise. Scanning electron microscope (SEM) images of samples were taken with field emission SEM (JEOL JSM-6700F).

3. Confocal white light reflection (CWLR) imaging for characterization of metal nanostructures

Following, we will discuss our proposed applications in (1) resolving gold nanoarrys, (2) distinguish the resonance energy difference between the isolated single and dimer gold nanoparticles' LSP and revealing the strength of the near-field coupling between individual gold nanospheres and their supporting SiO_2/Si substrate, and (3) correlating the polarization dependent CWLR images of single silver nanowires with the nanowire polarization dependent excitation of surface plasmon (SP).

3.1 CWLR imaging for characterization of gold nanoarrays

Gold nanoparticle arrays were fabricated by nanosphere lithography (Jensen et al., 1999) on cover glass substrates. Polystyrene (PS) microspheres with diameter 1 μm and 0.500 μm were used as masks, which will self-assemble into monolayer spheres on substrates. Gold thin film with thickness of about 0.050 μm was deposited by DC coater sputtering, and then the spheres were lifted off.

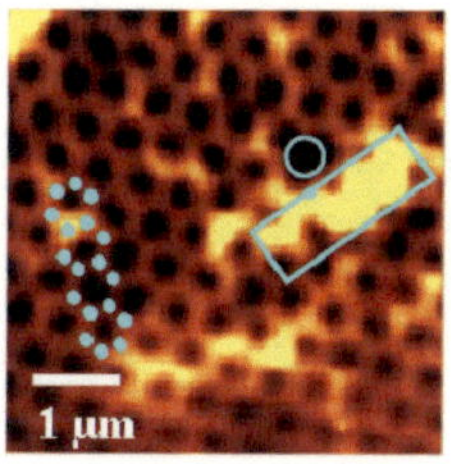

Fig. 4. The CWLR image at the wavelength of 0.480-0.520 μm for gold nanoarrays on cover glass which were fabricated by using 0.500 μm diameter PS as the lithographic mask.

Figure 4 gave a typical 5.0 x 5.0 μm^2 CWLR image for the obtained gold nanoarrays on cover glass fabricated by using 0.500 μm diameter PS as a lithographic mask. The image was extracted from the white light reflection intensity between the wavelength of 0.480-0.520 μm from the samples. Owing to smaller reflection of the cover glass substrate than that of gold particles, hexagonal bright rings in the image of Figure 4 correspond to gold nanoparticles while black areas correspond to the cover glass. The periodicity for the gold particle arrays, which is 0.500 μm, can be clearly resolved, demonstrating the high resolution of our technique. However, the gold particle size and centre-to-centre distance between two nearest gold particles is measured to be about 0.150 μm and 0.100 μm, respectively, by SEM images (not shown), which are out of the range of the system's spatial resolution. This can explain why the image for six gold particles in one hexagonal cell merges to form a hexagonal ring. The dots in Figure 4 work just as a guide for the eye labelling where the particle is. Meanwhile, different defects in the sample as indicated by the rectangular circles in Figure 4 were imaged as well, which reveals that the present imaging method can also be used to test the sample quality, similar to reports elsewhere (Ormonde et al., 2004), but here with much higher spatial resolution.

For comparison, in Figure 5, we have also given the CWLR images for gold nanoarrays on cover glass fabricated by using 1 μm diameter PS as a lithographic mask. Similar to Figure 4, owing to the smaller reflection of the substrate than metal particles, hexagonal bright dots in images Figures 5a to 5f correspond to gold particles while black areas correspond to the cover glass. From the images, we found that the image at 0.480-0.520 μm gives us the best resolution. All the images in this chapter were selected by this way. The gold particle size and centre-to-centre distance between two nearest gold particles is measured to be about 0.300 μnm and 0.200 μm, respectively, by SEM images (not shown). From images in Figures 5a to 5(f), the six gold nanoparticles in one hexagon cell can be resolved clearly as labelled by pink dots in Figure 5a for guiding. The size of these six nanoparticles in CWLR images is spatial resolution determined, which is ~0.410 μm. From these images, it is very easy to obtain the white light reflection spectra for the substrate and gold particles, respectively, as shown in Figure 5g. The contrast spectra are defined as

$$Contrast = (I_{sample} - I_{substrate}) / I_{substrate} \quad (2)$$

where I_{sample} and $I_{substrate}$ refer to the white light reflection intensity of the sample and the substrate, respectively. The result was shown as Figure 5h, which confirms that gold particles always have larger reflection than that of the substrate, as well as the role of SP in reflection, which is discussed in detail in the next section.

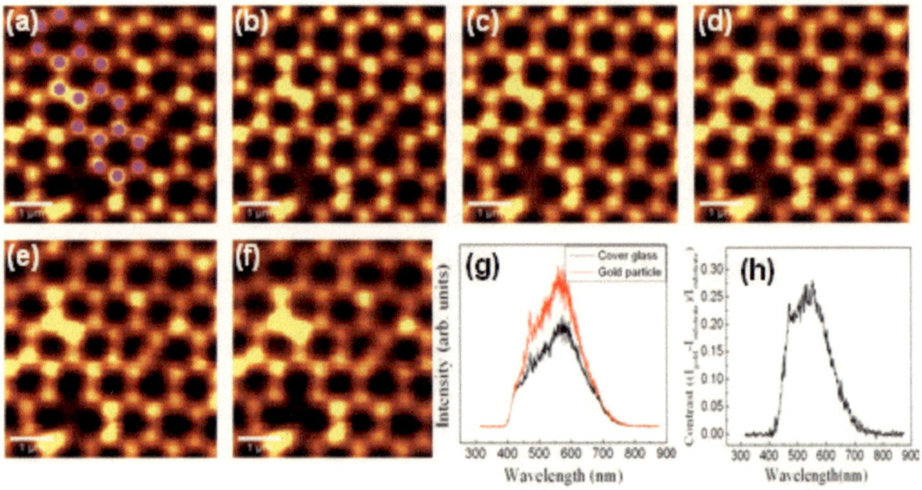

Fig. 5. The CWLR images (a-f) at different wavelength ranges for gold nanoarrays on cover glass fabricated by using 1 μm diameter PS as the lithographic mask while (g) and (h) are the reflected white light spectra and reflection contrast between the gold particle and substrate, respectively. Wavelength ranges for images (a) to (f) correspond to 0.440-0.480 μm, 0.480-0.520 μm, 0.520-0.560 μm, 0.560-0.600 μm, 0.600-0.640 μm and 0.640-0.680 μm, respectively.

3.2 CWLR imaging for characterization of individual gold nanospheres

Commercial gold nanospheres with a diameter of 50 ± 5 nm (Corpuscular Inc) deposited on 200 nm silicon dioxide (SiO_2) films were chosen for CWLR imaging as well. Their confocal white light reflection images were constructed by extracting the light intensity from the corresponding reflection spectra for a selected wavelength range, too. The image at the wavelength of 510-550 nm was shown in Figure 6b as one example. As can be seen from the SEM image shown in Figure 6a, it consists of three isolated single spheres and one dimer where the two spheres are almost in contact with each other. The four dark spots in Figure 6b represent the images of the gold nanospheres, which correspond well to the SEM image. The size of the dark spots is about 410 nm, determined by the spatial resolution of the imaging system. Considering the resolution limitation, it is reasonable that the white light reflection images can not distinguish between single sphere and dimer spheres. For comparison, Figure 6d presented the CWLR imaging results for the spheres by setting D2 to 100 μm while keeping other experimental conditions the same. It can be seen that the images from the spheres severely overlap, confirming the higher spatial resolution by using a 25 μm core diameter collection fiber again.

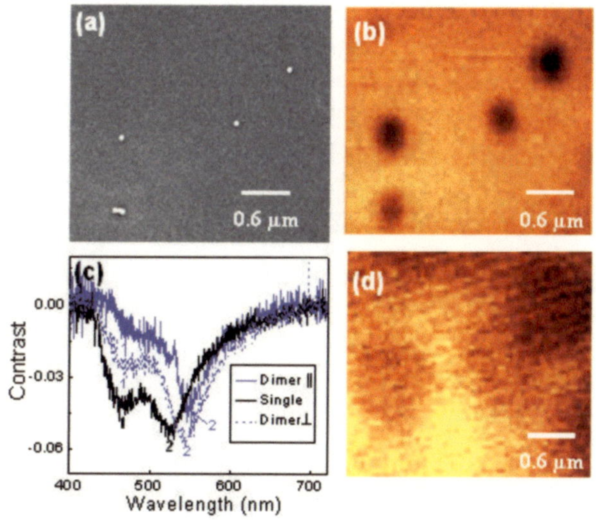

Fig. 6. SEM image (a) and CWLR images at wavelength 510-550 nm of gold nanospheres on 200 nm SiO_2 films with collection fiber core diameter 25 μm (b) and 100 μm (d). (c) The contrast spectra for single (black lines) and dimer spheres for incident light parallel (solid lines) and perpendicular (dotted lines) to the dimer axis. (Du et al., 2008).

Figure 6c plotted the contrast spectra for the single and dimer gold nanospheres after locating their positions. Two contrast dips (labelled 1 and 2, respectively) were observed from Figure 6. It is noticed that the position of dip 2 of isolated single spheres (at about 525 nm) coincides with that of the LSP of gold spheres with diameter 50 nm (Dijk et al., 2005). However, the same dip shows red shifts to 548 nm and 542 nm for the incident polarization parallel and perpendicular to the dimer axis, respectively. This results from the coupling effect between the two nanospheres of the dimer (Moores & Goettmann, 2006). The larger red shift for the parallel polarization than that of the perpendicular case is understandable considering the stronger coupling of the dimer for parallel polarization. Figure 6c also reveals a weak dip 1 for the single and dimer located at about 470 nm, which originates from the multi-polar SP excitation of the gold nanospheres. Firstly, SP excitation at about 470 nm has been observed for gold nanospheres with diameters close to 40 nm, although it was ascribed to false spectral lines that arose from using a 488 nm argon laser (Benrezzak et al., 2001). Secondly, similar multi-polar SP excitation has been reported for other isolated metal nanoparticles (Dijk et al., 2005). Moreover, the excitation of LSPs leads to the enhancement of the absorption of the nanospheres, which consequently has the effect of reducing the reflection intensity (Kawata, 2001), making the nanospheres dark in the corresponding images in Figure 6b. The different dip 2 position for the single and dimer spheres also reflects their different dipolar LSP resonant energies, further implying that the CWLR imaging method is capable of resolving the LSP energies for individual noble metal nanoparticles.

Then, to determine the decay length of the electromagnetic (EM) coupling between individual gold nanopspheres and its supporting substrate, different thicknesses (d) SiO_2 film on Si were obtained by annealing single-crystalline Si (d < 20 nm) in air or by RF sputtering SiO_2 (d > 20 nm) to serve as the substrate.

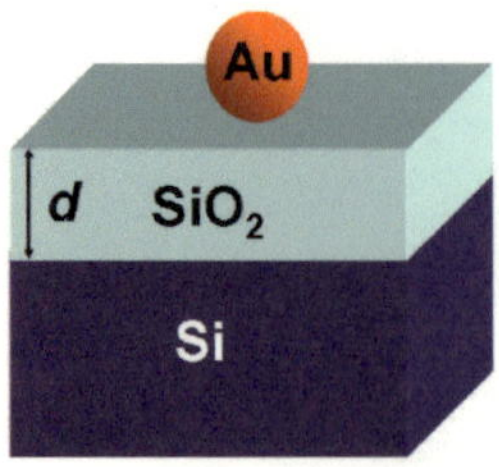

Fig. 7. Schematic diagram of the prepared gold nanospheres on SiO_2/Si substrate samples.

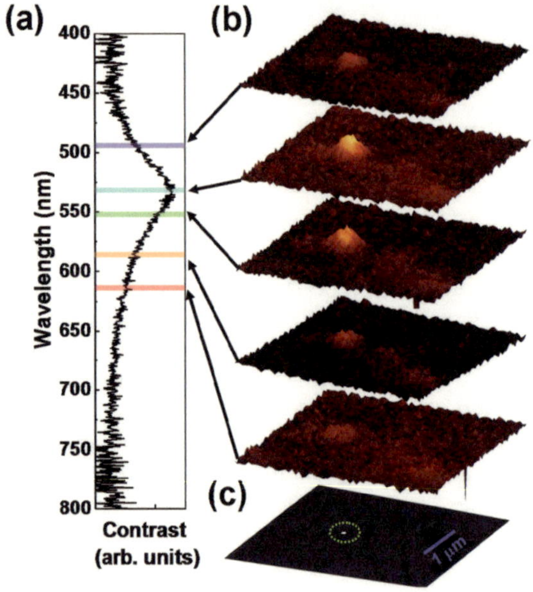

Fig. 8. Comparison between the CWLR contrast images (b) of an individual Au nanosphere on a SiO_2/Si substrate (with SiO_2 film 60 nm) at different selected wavelength regions labelled by the rectangular bars on the corresponding contrast spectra (a) along with the corresponding SEM image of the nanoshpere (c). (Du et al., 2010).

Figure 7 illustrated the schematic diagram of the prepared gold nanospheres on SiO_2/Si substrate samples and a typical SEM image was shown as Figure 8c. The distance between different individual nanospheres chosen for study herein was selected purposely to be larger than 1μm. Thus the EM coupling between them can be neglected. For convenience, herein the contrast is defined by the following Eq. (3), and hence their contrast images and the CWLR contrast spectra can be obtained.

$$Contrast = (I_{sub} - I_{AuNSP}) / I_{sub} \tag{3}$$

Where I_{AuNSP} and I_{sub} refer to the CWLR intensity from the positions of the Au NP and the substrate, respectively. Adopting Eq. (3), the CWLR contrast spectra for individual gold nanoparticles were obtained. A typical CWLR contrast spectra of an individual Au nanosphere on SiO_2/Si substrates was presented as Figure 8a along with its different CWLR images at different wavelength regions. The different wavelength regions constructing these contrast images were labelled by the rectangular bars on the left contrast spectra of Figure. 8a. Figure 8a presents an obvious peak at ~530 nm, which originates from the excitation of dipolar LSPs with the peak wavelengths corresponding to the dipolar LSP wavelengths (Du et al., 2008; Okamoto & Yamaguchi, 2003; Abe & Kajikawa, 2006; Pinchuk et al., 2004; Knight et al., 2009). The excitation of LSP enhances the light absorption by the nanosphere and reduces its reflection intensity (Kawata, 2001), then further leads to its higher contrast intensity relative to the substrate in the CWLR contrast images (Figure. 8b). As can be seen, with the selected wavelength region closer to the contrast peak position, the larger image contrast between the Au nanosphere and the substrate is obtained while the maximum contrast is reached at the LSP wavelength. Meanwhile, from the unequivocal one-to-one correspondence between the nanosphere in the contrast images (Figure. 8b) and its SEM image (Figure. 8c), we can locate the individual nanosphere we concern.

To explore the near-field coupling between individual gold nanospheres and their supporting substrate further, we have measured the contrast spectra between individual Au nanospheres on SiO_2/Si substrates with different thicknesses of SiO_2 on Si substrate. Several typical spectra were presented in Figure 9, in which the dipolar LSP peaks were guided by the dashed line. The second peak at shorter wavelength for cases of bare Si substrate and d = 6 nm corresponds to the multi-polar LSP excitation (Du et al., 2008). In what follows, we will mainly discuss the dipolar LSP (labelled as LSP for short in the following) wavelength (λ) behaviour as a function of the spacer SiO_2 film thickness d.

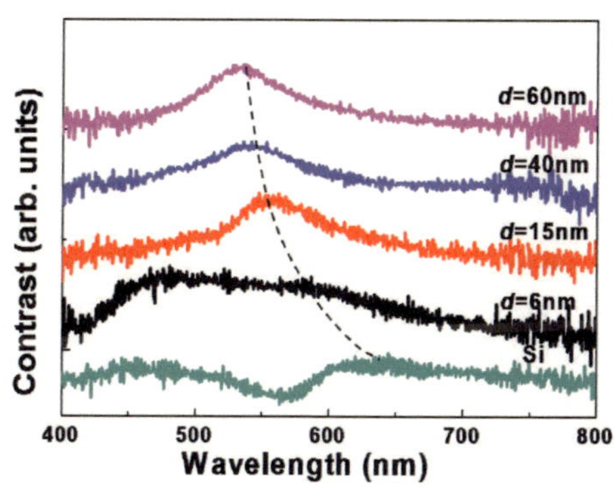

Fig. 9. Comparative CWLR contrast spectra between individual Au nanospheres on SiO_2/Si substrates (with SiO_2 film d nm) and on Si substrate. (Du et al., 2010).

Figure 9 reveals that the LSP wavelengths of all the samples show an obvious red-shift compared to the LSP resonant wavelength for an isolated Au nanosphere with diameter about 50 nm in air, which is at about 520 nm (Noguez, 2007). This originates from the different dielectric function of the substrate from that of air and agrees well with the literature reports (Okamoto & Yamaguchi, 2003; Abe & Kajikawa, 2006; Pinchuk et al., 2004). Moreover, the dashed line in Figure 9 demonstrates that the LSP wavelength blue-shifts with d increasing. To see this more clearly, we have plotted the function of the resonant wavelength λ of these LSP modes versus d in Figure 10. The experimental data of λ vs. d is noted to can be fitted with a single exponential decay function (the solid line) quite well. All these reflect the near-field EM interaction between the nanosphere and its supporting substrate, which dominates the corresponding LSP wavelength shift (Jain et al., 2007; Biring et al., 2008). Accordingly, the near-field EM coupling strength between them determines the magnitude of the dipolar LSP wavelength shift ($\Delta\lambda = \lambda - \lambda_0$) compared to the LSP wavelength (λ_0) of the isolated nanosphere case.

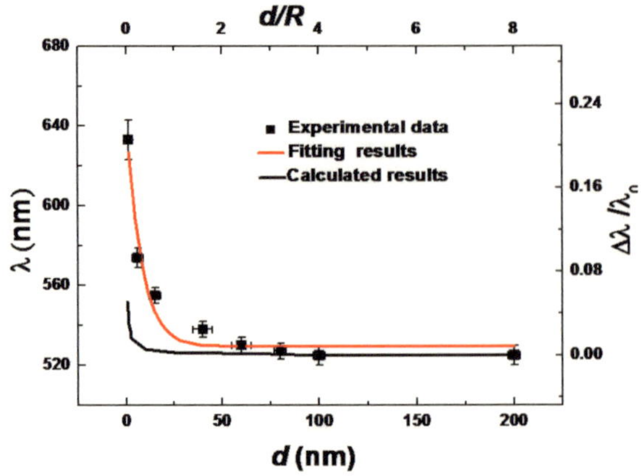

Fig. 10. The function of the LSP wavelength λ versus the spacer thickness *d* (left y-axis and bottom *x*-axis) and the function of the normalized LSP wavelength shift of $\Delta\lambda / \lambda_0$ versus the normalized spacer thickness *d/R* (right y-axis and top *x*-axis): the squared dots, dotted line and solid line correspond to the experimental data, calculated results and the single exponential decay fitting results, respectively). (Du et al., 2010).

By bringing a metal nanoparticle into the vicinity of a flat substrate, nonhomogeneous optical response due to the polarizability of the substrate material is expected. Generally, with an external EM field, a charge polarization on the NP can be induced, which further causes a charge distribution on the substrate. Under the quasi-static approximation, this charge distribution can be seen as the image charge distribution (Okamoto & Yamaguchi, 2006; Noguez, 2007) of the nanoparticle and it can in turn affect the local EM field around the nanoparticle, and further their optical responses. Adopting the method proposed elsewhere (Okamoto & Yamaguchi, 2006; Wind et al., 1987), the extinction spectra of the concerned nanospheres were calculated by including both the dipole and higher multiple image-charge effects along with the Fresnel reflection effect at the nanosphere-substrate interface. The corresponding obtained function of the LSP wavelength (λ) versus the spacer thickness (*d*) was plotted as the dashed line in Figure 10 as well for comparison. The calculation qualitatively verifies the decrease behaviour of the experimental data of λ versus *d* though it predicts a faster decay rate than the experimental results.

To quantify the near-field EM coupling strength between the individual nanosphere and its supporting substrate, the normalized LSP wavelength shift ($\Delta\lambda / \lambda_0$) was calculated as well by the quasi-static theory and for the experimental data. Experimentally, λ_0 is obtained by assuming that it equals to the LSP wavelength of the nanosphere on very thick SiO_2 film (take it as 200 nm). Through a simple linear transformation, both the experimental and theoretical data of λ versus *d* can be transferred into $\Delta\lambda / \lambda_0$ versus *d/R*. The obtained results were scaled according to the right *y*-axis (corresponds to the upper *x*-axis) in Figure 10. It can be seen that $\Delta\lambda / \lambda_0$ decreases with *d/R* and reaches zero for large enough *d/R*. This demonstrates that the strength of the near-field EM coupling owing to the substrate effect decreases with d/R and then vanishes with further increasing d/R. For thicker spacer ($d > 3R$), the influence of the image-charges owing to the substrate presence is ignorable and the LSP wavelength approaches to the limit of the individual nanosphere in the background. Meanwhile, the faster decay rate of the shift predicted by the quasi-static theory is resulted from the limitation of the theoretical model, which ignores the retardation effect while such effect is more pronounced for larger nanoparticles (Okamoto & Yamaguchi, 2003; Noguez, 2007). The solid line in the right y-axis of Figure 10 presents the single exponential decay fitting by Eq. (4)

$$y = a * \exp(-\frac{x}{t}) + y_0 \tag{4}$$

to the experimental data with a, t and y_0 as the fitting parameters. It reveals a decay length t about 0.30 in units of d/R and the fitting goodness equals to 0.97. The fitting results are interestingly noted to qualitatively agree with the 'plasmon ruler' scaling theory for the near-field EM coupling between two component noble metal nanoparticles of a dimer (Jain et al., 2007). It points out that the decay length is about 0.20 in units of the 'Gap/Diameter' regardless of the nature of the component metal nanoparticles of the dimer (Jain et al., 2007). The similarity of the decay length in magnitude is owing to that the image-charges induced in the substrate can be just replaced by the actual charges induced in the other particle for the dimeric nanoparticle's case. The deviation between their decay lengths is understandable considering the different polarizabilities between the dimer and our case. Thus, the near-field coupling strength between the individual Au nanosphere and the Si substrate is revealed to exponentially decrease with the spacer (SiO_2 film) thickness and the decay length is about 0.3 in units of d/R.

3.3 CWLR imaging for characterization of individual silver nanowires

Silver nanowire samples with a diameter about 100 nm were fabricated by a simple hydrothermal method (Wang et al., 2005) and deposited onto silicon for CWLR imaging. Two typical CWLR images of the silver nanowire were presented as Figures 11a and 11b, which clearly exhibits polarization dependent. By rotating 90 degree of the incident polarization, their contrast reverses from the comparison between Figures 11a and 11b. Even for the same bent nanowire, the different parts exhibits different contrast compared to the substrate. Owing to the larger reflectivity of silver than that of the substrate silicon, it is expected that the nanowires are brighter than the substrate under CWLR imaging system. However, this contradicts with the results of Figure 11, revealing that other factors besides material reflectivity contribute to the observed CWLR images.

It is known that the excitation of SPs of sliver nanowires is anisotropic (Schider et al., 2003), which is sensitive to the polarization direction of the incident light. This can account for the polarization dependence of the reflection images of Figure. 11, which also contribute to the different contrast for different parts of the same bent nanowire as well since for different parts of the same nanowire, the only difference lies in their different orientation relative to the incident light polarization direction. As the images shown in Figures 11a and 11b were extracted from the reflections in the range of 600 – 640 nm, it is overlapped with one of the SP mode (500 – 700 nm) (Kim et al., 2003; Mohanty et al., 2007) of the Ag nanowire, which experiences preferred excitation when the polarization of the incident light is parallel to the nanowire. Thus, this SP mode along the different material reflectivity contributes to the obtained polarization dependent CWLR images. It also demonstrates that the developed CWLR imaging system is able to correlate the polarization dependent CWLR images of single silver nanowires with the nanowire polarization dependent excitation of SP.

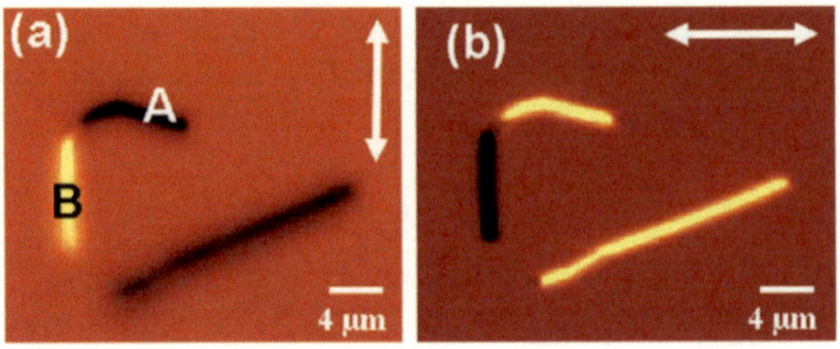

Fig. 11. The CWLR images at the wavelength of 600 - 640 nm for silver nanowires on silicon substrate. The double-direction arrows in the figure indicate the polarization direction of the incident light. (Du et al., 2008).

4. CWLR imaging for characterization of graphene

We also note that the proposed CWLR system is not limited to characterize metal nanostructures. It can also be extended to other samples, such as graphene sheet. In this part, we will show how the system was used to determine the number of graphene layers and to extract the corresponding refractive index.

The graphene's visibility strongly varies from one laboratory to another and it relies on experience of the observer though one can observe different colours/contrasts for graphene sheets of different thickness using the optical image with "naked eyes". Taking advantage of contrast spectra and image, this can be made quantitative and accurate. By combing Raman spectroscopy and optical images, graphene sheets with different layer numbers were first obtained as shown in Figure 12. Then, their contrast spectra were measured and plotted in Figure 12. For consistence with reference (Ni et al., 2007), the contrast spectra C(λ) were obtained by using C(λ)= (R0(λ)- R(λ))/ R0(λ), where R0(λ) is the reflection spectrum from the SiO_2/Si substrate while R(λ) is the reflection spectrum from graphene sheet. As revealed in Figure 12, the contrast spectrum of single layer graphene has a peak at about 550 nm, which makes the single layer graphene visible and is in green-orange range. Meanwhile, the peak position is almost unchanged with increasing number of layers up to ten. The contrast value for single layer graphene is about 0.09+0.005 and it increases with the number of layers, for example, 0.175+0.005, 0.255+0.010, and 0.330+0.015 for two, three, and four layers, respectively. For graphene of around ten layers, the contrast saturates and the contrast peak shifts towards red (samples a and b). For samples with larger number of layers (c to f), negative contrast occurs. This can easily be understood that these samples are so thick that the reflections from their surface become more intense than that from the substrate, resulting in negative value contrast.

Hence, different layer graphene gets different contrast value, which provides a standard for determining the number of layers for graphene. This can also be understood in terms of the

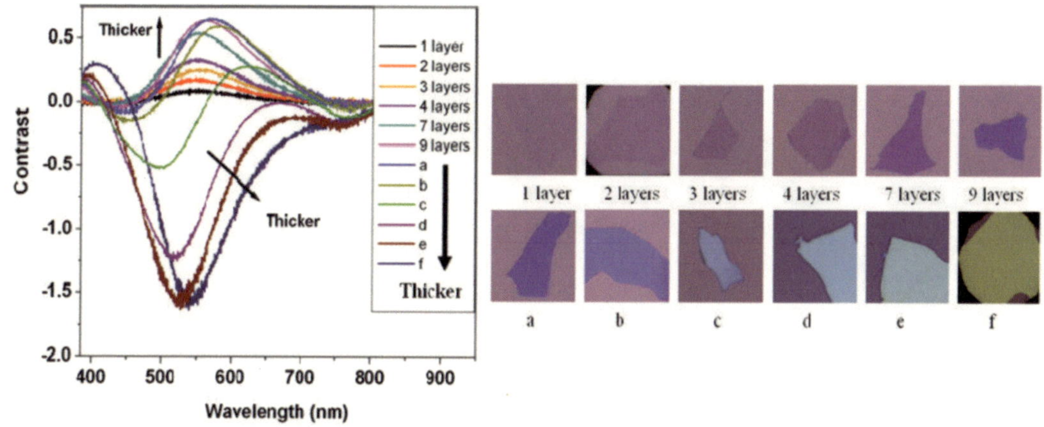

Fig. 12. The contrast spectra of graphene sheets with different thicknesses, together with the optical image of all the samples. Besides the samples with 1, 2, 3, 4, 7 and 9 layers, samples a, b, c, d, e and f are more than ten layers and the thickness increases from a to f. The arrows in the graph show the trend of curves in terms of the thicknesses of graphene sheets. (Ni et al., 2007).

Fresnel reflection theory. Consider the incident light from air ($n_0 = 1$) onto a graphene, SiO_2, and Si tri-layer system. The reflected light intensity from the tri-layer system can then be described by (Blake et al., 2007; Anders, 1967):

$$R(\lambda) = r(\lambda) r^*(\lambda) \quad (5)$$

$$r(\lambda) = \frac{r_a}{r_b} \quad (6)$$

$$r_a = (r_1 e^{i(\beta_1+\beta_2)} + r_2 e^{-i(\beta_1-\beta_2)} + r_3 e^{-i(\beta_1+\beta_2)} + r_1 r_2 r_3 e^{i(\beta_1-\beta_2)}) \quad (7)$$

$$r_b = (e^{i(\beta_1+\beta_2)} + r_1 r_2 e^{-i(\beta_1-\beta_2)} + r_1 r_3 e^{-i(\beta_1+\beta_2)} + r_2 r_3 e^{i(\beta_1-\beta_2)}) \quad (8)$$

where $r_1 = \frac{n_0 - n_1}{n_0 + n_1}, r_2 = \frac{n_1 - n_2}{n_1 + n_2}, r_3 = \frac{n_2 - n_3}{n_2 + n_3}$ are the reflection coefficients for different interfaces and $\beta_1 = 2\pi n_1 \frac{d_1}{\lambda}, \beta_2 = 2\pi n_2 \frac{d_2}{\lambda}$ are the phase difference when the light passes through the media which is determined by the path difference of two neighbouring interfering light beams. The thickness of the graphene sheet can be estimated as $d1 = N\Delta d$, where N represents the number of layers and Δd is the thickness of single layer graphene (Δd = 0.335 nm) (Kelly, 1981, Dresselhaus, 1996). The refractive index of graphene is used as a fitting parameter. The thickness of SiO_2, d_2, is 285 nm, with a maximum 5% error. The refractive index of SiO_2, n_2, is wavelength dependent (Palik, 1991). The Si substrate is considered as semi-infinite and the refractive index of Si, n_3, is also wavelength dependent (Palik, 1991). The reflection from SiO_2 background, R0(λ), was calculated by setting $n_1 = n_0 = 1$, and $d_1 = 0$.

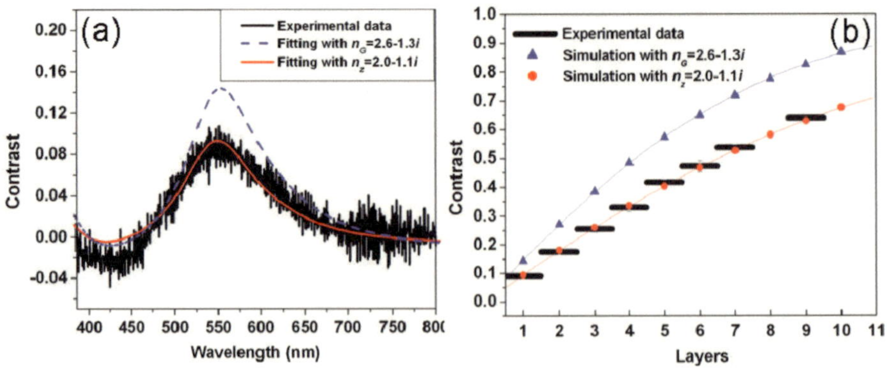

Fig. 13. (a) The contrast spectrum of experimental data (black line), the simulation result using n_z =2.0-1.1i (red line), and the simulation result using n_G = 2.6-1.3i (dash line). (b) The contrast simulated by using both n_G (blue triangles) and n_z (red circles), the fitting curve for the simulations (blue and red lines), and our experiment data (black thick lines), respectively, for one to ten layers of graphene. (Ni et al., 2007).

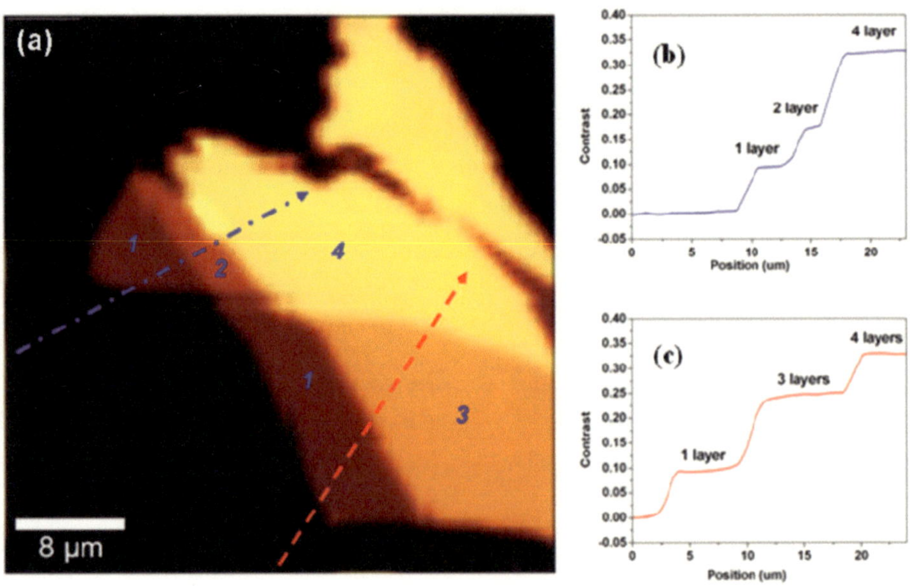

Fig. 14. (a) The contrast image of the sample. (b) and (c) The cross section of contrast image, which corresponds to the dash lines. (Ni et al., 2007).

The optimized simulation result is shown in Figure 13a reveals a refractive index of single layer graphene n_z = 2.0-1.1i, whereas the simulation result using the bulk graphite value of n_G (2.6-1.3i) shows large deviation from our experimental data. Using the optimized refractive index n_z, we have calculated the contrast of one to ten layers' graphene also as shown in Figure 13b, which agree well with the experimental data with the discrepancy being only 2%. By using this technique, the thickness of unknown graphene sheet can be determined directly by comparing the contrast value with the standard values shown in Figure 13b. Alternatively, it can be obtained using the following equation:

$$C = 0.0046+0.0925N-0.00255N^2 \quad (9)$$

where N ($\leq$10) is the number of layers of graphene.

In order to demonstrate the effectiveness of the contrast spectra in graphene thickness determination, we carried out the CWLR imaging, too. As shown in Figure 14a, distinct contrast for different thicknesses of graphene can be observed from the image. It is worth noting that the contrast image measurement can be done in a few minutes. Figure 14b and 14c show contrast along the two dash lines on the image. The contrast value for each thickness agrees well with those shown in Figure 13. Using Eq. (9), the N values along the two dash lines are calculated, where the N along the blue line is: 0.99, 1.93, and 3.83; and along the red line is: 0.98, 2.89 and 3.94. Again our results show excellent agreement.

Accordingly, by the proposed CWLR system, both the layer number of graphene and the refractive index of single layer graphene can be achieved. It does not need a single layer graphene as reference as that in Raman. It is also noted that the proposed system can be used to get information about the optical conductivity of some bilayer grapene sheet (Wang et al., 2010), which is not shown here.

5. Conclusion

In this chapter, we have proposed a far-field CWLR imaging system by combing a small aperture and a small collection fibre core diameter, which is fast, non-destructive and user friendly. It is demonstrated to provide a high spatial resolution about 410 nm, which is capable of resolve two nearest gold nanoparticles with the size and centre-to-centre distance in-between about 300 nm and 200 nm, respectively. Individual single, dimer gold nanospheres, silver nanowires, and graphene sheet were characterized by the imaging system as well. Apart from the dipolar LSP, excitation of multi-polar LSP of individual gold nanospheres was revealed. Compared to the resonance energy of single gold nanosphere, the resonance energy of the dimer is red-shifted due to the EM coupling between the two component nanospheres of the dimer. The near-field EM coupling effect between individual Au nanospheres and the supporting SiO_2/Si substrate was also studied by the CWLR imaging method, which reveals a decay length of 0.30 in units of d/R for the coupling strength, qualitatively agreeing well with the 'plasmon ruler' scaling theory. The anisotropic excitation of LSP of single silver nanowire was revealed to get contribution to the polarization dependent images besides their essential reflectivity difference from that of the substrate. It is also demonstrated that the CWLR spectra method provides a standard to identify the thickness of graphene sheet on Si substrate with ~300 nm SiO_2 capping layer, from which the refractive index (n_z = 2.0-1.1i) of graphene below ten layers can also be easily determined. As the CWLR imaging can be preformed at different wavelength, we also expect its other interesting applications such as biomaterial mapping and plasmonic studies in the future.

6. Acknowledgment

This work was financially supported by the Natural Science Foundation of China (No. 11004103), China Postdoctoral Science Foundation funded project (No. 20100471332), Jiangsu Planned Projects for Postdoctoral Research Funds (No. 0902016C), NUAA Research Funding (No. NS2010186), and NUAA Scientific Research Foundation.

7. References

Abe S.; Kajikawa K. (2006). Linear and Nonlinear Optical Properties of Gold Nanospheres Immobilized on a Metallic Surface. *Phys. Rev. B.*, Vol.74, No.3, pp. 035416, ISSN 1098-0121

Anders H. (1967). *Thin Films in Optics*, Focal Press, London

Benrezzak S.; Adam P. M.; Bijeon J. L.; Royer P. (2001). Observation of Nanometric Metallic Particles with an Apertureless Scanning Near-field Optical Microscope. *Surf. Sci.*, Vol. 491, No.1-2, pp. 195-207, ISSN 0039-6028

Biring S.; Wang H. H; Wang J. K; Wang Y. L. (2008). Light Scattering from 2D Arrays of Monodispersed Ag-nanoparticles Separated by Tunable Nano-gaps: Spectral Evolution and Analytical Analysis of Plasmonic coupling. *Opt. Express*, Vol.16, No.20, pp.15312-15324, ISSN 1094-4087

Blake P.; Hill E. W.; Neto A. H. C.; Novoselov K. S.; Jiang D.; Yang R.; Booth T. J.; Geim A. K. (2007). Making graphene visible. *Appl. Phys. Lett.*, Vol.91, No.6, pp. 063124, ISSN 0003-6951

Dixon A. E.; Damaskinos S.; Atkinson M. R. (1991). A Scanning Confocal Microscope for Transimission and Reflection Imaing. *Nature (London)*, Vol.351, No.6327, pp. 551-553, ISSN 0028-0836

Dijk M. A.; Lippitz M.; Orrit M. (2005). Far-Field Optical Microscopy of Single Metal Nanoparticles. *Acc. Chem. Res.*, 2005, Vol. 38, No. 7, pp. 594–601, ISSN 0001-4842

Dresselhaus M. S.; Dresselhaus G.; Eklund P. C. (1996). *Science of Fullerenes and Carbon Nanotubes*, Academic Press, ISBN 012-221820-5, San Diego, CA

Du C. L.; You Y. M.; Kasim J.; Ni Z. N.; Yu T.; Wong C. P.; Fan H. M.; Shen Z. X. (2008). Confocal White Light Reflection Imaging for Characterization of Metal Nanostructures, *Opt. Commun.* Vol. 281, No.21, pp. 5360-5363, ISSN 0030-4018

Du C. L.; You Y. M.; Kasim J.; Hu H. L.; Zhang X. J.; Shen Z. X. (2010). Near-field Coupling Effect between Individual Au Nanospheres and Their Supporting SiO_2/Si Substrate, *Plasmonics*, Vol. 5, No.2, pp. 105-109, ISSN 1557-1955

Grigorenko A. N.; Geim A. K.; Gleeson H. F.; Zhang Y.; Firsov A. A.; Khrushchev I. Y.;. Petrovic J. (2005). Nanofabricated Media with Negative Permeability at Visible Frequencies. *Nature,* Vol. 438, No. 7066, pp. 335-338, ISSN 0028-0836

Gütay L.; Bauer G. H. (2007). Spectrally Resolved Photoluminescence Studies on Cu(In,Ga)Se2 Solar Cells with Lateral Submicron Resolution. *Thin Solid Films,* Vol. 515, No. 15, pp. 6212-6216, ISSN 0040-6090

Jain P. K.; Huang W. Y.; El-Sayed M. A. (2007). On the Universal Scaling Behavior of the Distance Decay of Plasmon Coupling in Metal Nanoparticle Pairs: a Plasmon Ruler Equation. *Nano lett.* Vol.7, No.7, pp. 2080-2088, ISSN 1530-6984

Jensen T. R.; Duval M. L.; Kelly K. L.; Lazarides A. A.; Schatz G. C.; Duyne RP. Van. (1999). Nanosphere Lithography: Effect of the External Dielectric Medium on the Surface Plasmon Resonance Spectrum of a Periodic Array of Silver Nanoparticles. *J. Phys. Chem. B.*, Vol.103, No., pp. 9846-9853, ISSN 1089-5647

Kawata S. (2001). *Near-field Optics and Surface Plasmon Polaritons*, Springer, ISBN 3540415025, Verlag Berlin Heidelberg

Kelly, B. T. (1981). *Physics of Graphite*, Applied Science, ISBN 0853349606, London

Knight M. W.; Wu Y. P.; Lassiter J. B.; Nordlander P.; Halas N.J. (2009). Substrates Matter: Influence of an Adjacent Dielectric on an Individual Plasmonic Nanoparticle. *Nano Lett.* Vol.9, No. 5, pp.2188–2192, ISSN 1530-6984

Laurent G.; Félidj N.; Truong S. L.; Aubard J.; Lévi G.; Krenn J. R.; Hohenau A.; Leitner A.; Aussenegg F. R. (2005). Imaging Surface Plasmon of Gold Nanoparticle Arrays by Far-field Raman Scattering. *Nano Lett.*, Vol.5, No.2, pp. 253-258, ISSN 1530-6984

Laurent G.; Félidj N.; Grand J.; Aubard J.; Lévi G.; Hohenau A.; Aussenegg F. R.; Krenn J. R. (2006). Raman Scattering Images and Spectra of Gold Ring Arrays. *Phy. Rev. B.*, Vol.73, No.24, pp. 245417, ISSN 1098-0121

Lindfors K.; Kalkbrenner T.; Stoller P.; Sandoghdar V. (2004). Detection and Spectroscopy of Gold Nanoparticles Using Supercontinuum White Light Confocal Microscopy. *Phys. Rev. Lett.*, Vol.93, No.3, pp. 037401, ISSN 0031-9007

Mohanty P.; Yoon I.; Kang T.; Seo K. YT.; Varadwaj K.S.; Choi W. J.; Park Q. H.; Ahn J. P.; Suh Y. D.; Ihee H.; Kim B. (2007). Simple Vapor-phase Synthesis of Single-crystalline Ag Nanowires and Single-nanowire Surface-enhanced Raman Scattering. *J. Am. Chem. Soc.*, Vol.129, No.31, pp. 9576, ISSN 0002-7863

Moores A.; Goettmann F. (2006). The Plasmon Band in Noble Metal Nanoparticles: an Introduction to Theory and Applications. *New J. Chem.*, Vol.30, No.8, pp. 1121-1132, ISSN 1144-0546

Noguez C. (2007). Surface Plasmons on Metal Nanoparticles: the Influence of Shape and Physical Environment. *J. Phys. Chem. C* Vol. 111, No.10, pp. 3806-3819, ISSN 1932-7447

Novoselov, K. S.; Geim, A. K.; Morozov, S. V.; Jiang, D.; Zhang, Y.; Dubonos, S. V.; Grigorieva, I. V.; Firsov, A. A. (2004). Electric Field Effect in Atomically Thin Carbon Films. *Science,* Vol. 306, No. 5696, pp. 666-669, ISSN 1095-9203

Ni Z. H., Wang H. M., Kasim J., Fan H. M., Yu T., Wu Y. H., Feng Y. P., and Shen Z. X., (2007). Graphene Thickness Determination Using Reflection and Contrast Spectroscopy. *Nano Lett.* Vol.7, No.9, pp. 2758-2763, ISSN 1530-6984

Okamoto T; Yamaguchi I. (2003). Optical Absorption Study of the Surface Plasmon Resonance in Gold Nanoparticles Immobilized onto a Gold Substrate by Self-assembly Technique. *J. Phys. Chem. B.*, Vol.107, No.38, pp. 10321-10324, ISSN 1520-6106

Ormonde A. D; Hicks ECM.; Castillo J.; Van Duyne RP. (2004). Nanosphere Lithography: Fabrication of Large-area Ag Nanoparticle Arrays by Convective Self-assembly and Their Characterization by Scanning UV-visible Extinction Spectroscopy. *Langmuir,* Vol.20, No.16, pp.6927-6931, ISSN 0743-7463

Palik E. D. (1991). *Handbook of Optical Constants of Solids*, Academic Press, ISBN 0-12-544422-2, New York

Patel D. V.; McGhee C. N. J. (2007). *Clinical and Experimental Opbtbalmology,* Contemporary in Vivo Confocal Microscopy of the Living Human Cornea Using White Light and Laser Scanning Techniques: a Major Review. Vol.35, No.1, pp. 71-88, ISSN 1442-6404

Pinchuk A.; Hilger A.; Plessen G.; Kreibig U. (2004). Substrate Effect on the Optical Response of Silver Nanoparticles. *Nanotechnology,* Vol.15, No.12, pp. 1890-1896, ISSN 0957-4484

Rembe C.; Dräbenstedt A. (2006). Laser-scanning confocal vibrometer microscope: Theory and experiments. *Review of Scientific Instruments,* Vol. 77, No.8, pp. 083702, ISSN 0034-6748

Ribes A. C.; Damaskinos S.; Dixon A. E; Carver G. E; Peng C.; Fauchet P. M.; Sham T. K.; Coulthard I. (1995). Photoluminescence Imaging of Porous Silicon Using a Confocal Scanning laser Macroscope/Microscope. *Appl. Phys. Lett.,* Vol.66, No.18, pp.2321-2323, ISSN 0003-6951

Saijo Y.; Hozumi N.; Lee C.; Nagao M.; Kobayashi K.; Oakada N.; Tanaka N.; Filho E. S.; Sasaki H.; Tanaka M.; Yambe T. (2006). *Ultrasonics,* Vol. 44, pp. e51-e55, ISSN 0041-624X

Schider G.; Krenn J. R.; Hohenau A.; Ditlbacher H.; Leitner A.; Aussenegg F. R.; Schaich W. L.; Puscasu I.; Monacelli B.; Boreman G. (2003). Plasmon Dispersion Relation of Au and Ag Nanowires. *Phys. Rev. B.,* Vol. 68, No. 15, pp. 155427, ISSN 1098-0121

Singh B. K.; Hillier A. C. (2007). Multicolor Surface Plasmon Resonance Imaging of Ink Jet-Printed Protein Microarrays. *Anal. Chem.,* Vol.79, No.14, pp. 5124-5132, ISSN 0003-2700

Tao A.; Kim F.; Hess C.; Goldberger J.; He R. R.; Sun Y. G.; Xia Y. N.; Yang P. D. (2003). Langmuir-Blodgett Silver Nanowire Monolayers for Molecular Sensing Using Surface-enhanced Raman Spectroscopy. *Nano Letters,* Vol.3, No.9, pp. 1229-1233, ISSN 1530-6984

Veshapidze G.; Trachy M. L.; Shah M. H.; DePaola B. D. (2006). Reducing the Uncertainty in Laser Beam Size Measurement with a Scanning Edge Method. *Appl. Opt.,* Vol.45, No.32, pp. 8197-8199, ISSN 0003-6935

Verveer P. J.; Swoger J.; Pampaloni F.; Greger K.; Marcello M.; Stelzer E. H. K. (2007). High-Resolution Three-Dimensional Imaging of Large Specimens with Light Sheet–Based Microscopy. *Nature Methods,* Vol. 4, No. 4, pp. 311-313, ISSN 1548-7091

Wang Y. Y.; Ni Z. H.; Liu L.; Liu Y. H.; Cong C. X.; Yu T.; Wang X. J.; Shen D. Z.; and Shen Z. X. (2010). Stacking-dependent Optical Conductivity of Bilayer Graphene, *Acs Nano,* Vol.4, No.7, pp. 4074-4080, ISSN 1936-0851

Wang Z. H.; Liu J. W.; Chen X. Y.; Wan J. X.; Qian Y. T. (2004). A Simple Hydrothermal Route to Large-Scale Synthesis of Uniform Silver Nanowires. *Chem. Eur. J.,* Vol.11, No.1, pp. 160-163, ISSN 1521-3765

Wind M. M.; Vlieger J.; Bedeaux D. (1987). The Polarizability of a Truncated Sphere on a Substrate. *Physica A.,* Vol.141, pp. 33-57, ISSN 0378-4371

Youk Y. C.; Kim D. Y. (2006). A Simple Reflection-type Two-dimensional Refractive Index Profile Measurement Technique for Optical Waveguides. *Opt. Comm.,* Vol.262, No.2, pp. 206-210, ISSN 0030-4018

12

Time Resolved Camera: The New Frontier of Imaging Devices

Salvatore Tudisco
INFN - Laboratori Nazionali del Sud, Catania
Italy

1. Introduction

Time resolved imaging up to the single photon sensitivity is one of the most ambitious and important goals of photonics. Currently there are no commercial devices able to provide both the information on position (imaging) and arrival time of photons emitted by weak and ultra-weak sources. Only few and very expensive devices (Intensified CCD, electron bombarded CCD etc.) are able to reach the single photon detection threshold together with the possibility to collect photons in small time window (up to few ns). Unfortunately such devices are not able to provide any information on the arrival time of photons, fundamental for the recent developments on ultra-fast, time correlated optical sensing techniques.

Fluorescence-based imaging (both single and multi-photon) is the research field that has most influenced the development of fast and sensitive optical detectors. Examples of techniques in this class include Förster resonance energy transfer (FRET) (Jares-Erijman et al., 2003), fluorescence lifetime imaging microscopy (FLIM) (Becker et al., 2006), and fluorescence correlation spectroscopy (FCS) (Schwille et al., 1999). The success of these techniques, particularly FLIM, derives from the ability to characterize an environment based on the time domain behaviour of certain fluorophores with high resolution in space domain. This characterization can be done today with high levels of accuracy in 3D with minimal interference from the surroundings and almost no dependence on fluorophore concentration.

Many others scientific areas like astronomy, biophysics, biomedicine, nuclear and plasma physics etc. can benefit from a time resolved imaging device; it can improve the actual detection limits providing physical information otherwise inaccessible.

In astronomy and astrophysical science, one of the toughest problems affecting ground-based telescopes is the presence of the atmosphere, which distorts the spherical wave-front, creating phase errors in the image-forming ray paths. Even at the best sites, ground-based telescopes observing at visible wavelengths cannot achieve an angular resolution in the visible better than telescopes of 10 to 20 cm diameter, because of atmospheric turbulence alone. The cause is random spatial and temporal wave-front perturbations induced by turbulence in various layers of the atmosphere; one of the principal reasons for flying the Hubble Space Telescope was to avoid this image smearing. In addition, image quality is affected by permanent manufacturing errors and by long timescale wave-front aberrations introduced by mechanical, thermal, and optical effects in the telescope, such as defocusing, decentring, or mirror deformations generated by their supporting devices.

Adaptive optics (AO) and Natural Guide Star (NGS) are the solutions to these problems: a deformable mirror is inserted in the light path of the telescope, and its control signal is based on measurement of the incoming wave-front, performed by a suitable high-sensitivity and time resolved imaging detector. NGS adaptive optics suffers from the fact that it is not always possible to find a suitable NGS close enough to the portion of sky under investigation. Normally only 1% of the sky has an available NGS for current AO systems. The most promising way to overcome the lack of sufficiently bright natural reference stars is the use of artificial reference stars, also referred to as laser guide stars (LGS). These are patches of light created by the back-scattering of pulsed laser light by sodium atoms in the high mesosphere, or in the case of Rayleigh LGS, by molecules and particles located in the low stratosphere. Such an artificial reference star can be created as close to the astronomical target as desired, if will be available a wave-front sensor measuring the LGS wave-front in order to correct the atmospheric aberrations on the target object.

In plasma science, even 60 years after the invention of the laser, we witness a rapid development of systems generating electromagnetic pulses with extreme parameters such as duration, wavelength, peak power, and focused intensity.

The employment of solid-state laser materials allows the generation and subsequent amplification of light pulses as short as a few optical cycles only. When combined with the technique of chirped pulse amplification - CPA (Strickland et al. 1985) where the laser pulses are temporally stretched and recompressed before and after their amplification, table-top laser systems reaching peak powers of several tens or hundreds of Terawatt (1 TW = 10^{12} W) can be realized. At such huge intensities, the rapidly oscillating electric field of the laser pulse reaches peak values exceeding the atomic fields binding the electrons to the positively charged nucleus by several orders of magnitude. It is due to this fact that all kinds of matter when exposed to laser light shot under such extreme conditions are almost instantaneously ionized and a plasma – sometimes called the "4th state of matter" – is formed. Within such plasma, the interaction between the charged constituents mediated by the long-range Coulomb interaction governs the behaviour and the evolution of the plasma. This gives rise to a large magnitude of effects that makes the generation and application of plasmas a fascinating field of current research in physics.

In this context, the use of a time resolved imaging devices to get information on the spatial and temporal evolution of laser generated plasma is fundamental to improve the actual level of knowledge.

Thus far, in many time-resolved and/or high-sensitivity applications the detectors of choice have been photomultiplier tubes (PMTs) and multichannel plates (MCPs) (McPhate et al. 2005). While these devices can reach time uncertainties of a few tens of picoseconds (MCPs), usually are bulky, fragile, sensitive to electromagnetic disturbances (especially PMTs) and mechanical vibrations, require high supply voltages (2 – 3 kV) and are costly devices, particularly the high-performance models. Moreover multi-pixel images are not possible without bulky setups and expensive equipment. Thus, high sensitivity and/or time-resolved imaging has been relegated to applications requiring important investments for optical and detector equipment.

A number of solid-state solutions have been proposed as a replacement of MCPs and PMTs using conventional imaging processes. The challenge, though, has been to meet single photon sensitivity and low timing uncertainty. To address the sensitivity problem, cooled and/or intensified CCDs (Etoh et al. 2005) and ultra-low-noise CMOS APS architectures (Kawai et al. 2005) have been proposed. Multiplication of photo-generated charges by impact ionization has also been used in CCDs (Hynecek et al. 2001).

As an alternative to PMTs and MCPs, researchers have turned to solid-state photon counters based on avalanche photodiodes (APDs). In the last four decades, solid-state multiplication based photo-detectors have gradually evolved from relatively crude devices to the sophistication of today. Semiconductor APDs have the typical advantages of solid state devices (small size, low bias voltage, low power consumption, ruggedness and reliability, suitability to build integrated systems, etc.). Their quantum detection efficiency is inherently higher, particularly in the red and near infrared range. In APDs operating in linear mode the internal gain is not sufficient or barely sufficient to detect single photons. However, single photons can be efficiently detected by avalanche diodes operating in Geiger-mode, known as single-photon avalanche diodes (SPADs). Almost every imaging technology has one photo-detector device and the range of implementations is quite wide. In this context, SPADs have recently attracted significant interest thanks to their relative simplicity and ease of fabrication.

2. Single photon avalanche diodes

In the last decades, the possibility to build a silicon photo-sensor suitable for single photon counting applications was investigated by several research teams (Cova et al. 1983 and references therein). The original idea, firstly proposed by R.J. Mc Intyre since the 1960s, was to implement a semiconductor photodiode with characteristics suitable for the triggered avalanche operation mode, known as Geiger mode of operation, and therefore able to detect single photons.

Carriers generated by the absorption of a photon in the p–n junction, are multiplied by impact ionization thus producing an avalanche. APDs can reach timing uncertainties as low as a few tens of picoseconds thanks to the speed at which an avalanche evolves from the initial carrier pair forming in the multiplication region. An APD is implemented as photodiode reverse biased near or above breakdown, where it exhibits optical gains greater than one. When an APD is biased below breakdown it is known as proportional or linear APD. It can be used to detect clusters of photons and to determine their energy. When biased above breakdown, the optical gain becomes virtually infinite (see fig. 1). Thus, with relatively simple ancillary electronics, the APD becomes capable of detecting single photons. The APD operating in this regime is called single-photon avalanche diode - SPAD. If the primary carrier is photo-generated, the fast leading edge of the avalanche pulse marks the arrival time of the detected photon. After the avalanche is triggered, the current keeps flowing until the avalanche is quenched by lowering the bias voltage down to breakdown voltage or below. The bias voltage is then restored in order to detect another photon. This operation requires a suitable circuit that is usually referred to as a quenching circuit.

Individual detectors and detector arrays based on the SPAD technology have received renewed interest in recent years due to the versatility of their applications.

There are two main lines of research in silicon SPADs: one that advocates the use of highly optimized processes to boost performance and one that proposes to adapt SPAD design to existing processes to reduce cost and to maximize miniaturization. Both approaches have advantages and drawbacks.

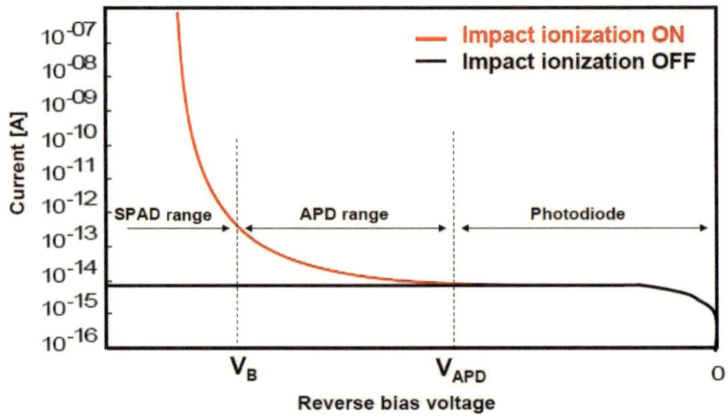

Fig. 1. Typical multiplication regions of reverse polarization junction.

The basic goals of the SPAD fabrication technology are to: *(i)* keep low the dark counting rate; *(ii)* keep low the afterpulsing probability; *(iii)* make uniform the electric field over the whole active area in order to have a photo-detection efficiency (PDE) independent from the absorption position; *(iv)* keep low the photo-generation of carriers outside the multiplication region in order to minimize the time uncertainty.

As it happens in PMTs, thermal generation effects produce current pulses even in the absence of illumination, and the Poissonian fluctuation of these dark counts represents the internal noise source of the detector. The dark pulses are due to carriers thermally generated in the SPAD junction, so that the count rate increases with the temperature, as does the dark current in ordinary photodiodes. The rate also increases with the excess bias voltage (EBV, which is the voltage over the breakdown) because of two effects: *i*) field-assisted enhancement of the emission rate from generation centers (Hurkx et al. 1996) and *ii*) increase of the avalanche triggering probability (Oldham et al. 1972).

The SPAD count rate includes also secondary pulses due to afterpulsing effects that in the case of dark count may strongly enhance the total measured rate.

During the avalanche some carriers are captured by deep levels in the junction depletion layer and subsequently released with a statistically fluctuating delay, whose mean value depends on the deep levels actually involved (Cova et al. 1991). Released carriers can retrigger the avalanche, generating afterpulses correlated with a previous avalanche pulse. The number of carriers captured during an avalanche pulse increases with the total number of carriers crossing the junction, that is, with the total charge of the avalanche pulse. Therefore afterpulsing increases with the delay of avalanche quenching and with the current intensity, which is proportional to the EBV (usually dictated by photon detection efficiency or time resolution requirements, or both (Cova et al. 1996)) so that the trapped charge per pulse first has to be minimized by minimizing the quenching delay. The situation is even more disadvantageous when the detector is cooled to reduce the dark-counting rate, since the release from trapping states becomes slower.

In conclusion, a really suitable technology for producing SPADs must not only reduce the generation and regeneration centers to very low concentration level, but also eliminate trapping level or at least minimize their concentration. The technological challenge is to design a process with such characteristics and still compatible with standard microelectronic industry processes.

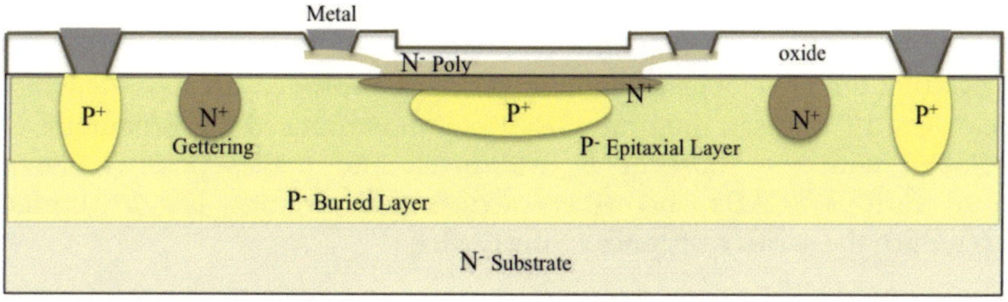

Fig. 2. Junction scheme of SPAD device

2.1 Fabrication and structure design

There exist several implementation styles for APDs, of which two are the most used. In the first style, known as reach-through APD, one builds a $p+$-π-p-n structure where π denotes very lightly p-doped (McIntyre, 1985). When reverse biased, the depletion region extends from the cathode to the anode. Thus, the multiplication region is deep in the $p/n+$ junction. Due to the depth of the multiplication region, this device is indicated for absorption of red and NIR photons up 1.1 μm (for silicon). Since the photoelectrons drift until the multiplication region, a larger timing uncertainty is generally observed.

The second implementation style is compatible with planar CMOS processes and it involves a shallow or medium depth p or n layer to form high-voltage pn junctions. Cova and others have investigated devices designed in this style since the 1970s, yielding a number of structures (Cova et al. 1981). All these structures have in common a pn junction and a zone designed to prevent premature edge breakdown. An example of the early structures is reported in the work of Zappa et al. (1997) $n+/p+$ enrichment in p-substrate was used, while premature edge breakdown was prevented by confining $p+$ enrichment in the centre of the device.

More recently, many authors have developed APDs, both in linear and Geiger mode, using dedicated planar and non planar processes, achieving superior performance in terms of sensitivity and noise (Kindt 1999).

The main disadvantage of using dedicated processes is generally the lack of libraries that can support complex functionalities and deep-submicrometre feature sizes, thus limiting array sizes. An interesting alternative is the use of a hybrid approach whereby the APD array and ancillary electronics are implemented in two different processes, each optimized for APD performance and speed, respectively. If the ancillary electronics is implemented in CMOS, high degrees of miniaturization are possible. The price to pay is increased fabrication complexity. The integration of SPADs in a low-cost CMOS process became feasible since 2003 (Sciacca et. al 2003).

An example of latest developed structures is shown in figure 2 (Sciacca et. al 2003). The *np* junction has an *n*+ shallow diffusion and a controlled Boron enrichment diffusion in the central zone of it. With respect to the outer abrupt *np* junction, the higher *p*+ doping concentration reduces the breakdown voltage in the central zone, which is the active area.

The process starts with a Si <100> *n*- substrate on which is grew a boron doped epitaxial layer with a *p*+ buried layer and with a *p*- doped layer. The reason to form a buried *pn* junction is twofold. First, the detector time-response is improved because the effect of photo-generated carriers diffusing in the undepleted region is reduced (G. Ripamonti et al. 1985). Second, isolation with the substrate is introduced and makes possible the monolithic integration of various SPADs and other devices and circuits. The *p*+ buried layer is necessary to reduce the series resistance of the device.

The *p*- layer must be thin enough to limit the photo-carrier diffusion effect above mentioned. A good trade off has to be found for this thickness, because if it is made too thin the edge breakdown occurs at a voltage not much higher than the breakdown voltage of the active area. The *p*+ sinkers are then created with a high-dose boron implantation step, in order to reduce the contact resistance of the anode and provide a low resistance path to the avalanche current.

The next step, a local gettering process, is a key step in the process because it guarantees a uniform defect concentration over the volume. At this point a heavy $POCl_3$ diffusion through an oxide mask is made on the topside of the wafer close to the device active area. Heavy phosphorus diffusions are well known to be responsible for transition metal gettering. Unfortunately, the well-known phosphorous predeposition on the backside of the wafer is not able to getter the distant active area of the device because metal diffusers (Pt, Au , Ti) diffuse too slowly during the final anneal. For this reason, if the gettering sites are created suitably close to the active region, a major improvement on dark counting rate is observed. The next step is the *p*+ enrichment diffusion obtained with a low energy boron implantation, producing a concentration peak, followed by a high temperature anneal and drive in (Lacaita et al. 1989).

The first generation of devices was fabricated with a deposited polysilicon cathode doped by Arsenic implantation and diffusion. The As+ ion implantation energy was carefully calculated in order to damage as little as possible the active area of the device; nevertheless, devices with very high dark-counting rate resulted. A remarkable improvement was obtained in the second generation by doping in situ the polysilicon. Further improvement was achieved in the third generation by accurately designing a Rapid Thermal Anneal to create a precisely controlled shallow Arsenic diffusion below the polysilicon in the *p*-epilayer.

An important issue for the SPAD quality is the uniformity of the electric field over the active area. If the electric field is not uniform, the PDE of the device becomes dependent on the absorption position over the active area. The lower the electric field the lower the PDE, the worst case being when the electric field is lower than the breakdown value.

2.2 Quenching strategy

After the avalanche is triggered, the current keeps flowing until the avalanche is quenched by lowering the bias voltage down to breakdown voltage or below. After a dead-time, the operative voltage must be restored in order to make the SPAD able to detect another photon. This operation requires a suitable electronics with the following tasks: (i) it senses the leading edge of the avalanche current; (ii) it generates a standard output pulse, synchronous with the current onset; (iii) it quenches the avalanche by lowering the bias below the breakdown voltage; (iv) it restores the photodiode voltage to the operating level. This circuit is usually referred to as quenching circuit. The two main quenching strategies are: passive and active (Cova et al. 1996). Recently, very promising results were obtained by using also a mixed approach (Mingguo et al. 2008).

In passive quenching the avalanche current itself is used to drop the voltage across the diode. This is generally accomplished via a ballast resistor R_L (normally few hundreds kΩ) placed on the anode or the cathode of the diode as shown in figure 3. The detection of the avalanche can be accomplished by measuring the current across a low resistivity path R_S. Pulse shaping may be performed using a discriminator.

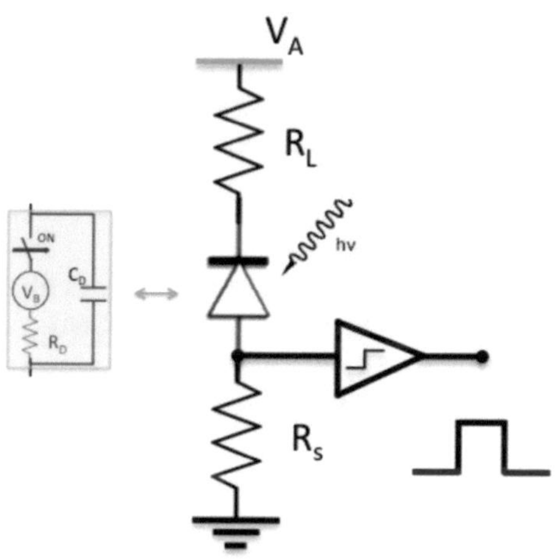

Fig. 3. Schematic representation of a passive quenching circuit. In the left part is shown the equivalent circuit of a SPAD detector.

Today modern technology gives also the possibility to produce SPAD detectors with integrate quenching mechanism based on a Metal-Resistor-Semiconductor structure. Precise resistive elements (or using the non-linear characteristics of a biased PMOS or NMOS, Niclass et al. 2005) are embedded for each individual micro-cell of the array and provide effective feedback for stabilization and quenching of the avalanche process also reducing the values of the stray capacitance from cathode to ground.

A circuit model, which emulates the evolution of the signal of a SPAD was developed in the 1960s to describe the behaviour of micro-plasma instability in silicon (McIntyre 1961). According to this model, the pre-breakdown state can be represented as a capacitance (junction capacitance, C_D) in series with the quenching resistor. Referring to Fig. 3(left), this state corresponds to the switch in the OFF condition. In steady state, the capacitance is charged at operating voltage $V_A > V_B$ where V_B is the breakdown voltage.

When a carrier traverses the high-field region, there is a certain probability, known as turn on probability, to initiate an avalanche discharge. If this happens, the new state of the system can be modelled adding to the circuit a voltage source V_B with a series resistor R_D in parallel to the diode capacitance (switch closed in Fig. 3(left)). R_D includes both the resistance of the neutral regions inside the silicon as well as the space charge resistance. C_D, originally charged at V_A, discharges through the series resistance down to the breakdown voltage with a time constant τ_D given by the product R_DC_D. It should be noted that the discharge current is initially limited by the build up of the avalanche process which can take some hundreds of ps.

As the voltage on C_D decreases, the current flowing through the quenching resistance, and as a consequence through the diode, tends to the asymptotic value of $(V_A-V_B)/(R_L+R_D)$. In this final phase, if R_L is high enough, the diode current is so low that a statistical fluctuation brings the instantaneous number of carriers flowing through the high-field region to zero, quenching the avalanche. The probability of such a fluctuation (turnoff probability) becomes significant when the diode current is below 10–20 mA (defined as latching current). The average time needed to stop the avalanche, when this condition is satisfied, is in the order of 1 ns. The latching current poses a strict limit on the lower value of R_L to some hundreds of kΩ. As the avalanche process is terminated, the switch is again open and the circuit is in its initial configuration. The capacitance charged at V_B, starts recharging to the bias voltage with a time constant C_DR_L, and the device becomes ready to detect the arrival of a new photon.

A photon that arrives during the very first part of recovery is almost certainly lost, since the avalanche triggering probability is almost negligible. Instead, subsequent photons have a progressively higher probability to trigger the SPAD. Unfortunately, SPAD triggering during recovery transition has mainly two deleterious effects due to the time-varying excess bias voltage.

Pulses having amplitude lower than the threshold of the discriminator are not sensed. Significant count losses are expected at higher counting rates. As a matter of fact, after each ignition, the detector has a dead-time which is not well-defined.

Time resolution is degraded for two reasons: the intrinsic time resolution of the SPAD is impaired when excess bias is reduced; additional jitter is introduced because pulses with different amplitudes cross the comparator threshold at different times.

In order to avoid the highlighted drawbacks of passive quenching circuit (PQC) and fully exploit the intrinsic performance of SPADs, a new approach was devised (Cova et al. 1996). The basic idea was to sense the rise of the avalanche pulse and react back on the SPAD, by forcing the quenching and reset transitions in short (few nanoseconds) times, with a controlled bias-voltage source. This approach was called active-quenching circuit (AQC) the literature on active quenching is extensive.

The simplest solution provides that the discriminator triggers also a driving stage that applies a quenching pulse, synchronous with the avalanche triggering and with a very low jitter. In order to quench the avalanche, the quenching pulse must be high enough to reduce the diode voltage below V_B. The detector is then kept off for a well-controlled hold-off time, at the end of which the driver swiftly restores the SPAD bias voltage to the operating level in order to be ready to detect another photon. During the reset transition, since spurious couplings and reflections could retrigger the discriminator, it can be latched off for the whole reset duration.

In this way the avalanche current intensity is kept constant due to the low-impedance of the driver biasing the detector, also the duration of the avalanche current pulse is constant and depends on the time taken by the avalanche signal to travel from the SPAD to the AQC, forth and back, and on the slope of the driver quenching pulse. By reducing this loop time, the afterpulsing will be limited, since the number of trapped carriers is linearly proportional to the avalanche pulse duration. Moreover, by using ACQ, the hold-off time is easily adjustable and both quenching and reset transitions are fast (tens of nanoseconds), thus minimizing the probability of non-standard avalanche triggering during the recovery transition. Nevertheless, recent works (Neri et al. 2010) demonstrates that with passive quenching strategy (by using hybrid Dead-Time models) it is also possible to evaluate the real amount of incident photon rate up to 10^7-10^8 cps.

In conclusion, even if ACQ are attractive, they usually require an extra complexity to a pixel, hindering the miniaturization, fundamental for imaging aim.

2.3 Performance and measurement techniques

In this section and in next, the attention will focus on SPADs with integrate quenching resistor. In contrast to APDs, to establish the performances are necessary a new set of parameters and measurement techniques.

A primary "static" characterization is made on-wafer by using temperature-controlled probe stations and semiconductor parameter analysers. By working in dark condition the SPAD reverse/forward I-V characteristics can be determined, with and without quenching resistor and as a function of the temperature. In this way we get information about leakage current, breakdown voltage, quenching resistor and hence on the overall device performance. For example the leakage current is linked to the generation of electrical carriers both in the bulk as well as on the surface depleted region around the junction. Then, low values of leakage current can be interpreted as the first evidence of the low defectivity of the diode and then of the low dark counting rate. The figure 4 reports the I-V characteristics of ST-Microelectronics SPAD (Mazzillo et al. 2008).

The operation of a SPAD with integrated quenching resistor can be easily investigated by measuring the current across a low resistivity path R_S ("dynamic" characterization). In figure 5a is reported the pulse of a single dark count event observed on a FBK SPAD.

The fast leading edge (rise time) is governed by the time that avalanche takes to spread all over the diode active area. The slower exponential decay (discharge time) as previous mentioned is instead determined by a constant time given by the product of the diode internal resistance (few kΩ) and the sum of parasitic and internal diode capacitances. After the breakdown current has been quenched, the diode will slowly recharge through the quenching resistor (see fig.5b).

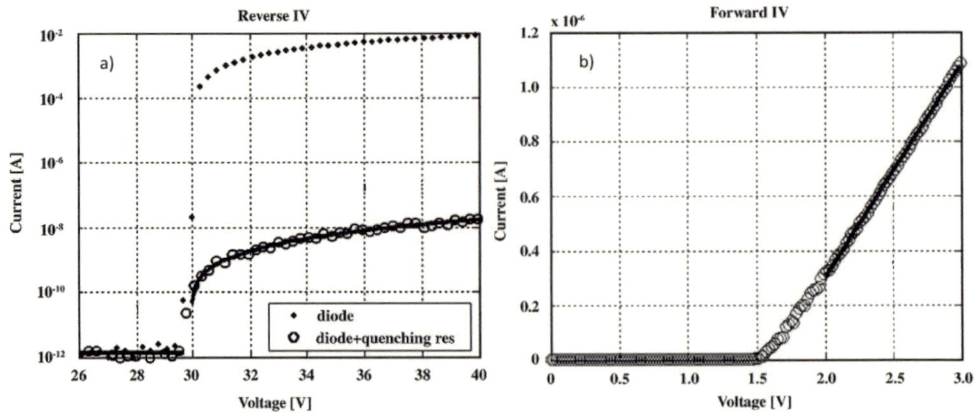

Fig. 4. I-V characteristics for a square shaped (30 μm side) ST-Microelectronics SPAD; *a)* reverse; *b)* forward.

Total charge and device gain can be extracted by integrating the pulse, converted into a current signal. The junction capacitance can be evaluated estimating the slope of the charge-EVB characteristic.

The *Dark Noise* is a function of EBV, temperature and detector area. Commonly it may vary from a minimum of tens of hertz to a few kilohertz and is a superposition of dark count and afterpulsing rate. In figure 5b the oscilloscope was triggered on a SPAD Geiger pulse (either dark count or afterpulsing event), while the device was kept in dark condition. The time delay between the leading edges of the reference pulse and a consecutive pulse, which can be due to dark count or afterpulsing, was measured. Then the average delay between two consecutive pulses was obtained. The inverse of this value represents the dark noise rate of the device.

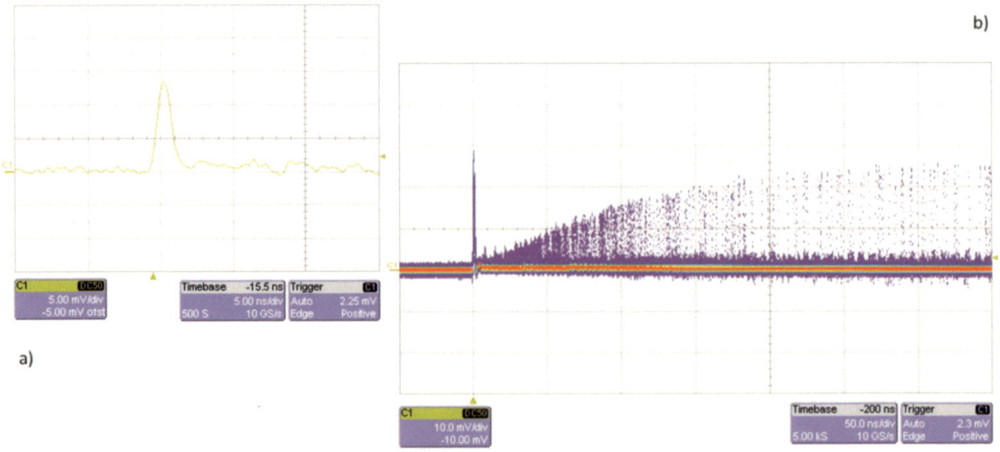

Fig. 5. Dynamic characterization of a circular FBK-IRST SPAD (30 μm diameter). The signals in *a*) and *b*) correspond to dark events when the device is respectively biased at 10 and 20 % of EBV. In *a*) is displayed the structure of a single pulse; *b)* reports the afterpulses distribution through a persistence image.

The distribution of the *Afterpulsing* events, as a function of the temperature and EBV, can be obtained by building the histogram of the time delay between two consecutive pulses within a time window of some hundreds of ns, being the probability of two independent dark counts events in such a time window negligible (the average time scale between two uncorrelated dark events is of the of the order of ms). The distribution is then normalized to the total number of collected events. The histogram is bell-shaped (Mazzillo et al. 2008) as a convolution of the exponentially decreasing probability of releasing a carrier from a trap and the exponentially increasing EBV due to the recharging phase of the depletion layer. The probability that a released carrier can trigger an avalanche event increases with time because during the recharging phase the voltage at the depletion layer approaches exponentially to the bias voltage. On the other hand the probability of release of the captured carriers decreases very quickly with time since this phenomenon is mainly due to traps that are located near the conduction and valence band edges. The afterpulsing probability is commonly a few percent.

The *Photon Detection Efficiency* (PDE) of SPAD detector is given by the product of two parameters: the quantum efficiency (QE) and the avalanche triggering probability (PT).

The QE represents the probability for a photon to generate an *e–h* pair in the active thickness of the device. It is given by the product of two factors: the transmittance of the dielectric layer on top of the silicon surface and the internal QE. Both are wavelength dependent. The former can be maximized, by implementing an anti-reflective coating. The second term represents the probability for a photon that has passed the dielectric layer to generate an *e–h* pair in the active thickness. In a conventional *n+/p/p+* diode, the active layer is roughly limited on top by the undepleted *n+* layer, whereas on the bottom by the *p+* layer used for the ohmic contact or by the highly doped substrate in case of epitaxial substrates. Indeed, when a pair is generated in those regions, there is a high probability for the electron and hole to recombine due to Auger or Shockley–Read–Hall processes (Tyagi 1991). For short wavelengths, the problem is focused in the top layer. As an example, a 420nm light is almost totally absorbed in the first 500 nm of silicon, which, for non-optimized fabrication processes, is usually well inside the undepleted layer.

There is a finite probability for a carrier to initiate an avalanche when passing through a high-field region. In case of a photogeneration event, two carriers are created travelling in opposite directions. Both contribute to the *PT* that can be evaluated from the following expression: $PT = Pe + Ph - PePh$. Where *Pe* and *Ph* are the electron and hole breakdown initiation probabilities. These terms can be calculated as a function of the generation position by solving two differential equations involving the carrier ionization rates. *Pe* and *Ph* depend on the impact ionization rates of electrons (α_n) and holes (α_p), respectively. These parameters are not well determined yet, and large discrepancies exist among the values extracted from the various models (Grant et al. 1973). Anyway, despite the differences in absolute values, some features are well established: (i) both coefficients increase with the electric field, (ii) the electron has an ionization rate higher than the hole (is about twice α_p), and (iii) their difference decreases with increasing fields.

This behaviour is reflected in the probabilities *Pe* and *Ph*. Thus, to maximize the triggering probability: (*i*) the photogeneration should happen in the p side of the junction in order for the electrons to pass the whole high field zone, and (*ii*) the bias voltage should be as high as possible.

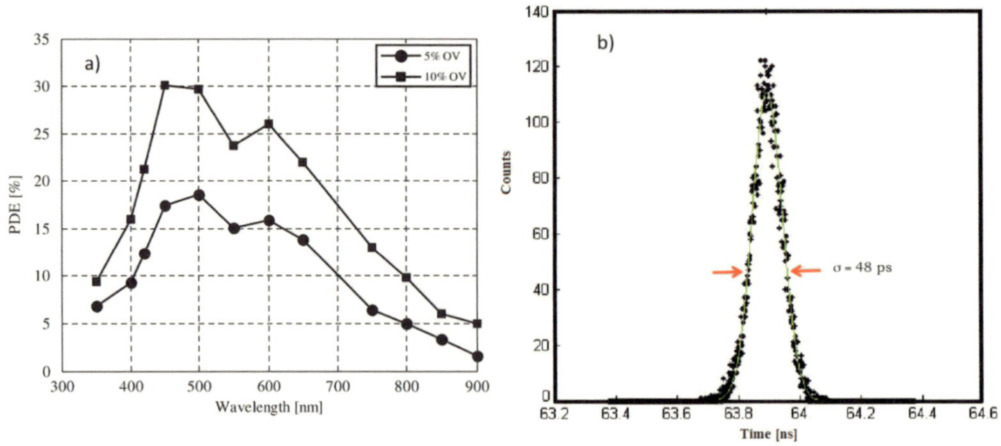

Fig. 6. *a)* PDE of ST-Microelectronics SPAD at 5 and 10% of EBV; *b*) Photon timing of passively quenched ST-Microelectronics SPAD.

The measurement principle adopted for the evaluation SPAD PDE is a direct comparison with a calibrated photodiode that receives the same photon flux of SPAD. An example of experimental set-up where, with high level of accuracy, are possible PDE measurements is discussed by Bonanno et al. 2005. In figure 6a is reported the typical PDE measured for the ST-Microelectronic devices.

The SPAD performance in *Photon Timing* is characterized by its time resolution curve, which is the statistical distribution of the delay between the true arrival time of the photon and the measured time, marked by the onset of the avalanche current pulse. These can be obtained by means of a time-correlated photon counting apparatus (Privitera at al. 2008), where optical pulses with short time duration (less than 30 - 40 ps) are used to illuminate the devices. After a careful check that the contribution of the laser pulse and the eventual electronic jitter to the SPAD response are negligible, by using a digital oscilloscope it is possible to build the histogram of the time delay between the laser reference pulse and the SPAD pulse.

In figure 5b is reported the time resolution curve obtained by using laser pulses of 408 nm wavelengths, 35 ps FWHM and 1 kHz repetition rate, on an ST-Microelectronic SPAD. It is consistent with the typical values reported in the literature for other devices.

3. Bidimensional arrays of SPADs

As discussed in the previous sections, SPAD arrays fabricated by standard planar silicon production processing, are at the moment considered an interesting topic of research because many photonic applications, not only scientific, would become available at much more reasonable cost.

The main requirement in the array fabrication is to guarantee electrical and optical isolation among pixels. In fact, each pixel must be operated independently from the others, independently on neighboring pixels and without disturbing them.

A typical way to achieve electrical isolation in silicon planar technology is to employ isolation diffusion, i.e. a thus electrically isolating the epitaxial layer, of the opposite doping sufficiently deep diffusion that reaches the substrate of the same doping type, type, as shown in Figure 7. However, the hardest problem to solve in SPAD array design is optical coupling. This effect is peculiar to the above-breakdown operation of SPADs.

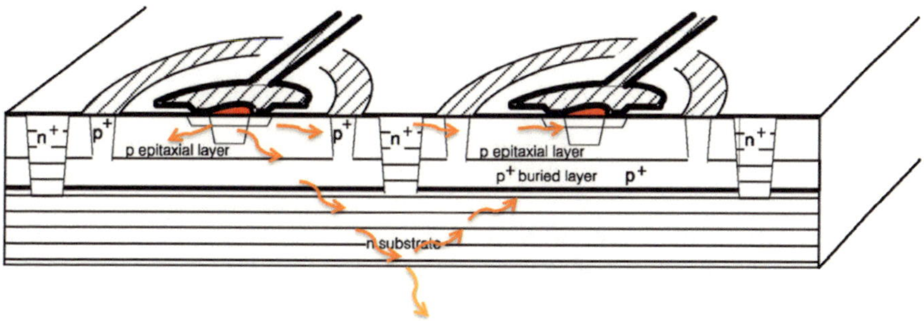

Fig. 7. Optical cross-talk representation between two neighbouring devices. The avalanche current in one pixel can trigger the process in surrounding pixels, due to photons emitted by hot carriers.

During the avalanche process each ionization impact causes high deceleration in electron drift, so the Bremsstrahlung process enables a secondary photons production inside the junction (the emission probability is estimated in 10^{-5} photons per carrier crossing the junction).

The optical cross talk is then activated by the emission of secondary photons, that can travel along the silicon bulk, transparent to the optical wavelength, and cross another diode that is triggering a correlated avalanche (Fig. 7). The time delay between the two produced signals is less than the measurable one and induced signals cannot be distinguished from the generated one. A further contribute to the optical cross talk was found in the indirect optical pat enabled by the optical reflection by the back part of the device.

In order to avoid the optical cross-talk, it is possible design and fabricate arrays that are optically and electrically isolated by deep thin trench technology. The trench process starts with a vertical etch (about 10 μm deep and 1μm large), a subsequent oxide deposition for complete electrical isolation. The process continues with tungsten fill to avoid optical crosstalk and ends with planarisation.

Nevertheless, all the cross-talk contributions previous mentioned, become important when the density of implemented elements is higher and the distances between neighbouring devices are smaller as required by imaging applications or by silicon PhotoMultiplier technology. They influence the total Dark counting rate of the device.

3.1 Silicon photomultiplier concept

A photomultiplier based on the silicon technology represents the new frontier of photodetection. SPADs integrated on the same substrate, with a common read-out, could satisfy such expectations. The main limitation of a single diode working in Gaiger mode is

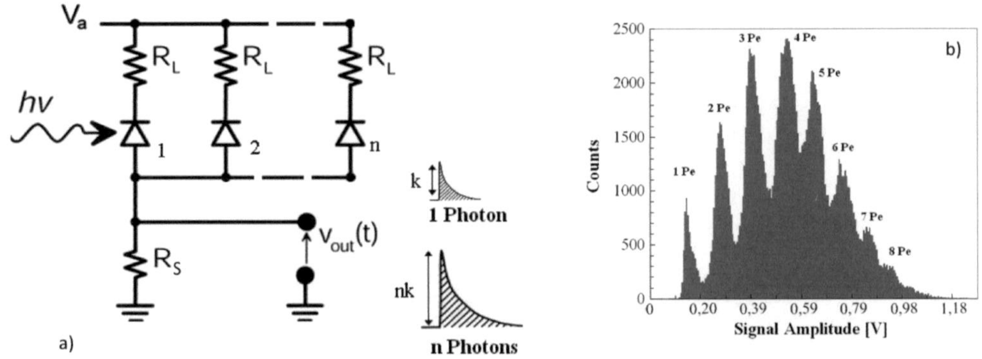

Fig. 8. *a*) Schematic representation of SiPM working principle. The amplitude of the out signals is proportional to the number of detected photons. In *b*) is reported the distribution.

that the output signal is the same regardless of the number of interacting photons. In order to partially overcome this limitation, the diode can be segmented in tiny micro-cells (each working in Gaiger mode) connected in parallel to a single output. Each element, when activated by a photon, gives the same current response, so that the output signal is proportional to the number of cells hit by a photon. The number of elements composing the device limits the dynamic range, and the probability that two or more photons hit the same micro-cell depends on the size of the micro-cell itself. This structure is called Silicon PhotoMultiplier (SiPM). Some of the advantages offered by the solid-state solution are: insensitivity to magnetic fields, ruggedness, compactness, low operating voltage and long lifespan. In addition, this technology facilitates the interconnection between the detector and the read-out electronics.

An interesting application of SiPMs is the detection of the light emitted by scintillators. Among the various types of scintillators, particular attention has recently been given to lutetium oxyorthosilicate (LSO) for its high light yield, short decay time and relatively good mechanical properties. These features are extremely useful, for example, in positron emission tomography (PET).

3.2 Image sensor

The identification of the fired pixel and the time at which the event occurred are the fundamental requirements in order to realize the time resolved imaging. These two features are almost simultaneously incompatible with the traditional reading techniques (CCD etc.). High-definition images require a large number of pixels and reading techniques designed to minimize the number of connections. On the other hand, these are almost incompatible with the information on the photon arrival time.

To avoid missing photon counts, a counter should be used for each pixel. However, large counters are not desirable due to the fill factor loss and/or extra time required to perform a complete readout of the contents of the chip. A possible solution to this problem is to access every pixel independently but sequentially using a digital random access scheme (Niclass et al. 2005).

In low light level applications one can use an event-driven readout, where the detector initiates and drives a column-wise detection process directly (Niclass et al. 2006). The drawback of this approach is that multiple photons cannot be detected simultaneously on the same column. In addition, the bandwidth of the column readout mechanism limits saturation levels of the entire column. These limitations are not problematic if the expected photon flux hitting the sensor is low.

An alternative approach for no low light level situations is the use of a latchless pipeline scheme. In this approach, the absorption of a photon causes the SPAD to inject a digital signal into a delay line that is then read externally. The timing of all injected pulses is evaluated so as to derive the time of arrival of the photon and the pixel of origin. This method allows detection of photons simultaneously over a column even though some restrictions apply on the timing of the optical set up. Note that in this design an effective gating mechanism is necessary to prevent photons detected outside a certain time window being interpreted as originating in a pixel other than the one responsible for that detection. The pixel access problem can be overcome if the photon time of arrival is performed in situ, i.e. on the pixel itself.

The main problem of the all discussed solutions is the physical size of the on-board ancillary circuitries (digital signal extractor, counters, memory elements, buffer readout, etc.), which drastically decrees the fill factor and increase the complexity of the chip manufacturing.

The alternative approach is to bring out the chip all the readout circuitries in order to increase the fill factor, simplify the matrix design and the manufacturing process (Tudisco et al. 2009).

A cross-wire readout scheme could be adopted to access every pixel independently; the avalanche signal of each diode is twist extracted from two contacts, these are shared one for all diodes of the same row and the other for all diodes of the same column. Figure 9 reports the scheme of two possible configurations. In such a way it is possible to deduce the hit position with a readout complexity which grows as $2n$ instead of n^2.

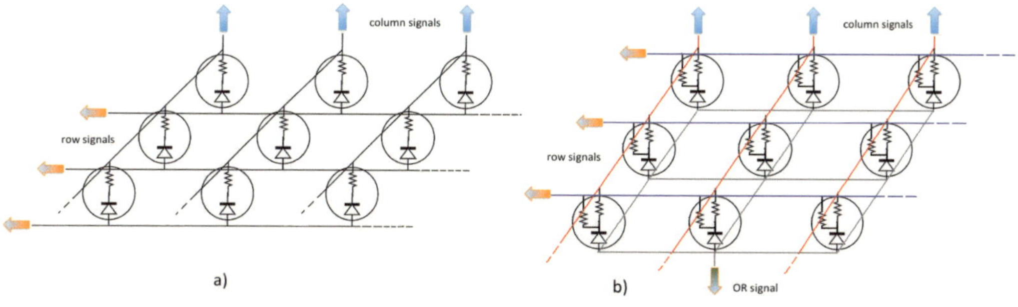

Fig. 9. Cross-wire readout scheme. In *a*) the avalanche signal of each pixel is collected from both side of the junction. In *b*) each pixel has two quenching resistors and a common anode contact.

If one of the devices "fires" in response to light applied to that device, it will generate a current pulse in one of corresponding row and column outputs. This allows the specific device to be determined unambiguously by its unique row-column signature.

The only real limitation of this spatial recognition approach is the uncertainty in the simultaneous detection of two photons. This uncertainty comes from events tagging and is related to the time jitter of row and column signals.

In the configuration of figure 9a the avalanche signal of each pixel is collected from both side of the junction, anode and cathode while the solution of figure 9b requires the integration of two quenching resistors for each diode, with advantages and disadvantages related to the existence of a common anode contact. The latter gives the possibility to collect the total OR-signal (like in SiPM configuration). Figure 10 shows the layout of some arrays made in collaboration with the FBK-IRST and currently under test and characterization.

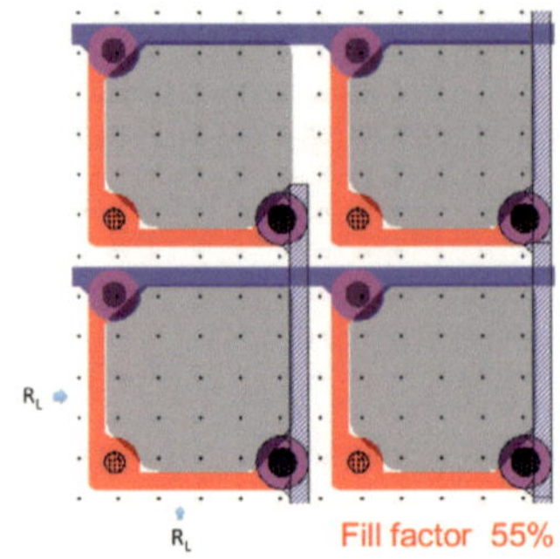

Fig. 10. Layout of FBK–IRST arrays.

Adopting the first solution was recently realised in collaboration with ST-Microelectronics an elementary first demonstrator of 10x10 elements (Tudisco et al. 2009).

The array layout has been designed with common anode and cathode bus. Metal strips used to contact the cathodes in each column are continuous while those used to contact the anodes in each row are interrupted in proximity of each cathode metal bus in order to avoid short-circuits with cathode metal strips. A high-doped silicon underpass guaranties the conductivity. The operating conditions were assured by the use of a single external quenching resistor for row (or column).

In such simplified configuration, respect to the figure 9a, the main disadvantage is the quenching of all devices in the row (or column) join to the slow recharge time coming from to use of externals quenching resistor. Nevertheless it remains a valid concept device.

In order to realize a first working prototype of "Time Resolved Camera" (TRC), row and column signals can be treated by standard nuclear electronics. For example, by using commercial VME modules (Constant Fraction Discriminators and multi-hits Time to Digital Converters, fig. 11c) it is possible to acquire and reconstruct the information concerning position and arrival time of photons.

An example of the achievable performance is reported in Figure 11b, in which the head of a Light Emission Diode (LED) was focus (trough a set of lens) on the top, alternative, of ccd/SPAD-array surface. The image was reconstructed after the subtraction of the dark count rate of each pixel. To maximize the image contrast we fix the black level of the colour scale at counting rate less than 500 Hz.

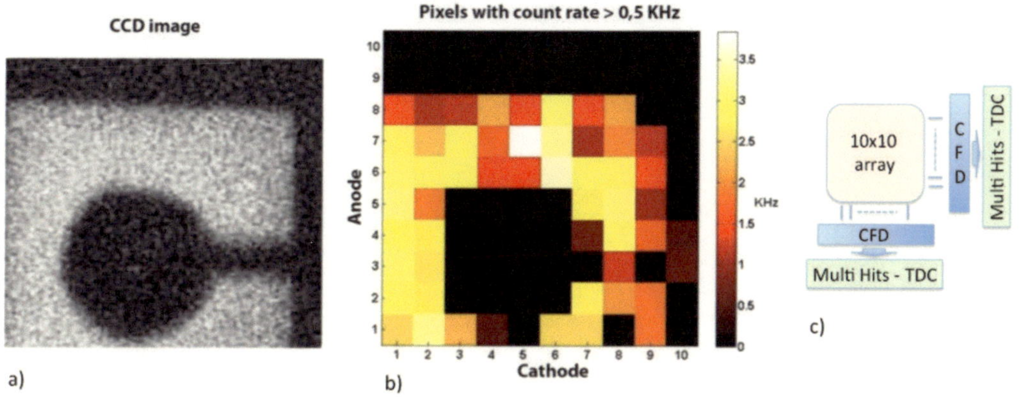

Fig. 11. *a*) CCD image from the Light Emission Diode. *b*) The same image obtained by 10x10 array. *c*) Schema Readout electronic.

The possibility to adjust the photon flux with a fine precision is a fundamental requirement in order to characterize a TRC. The use of LED source guarantees this possibility through the current limitation on it. With this regulation capability it is possible to collect image with a normal CCD camera (fig. 11a) at medium photon rate and test the SPAD matrix capability in all the sensor dynamic range, from single photon up to the saturation level. By using time modulated current signals it is also possible to test the time dependent characteristic of whole system.

TRC doesn't have frame time limitations like frame duration, frame rate, minimum time distance between frames etc. Every collected photons give his contribution to a continuous streaming of the photon source, the pictures can extracted putting together all the photons during a defined time, or can be collected in groups forming a sequence of video frames without any dead time between frames. The time and spatial positions of all detected photons are collected, and can be used for many correlation technique that use time and/or spatial correlation analysis.

3.2.1 Current status and prospective

The solution sketched in figure 9b is very promising. Arrays of several size and configuration (up to 170x170 elements) has been designed and manufactured in collaboration with FBK-IRST and are currently under test a characterization.

As previous mentioned, this solution guarantees the collection of the total OR signal (like in SiPM configuration) simplifying the ancillary readout circuitry. From the back, through the anodic contact, you can extract the photon arrival time information (e.g. through a single channel of Multi-Hit TDC) while the position information can be extracted from the front by using, for example, the scheme reported in figure 12b.

An alternative readout strategy currently under development is reported in figure 12a. A resistive path is realised on the row contacts and column contacts. By using the charge repartition principle it is possible to determine the row and column fired just sampling the two signals at the end of the chain. This solution drastically reduces the number of reading channels up to two for rows and two for columns.

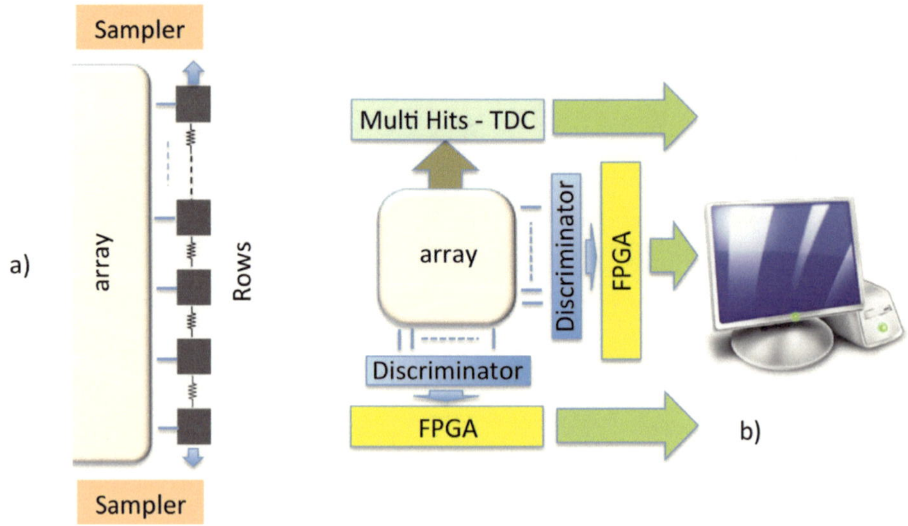

Fig. 12. *a*) Sketch of the alternative readout strategy through the rows resistive path and signals sampling. *b*) Possible readout schema of fig. 9b array.

4. Conclusion

In the last years, the R&D of SPADs technology has proceeded at great sped. The main challenge towards imaging devices is geometry and process optimization to yield high compactness while not excessively degrading performance.

On the other hand to make a Time Resolved Camera a compact object available on the market, many efforts have still to be made, in terms of identifying the best readout strategies and ancillary circuitries.

5. Acknowledgments

The author would like to acknowledge the partners of the collaboration, ST-Microelectronics and FBK-IRST and in particular G. Fallica and C. Piemonte and the 5th National Committee of INFN that funded this activity.

6. References

Jares-Erijman, E. A. and Jovin, T. M. (2003). *FRET imaging,* Nat. Biotechnol. Vol. 21 pp. 1387–1395, ISSN 1087-0156

Becker, W. et al., (2006). *Fluorescence lifetime images and correlation spectra obtained by multidimensional time-correlated single photon counting*, Microsc. Res. Tech. Vol. 69 pp. 186 - 195, ISSN 1059-910X

Schwille, P. et al., (1999). *Molecular dynamics in living cells observed by fluorescence correlation spectroscopy with one and two-photon excitation,* Biophys. J. Vol. 77 pp. 2251 - 2265, ISSN 0006-3495

Strickland, D. et al., (1985). *Compression of amplified chirped optica pulses,* Opt. Comunic. Vol. 56 pp. 219-221, ISSN 0030-4018

McPhate, J. et al., (2005) *Noiseless kilohertz-frame-rate imaging detector based on microchannel plates readout Medipix2 CMOS pixel chip, Proc. SPIE Vol.* 5881 pp. 88 - 97 ISBN 9780819458865

Etoh, T. G. et al. (2005) *Design of the PC-ISIS: photon-counting in-situ storage image sensor,* IEEE Workshop on CCDs and Advanced Image Sensors (Karuizawa, Japan) pp. 113–116 Available on from http://www.max.hi-ho.ne.jp/teranishi/

Kawai, N. et al. (2005) *A low-noise signal readout circuit using double-stage noise cancelling architecture for CMOS image sensors* IEEE Workshop on CCDs and Advanced Image Sensors (Karuizawa, Japan) pp. 27–30 Available on from http://www.max.hi-ho.ne.jp/teranishi/

Hynecek, J. et al. (2001) *Impactron – a new solid state image intensifier,* IEEE Trans. on Electron Devices Vol. 48 pp. 2238–41 ISSN 0018-9383

Cova, S. et al. (1981) *Towards picosecond resolution with single-photon avalanche diodes,* Rev. of Scie. Instrum. Vol. 52 pp. 408–412 ISSN 0034-6748

Cova, S. et al (1983) *A semiconductor detector for measuring ultraweak fluorescence decays with 70 ps FWHM resolution,* IEEE J. Quantum Electronics Vol. 19 pp. 630-634 ISSN 0018-9197

Cova, S. et al. (1989) *Trapping phenomena in avalanche photodiodes on nanosecond scale,* IEEE Electron Device Lett. Vol. 12, pp. 685–687 ISSN 0741-3106

Cova, S. et al. (1996) *Avalanche photodiodes and quenching circuits for single photon detection,* Appl. Opt. Vol. 35, pp. 1956–1976 ISSN 0003-6935.

Mc Intyre, R.J. (1961) *Theory of Microplasma Instability in Silicon,* J. Appl. Phys. Vol. 35 pp. 1370 ISSN 0021-8979

Mc Intyre, R.J. (1966) *Multiplication noise in uniform avalanche diodes,* IEEE Trans. on Electron Devices Vol. 13 pp. 164-168 ISSN 0018-9383

Mc Intyre, R.J. 1985 *Recent developments in silicon avalanche photodiodes,* Measurement Vol. 3 pp. 146–152 ISSN 0263-2241

Hurkx, G.A.M. et al. (1992) *A new analytical diode model including tunneling and avalanche breakdown,* IEEE Trans. Electron Devices, Vol. 39, pp ISSN 0018-9383

Oldham, W.G. et al. (1972) *Triggering phenomena in avalanche diodes," IEEE Trans. Electron Devices,* Vol. 19, pp. 1056–1060 ISSN ISSN 0018-9383

Zappa F et al. (1997) *Integrated array of avalanche photodiodes for single-photon counting,* IEEE ESSDERC (Stuttgart, Germany) pp 600–603

Kindt, W.J. (1999) *Geiger mode avalanche photodiode arrays for spatially resolved single photon counting,* PhD Thesis Delft University Press, ISBN 90-407-1845-8

Sciacca, et. al. (2003) *Silicon planar technology for single-photon optical detectors,* IEEE Trans. on Elect. Dev. Vol. 50, pp. 918-925 ISSN 0018-9383

Ripamonti, G.et al. (1985) *Carrier diffusion effects in the time-response of a fast photodiode,* Solid State Electron., Vol. 28, pp. 925–931, ISSN 0038-1101

Lacaita, A. et al. (1989) *Double epitaxy improves singlephoton avalanche diode performance,* Electron. Lett., Vol. 25, n°. 13 ISSN 0013-5194

Mingguo, L. et al. (2008) *Reduce Afterpulsing of single photon avalanche diodes using passive quenching with active reset*, IEEE Jour. of Quant. Electr. Vol. 44 n°5 1077-260X

Niclass, C et al. (2005) *Design and characterization of a CMOS 3-D image sensor based on single photon avalanche diodes*, IEEE J. Solid State Circuits Vol. 40 pp. 1847–1854 ISSN 0018-9200

Neri, L. et al. (2010) Generalization of DT equations for time dipendent sources, Sensors Vol. 10, pp. 10828-10836 ISSN 1424-8220

Mazzillo, M. et al. (2008) Single-photon avalanche photodiodes with integrated quenching resistor, Nucl. Instr. Metho. A Vol. 591 pp. 367–373 ISSN 0168-9002

Tyagi, M.S. (1991) *Introduction to Semiconductor Materials and Devices*, Wiley, ISBN 0471605603 9780471605607, New York.

Grant, W.N et al. (1973) Impact Ionization or Ionization Breakdown Gain and Coefficient Calculator Solid-state Electron. Vol. 16 pp. 1189 ISSN 0038-1101

Bonanno, G. et al (2005) *Electro-Optical Characteristics of the Single Photon Avalanche Diode (SPAD)*, Scientific Detectors for Astronomy, pp. 461 ISBN-1-4020-4329-5 Springer, Berlin, Dordrecht.

Privetara, S. et al. (2008) *Single Photon Avalanche Diodes: towards the large bidimensional arrays*, Sensors, Vol. 8 pp. 4636-4655 ISSN 1424-8220

Niclass, C. et al. (2005) *A single photon detector array with 64 × 64 resolution and millimetric depth accuracy for 3D imaging*, IEEE Int. Solid-State Circuits Conf. (ISSCC) (San Francisco, USA) pp 364–365

Niclass, C et al. (2006) *A 64 × 48 single photon avalanche diode array with event-driven readout*, Eur. Solid-State Circuits Conf. (ESSCIRC), Montreaux, Switzerland.

Tudisco, S. et al. (2009) *Bi-dimensional arrays of SPAD for time-resolved single photon imaging*, Nucl. Instr. Method. A Vol. 610 pp. 138-141 ISSN: 0168-9002

13

High-Resolution Near-Field Optical Microscopy: A Sub-10 Nanometer Probe for Surface Electromagnetic Field and Local Dielectric Trait

Jen-You Chu[1] and Juen-Kai Wang[2,3]
[1]*Material and Chemical Research Laboratories*
Industrial Technology Research Institute, Hsinchu
[2]*Center for Condensed Matter Sciences, National Taiwan University, Taipei*
[3]*Institute of Atomic and Molecular Sciences, Academia Sinica, Taipei*
Taiwan

1. Introduction

The first optical microscope was made by Zacharias Janssen in the late sixteenth century. Robert Hooke used his own microscope to observe bio-samples and draw these structures on the micrographia in the seventeenth century. August Köhler, in 1893, developed an illumination technique, allowing for even sample lighting and setting the corner stone for modern light microscopy. In 1930s, Zeiss laboratory invented the upright microscope that is a prototype of modern optical microscopy. Such earlier effort to extend human vision in examining tiny objects is however limited by wave diffraction, as pointed out by Helmholtz and Abbe [1]. Namely, in visible wavelength range, optical resolution is about 250 nm. Such resolution limit certainly does not satisfy the need for current development and progress of nanotechnology, in which objects within 100 nm portray unique properties and functions that are not predictable by direct extrapolation from their macro-counterparts. It is thus crucial to extend the optical resolution below the Abbe's limit if we want to further the role of optical microscopy in future technology advance.

Although both electron microscopy and scanning probe microscopy easily meet the resolution requirement for nanotechnology, optical microscopy provides two unique characteristics that are not possible by them. First, optical microscopy is a noninvasive characterization method and can therefore examine objects in ambient environment or through transparent condensed media. Second, the energy resolving power of an optical probe surpasses the other two microscopic techniques, greatly enhancing its species identification ability in conjunction with various spectroscopic probes. These two distinctive capabilities are promised to more clearly unravel the novel relationship between the size/shape of a nano-object and its property which is essential to the advancement of nanotechnology, if the spatial resolution of optical microscopy can reach sub-10 nanometer scale.

A dramatic improvement of optical resolution has become possible by the invention of aperture-type scanning near-field optical microscopy (SNOM) [2-5]. Viewing through a tiny pinhole, firstly proposed by Synge [6] in 1928, spurs a series of attempts to realize such concept. In 1984, two groups have simultaneously demonstrated it in the visible wavelength range [7-8]. Light through a sub-wavelength diameter aperture, which is evanescent in character, illuminates a sample within its near-field range. The optical response thus recorded while the sample is scanned laterally constructs a near-field optical image – scanning near-field optical microscope (SNOM). Its resolution depends on the diameter of aperture. Although breaking the Abbe's resolution limit, this new type of optical microscope holds three restrictions. The first one is that the resolution cannot be better than 50 nm, caused by the finite penetration into the optical field-confinement metal structure that defines the aperture. The second one is the trade-off between spatial resolution and collected signal intensity – the smaller the aperture is, the smaller the optical signal is collected [9]. The third one is that the throughput of the aperture is highly wavelength dependent, thus distorting the recorded optical spectrum. The aforementioned restrictions thus restrain the practice of aperture-type SNOM from the applications in nanophotonics and nanotechnology in general.

In contrast to the aperture-type SNOM, Wickramasinghe and coworkers in 1994 demonstrated an apertureless SNOM with a 10-nm resolution on the basis of sensing the dipole-dipole coupling between a tip end and a sample [10]. Later, Keilmann's team furthermore extended this approach to extract both amplitude and phase of such electromagnetic coupling simultaneously, concocting a comprehensive scattering-type SNOM (*s*-SNOM) [11-15], as depicted in Fig. 1. The collimated optical radiation impinges onto the site where the apex of an atomic force microscopic tip (AFM probe) is in close proximity of sample surface (enclosed by green dotted ellipse). Right-handed image of the figure 1 shows a blow-up view of the induced localized field at the tip apex. The restrictions of aperture-type SNOM are unleashed in such new-generation SNOM. Firstly, the resolution of s-SNOM is only limited by the tip radius. Secondly, the enhanced local field at the tip apex, owing to plasmon resonance, intensifies the electromagnetic interaction at the tip-sample system and thus boosts up the resulted scattered radiation for detection.

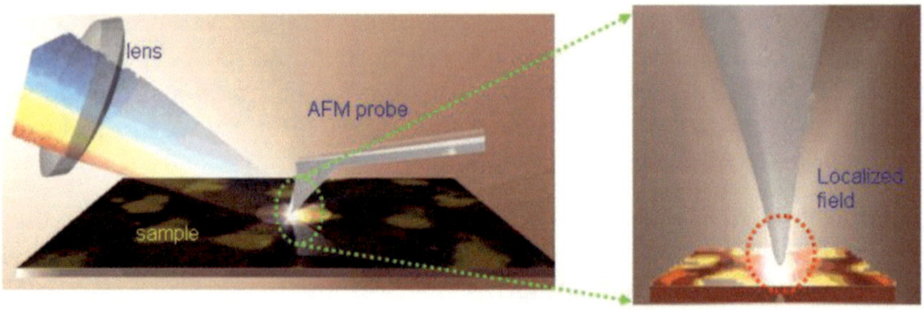

Fig. 1. Schematic view of scattering-type scanning near-field optical microscope.

In this chapter, we delineate the fundamental theory of *s*-SNOM (Section 2) and show its experimental setup (Section 3). Section 4 describes issues relevant to extracting near-field images. Its spatial resolution is demonstrated and discussed in Section 5. We show that *s*-SNOM is a powerful tool to unravel local optical properties in nanocomposite systems in Section 6. Furthermore in Section 7, we illustrate that it is possible to extract both amplitude and phase of surface plasmon polariton. The two cases above exemplify the great potential of *s*-SNOM to explore plasmonics and metamaterials, as well as nano-structured composite systems. Finally, Section 8 concludes this chapter.

2. Theory of *s*-SNOM

The field enhancement of a metallic tapered tip apex could be up to 1000-fold higher than the field of an incident optical wave that impinges onto it in an optimum condition for electromagnetic resonance [16]. As a consequence, the electromagnetic interaction between the tip and the incident field can be greatly altered owing to the presence of a near-by surface within the near-field enhancement zone, conferring an enhanced scattered radiation to the far field. Understanding the electromagnetic interaction taking place between the tip apex and the sample surface entails the detailed structure of the tip and solving the electromagnetic field in vicinity of the tip-surface composite system. Nonetheless, a simple model that can reveal the fundamental nature of such electromagnetic interaction would assist in understanding the limits of *s*-SNOM and in furthering its advancement and generalization. The first simplification is that only the response of the tip to the incident optical wave coming from the region around the tip apex is considered, whose dimension is much smaller than the wavelength of the incident wave, such that phase retardation is negligible over such region and thus a quasi-static treatment is justified [17-19]. Secondly, in spite that many multipole fields could be produced from the interaction between the incident wave and the tip apex of finite size and of complicated shape, only the dipole field from a sphere that approximates the tip apex is considered for it dominates the far-field radiation characteristics from the tip-surface system. Lastly, only a flat sample surface is considered for extracting the far-field scattered radiation, though a corrugated surface is expected to modify the outcome. The influence of the surface structure will be discussed later.

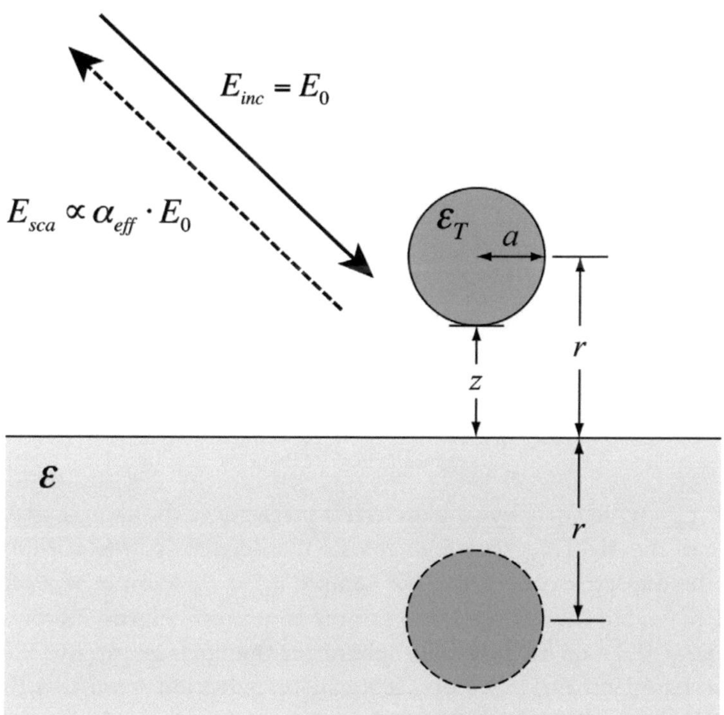

Fig. 2. Schematic of quasi-electrostatic dipole model illustrating the electromagnetic interaction happening in *s*-SNOM of side-illuminating geometry.

Figure 2 shows the geometric schematic of the interaction of a *p*-polarized incident electromagnetic wave with an electric field E_0 that impinges onto the surface with an incident angle θ_i. A sphere with a radius of a and a dielectric constant of ε_T that symbolizes the tip apex, residing at r above the surface of sample that has a dielectric constant of ε, in *s*-SNOM. The gap between the sphere and the surface is $z = r - a$. According to the Mie's theory of a sphere [20], the sphere illuminated by the incident light gives an induced point dipole with a dipole moment of p that is equal to $\alpha_T \cdot E_0$, where α_T is the polarizability of the sphere. As a consequence, such dipole at r above the surface induces an image dipole with a dipole moment of p' that resides at r directly below the surface. The image dipole creates another dipole field that together with the incident field produces the dipole moment p [21]. An effective polarizability, α_{eff}, can then be derived by solving the resultant equation self-consistently. Accordingly, the scattered radiation of the incident wave from such tip-surface system can be considered is produced by the combination of the dipole and image dipole. Namely, α_{eff} of such combination is given by

$$\alpha_{eff} = \alpha_T (1 + \beta \cos 2\theta_i) F(r, \theta_i), \tag{1}$$

where

$$F(r, \theta_i) = \frac{1}{1 - \beta \left(\frac{\varepsilon_T - 1}{\varepsilon_T + 2} \right) \left(\frac{1 + \cos^2 \theta_i}{8} \right) \left(\frac{a}{a + z} \right)^3}, \tag{2}$$

$$\alpha_T = 4\pi a^3 \frac{\varepsilon_T - 1}{\varepsilon_T + 2}, \tag{3}$$

and

$$\beta = (\varepsilon - 1)/(\varepsilon + 1). \tag{4}$$

z is the separation between the tip apex and the sample surface and is equal to $r - a$ and, furthermore, and β is the ratio between the dipole moments of the dipole of the tip apex and the image dipole. The scattering light field, given by

$$E_{sca} \propto \alpha_{eff} \times E_0, \tag{5}$$

directly reflects α_{eff} – which conveys the dielectric property of the sample and the tip-sample separation – and the field E_0 that induces it. Consequently, *s*-SNOM can extract the distribution of the dielectric constant of the sample if the tip-sample separation is fixed, as well as the optical field at the surface if the sample is uniform. Figure 3 shows the calculated scattering intensity $|E_0|^2$ of a chromium sphere on the surfaces of Au, SiO2 and Si as a function of r/a based on Eq. (5). Two facts can be extracted from this figure. First, the scattering intensity is sensitive to the dielectric constant of the sample, engendering *s*-SNOM to be a local probe of the dielectric property. Second, the scattering intensity bears a near-field nature – it decreases almost exponentially to a plateau as $r/a \le 1.5$. This feature thus makes s-SNOM in theory a near-field probe. As a final note to the near-field electromagnetic interaction mechanism, although more complete models have been proposed [22] the physical nature illustrated by the simple model still remains the main results from these complicated models.

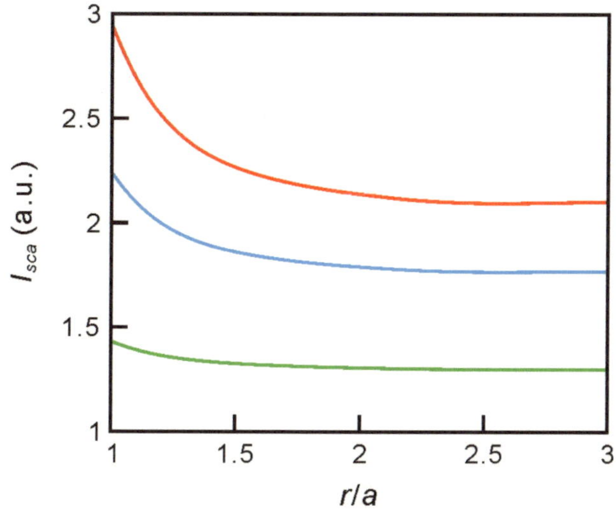

Fig. 3. Calculated scattering intensity, I_{sca}, of near-field interaction between a chromium sphere and three different samples — Au (red line), Si (blue line) and SiO_2 (green line) — *vs.* *z* based on quasi-static dipole model. *r* is the separation between the center of the sphere and the sample surface; *a* is the radius of the sphere.

For a *p*-polarized incident wave, the induced dipole at the sphere has a component normal to the surface, owing to the non-zero field component of the incident wave along the surface normal direction. The induced image dipole is also along the same direction, resulting in an effective polarizability that is larger than α_T. Furthermore, according to Eqs. (1) and (2), α_{eff} reaches the maximal value at $\theta_i = 60^\circ$ in most cases. Namely, at this incident angle, the scattered radiation is maximized. On the other hand, for an *s*-polarized incident wave, its field component is parallel to the surface, producing a point dipole that also follows the same direction. The image dipole beneath the sample surface is on the other hand along the opposite direction and thus produces a dipole field that counteracts the incident field. Effectively, it partially cancels the dipole above the surface. The resultant effective polarizability in this case is smaller than α_T. That is, the scattered radiation is much smaller than that produced in the case of the *p*-polarized wave excitation. The distinction can be thousand folds difference for a metallic tip [23].

3. Setup of *s*-SNOM

One major challenge facing the development of *s*-SNOM is how to separate the extremely weak near-field signal originated from the tip apex from the huge background signal originated from the giant scattering radiation from the shaft of the tip and the sample. This is caused by the fact that the illuminated region of the incident optical wave cannot be smaller than the one defined by the diffraction limit. This practical problem was solved by

Wickramasinghe *et al.* [10] and recently by Keilmann *et al.* [14-15]. In their approaches, an interferometric technique combining a dithering tip effectually extracts the weak near-field signal. Taking advantage of the nonlinear dependence of the scattering light field on the tip-sample separation, as shown in Eq. (2), the small variation on the tip-sample separation through the tip dithering or tapping produces many harmonic modulation terms in the scattering light field. While the first harmonic term is dominated by the large background that comes from the scattering from the tip shaft and the sample, the higher harmonic terms contain only the nature of the near-field electromagnetic interaction if the mechanical anharmonicity of the tip-dithering action is put under control [24-25]. The interferometry-based technique has two extra benefits. First, both amplitude and phase can in principle be extracted simultaneously. Second, the weak scattering radiation from the near-field zone can be amplified, which will be manifested below. There are two interferometry-based schemes: homodyne [23, 26] and heterodyne [27]. In the homodyne scheme, a coherent laser beam acts as the excitation light source and the local oscillator simultaneously. As the phase information of the scattering radiation is buried within the detected DC signal, its extraction is not straightforward. In the heterodyne scheme, one part of the laser beam is frequency shifted and, accordingly, the mixed signal between it and the other part is modulating in time, greatly facilitating the phase extraction. The shortcoming is that implementing multiple-wavelength excitation is formidable. The drawbacks of the two schemes have been overcome by pseudoheterodyne [28] and modified heterodyne schemes [29], separately. These interferometric schemes have been developed to extract both the amplitude and phase of the scattering contribution in the near-field region without being interfered by the background scattering [26-31]. Taking advantage of the nonlinear dependence of the scattering light field on the tip-sample separation, as shown in Eq. (2), the small variation on the tip-sample separation through tip dithering or tapping produces many harmonic modulation terms in the scattering light field. While the first-harmonic term is dominated by the large background, the higher-harmonic terms contain only the nature of the near-field electromagnetic interaction if the mechanical anharmonicity of the tip-dithering action is under control [24-25]. Here the modified heterodyne technique is delineated. For the homodyne scheme and its improved version, the authors refer to the publications by Keilmann *et al.* [23].

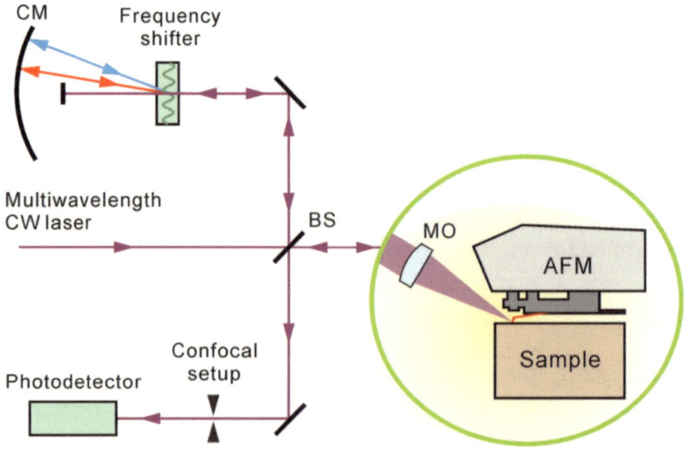

Fig. 4. Optical layout of modified heterodyne s-SNOM. BS, beamsplitter; MO, micro-objective lens; CM, curved mirror. Blue and red lines symbolize two excitation wavelengths.

Figure 4 shows the optical layout of modified heterodyne *s*-SNOM that is good for multiple-wavelength excitation. A CW laser – which can be a HeNe laser or a diode-pumped solid-state (DPSS) laser – serves as the coherent excitation light source. The laser beam is first split into two parts by a beamsplitter, BS. The transmitted part, as the excitation beam, is sent to the AFM setup in which a micro-objective lens (MO) focuses it into the tip apex with an incident angle of 60° and the field direction being in the incident plane (*p*-polarized excitation). The scattering radiation is collected backward with the same objective lens. As the AFM tip is dithering at a frequency of ~300 kHz (Ω), the collected scattering radiation is modulating at Ω and its harmonics. The reflected part, as the reference beam, is delivered to a frequency shifter that is an acousto-optical modulator. The diffracted beam that is shifted in frequency by Δ is reflected backward to the frequency shifter to experience another deflection and frequency shifting and thus follows the same path of the incoming beam. With this approach, although the reference beam with a different wavelength would be diffracted at a different angle, it should return back to the same path, without altering the optical alignment. The reference beam then combines with the collected scattering radiation at the beamsplitter. The resultant beam is sent to a high-speed photodetector through a confocal setup that is used to select only the region of the tip apex.

The detected signal by the photodetector is proportional to the square of the combined field of the frequency-shifted reference beam and the collected scattering radiation [27]:

$$\begin{aligned} I_{tot} &\propto \left| E_{ref} \exp\left[i(\omega_L - 2\Delta)t \right] + E_{sca} \exp\left[i(\omega_L + n\Omega)t + i\varphi \right] \right|^2 \\ &\propto I_{ref} + I_{sca} + 2\sqrt{I_{ref} I_{sca}} \cos\left[(2\Delta + n\Omega)t + \varphi \right] \end{aligned}, \tag{6}$$

where I_{sca} is the collected scattering intensity, I_{ref} is the intensity of the frequency-shifted reference beam, ω_L is the laser frequency, n is the harmonic order, and φ is the phase of the scattering radiation with respect to the reference beam. As the total detected signal is composed an AC signal modulated at a frequency of $2\Delta + n\Omega$, a lock-in amplifier referenced to such frequency retrieves the AC component. With the AFM tip is scanned over the sample surface, both I_{sca} and φ are thus retrieved, while the surface topology is simultaneously recorded. The system can be operated in two scanning modes: sample- and tip-scanning modes. In the sample-scanning mode, the sample is scanned by a piezo-driven *xy* stage while the tip is fixed in position. In the tip-scanning mode, the tip is, on the other hand, scanned while the sample is fixed.

4. Extracting near-field images

On the basis of the nonlinear behavior of the near-field interaction with respect to the tip-sample separation, as shown in Eq. (1), it is necessary to place the extracted optical signals of different harmonic orders in scrutiny to verify their exponential-decay dependences on such separation (Fig. 3). Figure 5(a) shows the extracted optical signals of first to fourth harmonic orders as a function of the tip-sample separation from a gold surface with nanometer-scaled gold particles. The scattering signal drops almost exponentially with the tip-sample separation, signifying the recorded third and fourth-harmonic signals are generated in near field. A fit of the fourth-harmonic signal to an exponential decay curve gives a decay length of 10 nm, reflecting the short-range electromagnetic interaction between the tip and the sample illustrated by Eq. (1) and demonstrating the near-field nature of the high harmonic-order signals. Figure 5(b) shows the topography image of rough Au coated silicon surface. The prominent features of this image show one-to-one correspondence with the images of the scattering optical signals of the second to fourth harmonic orders, as shown in Figs. 5(c) to (e). The interference-like ripples in Fig. 6(c) is an artifact that could be caused by the interference between the scattering light wave of the tip shaft and that of the sample, as similarly observed previously [23]. This assignment is because a slight nonlinear mechanical response of the tip dithering motion at a frequency of Ω provides a certain amount of sinusoidal component at 2Ω that then introduces some harmonic contribution at 2Ω into the elastic scattering optical signal from the tip shaft. That trait is also manifested itself in its non-decay dependence on the tip-sample separation, *z*, shown in Fig. 5(a), which is a direct evidence of the non-near field scattering contribution by the tip shaft.

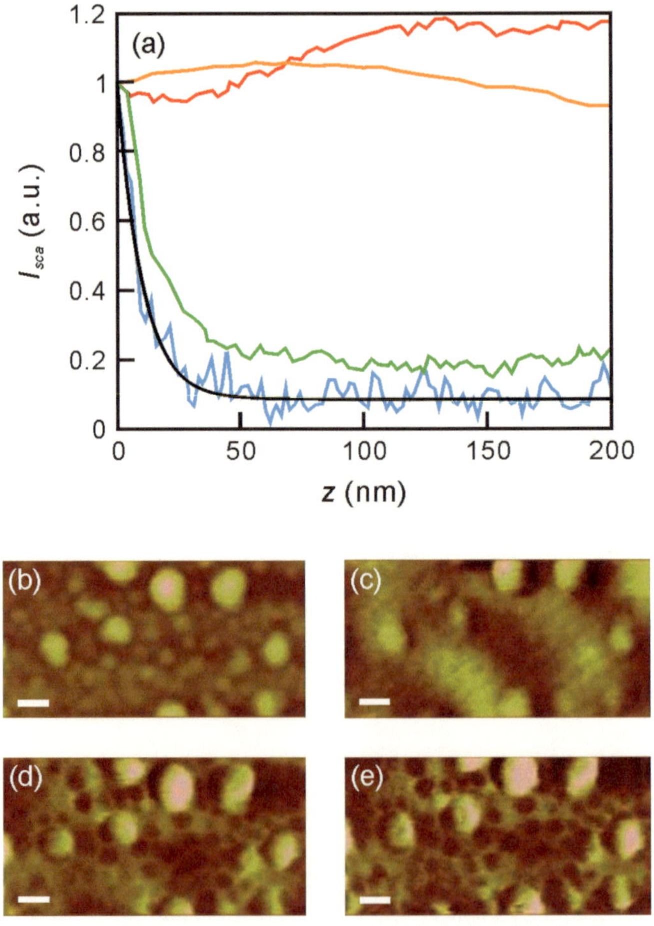

Fig. 5. (a) Collected *s*-SNOM signals of 1st (orange line), 2nd (red line), 3rd (green line), and 4th (blue line) harmonic orders on Au surface as a function of tip-sample separation, *z*; Scanned images of topography (b) and *s*-SNOM signal of 2nd (c), 3rd (d), and 4th (e) harmonic orders. Scale bars represent 100 nm. The black solid line is the fit to an exponential decay. The excitation wavelength is 632.8 nm.

Two more facts can be extracted from the near-field images of 3rd and 4th harmonic orders, shown in Figs. 5(d) and (e), are worthy of further discussion. Firstly, the near-field optical signal observed on large Au nanoparticles (diameter > 200 nm) is larger than that observed on small ones (diameter < 100 nm). This may be originated from the fact that the observed scattering signal from the tip-sample system can be enhanced by its plasmon resonance character [32]. As the plasmon resonance wavelength of a small Au nanoparticle is shorter than that of a large Au nanoparticle and thus is further away from the excitation wavelength of 632.8 nm [33], the corresponding effective polarizability for the small nanoparticle can be smaller than that for the large nanoparticle. Secondly, enhanced scattering signal is evident in the gap between these particles, which can be attributed to the following two causes. On the one hand, the gap mode existing between the two adjacent nanoparticles [34] induces localized strong field within and thus enhances optical scattering. The resonance wavelength of the gap mode is, on the other hand, red shifted and is closer to the excitation wavelength, thus increasing the corresponding effective polarizability and also enhancing the scattering radiation. These factors must be accounted for the interpretation of the extracted near-field images with the use of the excitation wavelength close to the electromagnetic resonance of the tip and the sample under scrutiny.

As to the geometry of the *s*-SNOM setup shown in Fig. 4, there are several advantages for such side-illumination geometry. In sum, except for the better laser excitation and signal collection efficiencies (60° incident angle and *p*-polarization excitation), the sample needs not be transparent. In contrast, for *s*-SNOM with a transmission-illumination geometry – in which the excitation laser beam illuminates the tip apex through the sample and along the tip shaft, transparent samples are must and furthermore a special polarization configuration is prepared for the excitation beam to enhance the field component at the tip apex along the tip shaft [35-37]. Nevertheless, for the side-illumination geometry, the cantilever that holds the tip shaft cannot obstruct the excitation laser beam from irradiating the tip apex. Namely, the tips must protrude from the front side of the cantilever. As a consequence, this geometry entails special AFM tips allowing for side illumination (such as AdvancedTEC tips from Nanosensor, Olympus Probes, Visible Apex tips from Bucker, etc.) or conventional tips modified by focused-ion beam.

Lastly, two differing considerations in choosing the tip material are given to the use of *s*-SNOM in probing local dielectric property and extracting local field. For the first application, the material under study is in general not near electromagnetic resonance with the incident optical wave, for which the scattering signal from the near-field interaction between the tip and the sample is often very weak. To enhance the scattering signal, the tip is preferably made by metal (silver or gold) that bears plasmon resonance with respect to the excitation laser beam. Equations (1) and (3) show that the resultant scattering signal can thus be amplified by the use of such metal tip. As the conventional AFM tips are made with the application of high precision etching process on crystalline silicon and silicon nitride, the shape of the tip apex can thus be controlled down to sub-20 nm scale. Coating of single or multiple layers of metals through thermal deposition can be applied to the tip to create a special dielectric property that is suitable for near-field applications [38]. The thickness of the coating has to be larger than the skin depth of the metal coating at the excitation wavelength to avoid the effect by the underneath tip material. However, silver and gold often tend to aggregate in tens of nanometer on the tip surface [39], creating non-continuous coating and thus bearing unstable optical response. Meticulous control in the metal coating process is necessity to increase its production yield. The resonance wavelength of a silver coated tip is about 450 nm while that of a gold one is about 600 nm. For such reason, the silver-coated tip is often used with the excitation wavelengths shorter than 600 nm, while the gold-coated tip is with the wavelengths longer than 600 nm. Furthermore, silver has a smaller optical loss than gold, but it can oxidize or sulfurize in ambient condition, altering its optical property in time. On the other hand, gold is rather stable in air, while it is softer than silver and thus does not endure during scanning probe operation. One has to place such factors into account while choosing tip coating for *s*-SNOM. For the applications of retrieving local field, the disturbance by the tip has to be minimized as much as possible. For such reason, the electromagnetic resonance of the tip needs to be avoided and therefore silicon or silicon nitride tips are commonly used directly. This is so because their electromagnetic resonance wavelengths are distant from the visible wavelength range. One concern has to be placed in the use of silicon tips. As silicon oxidizes in air, a thin silicon oxide coating (1-2 nm) can modify the optical response with respect to the excitation laser beam considerably, accordingly diminishing the scattering signal [38].

5. Spatial resolution

The resolution of *s*-SNOM by its operating principle depends on the size and the shape of the tip apex. In Fig. 5 above, the near-field amplitude image of the 4th harmonic order of the tip dithering frequency shows a one-to-one correspondence with the topography image. This match reflects that the spatial resolution of the recorded near-field image is comparable to that of the topography image. That is, the lateral resolution is at least 5 nm that is greatly dependent on the tip apex. On the other hand, the fit of the *s*-SNOM signal of the 4th harmonic order as a function of the tip-sample separation, Fig. 5(a), shows that it follows an exponential decay with a decay length of 9.8 nm. This result is an evidence for the vertical resolution of *s*-SNOM. Similar to other scanning probe microscopic techniques, many other factors can have great influences on the spatial resolution. Besides the tip shape – specifically the radius of the tip apex, the dielectric property of the sample under study and the geometric structure of the sample in vicinity of the tip apex are anticipated to play two additional key roles in *s*-SNOM on the basis of the tip-sample near-field interaction. Such considerations are worthy of further exploration.

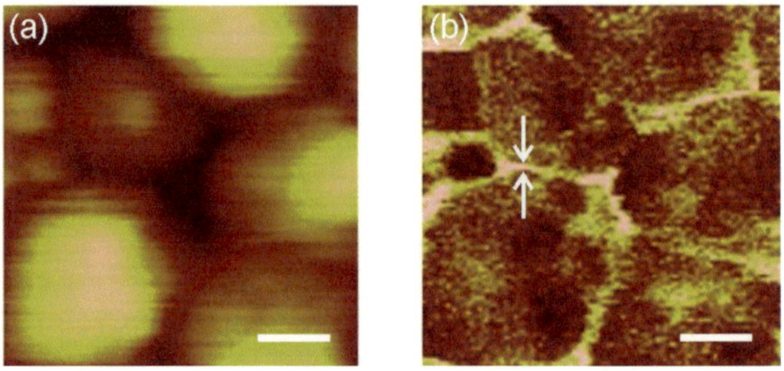

Fig. 6. Scanning topography (a) and *s*-SNOM optical amplitude (b) images of silver nanoparticle array embedded in anodic aluminum oxide. The gap between two counter-pointing arrows indicates the region where strong intensity resides. Scale bars represent 40 nm.

The following two examples illustrate how the field interaction and the geometric factors around the tip influence the spatial resolution of s-SNOM. The first sample is an array of silver nanoparticles embedded in aluminum oxide matrix. The silver nanoparticles were electrodeposited into a two-dimensional hexagonal ordered nanochannel array that was formed during anodization of smooth aluminum foil [40]. The gap between adjacent nanoparticles is about 5 nm and confirmed by electron microscopy. Figure 6 shows its topography and near-field intensity amplitude images recorded with a silicon tip with a radius of curvature of less than 10 nm. Notice that the close examination of the two images shows that a large intensity extends for ~6 nm between adjacent nanoparticles. In this observation, the nanoparticle array with a specific interparticle spacing serves as a ruler and confirms the lateral resolution of this s-SNOM measurement. An s-SNOM image of aggregated gold nanoparticles were reported previously [32] to portray the coherently oscillating nature of plasmon resonant of individual nanoparticles, expect that no confirmation of the gap by other means was provided. As an example, using a carbon nanotube attached on a silicon tip, Hillenbrand and his workers successfully recorded both amplitude and phase images of gold nanoparticles with a good spatial resolution, though no direct proof by electron microscopy was given [41]. The second example is a square array of annular trenches on a gold film of 200 nm thickness. These carved rings were made by focused ion beam. The diameters of the inner and outer circles are 250 and 330 nm, respectively; the depth is 200 nm; the spacing between adjacent rings is 600 nm. Figure 7 shows the topology and near-field amplitude scanned with sharp and blunt PtIr5-coated

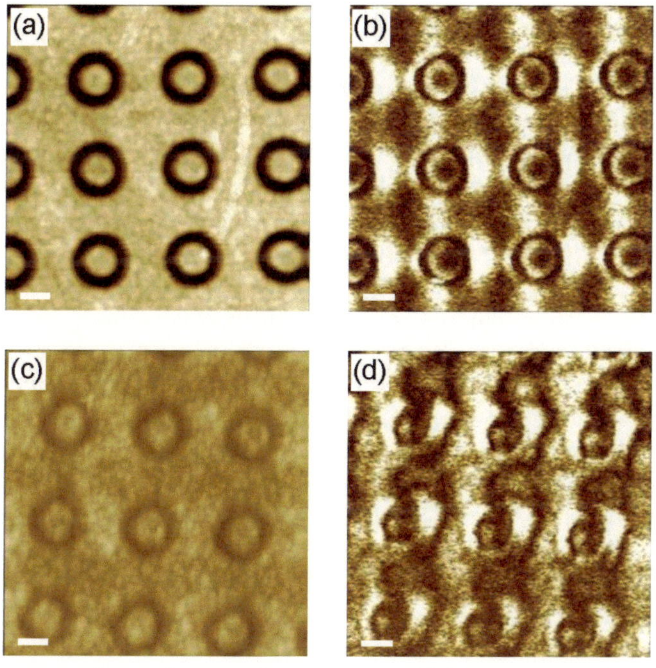

Fig. 7. Topology images of annular trench array on gold film made by focused ion beam recorded with (a) sharp and (c) blunt tips and their corresponding intensity amplitude images, (b) and (d), respectively. The diameters of the inner and outer circles are 250 and 330 nm, respectively. Scale bars represent 200 nm.

silicon tips. The recorded near-field amplitude image – Fig. 7(a) – with the sharp tip shows a good correspondence with the corresponding topology image – Fig. 7(b) – that reflects the examination with scanning electron microscopy. In contrast, the use of the blunt tip blurs the resultant topology image – Fig. 7(c) – and creates a peculiar near-field amplitude image – Fig. 7(d) – that is quite distinct from the one obtained with the sharp tip.

6. Applications in nanomaterials

In the sphere of nanomaterials, manifold materials are architecturally configured with nanometer-scale precision, aiming for specific and contingent, multiplex while coherently collaborative functions. It is requisite to comprehend their structural arrangement through innovative material processing methods and, even furthermore, their inter-correlated properties directly. The demonstration of *s*-SNOM being capable of resolving dielectric characteristic of surface in nanometer scale exemplifies one prototypical case. At first, such effort entails a sample that is made of distinct materials with known dielectric constants. Fischer's pattern [42] is chosen here to serve as the first example. Such pattern is made with the following nanosphere lithography procedure. First, polymethylmethacrylate (PMMA) spheres of 300 nm in diameter are self assembled into a hexagonal closely packed pattern on a silicon surface. Second, gold is thermally deposited to overlay such pattern with a gold film of 50 nm in thickness. Finally, the sample is immersed in an acetone solution to remove the PMMA spheres, yielding a hexagonal array of triangular gold disks. Some PMMA residues remain in the vicinity of the gold disks owing to incomplete dissolution of PMMA spheres. The resultant pattern sample thus bears three differing materials (silicon, gold and PMMA) that represent archetypal semiconductor, metal and polymer. The AFM topography image, shown in Fig. 8(a), only shows such hexagonally packed pattern devoid of material composition. In contrast, the scanned image of the scattering intensity of the third-harmonic order, $I_{sca}(3\Omega)$, is shown in Fig. 8(b) and clearly portrays three regions with distinct scattering intensities: triangular areas with the highest signal, round areas with a medium signal and peripheral areas surrounding the triangular areas with the smallest signal. The correspondence with the simultaneously recorded topography image suggests that the brightest areas stand for gold, the less bright areas signify silicon and the dimmest areas are indicative of PMMA. Such assignment can be confirmed with the quantitative comparison of their relative intensities with respect to the calculation according to the quasi-static dipole model, Eqs. (1)-(5). Figure 8(c) displays the calculated $I_{sca}(3\Omega)$ as a function of the real part of the dielectric constant of the sample, Re(ε). The measured scattering intensities of the three areas are also plotted in the figure according to the dielectric constants of the assigned materials. The experimental data points agree with the calculated values, if the imaginary part of the dielectric constant, Im(ε), is assumed to be -0.1 that is valid for these three lossless materials at 632.8 nm. This correspondence demonstrates the exploitation of *s*-SNOM to extract local dielectric property within sub-10 nanometers. The dielectric constant of material contains Lorentz and Drude contributions. The former one represents local oscillators coming from electronic and vibrational transitions, while the latter one reflects the influence of free carriers. Huber *et al.*, via taking advantage of the variation of the Lorentz contribution, used *s*-SNOM, operating at 10.7 μm, to resolve different materials of tungsten, aluminum, silicon and silicon oxide on the polished cross section of a pnp transistor [43]. More recently, they mapped the doping concentration distribution of a silicon transistor [44] with a resolution of 40 nm at 2.54 THz by the use of the Drude contribution: the carrier concentration influences the plasma contribution of the dielectric constant of silicon, yielding visible contrast to resolve the regions of different doping concentrations. This approach of recording local electrical property is non-contact and is better than the contact counterparts, such as surface capacitance microscopy, that often acquire interference results owing to electrical contact.

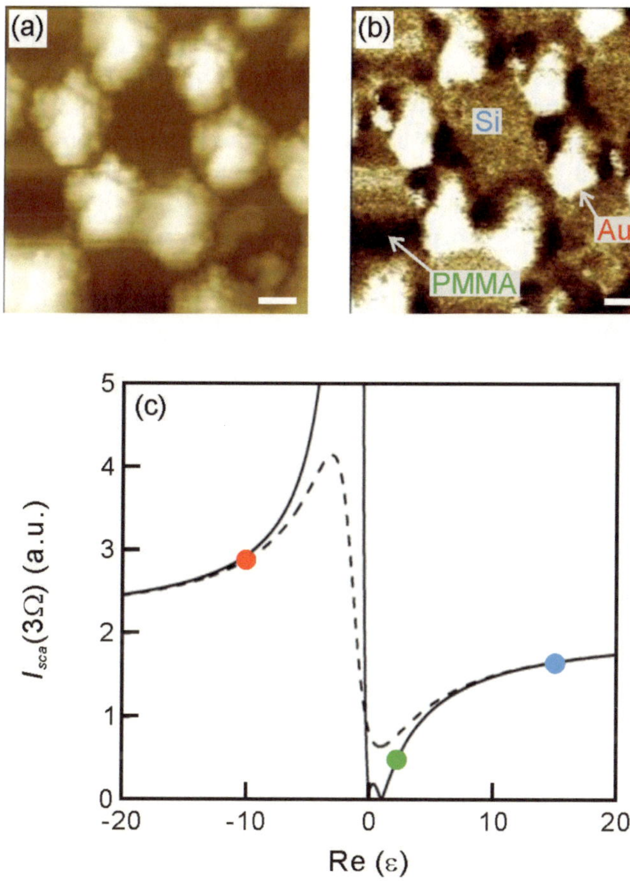

Fig. 8. Images of (a) topography and (b) near-field amplitude of the third-harmonic order, $I_{sca}(3\Omega)$, of a Fischer's pattern [10] excited at 632.8 nm; (c) comparison between calculated $I_{sca}(3\Omega)$ as a function of the real part of the dielectric constant of the sample, Re(ε), with the imaginary-part values of -0.1 (solid line) and -10 (dashed line) and measured values extracted from the corresponding regions of Si (blue dot), PMMA (green dot) and Au (red dot) in (b). Scale bars represent 100 nm.

The capability of resolving local dielectric property with s-SNOM has been utilized by us to examine nanometer-scale optical contrast of the phase-change layer of blue-ray recordable and erasable disks [45]. The quality of recording or reading data is greatly dependent on the detailed variation of optical characteristics within the recorded mark that is within 150 nm in size. Fast temperature quenching induces amorphous phase, while slow temperature cooling yields polycrystalline. The optical signature within such mark is not discernible with conventional aperture-typed SNOM owing to its >50 nm resolution, but would be possible with high-resolution *s*-SNOM. The near-field image of the AgInSbTe phase-change layer of a blue-ray disc obtained by the *s*-SNOM is shown in Fig. 9(a). The recorded marks are revealed by the prominent near-field optical contrast. According to the dielectric constants of crystalline and amorphous AgInSbTe at 632.8 nm (3.63+*i*21.2 and 8.14+*i*9.35, respectively) [46], the ratio between the calculated tip-induced scattering intensities of the crystalline and amorphous phases is 1.28 that that is very close to the experimentally determined value of 1.30. This observation indicates that the dark regions along the protruding track in Fig. 9(a) are amorphous AgInSbTe created during the recording process. This inference agrees with TEM observation. The scattering intensity profile of the recorded disc shows that the marks have the minimum width of ~160 nm by measured the cross section (dash lines), while no such feature is present in the non-recorded disc, shown in Fig. 9(b). Such prominent marks are not recognizable in the corresponding topographic image. In addition, 30-nm isolated crystalline domains within the amorphous recorded mark emerge in Fig. 9(a) and also were

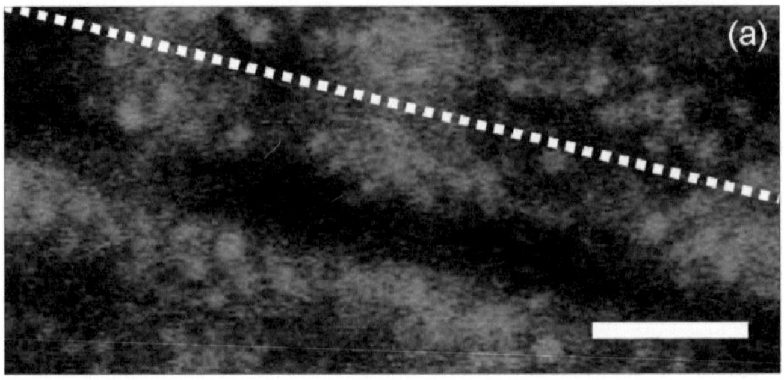

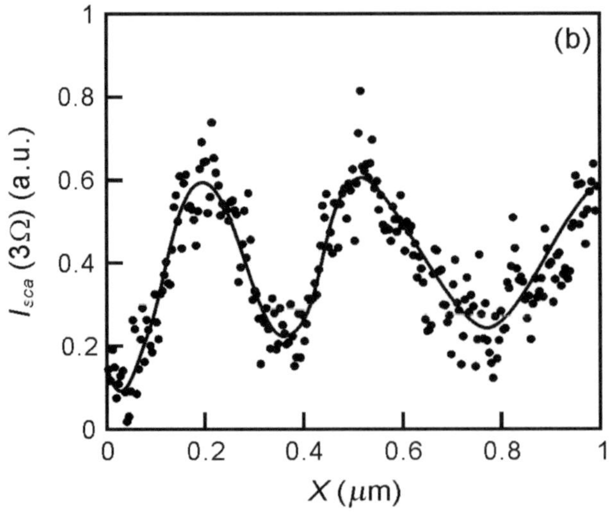

Fig. 9. (a) Image of the scattering intensity of third-harmonic order of BD-RE disc; (b) intensity profile along the track, marked by the dashed line in (a). Scale bar represents 200 nm.

observed by TEM measurement. This nanometer-scaled optical signature could potentially deteriorate the carrier-to-noise ratio in reading this optical storage medium. On the other hand, the studies of BD-RE with conductive AFM [47-48] only presented either conducting or non-conducting state of the phase-change layer, thus prohibiting from using this technique to investigate the intricate phase-change characteristics within the recorded marks. Compared with the result obtained with aperture-typed SNOM by Yoo and coworkers [50], its spatial resolution is not enough to extract the detailed variation, including the nanometer-scaled features within the recorded marks.

7. Applications in nanophotonics

In the sphere of nanophotonics, the energy of electromagnetic field is concentrated or controlled with some specially prepared nanostructures beyond the limit set by traditional optical theory where the optical mode density – on which the principles of optics rely – in the region of interest is assumed as in free space or only slightly perturbed from it. In contrast, the optical mode density in the content of nanophotonics has drastically distinct distribution that is governed by the nanostructured materials in proximity. Because of such distinct optical mode density, its understanding and therefore control are limited by itself that is beyond the traditional optical theory. Moreover, transferring its information content to our macroscopically sensible realm is also influenced greatly by such limitation, entailing unconventional experimental means to truthfully convey the optical information. The *s*-SNOM is representative of one of such many innovative techniques. According to Eq. (5), the scattering intensity induced by the tip-sample near-field interaction is also dependent on the field in the vicinity of such tip-sample complex. Specifically, the scattering radiation can be induced upon the presence of certain surface electromagnetic field. Surface plasmon wave (SPW) exemplifies such surface electromagnetic field, by which the evanescent wave propagates on the metal-dielectric or metal-vacuum interface [29, 43, 50]. In this section, the near-field characteristics of the SPW emanating from single hole or hole arrays of nanometer size made by focused ion beam on metal film are present.

The first example is single hole of 150 nm in diameter on a silver film of 200 nm in thickness. Figure 10 schematically shows the application of *s*-SNOM to such sample. In the case like that, three scattering field contributions are present in the thus obtained *s*-SNOM images. The first contribution is the outcome of the interaction between the incident wave and the tip apex that is in vicinity of the sample surface, producing the first scattering field, E_1. The second contribution is made by the SPW induced by the interaction of the incident wave and the hole. The thus induced SPW propagates to the tip apex, producing the second scattering field, E_2. The third contribution is made the SPW induced by the tip apex that is approached to the sample surface, which propagates to the hole, reflects back to the tip apex, and gives the third scattering field, E_3. The wave vector, $\boldsymbol{k}_{SPW}$, this surface plasmon wave follows the following dispersion relation, $|\boldsymbol{k}_{SPW}| = |\boldsymbol{k}_0| \cdot \sqrt{\varepsilon_1 \varepsilon_2/(\varepsilon_1 + \varepsilon_2)}$ [51], where $\boldsymbol{k}_0$ is the wave vector of the incident wave, and ε_1 and ε_2 is the dielectric constants of vacuum and silver, respectively. The detailed derivation of the resultant scattering wave as well as the corresponding one in inverse space is given in Ref. 50 and is not repeated here. According to the derived results, the interference between E_1 and E_2 produces two $|\boldsymbol{k}_{SPW}|$-radius circles centered at $\pm|\boldsymbol{k}_0|\sin\theta_i$. Furthermore, the interference between E_1 and E_3

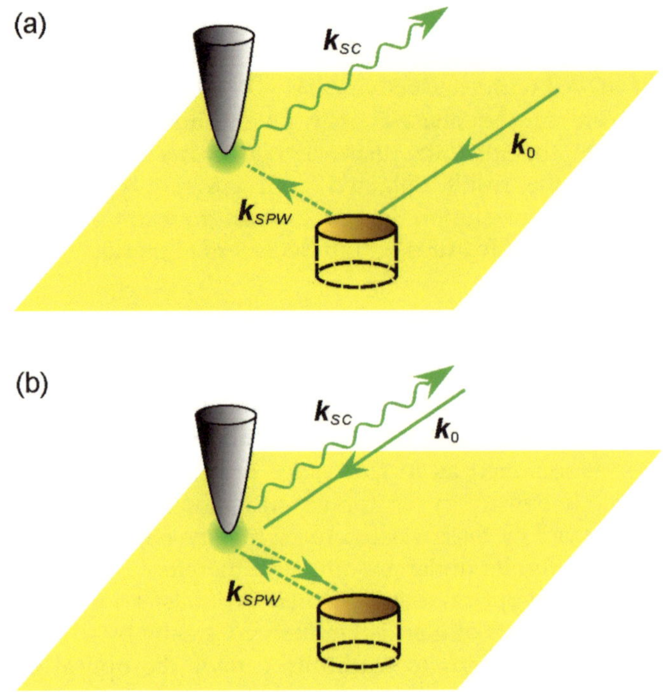

Fig. 10. Two scattering field contributions of surface plasmon waves generated from nanohole (a) and tip (b) during *s*-SNOM scanning of a circular hole on gold film. $\boldsymbol{k}_0$, the wave vector of the incident light wave; $\boldsymbol{k}_{SC}$, the wave vector of the scattering light wave; $\boldsymbol{k}_{SPW}$, the wave vector of the surface plasmon wave.

produces one $2|\boldsymbol{k}_{SPW}|$ -radius circle centered at origin. Such prediction is confirmed with the scanned intensity amplitude image of such sample, shown in Fig. 11. Notice that the intensity amplitude image, shown in Fig. 11(a), exhibits rather complex ring-like pattern around the nanohole. After performing fast Fourier transform (FFT), three circles emerge in the *k*-space image, Fig. 11(b), and match with the theoretical prediction above, supporting the interpretation of near-field images obtained by *s*-SNOM. The schematic diagram of the image in *k*-space is shown in Fig. 11(c). To further support the interpretation of the near-field amplitude image of single nanohole discussed above, numerical calculation based on finite-difference time-domain (FDTD) method was performed. The simulation, without considering the tip, was executed over a region of 16.8 μm × 12.8 μm with a mesh size of 20 nm and the dielectric function of silver was taken from Ref. 52. A Gaussian beam with a 4-μm waist was used to simulate the experimental condition. The calculated distribution of the field component normal to the surface in *k*-space only two $|\boldsymbol{k}_{SPW}|$ -radius circles, because the third contribution of the scattering field is induced by the tip apex and was not considered in the calculation. By removing the large circle of the recorded near-field image and subsequently transforming it back to *r* space, the resultant *r*-space image matches almost perfectly with the calculated result [50]. This consistency thus supports the theoretical analysis of the Fourier-transformed image above. Finally, the single-nanohole study above therefore demonstrates that examining near-field images in *k* space helps to identify the origins of different surface plasmon waves, allowing for in-depth investigation of the fundamental nature of surface electromagnetic waves that may exist around nanostructures [53]. The *k*-space examination can also be applied to ordered hole arrays and has been delineated in Ref. 50. For such case, each hole in the array has both E_2 and E_3 components, while only one E_1 is attributed to the tip apex. The resultant E_2 and E_3 are given by the sum of individual holes. Assuming that two successive scattering events is neglected because of the propagating loss of SPW, the total scattering field in *k*-space can be similarly derived [50], conferring a *k*-space pattern of the near-field image of the nanohole array that is the product of the *k*-space pattern of single nanohole and the structure factor $S(\boldsymbol{k}) = \sum_{m,n} \exp\left(ik_x ma + ik_y na\right)$. The corresponding Fourier-transformed images in k space obtained from the experimental data match very well with the predicted patterns [53]. One important implication emerges from this study. In the near-field study of surface plasmon polaritons with *s*-SNOM, the tip apex always acts as the plasmon inducer as well, making the resultant near-field images rather complicated. Performing Fourier analysis on the recorded near-field images facilitates the identification of the tip-induced contribution, setting a solid foundation for in-depth examination of the fundamental nature of surface plasmon polaritons.

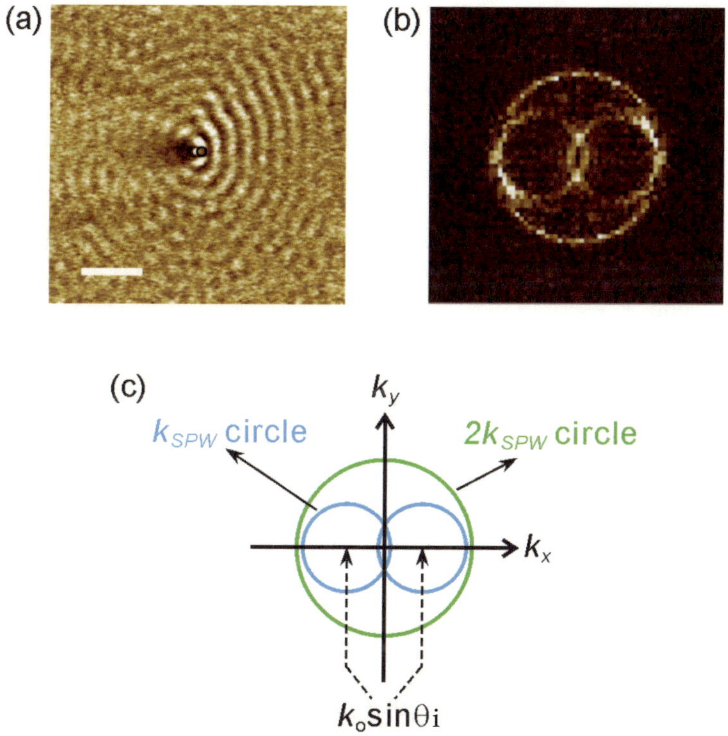

Fig. 11. (a) Scattering amplitude image and (b) Fourier transformed image single nanohole on gold film; (c) schematic Fourier-transformed image of different contributions. The hole diameter is 150 nm and the excitation wavelength is 532 nm. The hole is marked as open black circle. The scale bar represents 1 μm. k_0, wave vector of incident optical wave; k_{SPW}, wave vector of surface plasmon wave.

The aforementioned ordered hole arrays made on a metallic film can exhibit enhanced or suppressed optical transmission through such corrugated films at specific wavelengths [54]. Many applications, such as color filters, sensitive biodetectors based on fluorescence on such films, etc., have been projected. It has been proposed and experimentally verified that such anomalous properties are originated from the launching of SPW upon the excitation of an incident optical wave at those particular wavelengths [55], while examining such plasmon wave in high resolution is not made possible with conventional aperture-type SNOM. Figure 12 presents a closer view of topographic, near-field optical intensity and phase images on the edge of a square circular hole array. Notice that the SPW outside the array is constructed coherently by those emitting from individual holes and propagates away from the array, shown in Figs. 12(a) and (b), as the incident wavelength meets the criterion for exciting surface plasmon wave of an ordered array:

$$|\boldsymbol{k}_0|\sin\theta_i(\cos\varphi\hat{\boldsymbol{x}}+\sin\varphi\hat{\boldsymbol{y}})-\boldsymbol{k}_{SPW}=\frac{2\pi}{a}(m\hat{\boldsymbol{x}}+n\hat{\boldsymbol{y}}),\ (m,n=0,\pm1,\pm2,\cdots), \tag{7}$$

where φ is the azimuthal angle of $\boldsymbol{k}_0$ on the surface. If the incident wavelength is off the excitation condition, no such coherently constructed plasmon wave emanates from the edge of the hole array, shown in Figs. 12(c) and (d). The interior magnified near-field amplitude and phase images of a square elliptic hole array are shown in Figs. 12(f) and (g). In this case, the surface component of the incident light wave, $\boldsymbol{k}_{0,//}=\boldsymbol{k}_0-|\boldsymbol{k}_0|\cos\theta_i\hat{\boldsymbol{z}}$, is perpendicular to the long axis of the elliptic holes – $\hat{\boldsymbol{l}}$ (i.e., $\boldsymbol{k}_{0,//}\perp\hat{\boldsymbol{l}}$). As shown in Fig. 12(f), the regions of the localized field distribution manifest as the bright crescents located around the rim of the elliptical holes. This agrees with the expectation of the excitation of localized surface plasmon. In the corresponding phase image shown in Fig. 12(g), a 180° phase change takes place from the left crescent to the right crescent of the hole. The interpretation of these near-field results relies on the understanding that s-SNOM probes the field component normal to the surface in the present polarization configuration [29]. The two experimental observations above therefore indicate the nature of strong dipole oscillating field along the short axis of the elliptical hole – $\hat{\boldsymbol{s}}$. This interpretation agrees with two additional facts in Fig. 12: no vertical field component is observed and the phase appears randomly inside the hole. These results thus reflect the excited localized surface plasmon in the case that $\boldsymbol{k}_{0,//}\perp\hat{\boldsymbol{l}}$. In contrast, no dipole-like oscillating feature appears in the near-field images in the case that $\boldsymbol{k}_{0,//}\perp\hat{\boldsymbol{s}}$, reflecting weak plasmon excitation. A calculation result based on finite-different-time-domain method further confirms this interpretation [56].

8. Conclusions

In this chapter, we present a high-resolution near-field microscope – scattering-type SNOM (s-SNOM). Its basic operation principle is based on the near-field electromagnetic interaction between a scanning tip and the sample of interest. An interference-based heterodyne detection scheme is employed to extract both amplitude and phase of the extremely weak scattering radiation from the tip-sample complex. The local dielectric property of the sample – if the excitation wavelength is within the resonant excitation condition of the tip apex, and the local surface electromagnetic field – if the excitation wavelength is outside the resonance wavelength range of the tip apex, are readily extracted with this techniques. Considerations while exploiting such unique functions are discussed in detail. A few cases are presented to

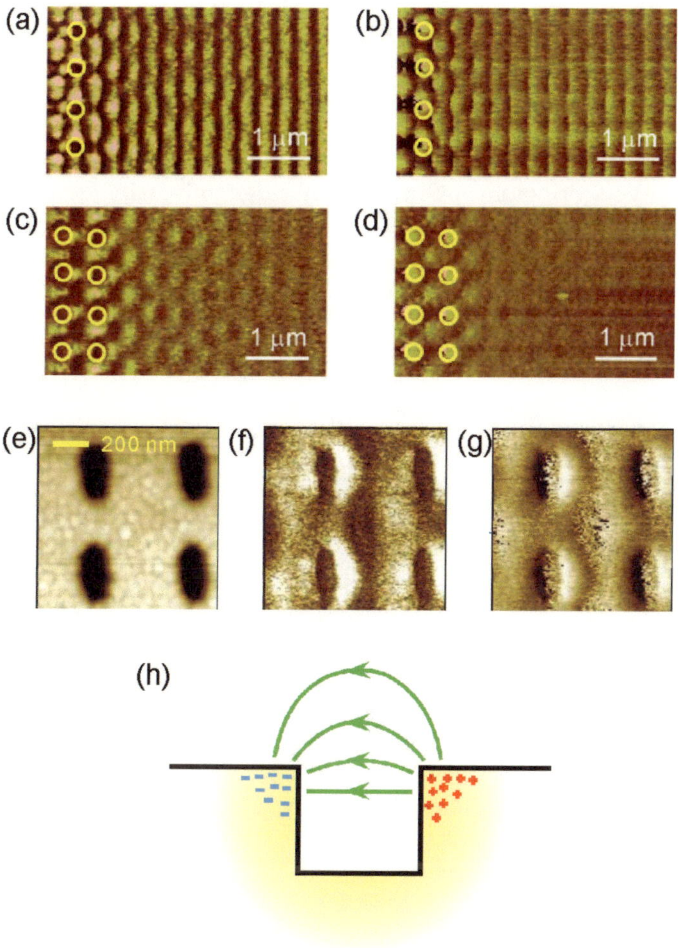

Fig. 12. Near-field intensity and phase images of a circular hole array with two excitation wavelengths (λ_{ex}). (a) Intensity and (b) phase images for λ_{ex} = 532 nm; (c) Intensity and (d) phase images for λ_{ex} = 632.8 nm. The yellow circles indicate the hole position. Topographic (e), near-field optical intensity (f) and phase images (g) around the rim of elliptical hole arrays with *s*-polarization excitation; (h) schematic cross-sectional view of charge distribution and electric force lines of a dipole oscillation.

exemplify these unique characteristics of this new-generation near-field microscopic technology. In particular, the extraction of local composition of the recorded marks of the phase-change layer of blue-ray recordable and erasable disks is presented. ~10-nm optical signatures within such recorded marks are identified and are correspondent to crystalline domains that are barely visible to high-resolution transmission electron microscopy. Furthermore, surface plasmon waves are revealed with *s*-SNOM and the effect caused by the tip on the acquired near-field images is carefully examined. The analysis in *k*-space domain successfully resolves this effect. The application of *s*-SNOM to nanomaterials and nanophotonics is still young. One can expect that its exploitation in these realms and even beyond can be furthered more so in near future.

9. References

[1] M. Born and E. Wolf, Principle of Optics (Cambridge university Press, UK, 1999).
[2] B. Hecht, B. Sick, U. P. Wild, V. Deckert, R. Zenobi, O. J. F. Martin and D. W. Pohl, *The J. of Chem. Phys.*, 112, 7761 (2000).
[3] *Near-Field Nano/Atom Optics and Technology*, edited by M. Ohtsu (Springer-Verlag, Berlin, 1998).
[4] D. Courjon, *Near-Field Microscopy and Near-Field Optics* (Imperial College Press, London, 2003).
[5] J. W. P. Hsu, *Mater. Sci. Engineer.*, 33, 1 (2001).
[6] E. H. Synge, *Philos. Mag.*, 6, 356 (1928).
[7] D. W. Pohl, W. Denk and M. Lanz, *Appl. Phys. Lett.*, 44, 651 (1984).
[8] A. Lewis, M. Isaacson, A. Harootunian and A. Muray, *Ultramicroscopy*, 13, 227 (1984).
[9] H. A. Bethe, *Phys. Rev.*, 66, 163 (1944).
[10] F. Zenhausern, Y. Martin, and H. K. Wickramasinghe, *Science*, 269, 1083 (1995).
[11] *Nano-Optics and Near-Field Optical Microscopy*, edited. by A. Zayats and D. Richard (Artech House, Boston, 2009)
[12] M. B. Raschke and C. Lienau, *Appl. Phys. Lett.*, 83, 5089 (2003).
[13] J. E. Hall, G. P. Wiederrecht, S. K. Gray, S. -H. Chang, S. Jeon, J. A. Rogers, R. Bachelot, and P. Royer, *Opt. Express*, 15, 4098 (2007).
[14] L. Gomez, R. Bachelot, A. Bouhelier, G. P. Wiederrecht, S. H. Chang, S. K. Grey, F. Hua, S. Jeon, J. A. Rogers, M. E. Castro, S. Blaize, I. Stefanon, G. Lerondel, and P. Loyer, *J. Opt. Soc. Am. B*, 23, 823 (2006).
[15] F. Keilmann, *Journal of Electron Microscopy*, 53, 187 (2004).
[16] M. I. Stockman, *Phys. Rev. Lett.*, 93, 137404-1 (2004).
[17] S. Patane, P. G. Gucciardi, M. Labardi, and M. Allegrini, Riv. Nuovo Cimento 27, 1 (2004).
[18] R. Hillenbrand, B. Knoll and F. Keilmann, *Journal of Microscopy*, 202, 77 (2000).
[19] B. Knoll and F. Keilmann, *nature*, 399, 134 (1999).
[20] C. F. Bohren, and D. R. Huffman, *Absorption and Scattering of Light by Small Particles* (Wiley, 1998).
[21] C. Meixner and P. R. Antonicwcz, Phys. Rev. B 13, 3276 (1976).
[22] A. Cvitkovic, N. Ocelic and R. Hillenbrand, *Opt. Express*, 15, 8550 (2007).
[23] F. Keilmann, and R. Hillenbrand, *Phil. Trans. R. Soc. Lond. A*, 362, 1 (2004).
[24] R. Hillenbrand, M. Stark and R. Guckenberger, *Appl. Phys. Lett.*, 76, 3478 (2000).
[25] A. Bek, R. Vogelgesang and K. Kern, *Appl. Phys. Lett.*, 87, 163115 (2005).
[26] B. Knoll and F. Keilmann, *Opt. Commun.*, 182, 321 (2000).
[27] F. Keilmann and R. Hillenbrand, *Phys. Rev. Lett.*, 85, 3029 (2000).
[28] N. Ocelic, A. Huber and R. Hillenbrand, *Appl. Phys. Lett.*, 89, 101124 (2006).

[29] J. Y. Chu, T. J. Wang, Y. C. Chang, M. W. Lin, J. T. Yeh, and J. K. Wang, *Ultramicroscopy*, 108, 314 (2008).
[30] M. B. Raschke, L. Molina, T. Elsaesser, D. H. Kim, W. Knoll and K. Hinrichs, *ChemPhysChem*, 6, 2197 (2005).
[31] Z. H. Kim, B. Liu and S. R. Leone, *J. Phys. Chem. B*, 109, 8503 (2005).
[32] Z. H. Kim and S. R. Leone, *Opt. Express*, 16, 1733 (2008).
[33] S. A. Maier, *Plasmonics: Fundamentals and Applications* (Springer-Verlag, Berlin, 2007).
[34] H. H. Wang, C. Y. Liu, S. B. Wu, N. W. Liu, C. Y. Peng, T. H. Chan, C. F. Hsu, J. K. Wang and Y. L. Wang, *Adv. Mat.*, 18, 491 (2006).
[35] R. Laddada, P. M. Adam, P. Royer and J. L. Bijeon, Opt. Eng., 37, 2142 (1998).
[36] G. P. Wiederrecht, *Eur. Phys. J. Appl. Phys.*, 28, 3 (2004).
[37] V.N. Konopsky, *Opt. Commun.*, 185, 83 (2000).
[38] D. Haefliger, J. M. Plitzko, and R. Hillenbrand, *Appl. Phys. Lett.*, 85, 4466 (2004).
[39] R. Gupta, M. J. Dyer, and W. A. Weimer, *J. of Appl. Phys.*, 92, 5264 (2002).
[40] A. P. Li, F. Muller, A. Birner, K. Nielsh and U. Gosele, *J. of Appl. Phys.*, 84, 6023 (1998).
[41] R. Hillenbrand, F. Keilmann, P. Hanarp, D. S. Sutherland and J. Aizpurua, *Appl. Phys. Lett.*, 83, 368 (2003).
[40] H. H. Wang, C. Y. Liu, S. B. Wu, N. W. Liu, C. Y. Peng, T. H. Chan, C. F. Hsu, J. K. Wang and Y. L. Wang, *Adv. Mat.*, 18, 491 (2006).
[41] A. J. Huber, F. Keilmann, J. Wittborn, J. Aizpurua, and R. Hillenbrand, *Nano lett.*, 8, 3766 (2008).
[42] U. C. Fischer, J. Heimel, H. J. Maas, M. Hartig, S. Hoeppener and H. Fuchs, *Surf. Interface Anal.*, 33, 75 (2002).
[43] A. Huber, D. Kazantsev, F. Keilmann, J. Wittborn, and R. Hillenbrand, Adv. Mater. 19, 2209 (2007).
[44] A. J. Huber, F. Keilmann, J. Wittborn, J. Aizpurua, and R. Hillenbrand, *Nano lett.*, 8, 3766 (2008).
[45] J. Y. Chu, S. C. Lo, S. C. Chen, Y. C. Chang, and J. K. Wang, *Appl. Phys. Lett.*, 95, 103105 (2009).
[46] J. M. Bruneau, B. Bechevet, B. Valon, E. Butaud, *Optical Data Storage Topical Meeting 1997*, 104 (1997).
[47] S. K. Lin, I. C. Lin and D. P. Tsai, *Opt. Express*, 14, 4452 (2006).
[48] A. J. G. Mank, A. E. Ton Kuiper, H. A. G. Nulens, B. Feddes, and G. Wei, *Jpn. J. Appl. Phys.*, 46, 5813 (2007)
[49] J. H. Yoo, J. H. Lee, S. Y. Yim, S. H. Park, M. D. Ro, J. H. Kim, I. S. Park, and K. Cho, *Opt. Express*, 12, 4467 (2004).
[50] Y. C. Chang, J. Y. Chu, T. J. Wang, M. W. Lin, J. T. Yeh, and J.-K. Wang, *Opt. Express*, 16, 740 (2008).
[51] H. Raether, *Surface Plasmons on Smooth and Rough Surfaces and on Gratings* (Springer-Verlag, Berlin, 1988).
[52] *Handbook of optical constants of solids*, edited by E. D. Palik, (Academic, Press, New York, 1985).
[53] L. Yin, V. K. Vlasko-Vlasov, J. Pearson, J. M. Hiller, J. Hua, U. Welp, D. E. Brown, and C. W. Kimball, *Nano Lett.*, 5, 1399 (2005).
[54] A. Degiron and T. W. Ebbesen, *J. Opt. A: Pure Appl. Opt.*, 7, S90 (2005).

[55] E. Devaux, T. W. Ebbesen, J.-C. Weeber and A. Dereux, *Appl. Phys. Lett.*, 83, 4936 (2003).

[56] T. J. Wang, J. Y. Chu, Y. C. Chang, M. W. Lin, J. T. Yeh, and J. K. Wang, unpublished.

14

LIDAR Atmospheric Sensing by Metal Vapor and Nd:YAG Lasers

Dimitar Stoyanov, Ivan Grigorov, Georgi Kolarov, Zahary Peshev and Tanja Dreischuh
Institute of Electronics, Bulgarian Academy of Sciences, Sofia Bulgaria

1. Introduction

LIDAR systems have widely been used for remote investigation of atmospheric parameters (Measures, 1984; Kovalev & Eichinger, 2004; Weitkamp, 2005). They are based on the so-called LIDAR (LIght Detection And Ranging) principle which consists in sending a laser pulse to the atmosphere and subsequent detecting of the radiation backscattered (at angle π) by atmospheric constituents or pollutants. As LIDAR is a time-of-flight technique, the return signal profile detected in the time domain contains range-resolved information about the atmospheric characteristics along the line of laser beam propagation. Advantages of the lidar measurement approaches, as compared to other available active techniques (e.g. radars), are the high spatial and temporal resolution, higher sensitivity and accuracy in sensing atmospheric particles, covering large observation areas, etc. These features make lidar systems powerful instruments for environmental measurements. At present, lidars find a variety of applications in different fields of the human activity. Along with the meteorology, atmospheric physics, and ecological monitoring, lidars are extensively used for volcanic and fire alerting, laser ranging, altimetry and bathymetry, lidar mapping and forestry, coastal morphology and hazards assessment in geology, as well as for many other applications in physics and astronomy, nuclear fusion, military, aviation, robotics, transportation, etc. There exists a variety of ground-based, air-borne and space-borne lidar systems distinguished by their types, schematics, regimes of operation, monitored parameters, constructions, etc. (Kovalev & Eichinger, 2004; Weitkamp, 2005). Among the most widely used systems are the one- or multi-wavelength aerosol lidars exploiting elastic scattering of light.

The present chapter describes the capabilities of LIDAR sensing techniques for atmospheric aerosol profiling by using elastic-scatter lidars based on metal vapor (MV) lasers, as well as on Nd:YAG lasers. First, a brief overview of the basic principles of the LIDAR remote sensing of the atmosphere is given. The single-scattering equations connecting return signal profiles, parameters of the experimental system and characteristics of the probed aerosols along the laser beam are presented in Sec.2, as well as some theoretical approaches for solving the lidar equations in the case of non-absorbing atmosphere (aerosol and molecular). General lidar schematics and methods for detection of lidar signals are also discussed. Special attention is paid to metal vapor and Nd:YAG lasers (Sec.3), and to advantages of using them in the lidar remote sensing of the atmosphere (Sec.4). Then, in Sec.5, a number of experimental results on lidar atmospheric sensing, obtained with these systems are presented. Reported are measurements focused on evaluations and range resolved profiling of a defined set of important optical characteristics of the atmospheric aerosol, such as backscattering and extinction coefficients, Ångström exponent, etc., as averaged over time so in their temporal evolution. This is in view of the close relation of these characteristics to the spatial distribution, concentration, size parameters, and dynamics of the atmospheric aerosol content. Also, the use of lidars for ecological measurements and detection of hidden aerosol pollution transport by lidar mapping, including the trans-border pollution transport over the Danube River, is shown. Special consideration is given to applications concerning remote sensing of different atmospheric phenomena such as volcanic ash and Saharan dust, to parallel observations with space-borne lidars, etc. In Sec.6, the role of multi-wavelength aerosol lidar probing in the mid-visible and near infrared (IR) ranges is underlined, as a powerful and reliable approach for atmospheric observations providing accurate range-resolved profiling of valuable atmospheric parameters with high spatial and temporal resolution.

2. LIDAR atmospheric sensing

LIDAR remote sensing of the atmosphere represents a complex activity joining together a variety of experimental equipments, measurement techniques, analytical methods, theoretical approaches, etc. Lidar sensing process includes the following principal stages:

- Emitting pulsed radiation into atmosphere;
- Receiving, detection, and recording of backscattered lidar signals;
- Lidar signal conversion and preprocessing;
- Data processing and profiling of major parameters of interest;
- Displaying, visualization, analysis and interpretation of obtained results.

The interpreted results from the lidar sensing can be used by different authorities (governmental, local, etc.) for decision making.

2.1 General lidar block-schematic

Regardless of the mentioned above diversity of lidars, there are some basic components common for all the systems, such as transmitter, receiver, and acquisition subsystems. A general block-diagram of a lidar is presented in Fig.1. One of the main parts of the lidar is the laser transmitter emitting pulsed radiation of appropriate power, spectral, spatial, and temporal characteristics. The laser beam is transmitted into the atmosphere by an opto-mechanical set-up. The latter comprises a set of optical elements (mirrors, splitters, etc.) for laser beam transportation and time-synchronization, an expanding telescope for minimising the output beam divergence, and mechanical mounts with precise translation and rotation mechanisms for beam steering and lidar adjustment.

The lidar receiver consists of optical receiving part and photo-electronic blocks. Backscattered radiation is collected by a telescope. Refractive and reflective telescopes of different types (Cassegrain, Newtonian, Schmidt, etc.), size, and configurations are typically used. A changeable properly-shaped and sized field-stop diaphragm, placed near the focal point of the telescope, provides angular spatial filtering of the backscattered radiation, forming by this manner the telescope's field of view in conformity with the laser beam divergence. An important part of the multiwavelength lidar receivers is the optical module for wavelength- and/or polarization separation and discrimination. It represents an optical assembly containing dichroic beam-splitters forming the lidar spectral channels for initial wavelength separation of the backscattered laser radiation. Narrow bandpass (1-3 nm FWHM) interference filters are usually placed in each of the spectral channels, providing the main fine spectral selection of the corresponding wavelengths and suppressing the solar radiation background, especially in day-time measurements. Residuals from the other laser wavelengths are further suppressed by using additional spectrally-selective optical elements (e.g. edgepass filters), leading to final transmission of less than 10^{-4}% in the blocking spectral regions. As a result, a good enough signal-to-noise ratio is normally achieved, allowing reliable detection of weak lidar signals and, hence, reaching high altitudes of lidar sounding. Lidar spectral channels can be equipped for measurement of the depolarization ratio of aerosol backscatter on defined wavelengths as it is important for aerosol particle shape characterization.

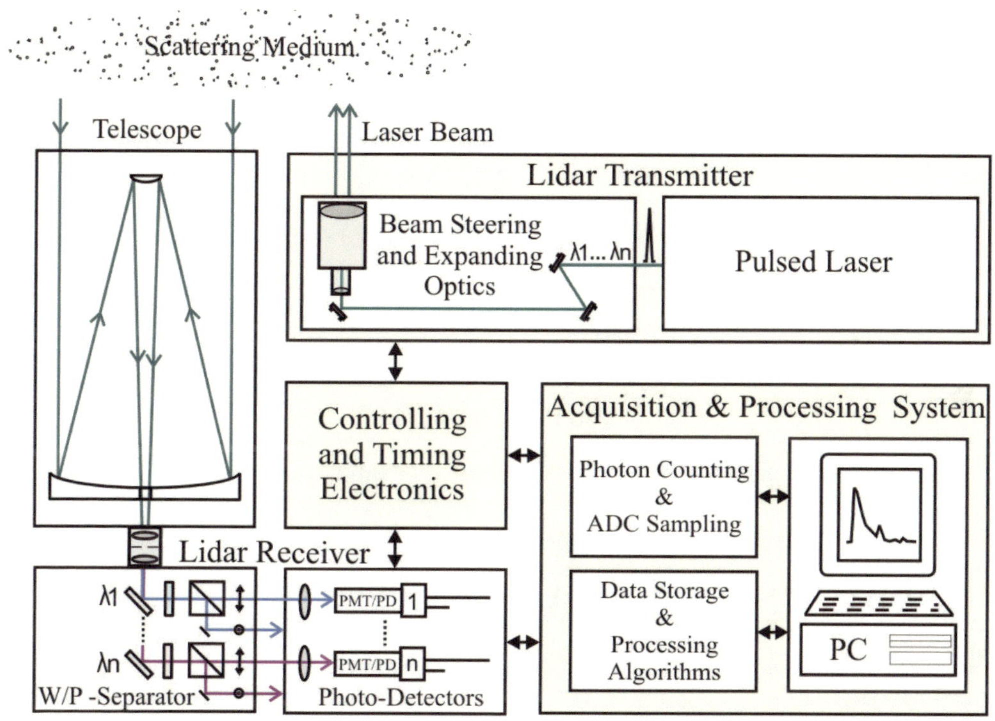

Fig. 1. General block diagram of a lidar system.

Spectrally selected optical lidar signals are detected and converted to electrical ones by using highly-sensitive photodetectors such as photomultipliers, avalanche photodiodes, and CCD-cameras. Further, the received signals enter the acquisition system which provides sampling and pre-processing of raw signals to standard lidar data. The acquisition system is designed to operate in either analog or photon-counting modes, depending on the lidar type, laser pulse energy, measurement tasks, etc. The acquired lidar profiles are stored and processed by specialized retrieving algorithms and software. The controlling and timing electronics provides the overall lidar operation.

2.2 Lidar equation

Basic aerosol parameters derived from lidar data are the aerosol backscattering and extinction coefficients. Theoretical models for retrieving their range profiles from raw lidar data are based on solving of so-called lidar equation describing the relation between the received range-resolved backscattered optical radiation power and atmospheric and system parameters. For a single-scattering elastic backscatter lidar (measuring backscattered light at the same wavelength as the sensing laser wavelength λ) the power $P(r)$, detected at a time t after the instant of pulse emission, is written as:

$$P(r) = P_0 \frac{c\tau}{2} A\varepsilon \frac{\gamma(r)}{r^2} \beta(r) \exp\left[-2\int_0^r \alpha(\rho)d\rho\right], \tag{1}$$

where $r=ct/2$ is the distance along the laser beam path, P_0 is the average power of a single laser pulse, τ is the pulse duration, A is the area of the primary receiving optics, ε is the overall system efficiency, $\beta(r)$ and $\alpha(r)$ are the backscattering and extinction coefficients, respectively, at wavelength λ. The term $\gamma(r)$ describes the overlap between the laser beam and the receiver field of view, being equal to 1 for ranges of complete overlap. To solve the lidar equation (1), it is useful to split the backscatter and extinction in molecular and aerosol terms:

$$\beta(r) = \beta_a(r) + \beta_m(r)\,; \quad \alpha(r) = \alpha_a(r) + \alpha_m(r)\,. \tag{2}$$

In Eq.(2) and further, the subscripts "a" and "m" stand for aerosol and molecular terms, respectively. We also assume negligible atmospheric absorption at a wavelength λ and $\gamma(r) = 1$. In next step we introduce the aerosol extinction-to-backscatter lidar ratio $S_a(r)$ as:

$$S_a(r) = \alpha_a(r) / \beta_a(r) \tag{3}$$

by analogy with the molecular extinction-to-backscatter ratio:

$$S_m = \alpha_m(r) / \beta_m(r) = 8\pi / 3\,. \tag{4}$$

$S_a(r)$ depends on range r through the size distribution, shape, and chemical composition of the aerosol particles. Substituting Eqs.(2) in Eq.(1), and bearing in mind Eqs.(3) and (4) , we obtain:

$$\begin{aligned} &P(r)r^2 \exp\left\{-2\int_0^r [S_a(\rho) - S_m]\beta_m(\rho)d\rho\right\} = \\ &= P_0 C[\beta_a(r) + \beta_m(r)]\exp\left\{-2\int_0^r [\beta_a(\rho) + \beta_m(\rho)]S_a(\rho)d\rho\right\} \end{aligned}. \tag{5}$$

In Eq.(5) the constant $C=A\varepsilon c\tau/2$ (so-called lidar constant) covers all system parameters. Next, following the steps described in Weitkamp, (2005), we take the logarithm of both sides of Eq.(5) and differentiate it with respect to r. Finally a differential equation of Bernoulli type is obtained. In a case when the values $\beta_a(r_0)$ of aerosol and $\beta_m(r_0)$ of molecular backscattering at a reference range r_0 are known, the lidar can be calibrated with these boundary conditions to determine the calibration constant P_0C. Then the solution of Eq.(5) can be written as follow:

$$\beta_a(r) = -\beta_m(r) + \frac{P(r)r^2 \exp\left\{-2\left[S_a(r) - S_m\right]\int_{r_0}^{r} \beta_m(\rho)d\rho\right\}}{\frac{P(r_0)r_0^2}{\beta_a(r_0)+\beta_m(r_0)} - 2\int_{r_0}^{r} S_a(\rho)P(\rho)\rho^2 \exp\left\{-2\left[S_a(\rho) - S_m\right]\int_{r_0}^{\rho} \beta_m(\rho')d\rho'\right\}d\rho}. \quad (6)$$

The molecular atmospheric backscattering coefficient $\beta_m(r)$ can be determined from the model of Standard atmosphere or calculated using meteorological data from a radiosonde. However, the aerosol backscattering $\beta_a(r)$ and extinction $\alpha_a(r)$ [or the aerosol lidar ratio $S_a(r)$], remain to be determined as two unknown variables from the values of one variable – the so-called range corrected signal $P(r)r^2$, obtained from the lidar measurement. This is not possible without using additional information about at least one of the unknown variables. When choosing the reference range r_0 near to the ground (and to the lidar), the aerosol backscattering $\beta_a(r_0)$ or extinction $\alpha_a(r_0)$ can be measured independently. After that, we can integrate Eq.(6) and retrieve the atmospheric backscattering coefficient profile. Unfortunately, there exists a mathematical instability in the calculations following Eq.(6) in sense that small errors in the determination of $\beta_a(r_0)$ at the reference distance r_0 or in the measured range corrected signal $P(r)r^2$, start to produce in few steps of integration procedure negative values for the denominator and, consequently, for $\beta_a(r)$. To solve this problem, Klett (Klett, 1981) and Fernald (Fernald, 1984) proposed an inverse integration of Eq.(6), starting from the far end of the lidar sounding path. Applying this idea and introducing $\beta_a(r_{\max}) = \beta_a^{\max}$ and $\beta_m(r_{\max}) = \beta_m^{\max}$, where r_{max} is the maximal distance, we can rewrite Eq.(6) as follows:

$$\beta_a(r) = -\beta_m(r) + \frac{P(r)r^2 \exp\left\{-2\left[S_a(r) - S_m\right]\int_{r}^{r_{\max}} \beta_m(\rho)d\rho\right\}}{\frac{P(r_{\max})r_{\max}^2}{\beta_a^{\max}+\beta_m^{\max}} + 2\int_{r}^{r_{\max}} S_a(\rho)P(\rho)\rho^2 \exp\left\{-2\left[S_a(\rho) - S_m\right]\int_{\rho}^{r_{\max}} \beta_m(\rho')d\rho'\right\}d\rho}. \quad (7)$$

The reference range r_{max} is chosen so that the aerosol backscatter coefficient is negligible compared to the molecular backscatter coefficient, i.e. $\beta_a(r_{max}) \ll \beta_m(r_{max})$. Normally such atmospheric conditions are observed in the upper troposphere. Thus, we can attach the upper backscattering value to the value of the molecular backscattering and calculate the backscatter profile in backward direction by Eq.(7), using the measured range-corrected lidar signal. Determination of the reference range r_{max} is a problem for some atmospheric conditions as cloudy atmosphere, intensive background, etc. (see Sec.6). This algorithm is now widely applicable in practice, assuming also invariant value for the aerosol lidar ratio S_a=const along the laser beam path in the atmosphere. The exact value of this constant is determined depending on the laser wavelength and also on *a priori* assumptions about the type of the observed aerosols. The influence of the assumption for constant aerosol lidar ratio on the results of calculated profiles of atmospheric backscattering coefficient is studied and described in Böckman et al., 2004. The backscattering profiles are calculated from numerical models of lidar returns in two stages: once using constant $S_a(r)$, and second – with variable profile of $S_a(r)$. As shown, the errors in the calculated aerosol backscatter profiles due to the variance of the aerosol lidar ratio at different atmospheric conditions could reach 25-30%. The conclusion thrusts on the strong recommendation to use all available *a priori* information about the atmospheric conditions and the observed aerosols to apply an adequate variable aerosol lidar ratio in lidar determination of the aerosol backscatter profiles.

2.3 Photon detection methods in lidar sensing

Photon detection (using photomultipliers, photodiodes, etc.) is a key operation in optical devices, including lidars (Gagliardi & Karp, 1976). The understanding of photon detection processes is of essential importance for the development and performance of the lidar. This process is described by the probability for a photoelectron emission, which is proportional to the square of the envelope of the classical electromagnetic field (the optical field intensity) on the photosensitive surface. The transformation of optical field into a photoelectron flux is a quantum process. The output photoelectron current is a random temporal process, corresponding to the random photon absorption by independent quantum systems of the photocathode and to the photoelectron emission. The concept for the randomness of the photoelectron current is always valid for the stochastic or determined optical fields. The output photoelectron current $I_{phe}(t)$ is described mathematically by a superposition of δ-pulses, each corresponding to a single process of photoelectron emission or

$$I_{phe}(t) \sim \sum_{k=1}^{N(0,t)} \delta(t - t_{phe,k}), \tag{8}$$

where k=1,…,$N(0,t)$ are the successive photoelectron numbers, $N(0,t)$ is the total number of photoelectrons and $t_{phe,k}$ are the so-called photoelectron arrival times. As seen, the information, carried by the optical field after the photodetection, is contained in the arrival times of photons (or photoelectrons). As the probability of photoelectron emission depends on the optical intensity, the increase of optical energy causes an increase of the number of photoelectrons per unit time and thus, decreasing the time intervals between adjacent photoelectrons. This is the so-called effect of photoelectron time-grouping, depending on the instant optical intensity. The optical field is transformed into a photoelectron flux of time-dependent intensity. Measuring the arrival times $t_{phe,k}$ of all photoelectrons one could, in principle, extract the whole information carried by the optical field.

Different effects in photon detectors prevent the extraction of the entire information from the received optical radiation. The output current $I_{out}(t)$ can be expressed now in the form:

$$I_{out}(t) \sim \sum_{k=1}^{N(0,t)} h_k(t - t_k), \tag{9}$$

where $h_k(t\text{-}t_k)$ are the output pulse functions of finite duration, depending on the photon detector parameters. They are usually called single electron pulses (SEPs) or photoevents. The SEPs are of fluctuating shapes, amplitudes, and arrival times t_k. The electric charge G_k of an individual k-th SEP is a fluctuating variable, depending mainly on the processes of the secondary emission. It is given by

$$G_k \sim q\int_{-\infty}^{\infty} h_k(t - t_k)dt\,, \tag{10}$$

where q is the electron charge. The SEP arrival times are expressed by $t_k=t_{phe,k}+t_{pd}+\Delta t_k$, where t_{pd} is the mean time delay and Δt_k are the centered time delay fluctuations or the jitter (transit time spread) of the photon detector. The mean time width τ_{pd} of the SEPs defines the receiving frequency bandwidth $\Delta\omega_{pd}\sim\pi/\tau_{pd}$ of the photon detector.

The parameters of $h_k(t-t_k)$ dramatically affect the photoreceiving process. It becomes impossible to measure the photoelectron arrival times at the photon detector output. It is due mainly to limitations imposed by the processing electronics. The temporal structure of the output current $I_{out}(t)$ provides successive extraction of the optical information from the flow of SEPs at the output in definite number of cases, strongly depending on the optical intensity. Four basic regimes (modes) of photodetection are typically recognized, requiring specific approaches to be applied in developing optical receivers and corresponding acquisition techniques. An approximate criterion for distinguishing these regimes is the number of photoevents (or SEPs) per the photon counter dead time τ_{dead} [as a rule $\tau_{dead}>(3–5)\tau_{pd}$]. Its appearance is due to the carrier restoration processes in transistors and diodes of electronic circuits after the photon detector (discriminators, amplifiers, etc). The dead time prevents the registration of two successive SEPs, separated by time intervals, shorter than τ_{dead}, causing nonlinearities in the detection process. The quoted above photodection modes (ordered by increasing of optical intensity) are as follows: single quantum (SQ) mode; photon counting (PC) mode; overlapping (OV) mode; and analog mode (see Fig.2).

In the so-called SQ mode, the time intervals between adjacent SEPs are quite larger than the dead time and, thus, the probability for appearance of adjacent SEPs, separated by intervals of the order of τ_{dead}, is minimized. This condition provides the measurement of individual arrival times of SEPs. Unfortunately, photon rates in this mode are very low, resulting in intolerably long accumulation times in many applications including the lidar sensing.

The photon counting is realized by conversion of SEPs into corresponding normalized electric pulses of standard amplitude (NSEPs) (photocounts). This transformation just causes the dead time effects. The normalized pulses are then counted (within some defined sampling intervals Δt) by standard electronic circuits. To count the normalized SEPs (i.e., to count photons) they have to be resolved in time. It is evident that minimum time intervals between the adjacent SEPs have to be longer than the dead time τ_{dead}. The maximum tolerable instantaneous photon count rate R_{NSEP} (in number of NSEPs per second) can be estimated approximately by the condition $(R_{\mathrm{NSEP}})_{\max}\tau_{dead}\leq 1$. Because of the time grouping effects of photoevents, a part of input SEPs will not be counted, if $(R_{\mathrm{NSEP}})_{\max}\tau_{dead}\sim 1$. That is why the values of $(R_{\mathrm{NSEP}})_{\max}\tau_{dead}\leq 0.1$ are more preferable to be satisfied in photon counting receivers in order to provide linear dependence of the counted NSEPs on the input optical intensity. The NSEPs are described by Poisson statistics as it is for the input photon flux. The dynamic range of PC receiver is very high. At higher intensities it is limited by the above condition $(R_{\mathrm{NSEP}})_{\max}\tau_{dead}\sim 1$. However, there are no limitations at low intensities. The only limitation here is the tolerable data accumulation (measurement) time. For these reasons the PC mode is widely used in modern lidar systems.

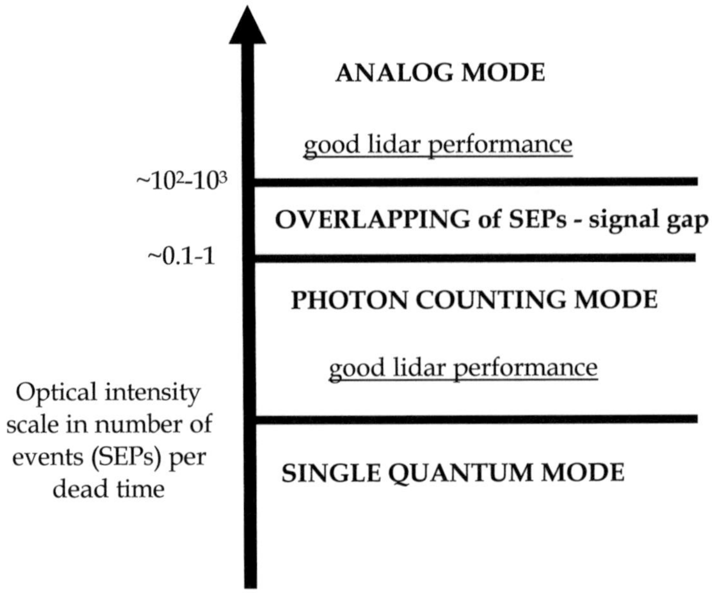

Fig. 2. Regimes of photon detection as a function of optical intensity.

The overlapping mode is typically recognized as a non-operational signal gap, where the performance of optical receivers is ineffective. In OV mode the output flow of SEPs (before the normalization) could be represented by a stochastic process, due to the random summation (at each moment) of a low number of SEPs of fluctuating individual charges, shapes, duration, and arrival times. The output current is not proportional to the optical intensity. The appearance of OV mode is due mainly to the electronics, when the photon counts are not resolved any more (due to the higher photon rates) in order to be counted. At the same time, due to the large fluctuations of the output amplitudes, signals could not be correctly sampled by analog-to-digital converters (ADCs). In OV mode the input optical intensity is higher than that in PC mode and this is why the OV mode is attractive for development of methods for providing the linear performance. Some novel techniques are reported (Stoyanov, 1997; Stoyanov et al., 2000). The analysis of OV mode is out of the scope of this chapter as it is not used in lidar sensing.

The analog mode is a basic regime of optical receiving (together with PC mode), widely used in lidar sensing. Here the number of SEPs at the photon detector output typically exceeds 10^2-10^3. In this mode, the fluctuations of SEP parameters can be neglected because of the averaging over a large number of SEPs at each moment. The output current is an analog signal of amplitude proportional to the input intensity and can be directly sampled by ADCs. The noises in the output current typically also display Poisson statistics, but because of the larger number of photons within each sampling interval it is transformed into Gaussian one.

The analog and PC regimes of optical detection impose special requirements to the lasers, used in lidars. Say, the analog mode is typically used with high pulsed power, low repetition rate lasers (as Nd:YAG lasers). The Nd:YAG lidars are also used in PC mode for large distances for which the number of received photons dramatically decreases. Here the measurement times are normally higher, due to the lower return intensities. In practice, the combined use of both regimes causes some problems as the gluing of both lidar profiles. The PC mode in lidar sensing can also be realized by lasers of lower pulsed power but at higher pulse repetition rates, providing high enough mean output power. In this case the above requirements to PC mode can be satisfied.

Assuming Poisson statistics (Sec.2.3) for the photon detector output signal fluctuations, the probability $W_{N(r)}$ of detecting $N(r)$ photons (or SEPs) from a distance r in PC and analog modes for a sampling interval Δt at a mean rate R_N is given by :

$$W_{N(r)} = (R_N \Delta t)^{N(r)} (N(r)!)^{-1} \exp(-R_N(r)\Delta t) . \tag{11}$$

For Poisson distributed lidar signals the noise variance $\sigma_N^2(r)$ is determined from the mean number of detected photons (SEPs) per sampling interval by the expression $R_N(r)\Delta t = \sigma_N^2(r)$. The contribution of Poisson noises with variance $\sigma_N^2(r)$ is essential in the total error balance of lidar measurements. Unfortunately, as it is clear from the lidar equation (Sec.2.2), a multitude of variables contribute to the total estimated error. In fact it is difficult to find an analytical expression for the errors of the calculated backscatter profiles by a classical way, differentiating the lidar equation. In the practice a roundabout approach is applied. It consists in two-step calculation of the backscatter coefficient profiles: i) using the measured values of the range corrected signal $P(r)r^2$; ii) adding some estimated deviation to the measured values $P(r)r^2 \pm \sigma_E(r)$, where $\sigma_E(r) \geq \sigma_N(r)$ is the estimated deviation profile. If no other useful information is available to estimate strictly $\sigma_E(r)$, at least the Poisson variance $\sigma_N(r)$ of lidar signals can be added for estimation of lower error limit and the error propagation along the backscatter calculus chain.

3. Metal vapor and Nd:YAG lasers: Basic parameters

The requirements for successful lidar atmospheric sensing impose strong limitations on the laser parameters as the pulse width, pulsed and mean powers, repetition frequency, operational wavelengths, stability, etc. That is why the number of laser types applied in lidars is limited. The Nd:YAG lasers are widely used in the most of lidar systems (Measures, 1984; Weitkamp, 2005) providing simultaneous sensing in analog and photon counting modes at typically 4 to 6 wavelengths (using harmonic generation techniques) in the IR, visible and near UV ranges (including Raman channels). Lasers emitting a set of basic wavelengths of proper parameters, say approximately equal output powers, are also of great importance for multiwave lidar atmospheric sensing. The use of such lasers can simplify the opto-mechanical lidar design.

The MV lasers eligible for lidar probing (in the sense of above requirements) are mainly lasing on two active media, namely copper (Cu) and gold (Au) vapors. They offer unique output parameters (Astadjov et al., 1988; Kim, 1991; Stoilov et al., 2000) attractive for development of lidars in the mid-visible range, capable to probe simultaneously the troposphere and stratosphere. These lasers emit pulses with mean power of up to 2 kW at relatively high repetition frequencies, normally ranging from 2 KHz to 100 KHz, depending on the laser type. The pulsed energy is substantially low (~0.1 mJ at 5-10 ns pulse duration). The combination of low pulse energy, high mean power, high repetition frequency and multiwavelength performance of MV lasers are their key advantages for application in lidar remote sensing. Typically, output powers of 1-3 W are sufficient for probing simultaneously the troposphere and stratosphere.

Most of the available Cu-vapor lasers operate using one of the two active substances - pure Cu or CuBr. They emit two basic spectral lines (at 510.6 nm and 578.2 nm) of close mean output powers. The beam and temporal characteristics at both lines are practically equal. These lasers provide Gaussian beams of low intensity fluctuations and simultaneous emission of pulses at both wavelengths. The CuBr lasers are more preferable for lidars, because of their very low readiness time (~5-15 min), due mainly to the quite lower working temperature of about 450 ^{0}C as compared to 1500 ^{0}C -1700 ^{0}C for lasers on pure Cu.

Lasers on Au-vapor emit radiation basically at a wavelength of 627.8 nm with temporal and beam parameters quite similar to these of pure Cu-vapor lasers. The combination of Cu- and Au-vapor lasers in a single laser tube is very attractive for lidar applications. Mean output powers of ~0.5 W for each of the three lines 510.6 nm, 578.2 nm and 627.8 nm will be sufficient to cover heights above 15 km in PC mode.

Conventional flashlamp-pumped Nd:YAG lasers, operating in Q-switching mode, provide pulses of 5-10 ns FWHM, at 10-100 Hz typical repetition rates, with extremely high pulse energy reaching more than 1 J at the fundamental wavelength (1064 nm) and up to hundreds of millijoules at the second (532 nm), third (355 nm), and fourth (266 nm) harmonics. These lasers can be designed for simultaneous operation at all the available wavelengths or at optional combinations of them, being by this manner powerful multi-wavelength sources of UV, visible, and IR light. They exhibit perfect shot-to-shot energy stability (instabilities of ≤±2%), long-term power drift of less than ±3%, and temporal jitter of ≤1 ns. In addition, Nd:YAG lasers possess excellent beam-quality characteristics such as Gaussian spatial profile, beam divergence down to 0.5 mrad, pointing stability of < 50 µrad, and polarization ratio > 80%. Diode-pumped Nd:YAG lasers, using bars of powerful laser diodes instead of flash-lamps, are also commercially available. They combine the features quoted above with compactness and high averaged power at kHz pulse frequencies.

4. Lidars on metal vapor and Nd:YAG lasers

Lidars on metal vapor and Nd:YAG lasers are used since mid 70-ties of the last century. The fast progress in development of Nd:YAG lasers of stable and very good output parameters, meeting the requirements for effective atmospheric probing, provided their wide applications in most of lidar sensing systems. In the first decade of the new century, the problems related to more precise range-resolved characterization of the atmosphere (aerosol and molecular content, clouds, air quality, atmospheric transport, etc.) became the most important challenge to the modern lidar systems and their future effective incorporation in the Global Atmospheric Watch (GAW) networks. An important approach for improving the characterization procedures is the combined use of lidars (and lidar networks) with some other instruments as sun-photometers, microwave (MW) radiometers, in-situ measurements, etc. In spite of the already proven synergy of this approach, the further improvement of the lidar atmospheric characterization remains an important task. In this sense, a possible solution could be the use of multiwave lidar sensing, especially in the visible range. That is because the typical aerosols, loading the troposphere and low stratosphere as dust, volcanic ash, thin clouds, etc., are of submicron and near micron size range. Thus, they are commensurable to the wavelengths in the visible range. The simultaneous use of laser radiations covering the green-to-red part of visible range can provide better description of the backscattering wavelength dependence on the aerosol particle sizes, especially in multimode size distributions. Efficient sources of radiation in this spectral domain are the mentioned above Cu- and Au- vapor lasers. The combination of lidars based on these lasers with Nd:YAG lidars operated at 1064 nm and 532 nm offers additional advantages to cover practically the mid-visible and near IR ranges, where the absorption effects can be neglected as a rule.

In Arshinov et al., (1983) the application of Cu-vapor lidar for measuring profiles of the atmospheric humidity and temperature is described. The Cu-vapor lidar system is used for detection of stratospheric aerosol layers at heights of up to 28 km (Kolarov et al., 1988). The scattering ratios (aerosol and molecular to molecular) exceeding 1.5 within the height range 22-28 km are measured at the wavelength of 510.6 nm (1-3 W mean power, 5 KHz repetition frequency, 19cm telescope diameter, and 10-100 s accumulation times) with a photon counting system. A complex system, containing Cu-vapor (510.6 nm) and Au-vapor (627.8 nm) lasers, combined with MW and IR radiometers and MW radar is described in Stoyanov et al., (1988). The two lasers are synchronized in time. The output powers are of 1 -3 W at 510.6 nm and 0.5 W at 627.8 nm, at pulse repetition frequencies 5-15 KHz. The experimentally demonstrated operational heights are 30 km for 510.6 nm and 22 km for 627.8 nm at 100 s accumulation times. The short accumulation times provided opportunities for studying some dynamic processes in the lower stratosphere. One of the most important results of these experiments is the demonstration of good lidar performance in the PC mode within the ranges from the planetary boundary layer (PBL) heights (~750 m) up to 30 km in the stratosphere. The first lidar and MW radiometers remote sensing experiment is performed using Cu-vapor laser of parameters as given above (Gagarin et al., 1987). The operational wavelengths of MW radiometers are 0.8 cm and 1.35 cm to be sensitive to the free water in cloud droplets and the water vapor. The measurement accuracy for the brightness temperature is below 0.1 K. The accuracy of determining cloud water content and water vapor is ~ 0.5% and 1%, respectively. In this schematic, the lack of range resolution of the radiometers is compensated by the good lidar resolution. In series of experiments the links between the time variations of the backscattered time-resolved lidar signals and the brightness temperatures at the radiometer wavelength are demonstrated and analyzed.

The aerosol lidar with CuBr-vapor laser developed at Laser Radars Lab of the Institute of Electronics (LRL-IE), Bulgarian Academy of Sciences, is shown in Fig.3a (Grigorov et al., 2010). The CuBr-laser generates pulses at high-repetition frequency of 13 kHz (10 kHz in the upgraded version), with duration of 10 ns at 510.6 nm and 578.2 nm. Laser beam is directed vertically upward. Two Cassegrain telescopes with 15 cm aperture and 2.25 m focal length receive the backscattered radiation at the two wavelengths. A registration in PC mode is applied. Received backscattered lidar signals are stored in the computer by means of a photon counting board providing spatial resolution of 15-30 m, in 1024 samples and averaging time of 1 min. The maximum height is 15 km, limited by the laser pulse repetition frequency. Under daytime conditions, the sounding height decreases to about 4-5 km, due to intensive sky illumination, reducing the signal-to-noise ratio (SNR). Each lidar measurement lasts about 3-4 hours and more. The lidar profiles, integrated over accumulation time of 1 min, are additionally averaged by summation of data from 30 profiles. Thus, the measurement time for each profile amounts to 30 min. The stored data are subsequently processed by Fernald's algorithm, using a program in MATLAB environment, developed in LRL-IE.

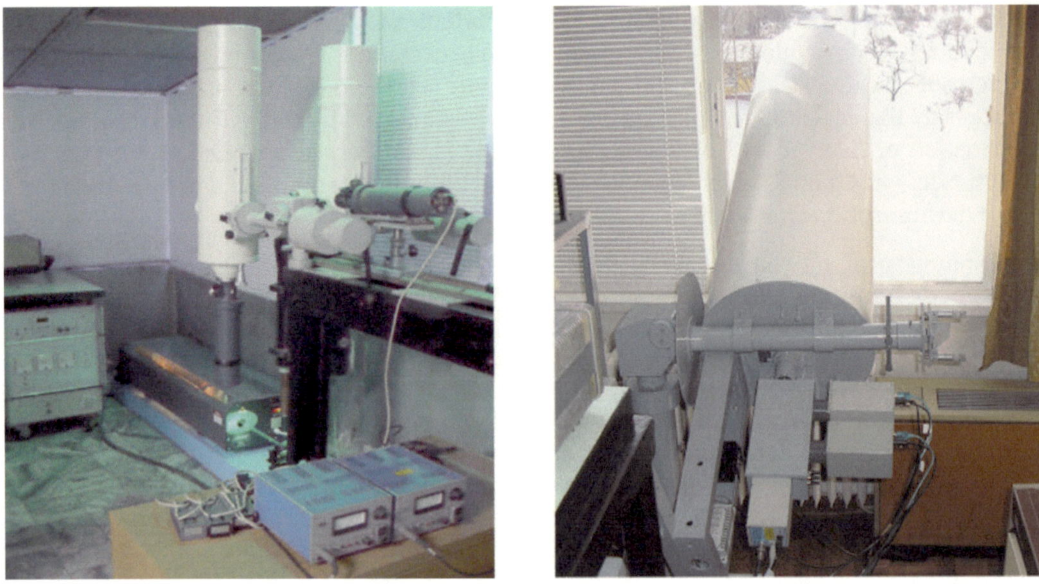

(a) (b)

Fig. 3. Photographs of the CuBr-vapor (a) and Nd:YAG (b) laser-based aerosol lidars at LRL-IE involved in the European Lidar Network measurement programs.

The Nd:YAG lidar system of the LRL-IE (Fig.3b) is a 3-channel combined aerosol-Raman lidar (Peshev et al., 2010). The laser provides output pulse energies of up to 1 J at 1064 nm and up to 120 mJ at 532 nm, at a repetition rate of 2-5 Hz, with pulse duration of 15 ns FWHM. The pulse power is of up to 70 MW at 1064 nm and up to 10 MW at 532 nm. The corresponding values of the averaged power for the two wavelengths are of up to 2 W and 0.25 W, respectively. The output beam divergence is of 2.5 mrad (total angle). These performance characteristics of the laser allow one to carry out nighttime and daytime lidar measurements. The optical part of the receiver contains a Cassegrain telescope (35 cm aperture; 2 m focal length) and a 3-channel spectrum-analyzing module based on dichroic beam-splitters, narrowband interference filters (1-3 nm FWHM), edge-pass filters, and neutral densities. The electronic part of the lidar receiving system consists of three compact photo-electronic modules. Each module comprises a photon detector (photomultiplier or avalanche photodiode), 10 MHz/14-bit ADC, high voltage power supply, and controlling electronics. The aerosol lidar channels operate in analog mode with 15 m range resolution. Receiving modules are connected to a computer by high-speed USB ports. The acquisition system is controlled by specialized software providing the accumulation, storage, and processing of lidar data. It allows for evaluating and plotting profiles of range-corrected lidar signals, aerosol backscattering coefficient, and estimation error.

5. Applications of LIDARS for remote atmospheric sensing

Atmospheric aerosols originate from natural and anthropogenic sources such as desert windstorms, forest fires, volcanic eruptions, sea spray, and combustion products of human activities. Aerosols, having different size distributions and chemical or physical properties, can affect the climate over large regions. Aerosol particles reflect the solar radiation, act as cloud condensation nuclei, modify the scattering properties and lifetime of clouds, influence the precipitation cycles, as well as atmospheric radiative and thermal balance, etc. Ejecting immense amounts of ashes and gases into the atmosphere, active volcanoes can strongly affect for long periods climate, ecology, aviation industry, agricultural activities, and human health over regions of up to global scale. This is why observations and alerting on volcanic aerosols in the atmosphere are of great importance. Lidars can provide real-time sensing of atmospheric aerosols over large areas with high spatial and temporal resolution.

The significance of lidar information, provided by a single lidar station is essentially enhanced if working in a lidar network. Such an idea combines together researches in different lidar networks as EARLINET (http://www.earlinet.org), MPLNET (http://mplnet.gsfc.nasa.gov/), AD-Net (http://www-lidar.nies.go.jp/AD-Net/), etc. The primary goal of the project EARLINET (European Aerosol Research Lidar Network) is the creation of a common database, banding the results of observation of lidar stations located in the European countries (Bösenberg et al., 2003). Main result of such cooperation is the establishment of a quantitative lidar dataset describing the aerosol vertical, horizontal, and temporal distribution, including its variability on a continental scale. Such a dataset could be a comprehensive data source to address the four-dimensional spatio-temporal distribution of aerosols on a global scale (Pappalardo et al., 2010).

Analyses of lidar data require additional information to improve the interpretations of both the type and origin of aerosol layers. Several regional models for simulation and prediction of the dust cycle in the atmosphere have been developed (Kallos et al., 1997; Ozsoy et al., 2001; Nickovic et al.; 2001; Perez et al., 2006a, 2006b). The Dust Regional Atmospheric Model (DREAM) (Nickovic et al.; 2001; Perez et al., 2006a, 2006b) provides reliable operational forecast maps of dust load and concentration in the atmosphere over the North Africa and Euro-Mediterranean region. The model is operated by the Barcelona Supercomputing Center (BSC), (http://www.bsc.es/projects/earthscience/DREAM/). Maps present the cloud coverage, wind directions and speeds, and dust loads. Another source of information about the origin of the aerosol layers offers the HYSPLIT (HYbrid Single-Particle Lagrangian Integrated Trajectory) model (Draxler & Hess, 1998; Draxler & Rolph, 2011). It represents a complete system for computing simple air parcel trajectories to complex dispersion and deposition simulations. The model can be run interactively through the READY system at the site http://www.arl.noaa.gov/HYSPLIT_info.php of the Air Resource Laboratory of NOAA, USA. Calculations of the forward/backward air mass trajectories give a plot of the path passed by the air mass for a chosen time period before arriving to the lidar station location.

The results of lidar applications in the remote sensing of the atmosphere, presented in this Section, are obtained in the LRL-IE using the two elastic backscatter lidar systems described in Sec.4. Some results are reported, concerning lidar mapping of aerosol fields over large industrial zones as one of the important applications of lidars in the regional ecological studies and expertise.

The Sofia lidar station, being one of the stations working in the frame of the EARLINET project, is involved in the following research activities:

- Regular lidar measurements of the atmosphere performed twice weekly;
- Observation of special phenomena, such as unusually high concentrations of aerosols in the troposphere (transportation of mineral dust from Sahara desert over the Mediterranean Sea to Europe, volcanic eruptions, formation of smoke layers resulting from forest or industrial fires, intense photochemical smog, etc.);
- Correlative measurements with space-borne lidars, in the frame of international cooperation.

5.1 Lidar mapping of aerosol pollutions over industrial regions

The monitoring of air pollution distribution over large industrial and urban zones is an important task for improving the quantitative and qualitative estimates of the pollution impact on the environmental conditions. The air pollution transport from local sources is connected with the motion of air masses driven by the wind. The high temporal and spatial resolution of lidar probing and the speed of measurements define the high efficiency of aerosol transport lidar mapping. The opportunity to scan areas of the order of 50-100 km^2 by a single scanning lidar system can provide valuable information for in-depth analysis of the pollution dynamics over broad regions containing a large number of potential local sources.

Results are presented below from a lidar mapping of air pollution distribution and transborder pollution transport over the Danube River in the region of Silistra (Bulgaria) – Kalarash (Romania). The measurements are performed using a scanning CuBr lidar system (λ=510.6 nm) applying photon counting detection. The main lidar parameters are similar to those described in Sec.4. The scanning system provides 1.6^0 scanning step from 0^0 to 360^0 in the horizontal plane with an angular elevation step of 1^0 within the range from -3^0 to 15^0. The scanning lidar is mounted on the roof of a high building located on the hill near the riverside of the Danube River. The total area scanned in successive measurement sessions in order to map different zones as urban and industrial ones, river ports, etc., is of more than 200 km^2. The scanning zone is divided into sectors, separately mapped by the lidar. The integration time at a given angle is chosen to be 1 min, providing maximum sounding distances to more than 12 km in horizontal direction. The measurement time for scanning a 30^0 angular sector is of about 30 min. Under conditions of measurements, one can accept the concept of approximately frozen aerosol fields. Lidar mapping measurements are supported by simultaneous in-situ measurements (including gas analysis) by a specialized transportable laboratory (Mitzev et al., 1995).

The processing of lidar data includes solutions of lidar equation for each angular direction, using the Klett's inversion method and estimation of the volume aerosol extinction coefficient. The latter is converted into aerosol mass concentration, by using our previously derived approximate empirical expression (Mitzev et al., 1995). Finally, the processed data in mass concentration profiles (stored in radial coordinates) are transformed to rectangular coordinates and attached to the geographical map (see Fig.4).

The lidar maps, created from these measurements contain valuable quantitative information about the spatial distribution of aerosol fields as well as their dominated paths of propagation. Fig.4 presents a lidar map of aerosol distribution (dark blue isolines of constant mass concentration) over an area of ~ 70 km^2 (12 km x 6 km). The Danube River and the Lake of Kalarash are displayed in blue. The contours of some geographical objects are given in red. Parts of urban regions of Kalarash and Silistra are presented in dark orange. The position of the scanning lidar is marked as well. As seen, two main areas of intensive aerosol loadings are well displayed on the lidar map. The upper map area shows the aerosol field distribution in the vicinity of the steel plant near Kalarash. This plant is one of the main pollution sources in the region of the investigation. The obtained lidar maps related to adjacent sectors show that the pollution emitted in this region is transported approximately parallel to the Danube River, probably not crossing the border. The aerosol plume in the mid-part of the image is identified as to be emitted from the cellulose plant located far from the mapped region. The pollution of this plume is supposed to pass through the river. This estimate is supported by the parallel gas analysis. The measured phenol concentration (typical for cellulose manufacturing) exceeds about 8 times the tolerable level in Silistra.

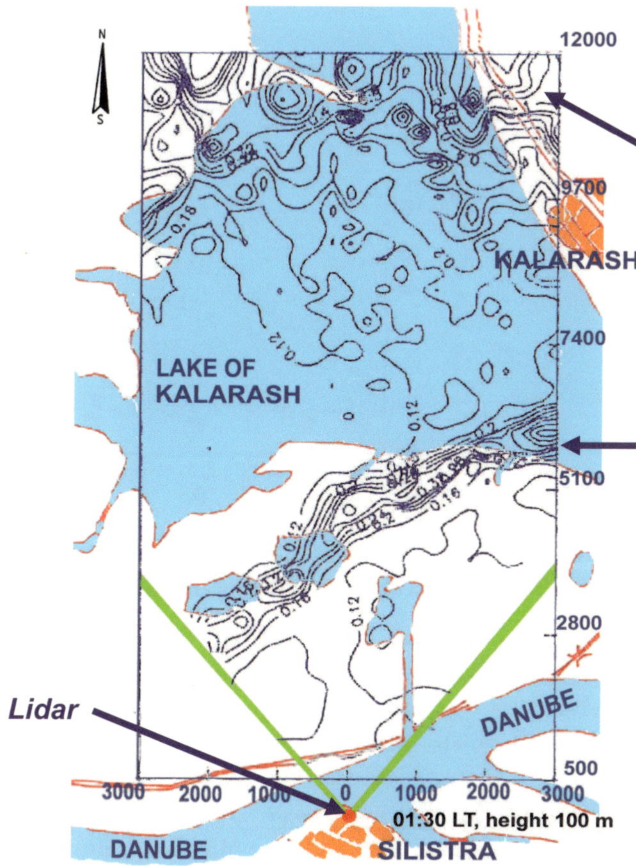

Fig. 4. Lidar map of aerosol pollution distribution over the Silistra - Kalarash region; 14.10.1992.

5.2 Lidar observations on Saharan dust loadings in the atmosphere

Numerous observations on Saharan dust presence in the atmosphere over Sofia are carried out by both lidars of LRL-IE, in order to follow the concentration, spreading, and temporal evolution of Saharan dust transported over European continent (Papayannis et al., 2008; Grigorov et al., 2009). Measurements are synchronized in time with the BCS-DREAM model forecasts for dust loadings and transport.

As an illustration, vertical profiles of the aerosol backscattering coefficient measured by the Nd:YAG lidar at 1064 nm and 532 nm on 4 November 2010, during a dust-transport event, are presented in Fig.5a. The altitude range 1-5 km above sea level (ASL) is only shown in order to zoom the profile part containing the Saharan dust layer. The latter is located in the range 2.8-4 km ASL, just above the PBL as typically. The color-coded DREAM dust loading forecast map for a time preceding the measurements is displayed in Fig.5c. As one can see on the map, a dust layer with density of about 0.2 g/m^2 has covered the lidar station region, in good correlation with the intense peaks of dust backscattering coefficient (exceeding $1x10^{-6}\ m^{-1}sr^{-1}$ at 532 nm) observed in Fig.5a.

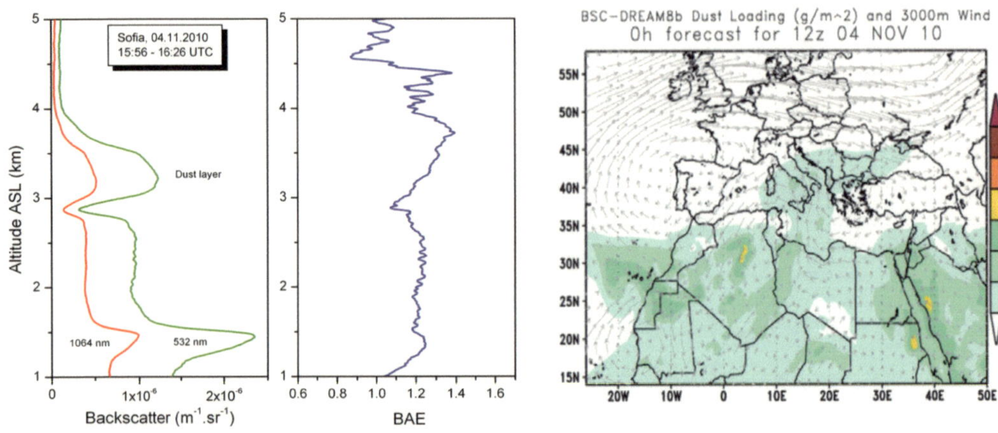

(a) (b) (c)

Fig. 5. Vertical profiles of the aerosol backscatter coefficient at the two lidar wavelengths (a), corresponding profile of BAE (b), and BSC-DREAM forecast map of dust loading (c).

The vertical profile of backscatter-related Ångström exponent (BAE) is shown in Fig.5b, corresponding to profiles in Fig.5a. The BAE values are nearly constant (~1.2) in the underlying PBL. They increase with height reaching 1.3-1.4 just in the dust layer. Such values (1-1.4) are typical for Saharan dust, implying sub-micron dust particle size domination.

Height-time-coordinate diagrams of the backscatter coefficient evolution are presented in Fig.6, as measured in successive time intervals by the CuBr and Nd:YAG lidars at 510.6 nm (a) and 532 nm (b), respectively. As obvious from both diagrams in Fig.6, the Saharan dust layer is well expressed, intense, and relatively stable in terms of height and thickness.

Nevertheless, one can perceive specific internal structure of density distribution evolving over time. The aerosol layer at 5-6 km ASL, observed by the upward-looking CuBr lidar (Fig.6a), is absent on the other diagram because the Nd:YAG lidar is operated at a slope angle of 58 degrees with respect to the zenith, receiving signals from different spatial domains.

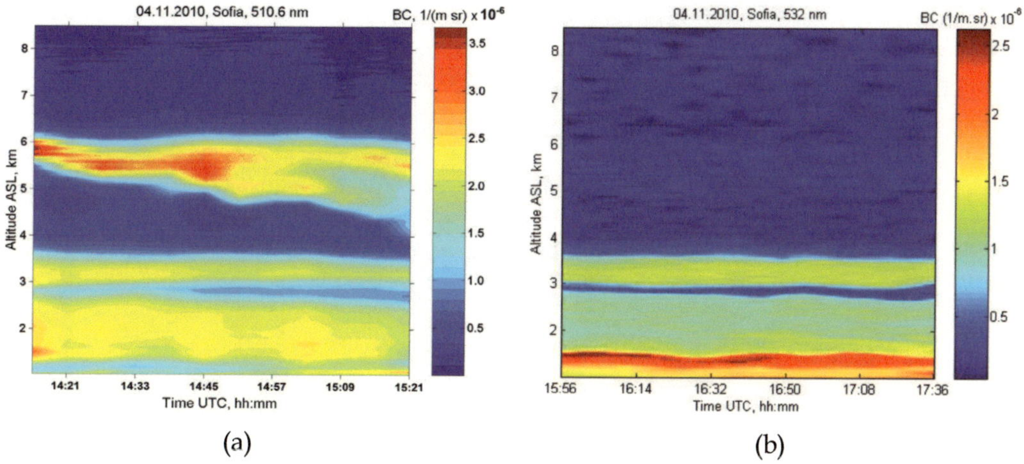

Fig. 6. Evolution diagrams of the aerosol backscatter coefficient at 510.6 nm (a) and 532 nm (b) as measured by the CuBr and Nd:YAG lidars, respectively, on 04.11.2010.

5.3 Detection of volcanic ashes

The eruption of Eyjafjallajokull volcano in Island on 14 April 2010 offered an opportunity lidar stations, participating in the European Lidar Network, to demonstrate the effectiveness of the lidar sensing for 4-dimensional characterization of the volcanic ash transport. The lidar monitoring of Eyjafjallajokull plumes spreading by the Sofia lidar station started on 18 April 2010 and finished on 25 May 2010 (Grigorov et al., 2011). Results of lidar measurements, performed on 22 April 2010, by using a CuBr lidar at a wavelength of 510.6 nm are given in Fig.7, showing presence of volcanic ash layer positioned at ~2.2 and 3 km altitudes AGL. The observed low limit of the layer frequently remains mixed with the PBL, at about 2-2.5 km altitude AGL.

The lidar observations are presented in two formats: as a single averaged vertical profile of the retrieved backscattering coefficient (Fig.7a) and as a map of the time evolution of the range-corrected lidar signal (RCS) (Fig.7b). The corresponding BSC-DREAM forecast map and the calculated HYSPLIT backward trajectories, proving the origin of the detected aerosols, are presented in Figs.7c and 7d, respectively.

On the plot of the backscattering coefficient, two peaks appear just at the top of the PBL, indicating the presence of aerosol layers at about 2.2 and 3 km altitude AGL. As it can be seen in Fig.7b, where the denser aerosol layers are color coded by orange-red colors, these two layers do not disappear during the whole period of measurement. The forecast map of BSC-DREAM concerning the Sahara dust transport (Fig.7c) shows an atmosphere free of desert dust over the Balkans at that time. In addition, the HYSPLIT backward trajectories (see Fig.7d), corresponding to altitudes of 1.5 km and 3 km AGL, cross the volcano site and/or European countries with volcanic ash atmospheric contamination. So, a conclusion can be drawn, that the detected two aerosol layers are due to the transport of the volcanic ash. The aerosol layers appearing at heights of about 8 km AGL are identified as cirrus clouds.

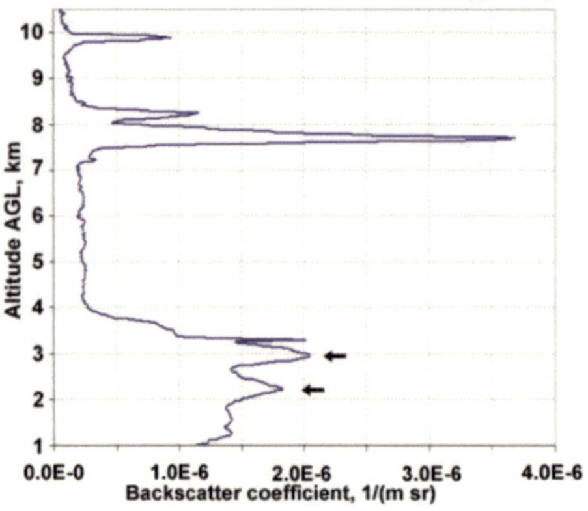

(a)

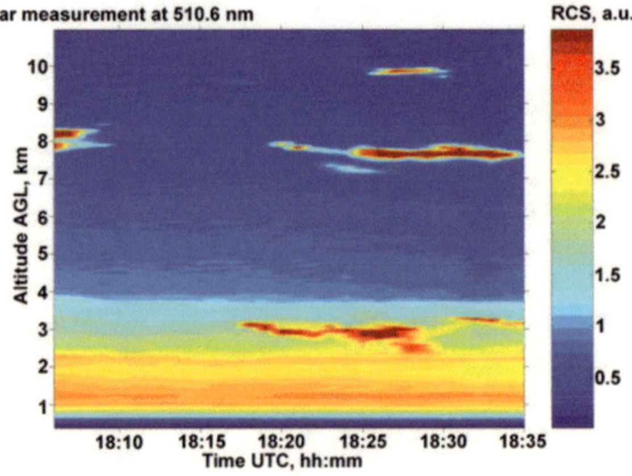

(b)

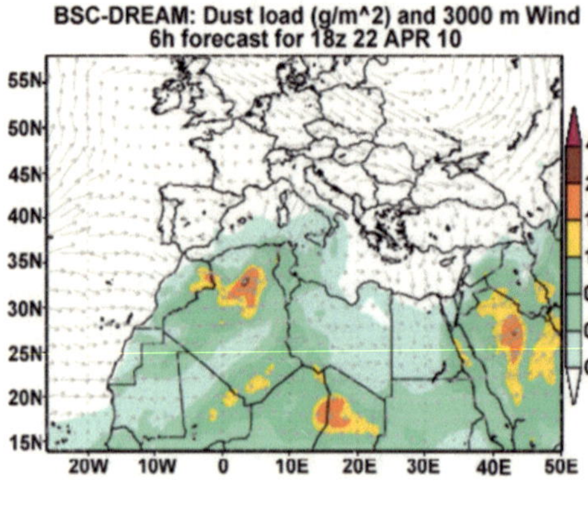

(c)

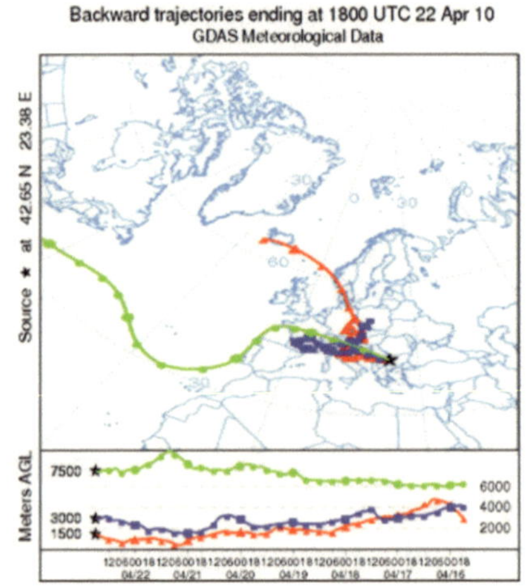

(d)

Fig. 7. Results of lidar measurements performed on 22.04.2010: a) averaged vertical profile of the retrieved aerosol backscattering coefficient; b) time evolution of range-corrected lidar signal (RCS); c) BSC-DREAM forecast map of Saharan dust load in the atmosphere; d) backward HYSPLIT air mass trajectories. The two peaks marked with arrows in Fig.7a are volcanic ash layers over Sofia.

5.4 Regular lidar atmospheric measurements

The specialized EARLINET database, resulting from the longtime monitoring of atmospheric aerosols by regular lidar measurements, contains a valuable information for atmospheric processes over Europe (Papayannis et al., 2008). It gives an opportunity for further improvement and validation of atmospheric models and retrieving algorithms applied for climatologic investigations.

Results of lidar measurements carried out by Sofia lidar station on 6 April 2009 are presented in Fig.8a. The color map represents the one-hour evolution of the retrieved aerosol backscattering coefficient based on lidar profiles with 5 min time averaging in the period 17:00-18:00 UT. The observed multi-layered aerosol structure can be explained analysing the meteorological situation using the corresponding BSC-DREAM dust load forecast map (Fig.8b). It shows that Sofia remains away from the Saharan dust flow. So, we suppose that the two aerosol layers, at 8 km and 9.5 km height, represent cirrus clouds. The layer at 3 km altitude is determined to be a residual aerosol layer, due to the decomposition of the PBL in the evening.

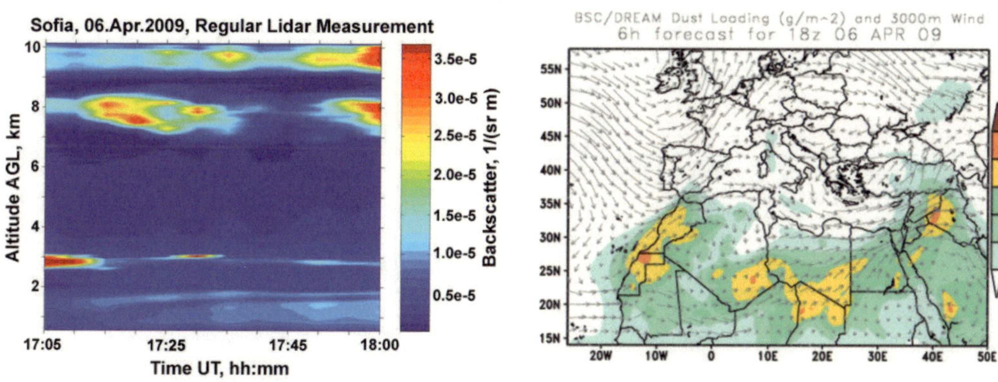

a) b)

Fig. 8. One-hour evolution diagram of the retrieved aerosol backscattering coefficient corresponding to lidar measurements carried out on 6 April 2009 (a) and the BSC-DREAM Saharan dust forecast map (b).

5.5 Correlative space-borne and ground-based lidar measurements

Atmospheric profiling by a network of ground-based lidar stations is an optimal approach for validation of results obtained by space-borne lidars, providing supporting data to fully exploit the information from satellite lidar missions. Such a mission is the Cloud-Aerosol Lidar and Infrared Pathfinder Satellite Observations (CALIPSO). The Cloud-Aerosol LIdar with Orthogonal Polarization (CALIOP), mounted on the CALIPSO satellite, is a Nd:YAG-laser-based lidar specially designed for aerosol and cloud monitoring. Several years correlative ground-based lidar measurements, performed by the EARLINET stations as synchonized with CALIPSO overpasses, contribute to the specialized database, illustrating the potential of the lidar network to provide a sustainable ground-based support for space-borne lidar missions (Pappalardo et al., 2009, 2010).

The Sofia CuBr-lidar group is involved in correlative measurements for CALIPSO since June 2006 (Grigorov et al., 2007). Results of mesurements performed on 28 April 2009 by the ground-based lidar and by the CALIOP lidar are presented in Figs.9a and 9b, respectively.

Vertical red lines on the plots indicate the time of satellite passage over Sofia. On the first plot two aerosol layers can be distinguished. The lower one, located at 1-1.5 km altitude, is due to air convection in the PBL. As seen on the corresponding forecast map of Sahara dust load (Fig.9c), the region covered by the dust flow is in the immediate vicinity of Sofia.

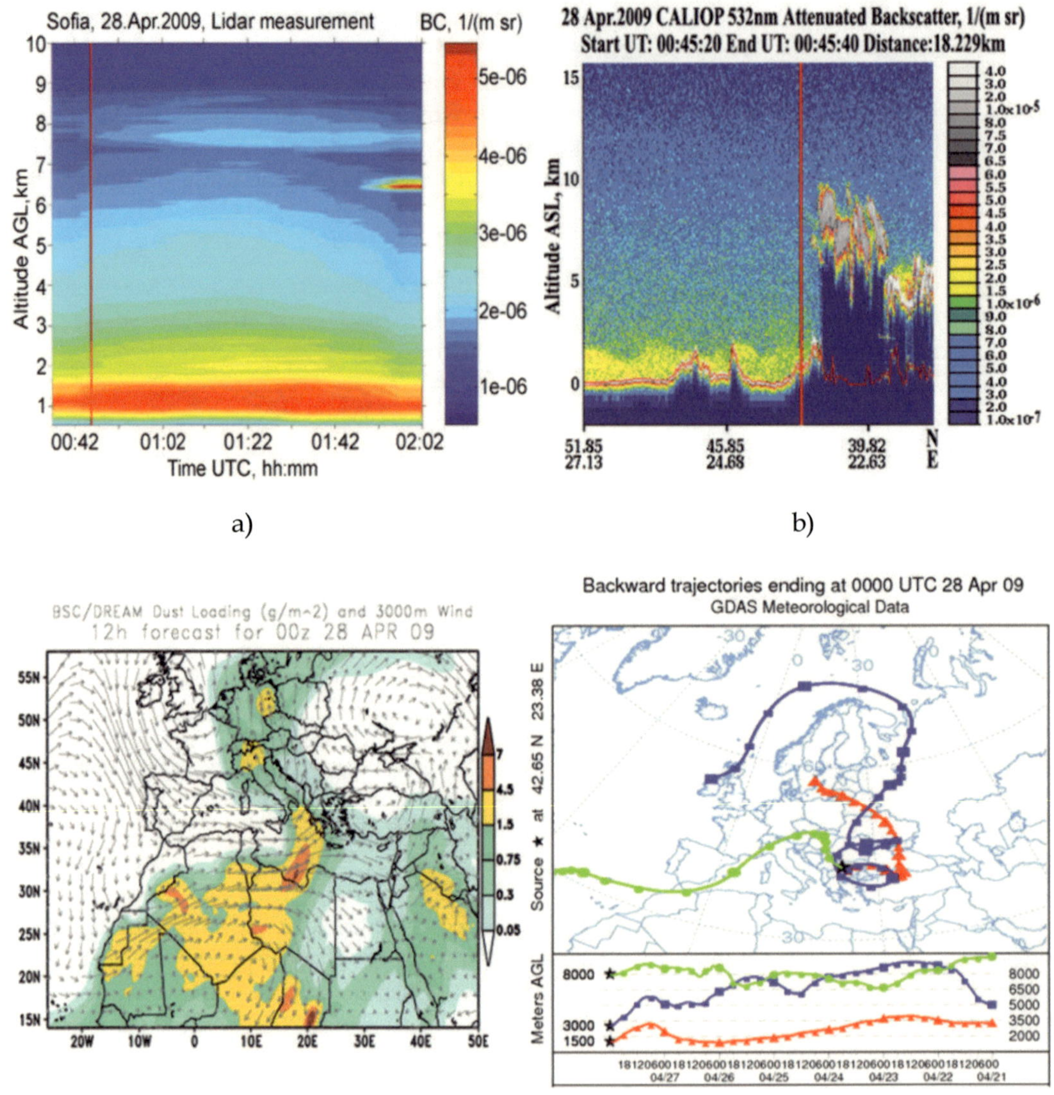

a) b)

c) d)

Fig. 9. Sofia-lidar-station and CALIPSO correlative measurements: a) ground-based CuBr lidar data; b) CALIPSO satellite lidar data; c) BSC-DREAM forecast map of Saharan dust load; d) HYSPLIT backward air mass trajectories. The vertical red lines on plots (a) and (b) show the satellite overpass time above Sofia.

Two of the calculated backward air mass trajectories (colored in red and blue on Fig.9d) originate from North Europe. Altough passing over regions loaded with Saharan dust, they are not related to aerosol layers perceptible by the lidar. The green trajectory has an origin above the Atlantic Ocean, indicating for probable transport of humid air which, mixed with the fine Saharan dust particles, can form the aerosol fields appearing at 7.5-8 km height. Similar aerosol layers are present at the same altitude on the plot of CALIPSO satellite lidar data. The satellite flies over Bulgaria from North-East to South-West. On the CALIOP lidar image, thick aerosol layers occur at 6-9 km altitudes close to the moment of satellite passage over Sofia (see Fig.9b). Those layers, moving from West, are observed by the ground-based lidar in about 1 hour (at ~ 01:50 h UTC, Fig.9a).

6. Multi-wavelength lidar sensing

Lidar sensing in the visible range is typically performed at the second harmonic of the Nd:YAG laser radiation - 532 nm, combined in some lidar systems with the Raman line of the nitrogen molecule at 607 nm. The interval between 532 nm and 1064 nm is not widely used in lidar sounding, mainly because of the dominating usage of Nd:YAG lasers. The US lidar network MPLNET makes use on laser emissions within approximately the same spectral bandwidth (523 nm or 527 nm). As clear from preceding considerations, the metal-vapor lasers cover partially the spectral interval mentioned above. In this Section we consider multi-wavelength lidar probing by MV and Nd:YAG lasers. Some useful theoretical simplifications, providing more correct determination of atmospheric backscattering coefficient profiles, and related experimental results are described and discussed.

6.1 Lidar equations in the case of multi-wavelength sensing by metal vapor lasers

Let us consider for simplicity the case of lidar sensing at two close wavelengths λ_1 and λ_2 (e.g. the ones emitted by a CuBr laser: λ_1=510.6 nm and λ_2=578.2 nm). The case of lidar sensing at more than two wavelengths can be analyzed using similar approach.

Assuming a vertical sounding and replacing further the distance r with the height h, the normalized range-corrected-signal profiles $S_1(h)$ and $S_2(h)$ for both wavelengths are given by

$$S_1(h) = C_1\gamma_1(h)\left[\beta_{a1}(h) + \beta_{m1}(h)\right]T_1(h) \ , \qquad (12a)$$

$$S_2(h) = C_2\gamma_2(h)\left[\beta_{a2}(h) + \beta_{m2}(h)\right]T_2(h) \ , \qquad (12b)$$

where C_1 and C_2 are the lidar constants; $\gamma_1(h)$ and $\gamma_2(h)$ are the geometrical overlapping functions; $\beta_{a1}(h)$ and $\beta_{a2}(h)$ are the aerosol backscattering profiles; $\beta_{m1}(h)$ and $\beta_{m2}(h)$ are the molecular backscattering profiles; $T_1(h) = \exp[-2\int_0^h \alpha_1(h')dh']$ and $T_2(h) = \exp[-2\int_0^h \alpha_2(h')dh']$ are the atmospheric transmissions. Let us choose some joint reference height h_0 for the two channels and denote by $S_1(h_0)$ and $S_2(h_0)$ the corresponding RCSs, as well as by $\beta_1(h_0)=[\beta_{a1}(h_0)+\beta_{m1}(h_0)]$ and $\beta_2(h_0)=[\beta_{a2}(h_0)+\beta_{m2}(h_0)]$ - the total scattering coefficients. Normalizing the lidar profiles $S_1(h)$ and $S_2(h)$ by their values at $h=h_0$, one can obtain (for $h > h_0$) the dimensionless lidar profiles:

$$L_1(h,h_0) = \frac{S_1(h)}{S_1(h_0)} = \frac{\beta_{a1}(h)+\beta_{m1}(h)}{\beta_1(h_o)}\Gamma_1(h,h_0); \quad \Gamma_1(h,h_0) = \frac{T_1(h)}{T_1(h_0)}\frac{\gamma_1(h)}{\gamma_1(h_0)} \qquad (13a)$$

$$L_2(h,h_0) = \frac{S_2(h)}{S_2(h_0)} = \frac{\beta_{a2}(h)+\beta_{m2}(h)}{\beta_2(h_o)}\Gamma_2(h,h_0); \quad \Gamma_2(h,h_0) = \frac{T_2(h)}{T_2(h_0)}\frac{\gamma_2(h)}{\gamma_2(h_0)} . \qquad (13b)$$

As seen, the normalized lidar profiles do not depend on lidar constants C_1 and C_2 and thus, on some lidar parameters as the emitted powers, receiver sensitivities, etc. The dependence on the overlapping functions can be minimized, if the reference height h_0 is chosen so that $\gamma_1(h) \sim \gamma_2(h) \sim 1$, for $h > h_0$. For non-absorbing atmosphere one could also accept the atmospheric transmissions for λ_1 and λ_2 to be close, i.e. $T_1(h)/T_1(h_0) \sim T_2(h)/T_2(h_0) \sim 1$, and thus, $\Gamma_1(h,h_0) \sim \Gamma_2(h,h_0) \sim 1$. As a result, the normalized profiles $L_1(h,h_0)$ and $L_2(h,h_0)$ will depend on the atmospheric parameters by the profiles of the aerosol and molecular backscattering coefficients. Thus, expressions (13a,b) can be written in the forms:

$$\beta_{a1}(h)+\beta_{m1}(h) = \beta_1(h_0)L_1(h,h_0), \tag{14a}$$

$$\beta_{a2}(h)+\beta_{m2}(h) = \beta_2(h_0)L_2(h,h_0). \tag{14b}$$

The molecular scattering coefficients are expressed by $\beta_{m1}(h) = B_m(h)\lambda_1^{-4}$ and $\beta_{m2}(h) = B_m(h)\lambda_2^{-4}$, where $B_m(h)$ does not depend on λ (Measures, 1984). Their ratio is given by:

$$\theta(\lambda_1,\lambda_2) = \beta_{m2}(h)/\beta_{m1}(h) = (\lambda_1/\lambda_2)^4 . \tag{15}$$

In the case of CuBr lidar, $\theta(\lambda_1,\lambda_2) = \theta \approx 0.6$.

By analogy with the molecular scattering, the wavelength dependence of the aerosol backscattering coefficients could be presented in the form:

$$\beta_{a1}(h,\lambda_1) = B_{a1}(h)\lambda_1^{-\eta(h)}; \quad \beta_{a2}(h,\lambda_1) = B_{a2}(h)\lambda_2^{-\eta(h)}, \quad h \geq h_0 , \tag{16}$$

where $B_{a1,2}(h)$ and $\eta(h)$ do not depend on the wavelength λ in broad spectral domains. The factor η typically varies within the range $0.57 \leq \eta \leq 1.8$ for different types of aerosol (Toriumi et al., 1994). Based on the initial supposition stating closeness of the two wavelengths, one can assume

$$B_{a1}(h) \approx B_{a2}(h) = B_a(h) \tag{17}$$

and the ratio $\mu(h,\lambda_1,\lambda_2)$ of backscattering coefficients $\beta_{a2}(h)$ and $\beta_{a1}(h)$ is expressed by a dependence similar to (15):

$$\mu(h,\lambda_1,\lambda_2) = \beta_{a2}(h)/\beta_{a1}(h) = (\lambda_1/\lambda_2)^{\eta(h)} . \tag{18}$$

The vertical profile of $\eta(h)$ used for characterizing aerosol types is given by

$$\eta(h) = \ln[\mu(h,\lambda_1,\lambda_2)]/\ln(\lambda_1/\lambda_2) = \ln[\beta_{a2}(h)/\beta_{a1}(h)] / \ln(\lambda_1/\lambda_2) . \tag{19}$$

The parameter $\eta(h)$ as defined in (19) is also called aerosol backscattering-related Ångström exponent (BAE) (Del Guasta, 2002; Kamei et al., 2006). It is involved by analogy with the Ångström exponent (Ångström, 1929, 1964; Shuster et al., 2005) which is normally expressed in terms of aerosol optical depth or extinction. Generally, $\eta(h)$ is a complex function and characteristics of the aerosol particle size distribution and mode volume fractions. In particular, the BAE values can be influenced by the relative humidity of the atmosphere (Del Guasta & Marini, 2000; Del Guasta, 2002). As calculated from lidar data, the BAE represents a range-resolved function, in contrast to the classical Ångström exponent, thus providing information about the range variations of the aerosol size distribution. In lidars emitting very different wavelengths (say, the first and the second harmonics), the large wavelength differences can cause large variations of the calculated factor $\eta(h)$, depending on shapes and mutual dispositions of the aerosol size distribution modes with respect to the laser wavelengths. This is due to the inequalities $B_{a1}(h) \neq B_{a2}(h)$, indicating for some difference in wavelength dependent scattering mechanisms. It is evident that the conditions (16, 19), when $B_{a1}(h) \approx B_{a2}(h)$ can be satisfied within some defined wavelength domain of bandwidth $\Delta\lambda$, depending on the aerosol size distribution and the aerosol composition. In these cases one can accept some similarity in the aerosol scattering mechanisms. In a clear atmosphere (single mode distribution) $\Delta\lambda$ can practically cover the entire visible and a part of infrared ranges. In the case of multimode size distribution, it can be quite narrower, depending also on the choice of both wavelengths (λ_1, λ_2). The validity of equations (16,19) is of great importance for extracting more and reliable information from the lidar sensing.

It is worth to discuss the opportunities for solving equations (14a,b) for the two wavelengths (λ_1, λ_2). An important requirement here is the normalized lidar profiles $L_1(h,h_0)$ and $L_2(h,h_0)$ to be well distinguished with respect to the noise. As seen, the number of arguments exceeds the number of equations. The parameters $\beta_1(h_0)$ and $\beta_2(h_0)$ could be determined by solving single wavelengths equations (16). The number of arguments could also be reduced using the links of both molecular scattering coefficients $\beta_{m1}(h)$ and $\beta_{m2}(h)$ [see Eq.(15)], as well as applying some well-known models for the standard molecular atmosphere. Using then the model for aerosol backscattering wavelength dependence present above, the number of arguments could also be additionally reduced. The application of this approach is out of the scope of this analysis.

The above analysis shows that the simultaneous multi-wavelength MVL lidar sensing in the mid-visible range, based on the application of the backscattering Ångström exponent profile, is attractive for characterizing vertical aerosol size distribution variations in the submicron and near-micron ranges. The combination with Nd:YAG lidar sensing (1064 nm, 532 nm) is a good approach for a reliable characterization of the most typical atmospheric aerosol loadings.

6.2 Estimation of aerosol-to-molecular scattering proportions

We present below some target experimental results obtained with the CuBr lidar system (Kolarov et al., 1995) emitting two basic wavelengths: λ_1=510.6 nm and λ_2=578.2 nm. To this purpose, let us define the profile of the ratio $R(h,h_0)$ of the two normalized lidar profiles $L_1(h,h_0)$ and $L_2(h,h_0)$:

$$R(h,h_0) = L_2(h,h_0)/L_1[h,h_0], \quad h \geq h_0 . \tag{20}$$

After some transformations, using the simplified expressions (15) we obtain:

$$R(h,h_0) = \left(\frac{\beta_1(h_0)}{\beta_2(h_0)}\right)\frac{\mu(h)\beta_{a1}(h) + \theta\beta_{m1}(h)}{\beta_{a1}(h) + \beta_{m1}(h)} . \tag{21}$$

The variation range of $R(h,h_0)$ for $h \geq h_0$ can be easily estimated. At the reference height h_0 one will obtain $R(h,h_0)=1$. Further, at a proper choice of the reference height h_0 the contribution of the molecular scattering could be neglected for some heights $h \geq h_0$ or $\beta_{a1,2}(h) >> \beta_{m1,2}(h)$ and then, $\mu(h) \sim R(h,h_0)\beta_2(h_0)/\beta_1(h_0)$ as well as $\mu(h_0) \sim \beta_{a2}(h_0)/\beta_{a1}(h_0)$. In the opposite case, one could neglect the contribution of the aerosol scattering with respect to the molecular scattering (Sec.2) above some height h_{max} ($h_{max}>h_0$), where $\beta_{a1,2}(h_{max}) << \beta_{m1,2}(h_{max})$. For this case an approximate estimate for the ratio $R(h_{max},h_0)$ can be also obtained. Using Eq.(21) one will obtain the estimate $R(h>h_{max},h_0) \sim \theta\beta_1(h_0)/\beta_2(h_0) \sim \theta$, if assuming $\beta_1(h_0)/\beta_2(h_0) \sim 1$ for lower heights [see Eqs.(16-19)]. Thus, the range of the height variations of the profile $R(h,h_0)$ will vary approximately within the borders:

$$\theta \leq R(h,h_0) \leq 1 \quad . \tag{22}$$

The use of the ratio $R(h,h_0)$ provides a clear and easy determination of the height h_{max} where one can accept an absence of an aerosol loading. In most of lidar systems it is typically accepted to attach the molecular scattering profile to the range-corrected profiles $S_1(h)$ and $S_2(h)$ at heights above 12-15 km. But, in the presence of intensive daily background this attachment could be even incorrect as the lidar profiles $S_1(h)$ and $S_2(h)$ can contain only noise at higher altitudes. The knowledge of the altitude h_{max} can provide the correct attachment of the molecular profiles in such cases. Moreover, this approach can provide well defined estimates of the aerosol scattering contribution in the point of attachment, if it is lower than h_{max} and thus, an attachment to the molecular scattering in the presence of an aerosol contribution. The measured ratio $R(h,h_0)$ can also be used for estimating the profile of aerosol-to-molecular backscattering coefficients.

Below, we demonstrate some experimental results concerning determination and analysis of calculated profiles of the ratio $R(h,h_0)$ for different aerosol loadings. The plots in Fig.10a present the two normalized lidar profiles: $L_1(h,h_0)$ (curve 1) and $L_2(h,h_0)$ (curve 2) at wavelengths λ_1 and λ_2, respectively. The reference height h_0 is equal to 2 km. Note that both normalized profiles are well distinguished. As expected, the profile for λ_1 is situated in the region of higher values than the profile for λ_2. This is due to the higher molecular scattering for shorter wavelengths. The ratio $R(h,h_{max},h_0)$ is larger than the lower estimate in relation (22) for height up to ~ 4 km. It becomes ~ 0.6 just at this altitude and above, therefore, h_{max} is of the order of 4 km. The attachment of the molecular scattering profile can be implemented at heights above 4 km. As evident, the height h_{max} for attachment of the Rayleigh profiles can be estimated here from lidar data only. This is essential for the accuracy of retrieved aerosol parameters, providing an opportunity to avoid sometimes the application of complicated set of calibration procedures (Bösenberg & Hoff, 2007).

The results in Fig.10b correspond to the case, when the calculated ratio $R(h,h_{max},h_0)$ is higher and approximately equal to ~ 0.8 in a large region of altitudes up to 10 km. This case can be characterized as a case of significant aerosol loading up to altitudes of 10 km. It must be noted that the both normalized profiles are approximately parallel and well distinguished as well. Some additional small aerosol contribution above 6km are well seen also in the profile of the ratio $R(h,h_{max},h_0)$. This fact could be explained by some changes of the aerosol size distribution at these altitudes, due to the presence of additional thin cloud loading having different aerosol scattering structure. For lower heights, the ratio $R(h,h_{max},h_0)$ tends to unity as expected. As seen, the ratio $R(h,h_{max},h_0)$ becomes very informative when the two wavelengths are closely disposed (but at well distinguished lidar profiles) so the uncertainties introduced by the parameters of two lidar channels can be minimized (see Eqs.13a,b). In this sense, the application of MVL in lidar atmospheric probing can be substantial for improving the measurement accuracy.

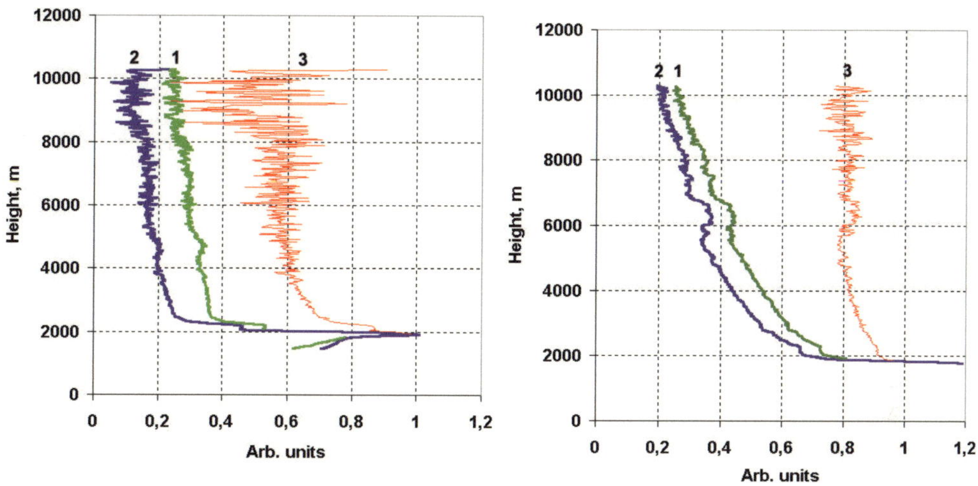

a) b)

Fig. 10. a) Normalized lidar profiles 1 and 2 (on both wavelengths, respectively) and the calculated ratio $R(h,h_{max},h_0)$ 3, measured in a clear atmosphere on 10.07.2008; b) Normalized lidar profiles 1 and 2 and the calculated ratio 3, measured in mixed (aerosol and molecular scattering) atmosphere on 17.07.2008.

6.3 Two-wavelength lidar probing of aerosol mode fractions over complex terrain

Results of lidar observations on atmospheric aerosols over a complex terrain (Fernando, 2010), representing adjoining city-, plain-, and mountain zones in Sofia region, are described below. A residential city zone is located along with the first 2 kilometers of the laser beam path. The distance range 2 - 5 km covers city outskirts and suburbs (plain zone) and the one from 5.5 km to about 9 km covers the mountain foot, slope, and ridge (mountain zone).

Lidar measurements (Fig.11) are carried out simultaneously at wavelengths 1064 nm and 532 nm (Peshev et al., 2011). Range profiles of evaluated aerosol backscattering coefficients (molecular component subtracted) on 28.01.2008, averaged over the period of measurements (17:10 h - 18:15 h GMT), are presented in Fig.12a. Over the plain area, the backscattering coefficient at 532 nm is permanently higher than the one at 1064 nm, starting with a ratio of about 1.5-2 and gradually decreasing to equalization close to the interface between the plain and mountain. Humps observed at the initial parts of the two profiles (city zone) are identified to be due to increased anthropogenic aerosol emissions. For the rest of the plain zone, profiles are quite smooth, without differentiated bulges.

Over the mountain zone, the situation is inverted - the backscattering at 1064 nm exceeds the one at 532 nm, both having values considerably higher than those for the plain zone, denoting presence of dense aerosol layers, most probably - water aerosol (fog or orographic clouds near the surface). The available meteorological data for the day support this conclusion.

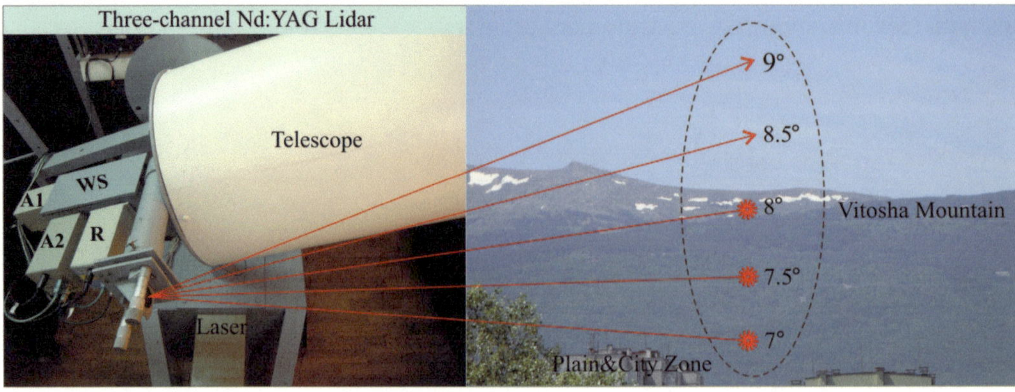

Fig. 11. Schematic view of the lidar experiment over complex terrain; WS - wavelength separator; A1 and A2 - aerosol channels at 1064 nm and 532 nm, respectively; R - Raman channel at 607 nm.

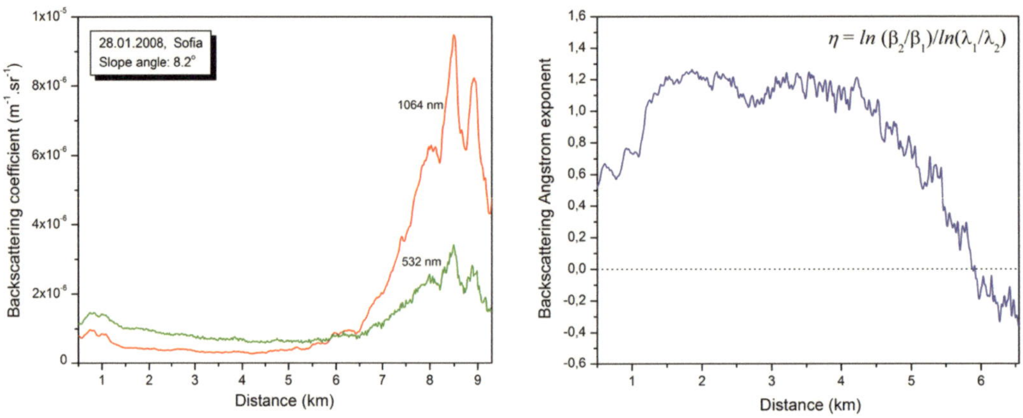

a) b)

Fig. 12. a) Time-averaged profiles of aerosol backscattering coefficient at the two wavelengths of measurement; b) Backscattering-related Ångström exponent profile.

The backscatter-related Ångström exponent defined in Eq.(19) is used as a qualitative indicator of aerosol particle properties. In Fig.12b, the range profile of BAE is presented, corresponding to aerosol backscatter profiles shown on Fig.12a. As one can see, the values of η vary within the range 0.5-1.25 for distances of up to 5 km (plain zone). These values indicate for prevalence of the fine-mode aerosol particle fraction, most probably of anthropogenic origin, as typical for urban areas under dry atmospheric conditions.

Around the plain-to-mountain interface zone located about 5.5 km away from the lidar station, BAE values approach zero indicating for prevailing contribution of the coarse-mode aerosol fraction.

The BAE takes gradually decreasing negative values for distances corresponding to the mountain zone (Fig.12b). Negative values of the extinction-related or backscatter-related Ångström exponent (Kamei, 2006; Lu, 2011; Guerrero-Rascado, 2009) occur in cases of large aerosol particles in the over-micron size range (e.g. large water droplets, ice particles, volcanic or fire ashes, partcle aggregates, etc.), as well as in case of large mode volume fractions of coarse aerosols with respect to those of fine aerosols. In these cases, the backscattering at longer wavelengths dominates over the one at shorter wavelengths, especially for substantially different wavelengths, as in Fig.12b.

7. Conclusions

Various lidar technologies have been developed since the first lidar demonstration more than 40 years ago. In this chapter we presented some of the basic applications of lidars in the atmospheric research. The high informativity of lidar probing is due to the strong interaction of optical waves with atmospheric particles - aerosols and molecules, combined with range-resolved acquisition of lidar signals. The fast spread of lidars all over the world in the last two decades led to their organizing in regional lidar networks as EARLINET (Europe), MPLNET (USA), AD-Net (Japan), etc., integrated now in the Global lidar network (GALION). As lidars are practically the only instruments for high resolution vertical atmospheric profiling, their further improvement is one of the most important tasks for the lidar community. One of the ways to enhance the quality of lidar output information and its significance for the global and regional atmospheric monitoring is the creation of complex multispectral lidar systems, capable to provide more detailed and reliable data for the retrieval of aerosol optical, microphysical, and radiative properties, etc. This approach requires the development of novel more effective inverse algorithms, as well. It is also expected the future lidar networks to operate in a close cooperation with other existing networks as the sun-photometer network, ground-level in-situ aerosol monitoring networks, satellite measurements (lidar and multispectral radiometers), radars, etc. The synergy resulting from such cooperation was demonstrated in a large number of experiments. The expected increase of the lidar station density all over the world is a good indication for their great significance in analysing local and global atmospheric processes and trends. In this connection we note the European project ACTRIS (Aerosols, Clouds, and Trace gases Research InfraStructure Network (www.actris.net)). Started in 2011 under 7-th Framework Program, it integrates the EARLINET, EUSSAR, CLOUDNET, and a new trace gas network into single-coordinated ground-based networks, with monitoring impact on the climate changes, air quality, and long-term transport of pollutants.

8. Acknowledgments

The results described here were funded partly by the Bulgarian National Science Fund under projects Ph-63, Ph-447, Ph-811 and the European Commission under the project EARLINET-ASOS, grant RICA-025991 EC FP6. The authors thank the NOAA Air Resources Laboratory (ARL) for the provision of the HYSPLIT backward trajectories. We also thank the Barcelona Supercomputing Center for Dust Regional Atmospheric Model (DREAM) and for the provision of Saharan dust forecast maps.

9. References

Ångström, A., (1929). On the atmospheric transmission of Sun radiation and on dust in the air, *Geogr. Ann.*, Vol.11, pp.156–166.

Ångström, A., (1964). The parameters of atmospheric turbidity, *Tellus*, Vol. 16, pp.64–75.

Arshinov, Yu.; Bobrovnikov, S.; Zuev, V. & Mitev, V. (1983). Atmospheric temperature measurements using a pure rotational Raman lidar, *Appl. Optics*, Vol.22, pp.2984-2987.

Astadjov, D.; Vuchkov, N. & Sabotinov, N. (1988). Parametric study of the CuBr laser with Hydrogen additivies, *IEEE J.Quant.Electron.*, Vol. QE-24, pp.1927-1935.

Böckmann, C.; Grigorov, I.; Hågård, A.; Horvat, M.; Iarlori, M.; Komguem, L.; Kreipl, S. & Frioud, M. (2004). Aerosol lidar intercomparison in the framework of the EARLINET project. 2. Aerosol backscatter algorithms, *Appl. Opt.*, Vol. 43 (4), pp. 977-989.

Bösenberg, J., et al. (2003). EARLINET: A European Aerosol Research Lidar Network to establish an aerosol climatology, Rep. 348, Max Planck Inst. für Meteorol., Hamburg, Germany.

Bösenberg, J. & Hoff, R., lead authors (2007). *Plan for the implementation of the GAW Aerosol Lidar Observation Network GALION*, GAW No.178, WMO, Hamburg, Germany. Available from ftp://ftp.wmo.int/Documents/PublicWeb/arep/gaw/gaw178-galion-27-Oct.pdf

Del Guasta, M. & Marini, S. (2000). On the retrieval of urbanaerosol mass concentration by a 532 and 1064 nm LIDAR. *J. Aeros. Sci.*, Vol.31, No.12, pp.1469–1488.

Del Guasta, M. (2002). Daily cycles in urban aerosols observed in Florence (Italy) by means of an automatic 532–1064 nm LIDAR. *Atm. Environ.*, Vol.36, No.17, pp.2853–2865.

Draxler, R.R. & Hess, G.D. (1998). An overview of the HYSPLIT_4 modeling system of trajectories, dispersion, and deposition. *Aust. Meteor. Mag.*, Vol. 47, 295-308.

Draxler, R.R. & Rolph, G.D. (2011). HYSPLIT (HYbrid Single-Particle Lagrangian Integrated Trajectory) Model access via NOAA ARL READY Website (http://ready.arl.noaa.gov/HYSPLIT.php). NOAA Air Resources Laboratory, Silver Spring, MD.

Fernald F., Analysis of atmospheric lidar observations: some comments, *Appl.Opt.* (1984) 23, (5), pp.852-853

Fernando, H. J. S. (2010). Fluid dynamics of urban atmospheres in complex terrain, *Ann. Rev. Fluid Mech.* Vol.42, pp.365-389

Gagarin, S.; Kalinkevich, A.; Kolarov, G.; Kutuza, B.; Mikhalev, M.; Mitsev, Ts.; Stoykova, E.; Stoyanov, D.; Ferdinandov, E. & Khaimov S. (1987). Investigation of the atmosphere by using microwave radiometric and lidar signals, *Atm. Ocean Optics*, Vol. 23, No. 2, pp. 121-129.

Gagliardi, R.M. & Karp, S. (1976). *Optical Communications*, Wiley, New York, USA.

Grigorov, I.V. & Kolarov, G.K. (2007). Measurements of atmospheric parameters using aerosol lidar, *JOAM*, Vol. 19, No. 11, pp. 3549-3552.

Grigorov, I.; Kolarov, G. & Stoyanov, D. (2009). Lidar remote monitoring of aerosol dust layers over Sofia, *Proc. ICEST'2009 XLIV Int. Scientific Conf.*, pp. 563-566.

Grigorov, I.; Kolarov, G. & Stoyanov, D. (2010). Remote monitoring of aerosol layers over Sofia in the frame of EARLINET-ASOS project, *AIP Conference Proceedings* ,Vol. 1203, 2010, DOI: 10.1063/1.3322514, pp. 585-590

Grigorov, I.; Stoyanov, D. & Kolarov, G. (2011). Lidar observation of volcanic dust layers over Sofia, *Proc. SPIE,* Vol. 7747, paper # 77470R.

Guerrero-Rascado, J. L.; Olmo, F.J.; Avilés-Rodríguez, I.; Navas-Guzmán, Pérez-Ramírez, F., Lyamani, D. H. & Arboledas, L. A. (2009). Extreme Saharan dust event over the southern Iberian Peninsula in september 2007: active and passive remote sensing from surface and satellite, *Atmos. Chem. Phys.,* Vol. 9, pp. 8453-8469, doi:10.5194/acp-9-8453-2009.

Kallos, G. et al. (1997). The regional weather forecasting system SKIRON: An overview, in Proceedings of the Symposium on Regional Weather Prediction on Parallel Computer Environments, edited by G. Kallos, V. Kotroni, and K. Lagouvardos, pp. 109–122, Univ. of Athens, Athens, Greece.

Kamei, A.; Sugimoto, N., Matsui, I., Shimizu, A. & Shibata, T. (2006). Volcanic Aerosol Layer Observed by Shipboard Lidar over the Tropical Western Pacific. *SOLA,* Vol.2, 001-004, doi:10.2151/sola.2006-001

Kim, J.J. (1991). Metal vapour lasers: a review of recent progress, *Opt. Quant. Electronics,* vol.23, pp.S469-S476.

Klett, J. (1981). Stable analytical inversion solution for processing lidar returns, *Appl.Opt.,* vol. 22, pp.211-220.

Kolarov, G.; Stoyanov, D.; Mitsev, Ts. & Againa, Ts. (1988). Sounding of the atmospheric aerosol by a lidar on copper vapour laser, *Atm. Optics,* Vol. 1, No. 6, pp. 125-126.

Kolarov, G.; Grigorov, I. & Stoyanov, D. (1995). Estimation of the ratio of aerosol to molecular backscattering by two closely disposed wavelengths using CuBr lidar sounding (510.6 nm, 578.2 nm), *Proc. SPIE,* Vol. 7027, paper # 702710.

Kovalev, V.A. & Eichinger, W.E. (2004). *Elastic Lidar: Theory, Practice, and Analysis Methods,* Wiley, New York, USA.

Lu, X; Jiang, Y.; Zhang, X.; Wang X. & Spinelli, N. (2011). Two-wavelength lidar inversion algorithm for determination of aerosol extinction-to-backscatter ratio and its application to CALIPSO lidar measurements, *J. Quantat. Spectr. & Radiat. Transf.,* Vol. 112, pp. 320–328.

Measures, R.M. (1984). *Laser Remote Sensing,* Wiley, New York, USA.

Mitzev, Tz.; Grigorov, I.; Kolarov, G. & Lolova, D. (1995). Investigation of transborder pollution by combining remote lidar sounding and stationary gas sampling, Proc.SPIE, vol.2506, pp.310-318.

Nickovic, S.; Papadopoulos, A.; Kakaliagou, O. & Kallos, G. (2001). Model for prediction of desert dust cycle in the atmosphere, *J. Geophys. Res.,* vol. 106, pp. 18,113- 18,129.

Ozsoy, E.; Kubilay, N.; Nickovic, S. & Moulin, C. (2001). A hemispheric dust storm affecting the Atlantic and Mediterranean (April 1994): Analyses, modelling, ground-based measurements and satellite observations, *J. Geophys. Res.,* Vol. 106, pp. 18, 439–18,460.

Pappalardo, G. et al. (2009). The EARLINET Contribution to the EarthCARE Mission, *EarthCARE 2009 Workshop,* 15.July.2009.

Pappalardo, G. et al. (2010). EARLINET correlative measurements for CALIPSO: First intercomparison results., *J. Geophys. Res.-Atmospheres,* Vol. 115, D00H19, doi: 10.1029/2009JD012147

Papayannis A. et al. (2008). Systematic lidar observations of Saharan dust over Europe in the frame of EARLINET (2000-2002), *J. Geophys. Res.,* Vol. 113, D10204.

Perez, C.; Nickovic, S.; Baldasano, J.M.; Sicard; M.; Rocadenbosch, F. & Cachorro, V. (2006a) . A long Saharan dust event over the western Mediterranean: Lidar, Sun photometer observations, and regional dust modelling *J. Geophys. Res.*, Vol. 111, D15214, doi:10.1029/2005JD006579.

Perez, C.; Nickovic, S.; Pejanovic, G.; Baldasano, J.M. & Ozsoy, E. (2006b). Interactive dust-radiation modeling: A step to improve weather forecasts, *J. Geophys. Res.*, Vol. 111, D16206, doi:10.1029/2005JD006717, 2006

Peshev, Z. Y.; Deleva, A. D., Dreischuh, T. N. & Stoyanov, D.V. (2010). Lidar measurements of atmospheric dynamics over high mountainous terrain, *AIP Conf. Proc.*, Vol. 1203, pp. 1108-1113.

Peshev, Z.Y.; Deleva, A.D., Dreischuh, T.N. & Stoyanov, D.V. (2011) Dynamical characteristics of atmospheric layers over complex terrain probed by two-wavelength lidar, *Proc. SPIE*, Vol. 7747, paper # 77470U.

Schuster, G. L.; Dubovik, O. & Holben, B. N. (2006). Angstrom exponent and bimodal aerosol size distributions, *J. Geophys. Res.*, Vol.111, D07207.

Stoilov, V.; Astadjov, D.; Vuchkov, N. & Sabotinov, N. (2000). High spatial intensity 10 W-CuBr laser with hydrogen additives, *Opt. Quant. Electr.*, vol.32, pp. 1209-1217.

Stoyanov, D.; Donchev, A.; Kolarov, G. & Mitsev, Ts. (1988). Copper and gold vapour laser radar for troposphere and stratospheric studies, *Atm. Optics*, Vol. 1, pp. 109-116.

Stoyanov, D. (1997). Counting of overlapped photon detector single pulses by analog/digital sampling and deconvolution, *Opt. Eng.*, Vol. 36, pp. 210-216.

Stoyanov D.; Vankov, O. & Kolarov, G. (2000). Measuring the arrival times of overlapped photo-events, *Nucl. Instrum. Methods Phys. Res. A*, Vol. 449, pp. 555-567.

Toriumi, R.; Tai, H.; Okumura, H. & Takeuchi, N. (1994). Wavelength dependence of aerosol optical parametres measuredby a tunable lidar, Abstracts of papers, 17th ILRC, July 25-29, Sendai, Japan, pp.91-92.

Weitkamp, C., Ed. (2005). *Lidar: Range-Resolved Optical Remote Sensing of the Atmosphere*, Springer Series in Optical Sciences, New York, USA.